SI Prefixes

Multiple	Exponential Form	Prefix	SI Symbol
1 000 000 000	10^9	giga	G
1 000 000	10^6	mega	M
1 000	10^3	kilo	k

Submultiple			
0.001	10^{-3}	milli	m
0.000 001	10^{-6}	micro	μ
0.000 000 001	10^{-9}	nano	n

Conversion Factors (FPS) to (SI)

Quantity	Unit of Measurement (FPS)	Equals	Unit of Measurement (SI)
Force	lb		4.4482 N
Mass	slug		14.5938 kg
Length	ft		0.3048 m

Conversion Factors (FPS)

$$1 \text{ ft} = 12 \text{ in. (inches)}$$
$$1 \text{ mi. (mile)} = 5{,}280 \text{ ft}$$
$$1 \text{ kip (kilopound)} = 1{,}000 \text{ lb}$$
$$1 \text{ ton} = 2{,}000 \text{ lb}$$

Engineering Mechanics

DYNAMICS

R. C. HIBBELER

FIFTH EDITION

Engineering Mechanics

DYNAMICS

MACMILLAN PUBLISHING COMPANY, INC.
NEW YORK
Collier Macmillan Publishers
LONDON

Macmillan Publishing Company
866 Third Avenue, New York, New York 10022

Collier Macmillan Canada, Inc.

LIBRARY OF CONGRESS CATALOGING IN PUBLICATION DATA

Hibbeler, R. C.
   Engineering mechanics—dynamics / R. C. Hibbeler. — 5th ed.
      p.   cm.
   Companion volume to Engineering mechanics—statics. These two works are also published together as Engineering mechanics—statics and dynamics.
   Includes index.
   ISBN 0-02-354661-1
   1. Dynamics. I. Title.
TA352.H5 1989
620.1'04—dc19   88-22105
                                                                  CIP

PRINTING: 3 4 5 6 7 8   YEAR: 0 1 2 3 4 5 6 7 8

Preface

The purpose of this book is to provide the student with a clear and thorough presentation of the theory and applications of engineering mechanics. To achieve this objective the author has by no means worked alone on this task, for to a large extent this book has been shaped by the comments and suggestions of more than a hundred reviewers in the teaching profession and many of the author's students who have used its previous editions.

Several changes have been made in preparing this new edition. Overall, particular attention has been given to developing accurate definitions and explanations of the concepts. A greater emphasis has been placed on drawing free-body diagrams, and the importance of selecting an appropriate coordinate system and associated sign convention for vector components is stressed when the equations of kinematics or the equations of motion are applied. Some rewriting of material has occurred; and in particular, relative motion analysis using Cartesian vectors and scalar components has been combined into a single section. Lastly, most of the problems in this edition are new, and there are more of them than in the previous edition. In this regard, extra care has been taken in the presentation and solution of the problems, and all the problem sets have been reviewed and the solutions checked by the author's colleagues to ensure both their clarity and numerical accuracy.

Organization and Approach. The contents of each chapter are organized into well-defined sections. Selected groups of sections contain an explanation of specific topics, illustrative example problems, and a set of homework problems. The topics within each section are placed into subgroups defined by boldface titles. The purpose of this is to present a structured method for introducing each new definition or concept, and to make the book convenient for later reference and review.

A "procedure for analysis" is given at the end of many sections of the book in order to provide the student with a review or summary of the material and a logical and orderly method to follow when applying the theory. As in the previous editions, the example problems are solved using this outlined method in order to clarify its numerical application. It is to be understood, however, that once the relevant principles have been mastered and enough confidence and judgment have been obtained, the student can then develop his or her own procedures for solving problems. In most cases it is felt that the first step in any procedure should require drawing a diagram. By doing so, the student forms the habit of tabulating the necessary data while focusing on the physical aspects of the problem and its associated geometry. If this step is correctly performed, applying the relevant equations of mechanics becomes somewhat methodical, since the data can be taken directly from the diagram. This step is particularly important when solving problems involving kinetics, and for this reason, application of the free-body diagram technique is strongly emphasized throughout the book.

Since mathematics provides a systematic means of applying the principles of mechanics, the student is expected to have prior knowledge of algebra, geometry, trigonometry, and, for complete coverage, some calculus. Vector analysis is introduced at points where it is most applicable. Its use often provides a convenient means for presenting concise derivations of the theory, and it makes possible a simple and systematic solution of many complicated three-dimensional problems. Occasionally, the example problems are solved using more than one method of analysis so that the student develops the ability to use mathematics as a tool whereby the solution of any problem may be carried out in the most direct and effective manner.

Problems. Numerous problems in the book depict realistic situations encountered in engineering practice. It is hoped that this realism will both stimulate the student's interest in engineering mechanics and provide a means for developing the skill to reduce any such problem from its physical description to a model or symbolic representation to which the principles of mechanics may be applied. As in the previous editions, an effort has been made to include some problems which may be solved using a numerical procedure executed on either a desk-top computer or programmable pocket calculator. Suitable numerical techniques along with associated computer programs are given in Appendix B. The intent here is to broaden the student's capacity for using other forms of mathematical analysis *without* sacrificing the time needed to focus on the application of the principles of mechanics. Problems of this type which either can or must be solved using numerical procedures are identified by a "square" symbol (■) preceding the problem number.

Throughout the text there is an approximate balance of problems using either SI or FPS units. Furthermore, in any set, an attempt has been made to arrange the problems in order of increasing difficulty. The answers to all but every fourth problem are listed in the back of the book. To alert the user to a problem without a reported answer, an asterisk (*) is placed before the problem number.

Contents. The book is divided into 11 chapters.* In particular, the kinematics of a particle is discussed in Chapter 12, followed by a discussion of particle kinetics in Chapter 13 (equation of motion), Chapter 14 (work and energy), and Chapter 15 (impulse and momentum). The concepts of particle dynamics contained in these four chapters are then summarized in a "review" section and the student is then given the chance to identify and solve a variety of different types of problems. A similar sequence of presentation is given for the planar motion of a rigid body: Chapter 16 (planar kinematics), Chapter 17 (equations of motion), Chapter 18 (work and energy), and Chapter 19 (impulse and momentum), followed by a summary and set of review problems for these chapters. If desired, it is possible to cover Chapters 12 through 19 in the following order with no loss in continuity: Chapters 12 and 16 (kinematics), Chapters 13 and 17 (equations of motion), Chapters 14 and 18 (work and energy), and Chapters 15 and 19 (impulse and momentum).

Time permitting, some of the material involving three-dimensional rigid-body motion may be included in the course. The kinematics and kinetics of this motion are discussed in Chapters 20 and 21, respectively. Chapter 22 (vibrations) may be included if the student has the necessary mathematical background. Sections of the book which are considered to be beyond the scope of the basic dynamics course are indicated by a star (★) and may be omitted. Note, however, that this more advanced material provides a suitable reference for basic principles when it is covered in other courses.

Acknowledgments. I have endeavored to write this book so that it will appeal to both the student and instructor. Through the years many people have helped in its development and I should like to acknowledge their valued suggestions and comments. Specifically, I wish to personally thank the following individuals who have contributed to this revision, namely, Professors Reginald Amory, Northeastern University; Robert D. Celmer, University of Hartford; C. H. Chang, University of Alabama; Major William Conner, U.S. Military Academy, West Point; Mitsunori Denda, Rutgers University; Frank D'Souza, Illinois Institute of Technology; Robert W. Fuessle, Bradley University; Sherman Goplen, North Dakota State University; John E. Griffith, University of Southern Florida; William Q. Gurley, University of Tennessee, Chattanooga; Edward E. Hornsey, University of Missouri, Rolla; Mohammad Khos-

*The first eleven chapters of this sequence form the contents of *Engineering Mechanics: Statics*.

rowjerdi, Western New England College; William M. Lee, U.S. Naval Academy, Annapolis; Captain Al Lewis, U.S. Military Academy, West Point; Will Liddell, Jr., Auburn University at Montgomery; John C. McWhorter, Mississippi State University; Nasser Moshrefi, Michigan Technical University; Larry Oline, University of Southern Florida; David Oglesby, University of Missouri, Rolla; John Peddieson, Tennessee Technical University; James R. Ritter, Old Dominion University; Robert Schmidt, University of Detroit; Jack Smith, U.S. Naval Academy, Annapolis; Robert Snell, Kansas State University; Henry L. Sundberg, Jr., Western New England College; Liang-Ning Tao, Illinois Institute of Technology; Theodore Tauchert, University of Kentucky; V. L. Vorsteveld, University of Vermont; Donald M. Wallace, Norwich University; William H. Walston, Jr., University of Maryland; Han-Chin Wu, University of Iowa.

Many thanks are also extended to all my students and to those in the teaching profession who have freely taken the time to send me their suggestions and comments. Since the list is too long to mention, it is hoped that those who have given help in this manner will accept this anonymous recognition. Furthermore, I greatly appreciate the personal support and attention given to me by my former editor Mr. Peter Gordon and currently, Mr. David Johnstone, along with the staff at Macmillan. Lastly, I should like to acknowledge the encouragement and assistance of my wife, Conny, during the time it has taken to prepare the manuscript for publication.

Russell Charles Hibbeler

Contents

Contents

13

Kinetics of a Particle: Force and Acceleration   77

14

Kinetics of a Particle: Work and Energy   125

15

Kinetics of a Particle: Impulse and Momentum   163

19

Planar Kinetics of a Rigid Body: Impulse and Momentum 369

Review 2: Planar Kinematics and Kinetics of a Rigid Body 397

20

Three-Dimensional Kinematics of a Rigid Body 411

21

Three-Dimensional Kinetics of a Rigid Body 439

22

Vibrations 485

APPENDIXES

A

Mathematical Expressions 521

B

Numerical and Computer Analysis 524

C

Vector Analysis 533

Answers 537

Index 547

12

Kinematics of a Particle

Engineering mechanics involves the study of both statics and dynamics. *Statics* is concerned with the equilibrium of bodies at rest or moving with constant velocity, whereas *dynamics* is concerned with bodies having accelerated motion. In this text, the subject of dynamics will be presented in two parts: *kinematics,* which treats only the geometrical aspects of motion, and *kinetics,* which is the analysis of the forces causing the motion. For simplicity particle dynamics will be discussed first, followed by topics in rigid-body dynamics presented in two and then three dimensions.

We will begin our study of dynamics by discussing particle kinematics. Recall that a *particle* has a mass but negligible size and shape. Therefore we must limit application to those objects that have dimensions that are of no consequence in the analysis of the motion. In most problems encountered, one is interested in bodies of finite size, such as rockets, projectiles, or vehicles. Such objects may be considered as particles, provided motion of the body is characterized by motion of its mass center and any rotation of the body is neglected.

This chapter begins with a study of the absolute motion of a particle, which is motion measured with respect to a fixed coordinate system. In this regard, motion along a straight line will be studied before the more general motion along a curved path. Afterward, the relative motion between two particles will be considered, using a translating coordinate system.

12.1 Rectilinear Kinematics

Rectilinear motion occurs when a particle moves along a straight-line path. The kinematics of the motion is characterized by specifying at any given instant the particle's position, velocity, and acceleration.

Position

The straight-line path of the particle can be defined using a single coordinate axis *s*, Fig. 12–1*a*. The origin *O* on the path is a fixed point, and from this point the *position vector* **r** is used to specify the position of the particle *P* at any given instant. For *rectilinear motion*, however, the *direction* of **r** is *always* along the *s* axis, and so it never changes. What will change is its magnitude and its sense or arrowhead direction. For analytical work it is therefore convenient to represent **r** by an *algebraic scalar s*, representing the *position coordinate* of the particle, Fig. 12–1*a*. The magnitude of *s* (and **r**) is the distance from *O* to *P*, usually measured in meters (m) or feet (ft) and the sense (or arrowhead direction of **r**) is defined by the algebraic sign of *s*. Although the choice is arbitrary, in this case *s* is positive since the coordinate axis is positive to the right of the origin. Likewise it is negative if the particle is located to the left of *O*.

Displacement

The *displacement* of the particle is defined as the *change* in its *position*. For example, if the particle moves from *P* to *P'*, Fig. 12–1*b*, the displacement is $\Delta \mathbf{r} = \mathbf{r}' - \mathbf{r}.$ Using algebraic scalars to represent $\Delta \mathbf{r}$, we also have $\Delta s = s' - s$. Here Δs is *positive* since the particle's final position is to the *right* of its initial position, i.e., $s' > s$. Likewise, if the final position is to the *left*, Δs is *negative*.

Since the displacement of a particle is a vector quantity, it should be distinguished from the distance the particle travels. Specifically, the *distance traveled* is a *positive scalar* which represents the total length of path traversed by the particle.

Velocity

If the particle moves through a displacement $\Delta \mathbf{r}$ from *P* to *P'* during the time interval Δt, Fig. 12–1*b*, the *average velocity* of the particle during this time interval is

$$\mathbf{v}_{avg} = \frac{\Delta \mathbf{r}}{\Delta t}$$

If we take smaller and smaller values of Δt, the magnitude of $\Delta \mathbf{r}$ becomes smaller and smaller. Consequently, the *instantaneous velocity* is defined as $\mathbf{v} = \lim_{\Delta t \to 0} (\Delta \mathbf{r}/\Delta t)$ or

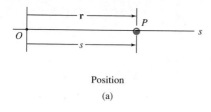

Position

(a)

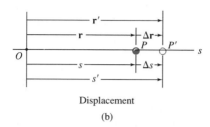

Displacement

(b)

Velocity

Fig. 12–1(a–c)   (c)

$$\mathbf{v} = \frac{d\mathbf{r}}{dt}$$

Representing **v** as an algebraic scalar, Fig. 12–1c, we can also write

$(\overset{+}{\rightarrow})$ $\qquad\qquad\boxed{v = \dfrac{ds}{dt}}\qquad\qquad$ (12–1)

Since Δt or dt is always positive, the sign used to define the *sense* of the velocity is the same as that of Δs (or ds). For example, if the particle is moving to the *right*, Fig. 12–1c, the velocity is *positive;* whereas if it is moving to the *left,* the velocity is *negative.* (This fact is emphasized here by the arrow written at the left of the above equation.) The *magnitude* of the velocity is known as the *speed* and is generally expressed in units of m/s or ft/s.

Occasionally, the term ''average speed'' is used. The *average speed* is *always* a positive scalar and is defined as the total distance traveled by a particle, s_T, divided by the elapsed time Δt, i.e.,

$$v_{\text{sp}} = \frac{s_T}{\Delta t}$$

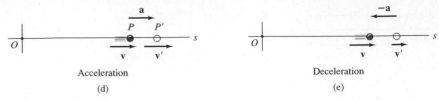

Fig. 12–1(d, e)

Acceleration

Provided the velocity of the particle is known at the two points P and P', the *average acceleration* of the particle during the time interval Δt is defined as

$$\mathbf{a}_{avg} = \frac{\Delta \mathbf{v}}{\Delta t}$$

Here $\Delta \mathbf{v}$ represents the difference in the velocity during the time interval Δt, i.e., $\Delta \mathbf{v} = \mathbf{v}' - \mathbf{v}$, Fig. 12–1d.

The *instantaneous acceleration* at time t is found by taking smaller and smaller values of Δt and corresponding smaller and smaller values of $\Delta \mathbf{v}$, so that $\mathbf{a} = \lim_{\Delta t \to 0} (\Delta \mathbf{v}/\Delta t)$ or, using algebraic scalars,

$(\overset{+}{\to})$
$$a = \frac{dv}{dt}$$
$(12\text{–}2)$

Taking the second time derivative of Eq. 12–1, we can also write

$(\overset{+}{\to})$
$$a - \frac{d^2 s}{dt^2}$$

Both the average and instantaneous acceleration can be either positive or negative. In particular, when the particle is *slowing down*, or its speed is decreasing, it is said to be *decelerating*. In this case, v' in Fig. 12–1e is *less* than v and so $\Delta v = v' - v$ will be negative. Consequently, a will also be negative and therefore it will act to the *left*, in the opposite *sense* to v. Also, note that when the *velocity* is *constant,* the *acceleration is zero* since $\Delta v = v - v = 0$. Units commonly used to express the magnitude of acceleration are m/s^2 or ft/s^2.

A differential relation involving the displacement, velocity, and acceleration along the path may be obtained by eliminating the time differential dt between Eqs. 12–1 and 12–2. Show that this yields

$(\overset{+}{\to})$
$$a \, ds = v \, dv$$
$(12\text{–}3)$

Constant Acceleration, $a = a_c$

When the acceleration is constant, each of the three kinematic equations $a_c = dv/dt$, $v = ds/dt$, and $a_c\, ds = v\, dv$ may be integrated to obtain formulas that relate a_c, v, s, and t.

Velocity as a Function of Time. Integrate $a_c = dv/dt$, assuming that initially $v = v_0$ when $t = 0$.

$$\int_{v_0}^{v} dv = \int_{0}^{t} a_c\, dt$$

$$v - v_0 = a_c(t - 0)$$

$(\overset{+}{\rightarrow})$

$$\boxed{v = v_0 + a_c t}$$
Constant Acceleration

$(12\text{--}4)$

Position as a Function of Time. Integrate $v = ds/dt = v_0 + a_c t$, assuming that initially $s = s_0$ when $t = 0$.

$$\int_{s_0}^{s} ds = \int_{0}^{t} (v_0 + a_c t)\, dt$$

$$s - s_0 = v_0(t - 0) + a_c(\tfrac{1}{2}t^2 - 0)$$

$(\overset{+}{\rightarrow})$

$$\boxed{s = s_0 + v_0 t + \tfrac{1}{2}a_c t^2}$$
Constant Acceleration

$(12\text{--}5)$

Velocity as a Function of Position. Either solve for t in Eq. 12–4 and substitute into Eq. 12–5, or integrate $v\, dv = a_c\, ds$, assuming that initially $v = v_0$ at $s = s_0$.

$$\int_{v_0}^{v} v\, dv = \int_{s_0}^{s} a_c\, ds$$

$$\tfrac{1}{2}v^2 - \tfrac{1}{2}v_0^2 = a_c(s - s_0)$$

$(\overset{+}{\rightarrow})$

$$\boxed{v^2 = v_0^2 + 2a_c(s - s_0)}$$
Constant Acceleration

$(12\text{--}6)$

The magnitudes and signs of s_0, v_0, and a_c, used in these equations, are determined from the chosen origin and positive direction of the s axis. As indicated by the arrows written at the left of these equations, here we have assumed positive quantities act to the right, in accordance with the coordinate axis s shown in Fig. 12–1.

It is important to remember that the above equations are useful *only when the acceleration is constant and when $t = 0$, $s = s_0$, $v = v_0$.* A common example of this motion occurs when a body falls freely toward the earth. If air resistance is neglected and the distance of fall is short, then the constant *downward* acceleration of the body when it is close to the earth is approximately 9.81 m/s^2 or 32.2 ft/s^2.*

PROCEDURE FOR ANALYSIS

Coordinate System. Whenever the kinematic equations are applied, it is *very important* first to establish a position coordinate s along the path, and to specify its *fixed origin* and positive direction. Because the path is *rectilinear*, the lines of direction of the particle's position, velocity, and acceleration *never change*. Therefore these quantities can be represented as algebraic scalars. For analytical work the sense of s, v, and a can then be determined from their *algebraic signs*. In the examples that follow, the positive sense will be indicated by an arrow alongside each kinematic equation as it is applied.

Kinematic Equations. Often a mathematical relationship between *any two* of the four variables a, v, s, and t can be established by observation or experiment. When this is the case, the relationships between the remaining variables can be obtained by either differentiation or integration, using the kinematic equations $a = dv/dt$, $v = ds/dt$, or $a\,ds = v\,dv$.* Since each of these equations relates *three* variables, then, if a variable is *known* as a function of another variable, a third variable can be determined by *choosing the kinematic equation which relates all three*. For example, suppose that the *acceleration* is known as a function of *position*, $a = f(s)$. The *velocity* can be determined from $a\,ds = v\,dv$ since $f(s)$ can be substituted for a to yield $f(s)\,ds = v\,dv$. Solution for v requires integration. Note that the velocity *cannot* be obtained by using $a = dv/dt$, since $f(s)\,dt = dv$ contains two variables, s and t, on the left side and so it *cannot* be integrated. Whenever integration is performed, it is important that the position and velocity be known at a given instant in order to evaluate either the constant of integration if an indefinite integral is used, or the limits of integration if a definite integral is used. Lastly, keep in mind that Eqs. 12–4 through 12–6 have only a limited use. *Never apply these equations unless it is absolutely certain that the acceleration is constant.*

*Some standard differentiation and integration formulas are given in Appendix A.

*The proof is given in Example 13–2.

Example 12–1

The car in Fig. 12–2 moves in a straight line such that for a short time its velocity is defined by $v = (9t^2 + 2t)$ ft/s, where t is in seconds. Determine its position and acceleration when $t = 3$ s.

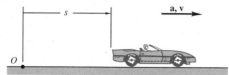

Fig. 12–2

SOLUTION

Coordinate System. The position coordinate extends from the fixed origin O to the car, positive to the right.

Position. The car's velocity is given as a function of time so that its position can be determined from $v = ds/dt$, since this equation relates v, s, and t. Noting that $s = 0$ when $t = 0$, we have*

$(\xrightarrow{+})$

$$v = \frac{ds}{dt} = (9t^2 + 2t)$$

$$\int_0^s ds = \int_0^t (9t^2 + 2t)\, dt$$

$$s \Big|_0^s = 3t^3 + t^2 \Big|_0^t$$

$$s = 3t^3 + t^2$$

When $t = 3$ s,

$$s = 3(3)^3 + (3)^2 = 90 \text{ ft} \qquad\qquad \textbf{\textit{Ans.}}$$

Acceleration. Knowing the velocity as a function of time, the acceleration is determined from $a = dv/dt$, since this equation relates a, v, and t.

$(\xrightarrow{+})$

$$a = \frac{dv}{dt} = \frac{d}{dt}(9t^2 + 2t)$$

$$= 18t + 2$$

When $t = 3$ s,

$$a = 18(3) + 2 = 56 \text{ ft/s}^2 \rightarrow \qquad\qquad \textbf{\textit{Ans.}}$$

Note that the formulas for constant acceleration *cannot* be used to solve this problem.

*The *same result* can be obtained by evaluating a constant of integration C rather than using definite limits on the integral. For example, integrating $ds = (9t^2 + 2t)\, dt$ yields $s = 3t^3 + t^2 + C$. Using the condition that at $t = 0$, $s = 0$, then $C = 0$.

Example 12–2

A small projectile is fired vertically *downward* into a fluid medium with an initial velocity of 60 m/s. If the projectile experiences a deceleration which is equal to $a = (-0.4v^3)$ m/s², where v is measured in m/s, determine the velocity v and position s four seconds after the projectile is fired.

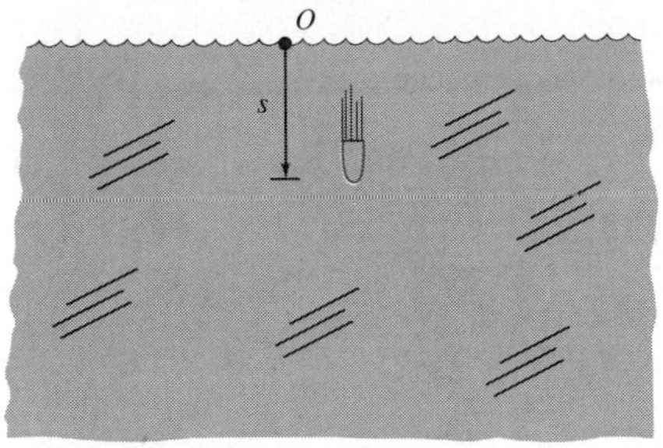

SOLUTION

Coordinate System. Since the motion is downward, the position coordinate is positive downward, with origin located at O, Fig. 12–3.

Velocity. The acceleration is given as a function of velocity so that the velocity can be obtained from $a = dv/dt$, since this equation relates v, a, and t. (Why not use $v = v_0 + a_c t$?) Separating the variables and integrating, with $v_0 = 60$ m/s when $t = 0$, yields

$(+\downarrow)$

$$a = \frac{dv}{dt} = -0.4v^3$$

Fig. 12–3

$$\int_{60}^{v} \frac{dv}{-0.4v^3} = \int_{0}^{t} dt$$

$$\frac{1}{-0.4}\left(\frac{1}{-2}\right)\frac{1}{v^2}\bigg|_{60}^{v} = t - 0$$

$$\frac{1}{0.8}\left[\frac{1}{v^2} - \frac{1}{(60)^2}\right] = t$$

$$v = \left\{\left[\frac{1}{(60)^2} + 0.8t\right]^{-1/2}\right\} \text{ m/s}$$

Here the positive root is taken, since the projectile is moving downward. When $t = 4$ s,

$$v = 0.559 \text{ m/s} \downarrow \qquad\qquad \textbf{\textit{Ans.}}$$

Position. Knowing the velocity as a function of time, we obtain the position from $v = ds/dt$, since this equation relates s, v, and t. Using the initial condition $s = 0$, when $t = 0$, we have

$(+\downarrow)$

$$v = \frac{ds}{dt} = \left[\frac{1}{(60)^2} + 0.8t\right]^{-1/2}$$

$$\int_{0}^{s} ds = \int_{0}^{t} \left[\frac{1}{(60)^2} + 0.8t\right]^{-1/2} dt$$

$$s = \frac{2}{0.8}\left[\frac{1}{(60)^2} + 0.8t\right]^{1/2}\bigg|_{0}^{t}$$

$$s = \frac{1}{0.4}\left\{\left[\frac{1}{(60)^2} + 0.8t\right]^{1/2} - \frac{1}{60}\right\} \text{ m}$$

When $t = 4$ s,

$$s = 4.43 \text{ m} \qquad\qquad \textbf{\textit{Ans.}}$$

Example 12–3

A boy tosses a ball in the vertical direction off the side of a cliff, as shown in Fig. 12–4. If the initial velocity of the ball is 15 m/s upward, and the ball is released 40 m from the bottom of the cliff, determine the maximum height s_B reached by the ball and the speed of the ball just before it hits the ground. During the entire time the ball is in motion, it is subjected to a constant downward acceleration of 9.81 m/s² due to gravity. Neglect the effect of air resistance.

SOLUTION

Coordinate System. The origin O for the position coordinate s is taken at ground level with positive upward, Fig. 12–4.

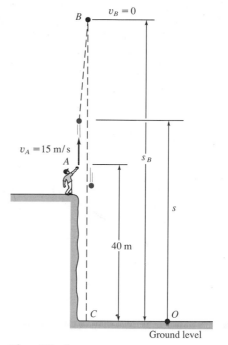

Fig. 12–4

Maximum Height. At the maximum height $s = s_B$ the velocity $v_B = 0$. Since the ball is thrown *upward* when $t = 0$, it is subjected to a velocity $v_A = +15$ m/s (positive since it is in the same sense as positive displacement). For the entire motion, the acceleration is *constant* such that $a_c = -9.81$ m/s² (negative since it acts in the *opposite* sense to positive velocity or positive displacement). Since a_c is *constant* throughout the entire motion, the position may be related to the velocity at the two points A and B on the path as follows:

$$(+\uparrow) \qquad v_B^2 = v_A^2 + 2a_c(s_B - s_A)$$
$$0 = (15)^2 + 2(-9.81)(s_B - 40)$$
$$s_B = 51.5 \text{ m} \qquad\qquad Ans.$$

Velocity. To obtain the velocity of the ball just before it hits the ground, Eq. 12–6 can be applied between points B and C, Fig. 12–4,

$$(+\uparrow) \qquad v_C^2 = v_B^2 + 2a_c(s_C - s_B)$$
$$= 0 + 2(-9.81)(0 - 51.5)$$
$$v_C = -31.8 \text{ m/s} = 31.8 \text{ m/s} \downarrow \qquad Ans.$$

The negative root was chosen since the ball is moving *downward*.

Similarly, Eq. 12–6 may also be applied between points A and C, i.e.,

$$(+\uparrow) \qquad v_C^2 = v_A^2 + 2a_c(s_C - s_A)$$
$$= 15^2 + 2(-9.81)(0 - 40)$$
$$v_C = -31.8 \text{ m/s} = 31.8 \text{ m/s} \downarrow \qquad Ans.$$

Note: It should be realized that the ball is subjected to a *deceleration* from A to B of 9.81 m/s², and then from B to C it is *accelerated* at this rate. Furthermore, even though the ball momentarily comes to *rest* at B ($v_B = 0$) the acceleration at B is 9.81 m/s² downward!

Example 12–4

A metallic particle is subjected to the influence of a magnetic field such that it travels vertically downward through a fluid that extends from plate A to plate B, Fig. 12–5. If the particle is released from rest at C, $s = 100$mm, and the acceleration is measured as $a = (4s)$ m/s^2, where s is in meters, determine the velocity of the particle when it reaches plate B, $s = 200$ mm, and the time it needs to travel from C to plate B.

SOLUTION

Coordinate System.   As shown in Fig. 12–5, s is taken positive downward, measured from plate A.

Velocity.   Knowing the acceleration as a function of position, the velocity as a function of position can be obtained by using $v\,dv = a\,ds$. Why not use the formulas for constant acceleration? Realizing that $v = 0$ at $s = 100$ mm $= 0.1$ m, we have

$(+\downarrow)$

$$v\,dv = a\,ds$$

$$\int_0^v v\,dv = \int_{0.1}^s 4s\,ds$$

$$\tfrac{1}{2}v^2 \Big|_0^v = \frac{4}{2}s^2 \Big|_{0.1}^s$$

$$v = 2(s^2 - 0.01)^{1/2} \qquad (1)$$

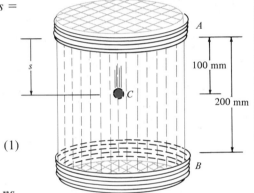

Fig. 12–5

At $s = 200$ mm $= 0.2$ m,

$$v_B = 0.346 \text{ m/s} = 346 \text{ mm/s} \downarrow \qquad \textit{Ans.}$$

The positive root is chosen since the particle is traveling downward, i.e., in the $+s$ direction.

Time.   The time for the particle to travel from C to plate B can be obtained using $v = ds/dt$ and Eq. (1), where $s = 0.1$ m when $t = 0$.

$(+\downarrow)$

$$ds = v\,dt$$

$$= 2(s^2 - 0.01)^{1/2}\,dt$$

$$\int_{0.1}^s \frac{ds}{(s^2 - 0.01)^{1/2}} = \int_0^t 2\,dt$$

$$\ln\left(s + \sqrt{s^2 - 0.01}\right) \Big|_{0.1}^s = 2t \Big|_0^t$$

$$\ln\left(s + \sqrt{s^2 - 0.01}\right) + 2.30 = 2t$$

At $s = 200$ mm $= 0.2$ m,

$$t = \frac{\ln\left(0.2 + \sqrt{(0.2)^2 - 0.01}\right) + 2.30}{2} = 0.657 \text{ s} \qquad \textit{Ans.}$$

Example 12–5

A particle moves along a horizontal straight line such that its velocity is given by $v = (3t^2 - 6t)$ m/s, where t is the time in seconds. If it is initially located at the origin O, determine the distance traveled during the time interval $t = 0$ to $t = 3.5$ s, the average velocity, and the average speed of the particle during this time interval.

SOLUTION

Coordinate System. Here we will assume positive motion to the right, measured from a fixed origin O.

Distance Traveled. Since the velocity is related to time, the position related to time may be found by integrating $v = ds/dt$ with $t = 0$, $s = 0$.

$(\overset{+}{\rightarrow})$

$$ds = v\, dt$$

$$= (3t^2 - 6t)\, dt$$

$$\int_0^s ds = 3 \int_0^t t^2\, dt - 6 \int_0^t t\, dt$$

$$s = (t^3 - 3t^2)\ \text{m} \qquad\qquad (1)$$

In order to determine the distance traveled in 3.5 s, it is necessary to investigate the path of motion. The graph of the velocity function, Fig. 12–6a, reveals that for $0 < t < 2$ s the velocity is *negative*, which means the particle is traveling to the *left*, and for $t > 2$ s the velocity is *positive*, and hence the particle is traveling to the *right*. Also, $v = 0$ at $t = 2$ s. The particle's position when $t = 0$, $t = 2$ s, and $t = 3.5$ s can be computed from Eq. (1). This yields

$$s|_{t=0} = 0, \qquad s|_{t=2s} = -4\ \text{m}, \qquad s|_{t=3.5s} = 6.12\ \text{m}$$

The path is shown in Fig. 12–6b. Hence, the distance traveled in 3.5 s is

$$s_T = 4 + 4 + 6.12 = 14.1\ \text{m} \qquad\qquad Ans.$$

Velocity. The *displacement* from $t = 0$ to $t = 3.5$ s is

$$\Delta s = s|_{t=3.5s} - s|_{t=0} = 6.12 - 0 = 6.12\ \text{m}$$

so that the average velocity is

$$v_{avg} = \frac{\Delta s}{\Delta t} = \frac{6.12}{3.5 - 0} = 1.75\ \text{m/s} \rightarrow \qquad\qquad Ans.$$

The average speed is defined in terms of the *distance traveled* s_T. Hence,

$$(v_{sp})_{avg} = \frac{s_T}{\Delta t} = \frac{14.12}{3.5 - 0} = 4.03\ \text{m/s} \qquad\qquad Ans.$$

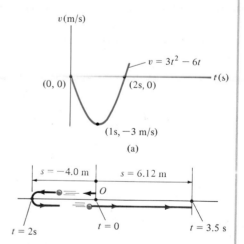

v(m/s)

$v = 3t^2 - 6t$

$(0, 0)$ $(2s, 0)$ t(s)

$(1s, -3$ m/s$)$

(a)

$s = -4.0$ m $s = 6.12$ m

O

$t = 2s$ $t = 0$ $t = 3.5$ s

(b)

Fig. 12–6

PROBLEMS

12–1. If a ball is thrown upwards 8 m into the air before it starts to fall, determine its initial velocity and the total time of flight.

12–2. A train is traveling along a straight track at 56 ft/s when its brakes are applied, subjecting it to a constant deceleration. If it comes to a stop in 800 ft, determine its deceleration and the time needed to make the stop.

12–3. A particle travels along a straight-line path such that in 4 s it moves from an initial position $s_A = -8$ m to a position $s_B = +3$ m. Then in another 5 s it moves from s_B to $s_C = -6$ m. Determine the particle's average velocity and average speed during the 9-s time interval.

*12–4. A truck, traveling along a straight road at 20 km/h, increases its speed to 120 km/h in 15 s. If its acceleration is constant, determine the distance traveled.

12–5. A balloon is traveling upwards at a constant speed of 6 m/s. When a bag of sand falls from it, it strikes the ground in 3 s. Determine the height of the balloon at the moment the bag is released. Also determine the highest point reached by the bag of sand.

12–6. A particle is moving along a straight line such that its position is given by $s = (4t - t^2)$ ft, where t is in seconds. Determine the distance traveled from $t = 0$ to $t = 5$ s, the average velocity, and the average speed of the particle during this time interval.

12–7. A sphere is fired downwards into a medium with an initial speed of 27 m/s. If it experiences a deceleration of $a = (-6t)$ m/s,^{2} where t is in seconds, determine the distance traveled before it stops.

*12–8. A particle travels along a straight line with a constant acceleration. When $s = 4$ ft, $v = 3$ ft/s and when $s = 10$ ft, $v = 8$ ft/s. Determine the velocity as a function of position.

12–9. A particle travels to the right along a straight path with a velocity $v = [10/(2 + s^2)]$ m/s, where s is in meters. Determine its position when $t = 4$ s if $s = 3$ m when $t = 0$.

12–10. A particle travels in a straight line such that for a short time 4 s $\leq t \leq$ 8 s its motion is described by $v = (4/a)$ ft/s, where a is in ft/s^2. If $v = 10$ ft/s when $t = 4$ s, determine the particle's acceleration when $t = 6$ s.

12–11. Tests reveal that a normal driver can *react* to a situation in 0.75 s before beginning to avoid a collision. It takes about 3 s for a driver having 0.1% alcohol in his system to do the same. If two such drivers are traveling on a straight road at 30 mph (44 ft/s) and their cars can decelerate at 2 ft/s^2, determine the shortest stopping distance d for each from the moment they see the pedestrians at A. *Moral:* If you must drink, please don't drive!

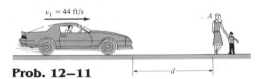

$v_1 = 44$ ft/s

A

Prob. 12–11 $\vert\!\leftarrow\!\!-d\!-\!\!\rightarrow\!\vert$

■*12–12. A particle moves along a straight-line path with an acceleration of $a = 5/(3s^{1/3} + s^{5/2})$, where s is in meters. Determine the particle's velocity when $s = 2$ m, if it starts from rest when $s = 1$ m. Use Simpson's rule to evaluate the integral.

■12–13. A projectile, initially at the origin, moves vertically downward along a straight-line path through a fluid medium such that its velocity is defined as $v = 3(8e^{-t} + t)^{1/2}$ m/s, where t is in seconds. Plot the position s of the projectile during the first 2 s. Use the Runge-Kutta method to evaluate s with incremental values of $h = 0.25$ s.

12–14. The acceleration of a rocket traveling upwards is given by $a = (6 + 0.02s)$ m/s^2, where s is in meters. Determine the rocket's velocity when $s = 2$ km and the time needed to reach this elevation. Initially, $v = 0$ and $s = 0$ when $t = 0$.

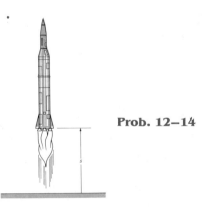

Prob. 12–14

12–15. The position of a particle traveling along a straight path is defined by $s = (t^3 - 9t^2 + 15t)$ ft, where t is in seconds. Determine the position and the total distance traveled in $t = 6$ s. *Hint:* To determine the total distance traveled, plot the path and calculate the displacements between the times when the particle stops.

*12–16. A motorcycle starts from rest and travels on a straight road with a constant acceleration of 10 ft/s² for 8 s, after which it maintains a constant speed for 2 s. Finally it decelerates at 14 ft/s² until it stops. Determine the total distance traveled and the average speed.

12–17. The position of a particle along a straight line is given by $s = (1.5t^3 - 13.5t^2 + 22.5t)$ ft, where t is in seconds. Determine the position of the particle when $t = 6$ s and the total distance it travels during the 6-s time interval. *Hint:* Plot the path to determine the total distance traveled.

12–18. A particle moving along a straight line is subjected to a deceleration $a = (-2v^3)$ m/s², where v is in m/s. If it has a velocity $v = 8$ m/s and a position $s = 10$ m when $t = 0$, determine its velocity and position when $t = 4$ s.

12–19. Plane B is traveling a distance d ahead of plane A. Both planes are traveling at 550 ft/s when the pilot of B suddenly causes his plane to decelerate at 12 ft/s². It takes the pilot of plane A 0.75 s to react and then he decelerates at 15 ft/s². Determine the minimum distance d between the planes so as to avoid a collision. Neglect the size of the planes.

Prob. 12–19

*12–20. The acceleration of a particle along a straight line is defined by $a = (2t - 9)$ m/s², where t is in seconds. At $t = 0$, $s = 1$ m and $v = 10$ m/s. When $t = 9$ s, determine (a) the particle's position, (b) the total distance traveled, and (c) the velocity. Assume the positive direction is to the right.

12–21. At $t = 0$ bullet A is fired vertically with an initial (muzzle) velocity of 450 m/s. When $t = 3$ s, bullet B is fired upwards with a muzzle velocity of 600 m/s. Determine the time t, after A is fired, as to when bullet B passes bullet A. At what elevation does this occur?

12–22. A stone is observed to fall past a 1.25-m-high window in 0.2 s. Determine (a) the average speed of the stone while it is in view, (b) the velocity of the stone when it reaches the bottom ledge, and (c) the height above the top of the window from which it fell from rest.

12–23. A particle is moving along a straight line with an initial velocity of 6 m/s when it is subjected to a deceleration of $a = (-1.5v^{1/2})$ m/s², where v is in m/s. Determine how far it travels before it stops. How much time does this take?

*12–24. Car A starts from rest at $t = 0$ and travels along a straight road with a constant acceleration of 6 ft/s² until it reaches a speed of 80 ft/s. Afterwards it maintains this speed. Also, when $t = 0$, car B located 6000 ft down the road is traveling towards A at a constant speed of 60 ft/s. Determine the time and distance traveled by car A when they pass each other.

12–25. A bicycle travels along a straight line with an acceleration of $a = (2 - v^2/500)$ ft/s², where the velocity v is in ft/s. If it starts from rest, determine its maximum speed. How far must it travel to reach a speed of 8 ft/s?

12–26. Two particles A and B start from rest at the origin $s = 0$ and move along a straight line such that $a_A = (6t - 3)$ ft/s² and $a_B = (12t^2 - 8)$ ft/s², where t is in seconds. Determine the distance between them when $t = 4$ s and the total distance each has traveled in $t = 4$ s.

12–27. As a body is projected to a high altitude above the earth's *surface*, the variation of the acceleration of gravity with respect to altitude y must be taken into account. Neglecting air resistance, this acceleration is determined from the formula $a = -g_o[R^2/(R + y)^2]$, where g_o is the constant gravitational acceleration at sea level, R is the radius of the earth, and the positive direction is measured upward. If $g_o = 9.81$ m/s² and $R = 6356$ km, determine the minimum initial velocity (escape velocity) at which a projectile should be shot vertically from the earth's surface so that it does not fall back to the earth. *Hint:* This requires that $v = 0$ as $y \rightarrow \infty$.

*12–28. Accounting for the variation of gravitational acceleration a with respect to altitude y (see Prob. 12–27), derive an equation that relates the velocity of a freely falling particle to its altitude. Assume that the particle is released from rest at an altitude y_o from the earth's surface. With what velocity does the particle strike the earth if it is released from rest at an altitude $y_o = 500$ km? Use the numerical data in Prob. 12–27.

12.2 Graphical Solutions

When a particle's motion during a time period is erratic, it may be difficult to obtain a continuous mathematical function to describe its position, velocity, or acceleration. Instead, the motion may best be described graphically using a series of curves. If this graph describes the behavior of any two of the variables, a, v, s, t, a graph describing the behavior of the other variables can be established by using the kinematic equations $a = dv/dt$, $v = ds/dt$, $a\ ds = v\ dv$. The following situations frequently occur.

Given the *s-t* Graph, Construct the *v-t* Graph

If the position of a particle can be *experimentally determined* during a time period t, the *s-t* graph for the particle can be plotted, Fig. 12–7a. Since *v = ds/dt, the v-t graph can be established by measuring the slope (ds/dt) of the s-t graph at various times and plotting the results.* (In a graphical sense, the slope is measured with a ruler and protractor.) For example, measurement of the slopes v_0, v_1, v_2, and v_3 at the intermediate points (t_0, s_0), (t_1, s_1), (t_2, s_2), and (t_3, s_3) on the *s-t* graph, Fig. 12–7a, gives corresponding points on the *v-t* graph shown in Fig. 12–7b.

It may also be possible to establish the *v-t* graph *mathematically,* provided the curves of the *s-t* graph can be expressed in the form of an equation $s = f(t)$. Corresponding equations describing curves of the *v-t* graph are then determined by *differentiation* since $v = ds/dt = d\ (f(t))/dt$.

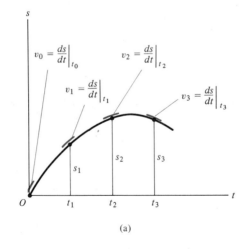

(a)

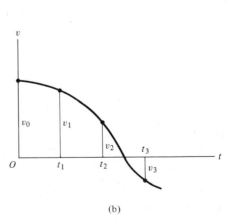

(b)

Fig. 12—7

Given the *v-t* Graph, Construct the *a-t* Graph

When the particle's *v-t* graph is known, as in Fig. 12–8*a*, the acceleration can be determined at any instant using $a = dv/dt$. Hence, *the a-t graph can be established by measuring the slope (dv/dt) of the v-t graph at various times and plotting the results*. For example, measurement of the slopes a_0, a_1, a_2, and a_3 at the intermediate points (t_0, v_0), (t_1, v_1), (t_2, v_2), and (t_3, v_3) on the *v-t* graph, Fig. 12–8*a*, yields corresponding points on the *a-t* graph shown in Fig. 12–8*b*.

Any segmented curves of the *a-t* graph can also be determined *mathematically* provided the equations of the corresponding curves of the *v-t* graph are known, $v = g(t)$. This is done by simply taking the *derivative* since $a = dv/dt = d\,(g(t))/dt$.

Since differentiation reduces a polynomial of degree *n* to that of degree *n-1*, then from the above explanation, if the *s-t* graph is parabolic (second-degree curve), the *v-t* graph will be a sloping line (first-degree curve), and the *a-t* graph will be a constant or a horizontal line (zero-degree curve).

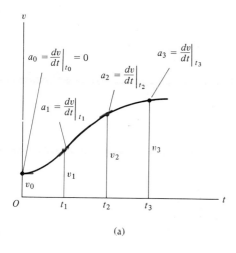

(a)

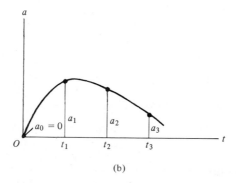

Fig. 12–8

(b)

Example 12–6

A car moves along a straight road such that its position is described by the graph shown in Fig. 12–9a. Construct the v-t and a-t graphs for the time period $0 \leqslant t \leqslant 30$ s.

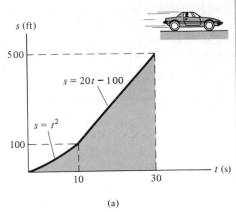

(a)

SOLUTION

v-t Graph. Since $v = ds/dt$, the v-t graph can be determined by differentiating the equations defining the s-t graph, Fig. 12–9a. We have

$$0 \leqslant t < 10 \text{ s}; \quad s = t^2 \quad v = \frac{ds}{dt} = 2t$$

$$10 \text{ s} < t \leqslant 30 \text{ s}; \quad s = 20t - 100 \quad v = \frac{ds}{dt} = 20$$

The results are plotted in Fig. 12–9b. We can also obtain specific values of v by measuring the *slope* of the s-t graph at a given instant. For example, at $t = 20$ s, the slope of the s-t graph is determined from the straight line from 10 s to 30 s, i.e.,

$$t = 20 \text{ s}; \quad v = \frac{\Delta s}{\Delta t} = \frac{500 - 100}{30 - 10} = 20 \text{ ft/s}$$

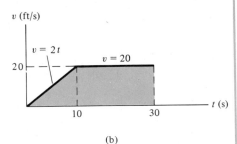

(b)

a-t Graph. Since $a = dv/dt$, the a-t graph can be determined by differentiating the equations defining the lines of the v-t graph. This yields

$$0 \leqslant t < 10 \text{ s}; \quad v = 2t \quad a = \frac{dv}{dt} = 2$$

$$10 < t \leqslant 30 \text{ s}; \quad v = 20 \quad a = \frac{dv}{dt} = 0$$

The results are plotted in Fig. 12–9c. Show how to determine the specific value of the car's acceleration when $t = 5$ s by measuring the slope of the v-t graph.

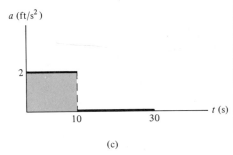

(c)

Fig. 12–9

Given the *a-t* Graph, Construct the *v-t* Graph

If the *a-t* graph is given, Fig. 12–10a, the *v-t* graph may be constructed using the equation $a = dv/dt$, written in integrated form as $\Delta v = \int a\, dt$. In this case, *the change in the particle's speed during a period of time is equal to the area under the a-t graph during the same time period*, Fig. 12–10b. (In a graphical sense, any *small* area may be approximated as a trapezoid or rectangle.) Using this method, one begins with knowing the particle's initial velocity v_0 and adding (algebraically) to this small area increments Δv determined from the *a-t* graph. In this manner, one determines successive points, $v_1 = v_0 + \Delta v$, etc., for the *v-t* graph. Notice that an algebraic addition of area is necessary, since areas lying above the *t* axis correspond to an increase in *v* ("positive" area), whereas those lying below the *t* axis indicate a decrease in *v* ("negative" area).

If the curves of the *a-t* graph can be described by a series of equations, then each of these equations may be *integrated* to yield equations describing the corresponding curves of the *v-t* graph. Hence, if the *a-t* graph is linear (first-degree curve), the integration will yield a *v-t* graph that is parabolic (second-degree curve), etc.

Given the *v-t* Graph, Construct the *s-t* Graph

When the *v-t* graph is given, Fig. 12–11a, it is possible to determine the *s-t* graph using $v = ds/dt$, written in integrated form as $\Delta s = \int v\, dt$. In this case *the particle's displacement during a period of time is equal to the area under the v-t graph during the same time period*, Fig. 12–11b. In the same manner as stated above, one begins by knowing the particle's initial position s_0 and adding (algebraically) to this small area increments Δs determined from the *v-t* graph.

If it is possible to describe the segments of the *v-t* graph by a series of equations, then each of these equations may be *integrated* to yield equations describing the *s-t* graph.

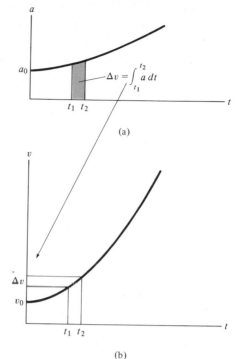

Fig. 12–10

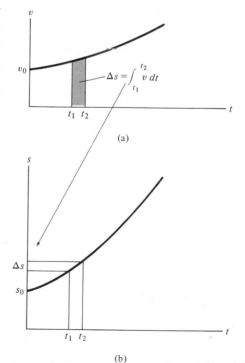

Fig. 12–11

17

Example 12-7

The rocket sled in Fig. 12-12a starts from rest and travels along a straight track such that it accelerates at a constant rate for 10 s and then decelerates at a constant rate. Draw the v-t and s-t graphs and determine the time t' needed to stop the sled. How far has the sled traveled?

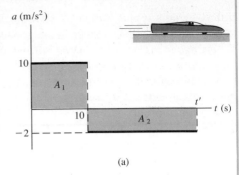

(a)

SOLUTION

v-t Graph. Since $dv = a\,dt$, the v-t graph is determined by integrating the straight-line segments of the a-t graph. Using the initial condition $v = 0$ when $t = 0$, we have

$$0 \leqslant t < 10 \text{ s}; \quad a = 10; \quad \int_0^v dv = \int_0^t 10\,dt$$
$$v = 10t$$

When $t = 10$ s, $v = 10(10) = 100$ m/s. Using this as the initial condition for the next time period, we have

$$10 \text{ s} < t \leqslant t'; \quad a = -2; \quad \int_{100}^v dv = \int_{10}^t -2\,dt$$
$$v = -2t + 120$$

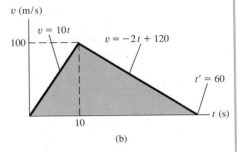

(b)

When $t = t'$ we require $v = 0$. This yields

$$t' = 60 \text{ s} \qquad \textit{Ans.}$$

The results are shown in Fig. 12-12b.

s-t Graph. Since $ds = v\,dt$, integrating the equations of the v-t graph yields the corresponding equations of the s-t graph. Using the initial condition $s = 0$ when $t = 0$, we have

$$0 \leqslant t \leqslant 10 \text{ s}; \quad v = 10t; \quad \int_0^s ds = \int_0^t 10t\,dt$$
$$s = 5t^2$$

When $t = 10$ s, $s = 5(10)^2 = 500$ m. Using this initial condition,

$$10 \leqslant t \leqslant 60 \text{ s}; \quad v = -2t + 120; \quad \int_{500}^s ds = \int_{10}^t (-2t + 120)\,dt$$
$$s - 500 = -t^2 + 120t - (-(10)^2 + 120(10))$$
$$s = -t^2 + 120t - 600$$

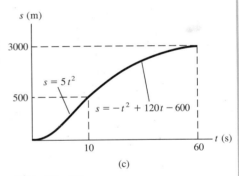

(c)

Fig. 12-12

When $t' = 60$ s, the position is

$$s = -(60)^2 + 120(60) - 600 = 3000 \text{ m} \qquad \textit{Ans.}$$

The s-t graph is shown in Fig. 12-12c. Note that a direct solution for s when $t' = 60$ s is possible, since the *triangular area* under the v-t graph would yield the displacement $\Delta s = s - 0$ from $t = 0$ to $t' = 60$ s. Hence,

$$\Delta s = \tfrac{1}{2}(60)(100) = 3000 \text{ m} \qquad \textit{Ans.}$$

Given the *a*-*s* Graph, Construct the *v*-*s* Graph

In some cases an *a*-*s* graph is known, so that points on the *v*-*s* graph can be determined by using $v \, dv = a \, ds$. Integrating this equation between the limits $v = v_1$ at $s = s_1$ and $v = v_2$ at $s = s_2$, we have $\frac{1}{2}(v_2^2 - v_1^2) = \int_{s_1}^{s_2} a \, ds$. Thus, small segments of area under the *a*-*s* graph, $\int_{s_1}^{s_2} a \, ds$, shown colored in Fig. 12–13*a*, equal one-half the difference in the squares of the speed, $\frac{1}{2}(v_2^2 - v_1^2)$. By approximation of the area, $\int_{s_1}^{s_2} a \, ds$, it is possible to compute the value of v_2 at s_2 if an initial value of v_1 at s_1 is known, i.e., $v_2 = (2 \int_{s_1}^{s_2} a \, ds + v_1^2)^{1/2}$, Fig. 12–13*b*. The *v*-*s* graph can be constructed in this manner starting from the initial velocity v_0.

Another way to construct the *v*-*s* graph is to first determine the equations which define the various curves of the *a*-*s* graph. Then the corresponding equations defining the curves of the *v*-*s* graph can be obtained directly from integration, using $v \, dv = a \, ds$.

Given the *v*-*s* Graph, Construct the *a*-*s* Graph

If the *v*-*s* graph is known, the acceleration *a* at any position *s* can be determined using the following graphical procedure. At any point $P(s, v)$, Fig. 12–14*a*, the slope dv/ds of the *v*-*s* graph is determined. Since $a \, ds = v \, dv$, then $a = v(dv/ds)$, Fig. 12–14*b*. Of course, one must work with a consistent set of units when using this procedure; for example, if *v* is measured in m/s and *s* in meters, then *a* will be calculated in m/s^2.

We can also determine the curves describing the *a*-*s* graph analytically, provided the equations of the corresponding curves of the *v*-*s* graph are known. As above, this requires application of $a \, ds = v \, dv$.

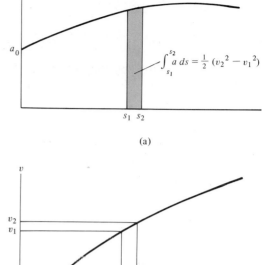

$$\int_{s_1}^{s_2} a \, ds = \frac{1}{2}(v_2{}^2 - v_1{}^2)$$

(a)

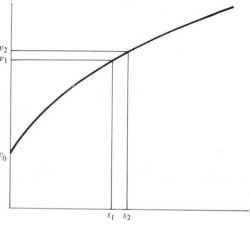

(b)

Fig. 12–13

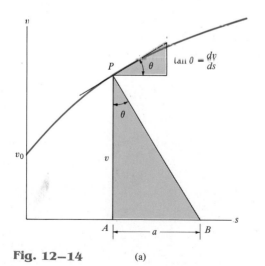

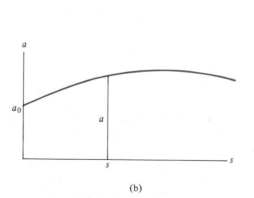

Fig. 12–14   (a)   (b)

Example 12–8

The v-s graph describing the motion of a motorcycle is shown in Fig. 12–15a. Construct the a-s graph of the motion and determine the time needed for the motorcycle to reach the position $s = 400$ ft.

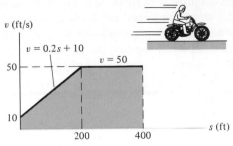

(a)

SOLUTION

a-s Graph. Since the equations of the v-s graph are given, the a-s graph can be determined using the equation $a\,ds = v\,dv$, which yields

$0 \leqslant s < 200$ ft; $v = 0.2s + 10$

$$a = v\frac{dv}{ds} = (0.2s + 10)\frac{d}{ds}(0.2s + 10)$$

$$= 0.04s + 2$$

200 ft $< s \leqslant 400$ ft; $v = 50$; $a = v\dfrac{dv}{ds} = (50)\dfrac{d}{ds}(50)$

$$= 0$$

The results are plotted in Fig. 12–15b.

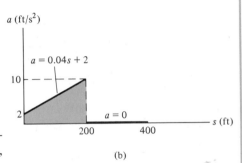

(b)

Fig. 12–15

Time. The time can be obtained using the v-s graph and $v = ds/dt$, because this equation relates v, s, and t. For the two segments of motion,

$0 \leqslant s < 200$ ft; $v = 0.2s + 10$; $dt = \dfrac{ds}{v} = \dfrac{ds}{0.2s + 10}$

$$\int_0^t dt = \int_0^s \frac{ds}{0.2s + 10}$$

$$t = 5 \ln (0.2s + 10) - 5 \ln 10$$

At $s = 200$ ft, $t = 5 \ln (0.2(200) + 10) - 5 \ln 10 = 8.05$ s.

200 ft $< s \leqslant 400$ ft; $v = 50$; $dt = \dfrac{ds}{v} = \dfrac{ds}{50}$

$$\int_{8.05}^t dt = \int_{200}^s \frac{ds}{50}$$

$$t - 8.05 = \frac{s}{50} - 4$$

$$t = \frac{s}{50} + 4.05$$

Therefore, at $s = 400$ ft,

$$t = \frac{400}{50} + 4.05 = 12.0 \text{ s} \qquad\qquad \textit{Ans.}$$

PROBLEMS

12–29. The bobsled moves down along the straight course such that its *v-t* graph is as shown. Construct the *s-t* and *a-t* graphs for the same 50-s time interval. When $t = 0$, $s = 0$.

12–30. A rocket car starting from rest travels along a straight road and for 10 s has an acceleration as shown. Construct the *v-t* graph that describes the motion and find the distance traveled in 10 s.

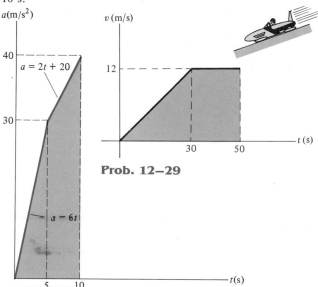

Prob. 12–30

12–31. A motorcycle starts from rest at $s = 0$ and travels along a straight road with the speed shown by the *v-t* graph. Determine the total distance the motorcycle travels until it stops when $t = 15$ s. Also plot the *a-t* and *s-t* graphs.

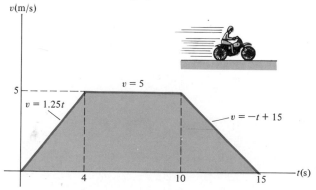

Prob. 12–31

12–29 graph: **Prob. 12–29**

*12–32. An airplane lands on the straight runway, originally traveling at 110 ft/s when $s = 0$. If it is subjected to the decelerations shown, determine the time t' needed to stop the plane and construct the *s-t* graph for the motion.

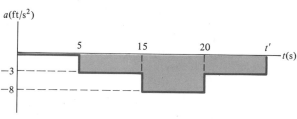

Prob. 12–32

12–33. The *s-t* graph for a train has been experimentally determined. From the data, construct the *v-t* and *a-t* graphs for the motion; $0 \leqslant t \leqslant 40$ s. For $0 \leqslant t \leqslant 30$ s, the curve is $s = (0.4t^2)$ m, where t is in seconds.

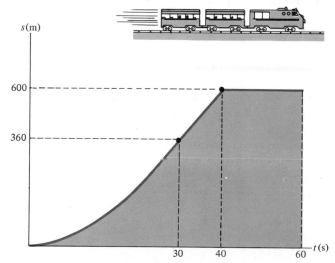

Prob. 12–33

12–34. A train starts from rest when $t = 0$ and $s = 0$. It travels along a straight track with a constant acceleration of 0.75 ft/s². When $t = 15$ s it maintains a constant speed until it has traveled a total distance of 500 ft. Determine its maximum speed and draw the *s-t* graph for the motion.

21

12–35. A two-stage rocket is fired vertically from rest at $s = 0$ with an acceleration as shown. After 30 s the first stage A burns out and the second stage B ignites. Plot the v-t and s-t graphs which describe the motion of the second stage for $0 \leqslant t \leqslant 60$ s.

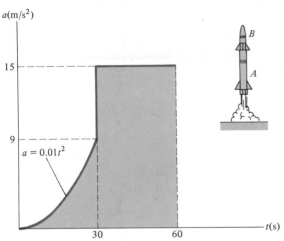

Prob. 12–35

*12–36. The v-t graph for the motion of a car as it moves along a straight road is shown. Draw the a-t graph and determine the maximum acceleration during the 30-s time interval. The car starts from rest at $s = 0$.

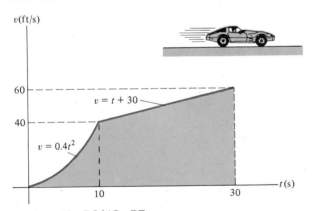

Probs. 12–36/12–37

12–37. The v-t graph for the motion of a car as it moves along a straight road is shown. Draw the s-t graph and determine the average speed and the distance traveled for the 30-s time interval. The car starts from rest at $s = 0$.

12–38. The race car travels from rest along a straight road and reaches a speed of 26 m/s in 8 s as shown on the v-t graph. The flat part of the graph is caused by shifting gears. Draw the a-t graph and determine the maximum acceleration of the car.

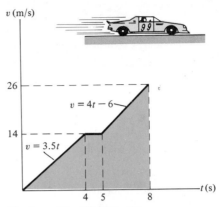

Probs. 12–38/12–39

12–39. The race car travels from rest at $s = 0$ along a straight road and reaches a speed of 26 m/s in 8 s as shown on the v-t graph. The flat part of the graph is caused by shifting gears. Draw the s-t graph and determine the total distance traveled.

*12–40. A particle is moving at $v = (4s)$ ft/s, where s is given in ft. Draw the v-t and a-t graphs for the particle for $0 \leqslant t \leqslant 5$ s. Initially, $s_0 = 10$ ft at $t = 0$.

12–41. Starting from rest at $s = 0$, a boat travels in a straight line with an acceleration as shown by the a-s graph. Determine the boat's speed when $s = 40$, 90, and 200 ft.

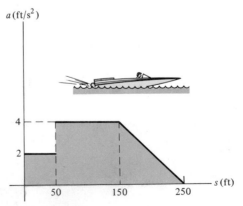

Prob. 12–41

12–42. The car is originally traveling at 15 ft/s when it is subjected to the motion shown by the *a-t* graph. Determine the car's maximum speed and the time t' when it stops.

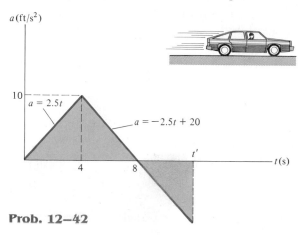

Prob. 12–42

12–43. The *v-s* graph for an airplane traveling on a straight runway is shown. Determine the acceleration of the plane at $s = 100$ m and $s = 150$ m. Draw the *a-s* graph.

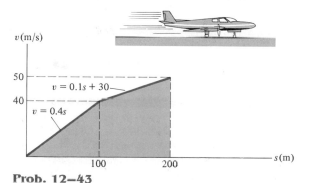

Prob. 12–43

*12–44.** Two rockets start from rest at the same ground elevation. Rocket *A* accelerates vertically at 20 m/s^2 for 12 s and then maintains a constant speed. Rocket *B* accelerates at 15 m/s^2 until reaching a constant speed of 150 m/s and then maintains this speed. Construct the *a-t*, *v-t*, and *s-t* graphs for each rocket until $t = 20$ s. What is the distance between the rockets when $t = 20$ s?

12–45. The car starts from rest at $s = 0$ and is subjected to an acceleration as shown by the *a-s* graph. Draw the *v-s* graph and determine the time needed to travel 200 ft.

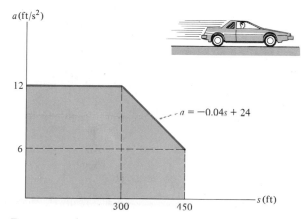

Prob. 12–45

12–46. A freight train starts from rest when $t = 0$ and travels on a straight track with a constant acceleration of 0.75 ft/s^2. When $t = t'$ it maintains a constant speed so that when $t = 160$ s the train has traveled 1500 ft. Determine the time t' and draw the *v-t* graph for the motion.

12.3 General Curvilinear Motion

When a particle moves along a curved path, the motion is called *curvilinear motion*. Because the path is often represented in three dimensions, *vector analysis* will be used to formulate the particle's position, velocity, and acceleration.* In this section the general aspects of curvilinear motion are discussed and in subsequent sections three types of coordinate systems often used to describe this motion will be introduced.

*A summary of some of the important concepts of vector analysis is given in Appendix C.

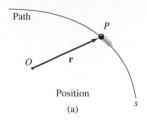

Position

(a)

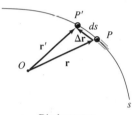

Displacement

(b)

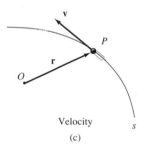

Velocity

(c)

Fig. 12–16(a–c)

Position

Consider a particle located at point P on a space curve defined by the path function s, Fig. 12–16a. The position of the particle, measured from a fixed point O, will be designated by the *position vector* $\mathbf{r} = \mathbf{r}(t)$. This vector is a function of time since, in general, both its magnitude and direction change as the particle moves along the curve.

Displacement

Suppose that during a small time interval Δt the particle moves a distance Δs along the curve to a new position P', defined by $\mathbf{r}' = \mathbf{r} + \Delta \mathbf{r}$, Fig. 12–16b. The *displacement* $\Delta \mathbf{r}$ represents the change in the particle's position and is determined by vector subtraction; i.e., $\Delta \mathbf{r} = \mathbf{r}' - \mathbf{r}$.

Velocity

During the time Δt, the *average velocity* of the particle is defined as

$$\mathbf{v}_{\text{avg}} = \frac{\Delta \mathbf{r}}{\Delta t}$$

The *instantaneous velocity* is determined from this equation by letting $\Delta t \to 0$ and consequently the direction of $\Delta \mathbf{r}$ approaches the *tangent* to the curve at point P. Hence, $\mathbf{v} = \lim\limits_{\Delta t \to 0} (\Delta \mathbf{r}/\Delta t)$ or

$$\mathbf{v} = \frac{d\mathbf{r}}{dt} \qquad (12\text{--}7)$$

Since $d\mathbf{r}$ will be tangent to the curve at P, the *direction* of $\mathbf{v}$ is *also tangent to the curve*, Fig. 12–16c. The *magnitude* of $\mathbf{v}$, which is called the *speed*, may be obtained by noting that the magnitude of the displacement $\Delta \mathbf{r}$ is the length of the straight line segment from P to P', Fig. 12–16b. Realizing that this length, Δr, approaches the arc length Δs as $\Delta t \to 0$, we have $v = \lim\limits_{\Delta t \to 0} (\Delta r/\Delta t) = \lim\limits_{\Delta t \to 0} (\Delta s/\Delta t)$, or

$$v = \frac{ds}{dt} \qquad (12\text{--}8)$$

Thus, the *speed* may be obtained by differentiating the path function s with respect to time.

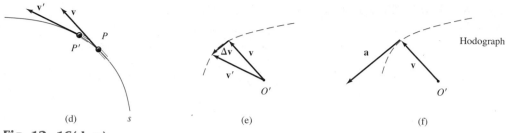

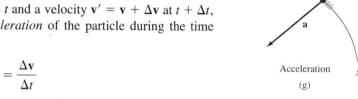

Fig. 12–16(d–g)

Acceleration

If the particle has a velocity $\mathbf{v}$ at time t and a velocity $\mathbf{v}' = \mathbf{v} + \Delta\mathbf{v}$ at $t + \Delta t$, Fig. 12–16d, then the *average acceleration* of the particle during the time interval Δt is

$$\mathbf{a}_{\text{avg}} = \frac{\Delta\mathbf{v}}{\Delta t}$$

where $\Delta\mathbf{v} = \mathbf{v}' - \mathbf{v}$. To study this time rate of change, the two velocity vectors in Fig. 12–16d are plotted in Fig. 12–16e such that their tails are located at the fixed point O' and their heads reach points on the dashed curve. This curve is called a *hodograph*, and when constructed, it describes the locus of points for the head of the velocity vector in the same manner as the *path s* describes the locus of points for the head of the position vector, Fig. 12–16a.

To obtain the *instantaneous acceleration*, let $\Delta t \to 0$ so that in the limit $\Delta\mathbf{v}$ becomes *tangent to the hodograph* and we have $\mathbf{a} = \lim\limits_{\Delta t \to 0} (\Delta\mathbf{v}/\Delta t)$, or

$$\mathbf{a} = \frac{d\mathbf{v}}{dt} \qquad (12\text{–}9)$$

Using Eq. 12 7, we can also write

$$\mathbf{a} = \frac{d^2\mathbf{r}}{dt^2}$$

By definition of the derivative, $\mathbf{a}$ acts *tangent to the hodograph*, Fig. 12–16f, and therefore, *in general*, $\mathbf{a}$ *is not tangent to the path of motion*, Fig. 12–16g. To clarify this point, realize that $\Delta\mathbf{v}$ and consequently $\mathbf{a}$ must account for the change made in *both* the magnitude *and* direction of the velocity $\mathbf{v}$ as the particle moves from P to P', Fig. 12–16d. Just a magnitude change increases (or decreases) the "length" of $\mathbf{v}$, and this in itself would allow $\mathbf{a}$ to remain tangent to the path. However, in order for the particle to follow the curve, the directional change always swings the velocity vector toward the "inside" or "concave side" of the curve, and therefore $\mathbf{a}$ cannot remain tangent to the path.

12.4 Curvilinear Motion: Rectangular Components

T2

Occasionally the motion of a particle is expressed in terms of its three rectangular components, which are measured from the origin of a fixed x, y, z frame of reference.

Position

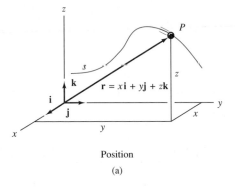

Position

(a)

If at a given instant the particle P is at point (x, y, z) on the curved path s, Fig. 12–17a, its location is then defined by the *position vector*

$$\boxed{\mathbf{r} = x\mathbf{i} + y\mathbf{j} + z\mathbf{k}} \qquad (12\text{–}10)$$

Because of the particle motion and the shape of the path, the x, y, z components of $\mathbf{r}$ are generally all functions of time; i.e., $x = x(t)$, $y = y(t)$, and $z = z(t)$, so that $\mathbf{r} = \mathbf{r}(t)$.

In accordance with the discussion in Appendix C, the *magnitude* of $\mathbf{r}$ is *always positive* and defined from Eq. C–3 as

$$r = \sqrt{x^2 + y^2 + z^2}$$

The *direction* is specified by the components of the unit vector $\mathbf{u}_r = \mathbf{r}/r$.

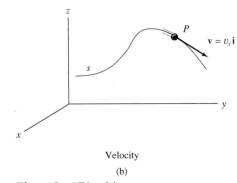

Velocity

(b)

Fig. 12–17(a, b)

Velocity

The first time derivative of $\mathbf{r}$ yields the velocity $\mathbf{v}$ of the particle. Hence,

$$\mathbf{v} = \frac{d\mathbf{r}}{dt} = \frac{d}{dt}(x\mathbf{i}) + \frac{d}{dt}(y\mathbf{j}) + \frac{d}{dt}(z\mathbf{k})$$

When taking the derivative, it is necessary to account for changes in *both* the magnitude and direction of the vector's components. From Eq. C–23, the derivative of the $\mathbf{i}$ component of $\mathbf{v}$ is therefore

$$\frac{d}{dt}(x\mathbf{i}) = \frac{dx}{dt}\mathbf{i} + x\frac{d\mathbf{i}}{dt}$$

The second term on the right side is zero, since the x, y, z reference frame is *fixed*, and therefore the *direction* (and the *magnitude*) of $\mathbf{i}$ does not change with time. Differentiation of the $\mathbf{j}$ and $\mathbf{k}$ components may be carried out in a similar manner, which yields the final result,

$$\boxed{\mathbf{v} = \frac{d\mathbf{r}}{dt} = v_x\mathbf{i} + v_y\mathbf{j} + v_z\mathbf{k}} \qquad (12\text{–}11)$$

where

$$v_x = \dot{x}$$
$$v_y = \dot{y}$$
$$v_z = \dot{z}$$

(12–12)

The "dot" notation $\dot{x}$, $\dot{y}$, $\dot{z}$ represents the first time derivatives of the parametric equations $x = x(t)$, $y = y(t)$, and $z = z(t)$, respectively.

The velocity has a *magnitude* defined as the *positive* value of

$$v = \sqrt{v_x^2 + v_y^2 + v_z^2}$$

and a *direction* that is specified by the components of the unit vector $\mathbf{u}_v = \mathbf{v}/v$. This direction is *always tangent* to the path, as shown in Fig. 12–17b.

Acceleration

The acceleration of the particle is obtained by taking the first time derivative of Eq. 12–11 (or the second time derivative of Eq. 12–10). Using dots to represent the derivatives of the components, we have

$$\mathbf{a} = \frac{d\mathbf{v}}{dt} = a_x\mathbf{i} + a_y\mathbf{j} + a_z\mathbf{k}$$

(12–13)

where

$$a_x = \dot{v}_x = \ddot{x}$$
$$a_y = \dot{v}_y = \ddot{y}$$
$$a_z = \dot{v}_z = \ddot{z}$$

(12–14)

Hence, a_x, a_y, and a_z represent, respectively, the first time derivatives of the time functions v_x, v_y, and v_z, or the second time derivatives of the time functions x, y, and z.

The acceleration has a *magnitude* defined as the *positive* value of

$$a = \sqrt{a_x^2 + a_y^2 + a_z^2}$$

and a *direction* specified by the components of the unit vector $\mathbf{u}_a = \mathbf{a}/a$. Since $\mathbf{a}$ represents the time rate of *change* in velocity, in general $\mathbf{a}$ will *not* be tangent to the path traveled by the particle, Fig. 12–17c.

PROCEDURE FOR ANALYSIS

Coordinate System. A rectangular coordinate system is often used to solve problems when the motion can be expressed in terms of its x, y, z components.

Kinematic Quantities. Since *rectilinear motion* occurs along each axis, a description of the motion of each component can be determined using $v = ds/dt$, $a = dv/dt$, and $a \, ds = v \, dv$ as outlined in Sec. 12.1 and formalized above. Once the components have been specified, the resultant motion is determined from the Pythagorean Theorem.

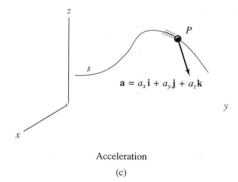

Acceleration

(c)

Fig. 12–17c

27

Example 12–9

At any instant the position of the kite in Fig. 12–18a is defined by the coordinates $x = (30t)$ ft and $y = (9t^2)$ ft, where t is given in seconds. Determine (a) the equation which describes the path and the distance of the kite from the boy when $t = 2$ s, (b) the magnitude and direction of the velocity when $t = 2$ s, and (c) the magnitude and direction of the acceleration when $t = 2$ s.

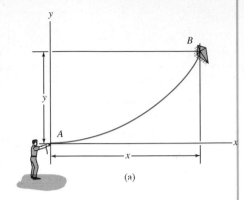

(a)

SOLUTION

Position. The equation of the path is determined by eliminating t from the expressions for x and y, i.e., $t = x/30$ so that $y = 9(x/30)^2$ or

$$y = x^2/100 \qquad \textit{Ans.}$$

This is the equation of a parabola, Fig. 12–18a. When $t = 2$ s,

$$x = 30(2) = 60 \text{ ft}, \qquad y = 9(2)^2 = 36 \text{ ft}$$

The distance from A to B is therefore

$$r = \sqrt{(60)^2 + (36)^2} = 70.0 \text{ ft} \qquad \textit{Ans.}$$

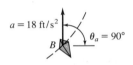

(b)

Velocity. Using Eqs. 12–12, the components of velocity when $t = 2$ s are

$$v_x = \frac{d}{dt}(30t) = 30 \text{ ft/s} \rightarrow$$

$$v_y = \frac{d}{dt}(9t^2) = 18t \Big|_{t=2\text{s}} = 36 \text{ ft/s} \uparrow$$

When $t = 2$ s, the magnitude of velocity is

$$v = \sqrt{(30)^2 + (36)^2} = 46.9 \text{ ft/s} \qquad \textit{Ans.}$$

The direction is tangent to the path, Fig. 12–18b, where

$$\theta_v = \tan^{-1}\frac{36}{30} = 50.2° \qquad \textit{Ans.}$$

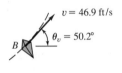

(c)

Fig. 12–18

Acceleration. The components of acceleration are determined from Eqs. 12–14,

$$a_x = \frac{d}{dt}(30) = 0$$

$$a_y = \frac{d}{dt}(18t) = 18 \text{ ft/s}^2 \uparrow$$

Here the acceleration is constant for all time and thus

$$a = \sqrt{(0)^2 + (18)^2} = 18 \text{ ft/s}^2 \qquad \textit{Ans.}$$

The direction, as shown in Fig. 12–18c, is

$$\theta_a = \tan^{-1}\frac{18}{0} = 90° \qquad \textit{Ans.}$$

Example 12–10

The motion of a bead B sliding down along the spiral path shown in Fig. 12–19 is defined by the position vector $\mathbf{r} = \{0.5 \sin (2t)\mathbf{i} + 0.5 \cos (2t)\mathbf{j} - 0.2t\mathbf{k}\}$ m, where t is given in seconds and the arguments for sine and cosine are given in radians (π rad $= 180°$). Determine the location of the bead when $t = 0.75$ s and the magnitudes of the bead's velocity and acceleration at this instant.

SOLUTION

Position. Evaluating $\mathbf{r}$ when $t = 0.75$ s yields

$$\mathbf{r}|_{t=0.75\,\text{s}} = \{0.5 \sin (1.5 \text{ rad})\mathbf{i} + 0.5 \cos (1.5 \text{ rad})\mathbf{j} - 0.2(0.75)\mathbf{k}\} \text{ m}$$
$$= \{0.499\mathbf{i} + 0.035\mathbf{j} - 0.150\mathbf{k}\} \text{ m} \qquad \textit{Ans.}$$

The magnitude of this vector represents the distance of the bead from the origin O, which is

$$r = \sqrt{(0.499)^2 + (0.035)^2 + (-0.150)^2} = 0.522 \text{ m} \qquad \textit{Ans.}$$

The direction of $\mathbf{r}$ is obtained from the components of the unit vector

$$\mathbf{u}_r = \frac{\mathbf{r}}{r} = \frac{0.499}{0.522}\mathbf{i} + \frac{0.035}{0.522}\mathbf{j} - \frac{0.150}{0.522}\mathbf{k}$$
$$= 0.956\mathbf{i} + 0.067\mathbf{j} - 0.287\mathbf{k}$$

Hence, the coordinate direction angles α, β, and γ, Fig. 12–19, are

$$\alpha = \cos^{-1} (0.956) = 17.1° \qquad \textit{Ans.}$$
$$\beta = \cos^{-1} (0.067) = 86.2° \qquad \textit{Ans.}$$
$$\gamma = \cos^{-1} (-0.287) = 106° \qquad \textit{Ans.}$$

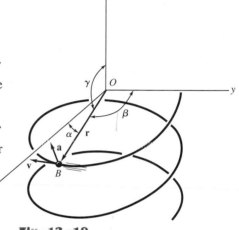

Fig. 12–19

Velocity. The velocity is defined by

$$\mathbf{v} = \frac{d\mathbf{r}}{dt} = \{1 \cos (2t)\mathbf{i} - 1 \sin (2t)\mathbf{j} - 0.2\mathbf{k}\} \text{ m/s}$$

Hence, when $t = 0.75$ s the magnitude of velocity, or speed, is

$$v = \sqrt{v_x^2 + v_y^2 + v_z^2}$$
$$= \sqrt{(1 \cos (1.5 \text{ rad}))^2 + (-1 \sin (1.5 \text{ rad}))^2 + (-0.2)^2}$$
$$= 1.02 \text{ m/s} \qquad \textit{Ans.}$$

The velocity is tangent to the path as shown in Fig. 12–19.

Acceleration. The bead's acceleration $\mathbf{a}$, which is shown in Fig. 12–19, is *not* tangent to the path. Show that

$$\mathbf{a} = \frac{d\mathbf{v}}{dt} = \{-2 \sin (2t)\mathbf{i} - 2 \cos (2t)\mathbf{j}\} \text{ m/s}^2$$

and when $t = 0.75$ s the magnitude is $a = 2$ m/s^2.

12.5 Motion of a Projectile

Free-flight motion of a projectile is often studied in terms of its rectangular components since the projectile's acceleration *always* acts in the vertical direction. To illustrate the concepts involved in the kinematic analysis, consider a projectile launched at point (x_0, y_0), as shown in Fig. 12–20. The path is defined in the x-y plane such that the initial velocity is $\mathbf{v}_0$, having components $(\mathbf{v}_x)_0$ and $(\mathbf{v}_y)_0$. When air resistance is neglected, the only force acting on the projectile is its weight, which creates a *constant downward acceleration* of approximately $a_c = g = 9.81$ m/s^2 or $g = 32.2$ ft/s^2.* Hence, $a_y = -g$ and $a_x = 0$.

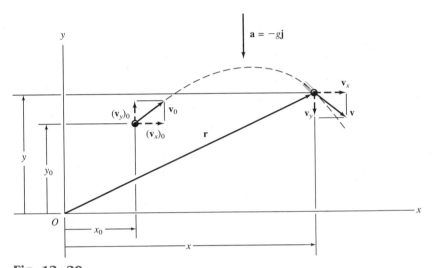

Fig. 12–20

Horizontal Motion

Since the component of acceleration in the x direction is $a_x = 0$, application of the constant acceleration equations, 12–4 to 12–6, yields

$$(\xrightarrow{+})\, v = v_0 + a_c t; \qquad\qquad v_x = (v_x)_0$$
$$(\xrightarrow{+})\, x = x_0 + v_0 t + \tfrac{1}{2} a_c t^2; \qquad x = x_0 + (v_x)_0 t$$
$$(\xrightarrow{+})\, v^2 = v_0^2 + 2 a_c (s - s_0); \qquad v_x = (v_x)_0$$

The first and last equations indicate that *the horizontal component of velocity remains constant throughout the motion.*

Vertical Motion

Since the positive y axis is directed upward, the component of acceleration in the y direction is $a_y = -g$. Applying Eqs. 12–4 to 12–6, we get

*This assumes that the earth's gravitational field does not vary with altitude (see Example 13–2).

$$(+\uparrow)v = v_0 + a_c t; \qquad\qquad v_y = (v_y)_0 - gt$$
$$(+\uparrow)y = y_0 + v_0 t + \tfrac{1}{2}a_c t^2; \qquad y = y_0 + (v_y)_0 t - \tfrac{1}{2}gt^2$$
$$(+\uparrow)v^2 = v_0^2 + 2a_c(y - y_0); \qquad v_y^2 = (v_y)_0^2 - 2g(y - y_0)$$

Recall that the last equation can be formulated on the basis of eliminating the time t between the first two equations, and therefore *only two of the above three equations are independent of one another*.

To summarize, problems involving the motion of a projectile can have at most three unknowns since only three independent equations can be written. These equations consist of one equation in the horizontal direction and two in the vertical direction. Scalar analysis can be used here because the component motions along the x and y axes are *rectilinear*. Once $\mathbf{v}_x$ and $\mathbf{v}_y$ are obtained, realize that the resultant velocity $\mathbf{v}$ is defined by the *vector sum* as shown in Fig. 12–20.

PROCEDURE FOR ANALYSIS

Using the above results, the following procedure provides a method for solving problems concerning free-flight projectile motion.

Coordinate System. Establish the fixed x, y coordinate axes and sketch the trajectory of the particle. Between any *two points* on the path specify the given problem data and the *three unknowns*. In all cases the acceleration of gravity acts downward. The particle's initial and final velocities should be represented in terms of their x and y components. Remember that positive and negative position, velocity, and acceleration components always act in accordance with the associated coordinate directions.

Kinematic Equations. Depending upon the known data and what is to be determined, a choice should be made as to which three of the following four equations should be applied between the two points on the path to obtain the most direct solution to the problem.

Horizontal Motion. The *velocity* in the horizontal or x direction is *constant*, i.e., $(v_x) = (v_x)_0$, and

$$x = x_0 + (v_x)_0 t$$

Vertical Motion. In the vertical or y direction *only two* of the following three equations can be used for solution.

$$v_y = (v_y)_0 + a_c t$$
$$y = y_0 + (v_y)_0 t + \tfrac{1}{2}a_c t^2$$
$$v_y^2 = (v_y)_0^2 + 2a_c(y - y_0)$$

The following examples numerically illustrate application of this procedure.

Example 12–11

A cannon ball is fired from point A with a horizontal muzzle velocity of 120 m/s, as shown in Fig. 12–21. If the cannon is located at an elevation of 60 m above the ground, determine the time for the cannon ball to strike the ground and the range R.

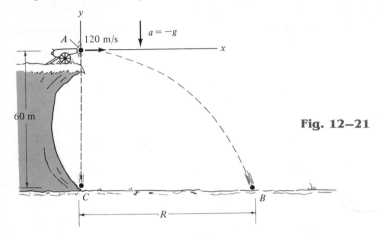

Fig. 12–21

SOLUTION

Coordinate System. The origin of coordinates is established at the initial point A, Fig. 12–21. The initial velocity of the cannon ball has components $(v_A)_x = 120$ m/s and $(v_A)_y = 0$. Also, between points A and B the acceleration is $a_y = -9.81$ m/s^2. Since $(v_B)_x = (v_A)_x = 120$ m/s, the three unknowns are $(v_B)_y$, R, and the time of flight t.

Vertical Motion. The vertical distance from A to B is known, and therefore we can obtain a direct solution for t by using the equation

$(+\uparrow)$
$$y = y_0 + (v_y)_0 t + \tfrac{1}{2}a_c t^2$$
$$-60 = 0 + 0 + \tfrac{1}{2}(-9.81)t^2$$
$$t = 3.50 \text{ s} \qquad Ans.$$

This calculation also indicates that if an object is released from rest at A at the instant the cannon is fired (horizontally), it will strike the ground at C the same instant the cannon ball strikes the ground at B.

Horizontal Motion. Since t has been calculated, R is determined as follows:

$(\xrightarrow{+})$
$$x = x_0 + (v_x)_0 t$$
$$R = 0 + 120(3.50)$$
$$R = 420 \text{ m} \qquad Ans.$$

Example 12–12

A ball is thrown from a position 5 ft above the ground to the roof of a 40-ft-high building, as shown in Fig. 12–22. If the initial velocity of the ball is 70 ft/s, inclined at an angle of 60° from the horizontal, determine the range or horizontal distance R from the point where the ball is thrown to where it strikes the roof.

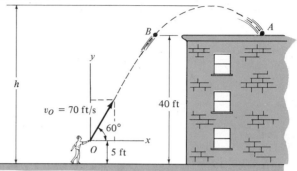

Fig. 12–22

SOLUTION

Coordinate System. When the motion is analyzed between points O and A, the three unknowns are represented as the horizontal distance R, time of flight t_{OA}, and vertical component of velocity $(v_y)_A$. With the origin of coordinates at O, Fig. 12–22, the ball's initial velocity has components of

$$(v_x)_O = 70 \cos 60° = 35 \text{ ft/s} \rightarrow$$
$$(v_y)_O = 70 \sin 60° = 60.62 \text{ ft/s} \uparrow$$

Also, $(v_x)_A = (v_x)_O = 35$ ft/s and $a_y = -32.2$ ft/s^2.

Horizontal Motion

$(\xrightarrow{+})$
$$x_A = x_O + (v_x)_O t_{OA}$$
$$R = 0 + 35t_{OA} \qquad (1)$$

Vertical Motion. Relating t_{OA} to the initial and final elevations of the ball,

$(+\uparrow)$
$$y_A = y_O + (v_y)_O t_{OA} + \tfrac{1}{2}a_c t_{OA}^2$$
$$(40 - 5) = 0 + 60.62t_{OA} + \tfrac{1}{2}(-32.2)t_{OA}^2$$

Solving for the two roots, using the quadratic formula, we have $t_{OB} = 0.712$ s and $t_{OA} = 3.05$ s. The first root (shortest time) is designated as t_{OB}, since it represents the time needed for the ball to reach point B, which has the same elevation as point A, Fig. 12–22. Substituting t_{OA} into Eq. (1) and solving for R yields

$$R = 35(3.05) = 107 \text{ ft} \qquad \qquad \textit{Ans.}$$

Using only one equation, can you show that the maximum height reached by the ball, Fig. 12–22, is $h = 62.1$ ft?

Example 12–13

When a ball is kicked from A as shown in Fig. 12–23, it just clears the top of a wall at B as it reaches its maximum height. Knowing that the distance from A to the wall is 20 m and the wall is 4 m high, determine the initial speed at which the ball was kicked. Neglect the size of the ball.

Fig. 12–23

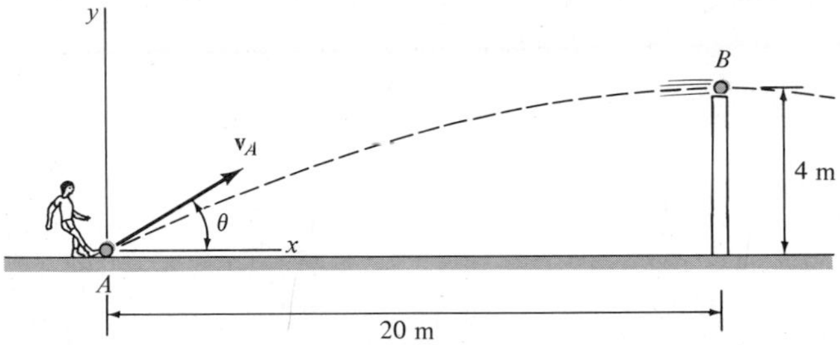

SOLUTION

Coordinate System. The three unknowns are represented by the initial speed v_A, angle of inclination θ, and the time t_{AB} to travel from A to B, Fig. 12–23. At the highest point, B, the velocity $(v_y)_B = 0$, and $(v_x)_B = (v_x)_A = v_A \cos \theta$.

Horizontal Motion

$(\xrightarrow{+})$
$$x_B = x_A + (v_x)_A t_{AB}$$
$$20 = 0 + (v_A \cos \theta)t_{AB} \qquad (1)$$

Vertical Motion

$(+\uparrow)$
$$(v_y)_B = (v_y)_A + a_c t_{AB}$$
$$0 = v_A \sin \theta - 9.81 t_{AB} \qquad (2)$$

$(+\uparrow)$
$$(v_y)_B^2 = (v_y)_A^2 + 2a_c[y_B - y_A]$$
$$0 = v_A^2 \sin^2 \theta + 2(-9.81)(4 - 0) \qquad (3)$$

To obtain v_A, eliminate t_{AB} from Eqs. (1) and (2), which yields

$$v_A^2 \sin \theta \cos \theta = 196.2 \qquad (4)$$

Solve for v_A^2 in Eq. (4) and substitute into Eq. (3), so that

$$\frac{\sin \theta}{\cos \theta} = \tan \theta = \frac{2(9.81)(4)}{196.2} = 0.4$$

$$\theta = \tan^{-1}(0.4) = 21.8°$$

Then using Eq. (4), the required initial speed is

$$v_A = \sqrt{\frac{196.2}{(\sin 21.8°)(\cos 21.8°)}} = 23.9 \text{ m/s} \qquad \textit{Ans.}$$

PROBLEMS

12–47. A car travels east 2 km for 5 minutes, then north 3 km for 8 minutes, and then west 4 km for 10 minutes. Determine the total distance traveled and the displacement of the car. Also, what is the magnitude of the average velocity and the average speed?

*12–48. If the velocity of a particle is defined as $\mathbf{v}(t) = \{1.5t^2\mathbf{i} + 1.8t\mathbf{j} + t^3\mathbf{k}\}$ m/s, where t is in seconds, determine the displacement of the particle from $t = 1$ s to $t = 3$ s.

12–49. If the velocity of a particle is defined as $\mathbf{v}(t) = \{1.5t^2\mathbf{i} + 1.8t\mathbf{j} + t^3\mathbf{k}\}$ m/s, determine the magnitude and coordinate direction angles α, β, γ of the particle's acceleration when $t = 2$ s.

12–50. A particle is subjected to an acceleration $\mathbf{a} = \{9t^2\mathbf{i} + 12t^3\mathbf{j} - 6t\mathbf{k}\}$ ft/s^2. Determine the particle's position (x, y, z) when $t = 2$ s. When $t = 0$, the particle is located at point $\mathbf{r}_0 = \{2\mathbf{i} - 1\mathbf{J} + 2\mathbf{k}\}$ ft and has a velocity of $\mathbf{v}_0 = \{2\mathbf{i} - 6\mathbf{j} + 5\mathbf{k}\}$ ft/s.

12–51. A particle travels along the path $y = x^2$, where x and y are in meters. If the particle's component of velocity in the y direction is always $v_y = 3$ m/s, determine the magnitude of the particle's velocity when $t = 2$ s. When $t = 0$ the particle is at point (1 m, 1 m). How far from the origin is the particle when $t = 2$ s?

*12–52. A particle travels with a constant speed v around a helix defined by the parametric equations $x = (5 \cos 2t)$ m, $y = (5 \sin 2t)$ m, $z = (3t)$ m, where t is in seconds. Determine the particle's acceleration as a function of time.

12–53. The position of a particle is defined by $\mathbf{r} = \{5 \cos 2t\mathbf{i} + 4 \sin 2t\mathbf{j}\}$ m, where t is in seconds and the arguments for the sine and cosine are given in radians. Determine the magnitudes of the velocity and acceleration of the particle when $t = 1$ s. Also, prove that the path of the particle is elliptical.

12–54. The two particles A and B start at the origin O and travel in opposite directions along the circular path. If A and B travel at a constant speed of $v_A = 0.7$ m/s and $v_B = 1.5$ m/s, respectively, determine in $t = 2$ s, (a) the displacement of each particle, (b) the position vector to each particle, and (c) the distance between the particles.

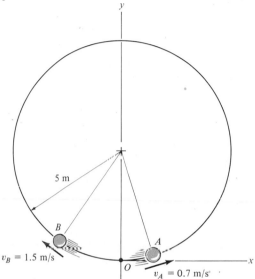

Prob. 12–54

12–55. The curvilinear motion of a particle is defined by $x = 5 \sin (4t)$, $y = 8 \cos (3t) + 2$, and $z = 14t^3$, where the x, y, z position is given in millimeters, the time in seconds, and the arguments for the sine and cosine are given in radians. Determine the magnitudes and coordinate direction angles α, β, γ of the particle's velocity and acceleration when $t = 2$ s.

*12–56. A particle travels along a path such that its position is $\mathbf{r} = \{(5 \sin t + 3)\mathbf{i} + (5 \cos t)\mathbf{j}\}$ ft, where t is in seconds and the arguments for the sine and cosine are in radians. Find the equation $y = f(x)$ which describes the path, and show that the magnitudes of the particle's velocity and acceleration are constant. What are these magnitudes, and what is the magnitude of the particle's displacement from $t = 1$ s to $t = 3$ s?

12–57. The position of particles A and B is described by the vectors $\mathbf{r}_A = \{4.5t\mathbf{i} + 13.5t(2 - t)\mathbf{j}\}$ m and $\mathbf{r}_B = \{4.5(t^2 - 2t + 2)\mathbf{i} + (4.5t - 9)\mathbf{j}\}$ m, respectively, where t is in seconds. Determine the point where the particles collide and their speeds just before the collision. How long does it take before the collision occurs?

12–58. The motion of particles A and B is described by the position vectors $\mathbf{r}_A = \{2t\mathbf{i} + (t^2 - 1)\mathbf{j}\}$ ft and $\mathbf{r}_B = \{(t + 2)\mathbf{i} + (2t^2 - 5)\mathbf{j}\}$ ft, respectively, where t is in seconds. Determine the point where the particles collide and their speeds just before the collision.

12–59. A particle moves along the path $\mathbf{r} = \{4t^4\mathbf{i} + (3t + 8t^3)\mathbf{j}\}$ in., where t is in seconds. Determine the magnitudes of the particle's velocity and acceleration when $t = 2$ s. Also determine the equation $y = f(x)$ of the path.

*12–60. The position of a particle is defined by $\mathbf{r} = \{\theta^3\mathbf{i} + \sin 2\theta\mathbf{j} + \cos^2 \theta\mathbf{k}\}$ ft, where θ is in radians. If $\theta = (2t^2)$ rad, where t is in seconds, determine the velocity $\mathbf{v}$ and acceleration $\mathbf{a}$ when $t = 1$ s. Express $\mathbf{v}$ and $\mathbf{a}$ as Cartesian vectors.

12–61. A particle is moving along the curve $y = x - (x^2/400)$, where x and y are in ft. If the velocity component in the x direction is $v_x = 2$ ft/s and remains *constant*, determine the magnitudes of the velocity and acceleration when $x = 20$ ft.

12–62. A particle moves along a hyperbolic path $x^2/16 - y^2 = 28$. If the x component of velocity is $v_x = 4$ m/s and remains constant, determine the magnitude of the particle's velocity and acceleration when it is at point (32 m, 6 m).

12–63. A particle moves along the curve $y = e^{2x}$ such that its velocity has a constant magnitude of $v = 4$ ft/s. Determine the x and y components of velocity when the particle is at $y = 5$ ft.

*12–64. The particle travels along the path defined by the parabola $y = 0.5x^2$. If the component of velocity along the x axis is $v_x = (5t)$ ft/s, where t is in seconds, determine the particle's distance from the origin O and the magnitude of acceleration when $t = 1$ s. When $t = 0$, $x = 0$, $y = 0$.

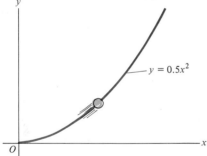

Prob. 12–64

■12–65. A particle is traveling with a velocity of $\mathbf{v} = \{3\sqrt{t}e^{-0.2t}\mathbf{i} + 4e^{-0.8t^2}\mathbf{j}\}$ m/s, where t is in seconds. Determine the magnitude of the particle's displacement from $t = 0$ to $t = 3$ s. Use Simpson's rule to evaluate the integrals. What is the magnitude of the particle's acceleration when $t = 2$ s?

12–66. It is observed that the skier leaves the ramp A and then strikes the ground at B in $t_{AB} = 5$ s. Determine his initial speed v_A and the launch angle θ_A.

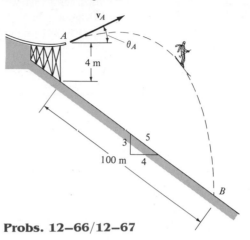

Probs. 12–66/12–67

12–67. It is observed that the skier leaves the ramp A at an angle $\theta_A = 25°$ with the horizontal. If he strikes the ground at point B, determine his initial speed v_A and the time of flight t_{AB}.

*12–68. The boy at A attempts to throw a ball over the roof of a barn with an initial speed of $v_A = 15$ m/s. Determine the angle θ_A at which the ball must be thrown so that it reaches its <u>highest</u> altitude at C. Also, find the distance d where he should stand to make the throw.

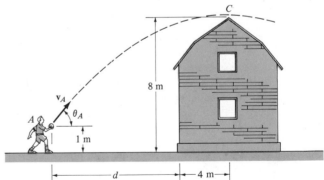

Probs. 12–68/12–69

12–69. The boy at A attempts to throw a ball over the roof of a barn such that it is launched at an angle $\theta_A = 40°$. Determine the minimum speed v_A at which he must throw the ball so it reaches its highest altitude at C. Also, find the distance d where the boy must stand so that he can make the throw.

12–70. The pitcher throws the baseball horizontally with a speed of 150 ft/s from a height of 5 ft. If the batter is 60 ft away, determine the time needed for the ball to arrive to the batter and the height h at which it passes the batter.

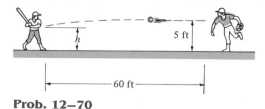

Prob. 12–70

12–71. A golf ball is struck with a velocity of 80 ft/s as shown. Determine the distance d to where it will land.

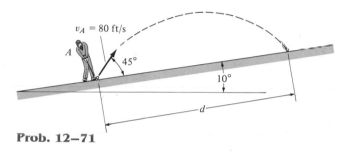

Prob. 12–71

*12–72. A ball is thrown from the ground and in 3 s it is caught by someone on top of a building, at a point which has coordinates $x = 100$ ft, $y = 40$ ft, measured from the point where the ball is thrown. Determine the velocity and angle θ at which it is thrown.

12–73. Measurements of a shot which was recorded on a videotape during a basketball game are shown. The ball passed through the hoop even though it barely cleared the hands of player B who attempted to block it. Neglecting the size of the ball, determine the magnitude v_A of its initial velocity and the height h of the ball when it passes over player B.

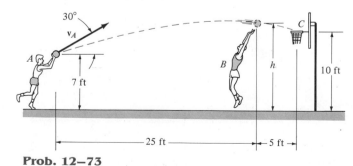

Prob. 12–73

12–74. The box is traveling at 3 m/s when it leaves the end of the chute at A. Determine the time of flight from A to B and the range R of the trajectory.

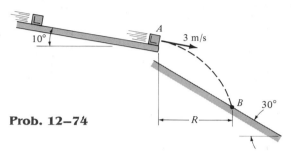

Prob. 12–74

12–75. The ball at A is kicked with a speed of $v_A = 80$ ft/s and an angle of $\theta_A = 30°$. Determine the point $(x, -y)$ where it strikes the ground. Assume the ground has the shape of a parabola as shown.

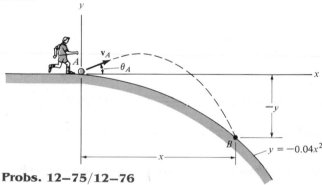

Probs. 12–75/12–76

*12–76. The ball at A is kicked such that $\theta_A = 30°$. If it strikes the ground at B having coordinates $x = 15$ ft, $y = -9$ ft, determine the speed at which it is kicked and the speed at which it strikes the ground.

12–77. The man at A wishes to throw two darts at the target at B so that they arrive at the *same time*. If each dart is thrown with a speed of 10 m/s, determine the angles θ_C and θ_D at which they should be thrown and the time between each throw. Note that the first dart must be thrown at $\theta_C(>\theta_D)$, then the second dart is thrown at θ_D.

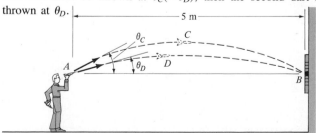

Prob. 12–77

37

12–78. The girl at A can throw a ball at $v_A = 10$ m/s. Calculate the maximum possible range R_{max} and the associated angle θ at which it should be thrown. Assume the ball is caught at B at the same elevation from which it is thrown.

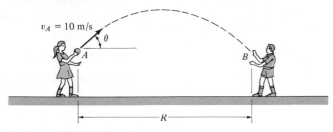

Probs. 12–78/12–79

12–79. Show that the girl at A can throw the ball to the boy at B by launching it at equal angles measured up or down from a $45°$ inclination. If $v_A = 10$ m/s, determine the range R if this value is $15°$, i.e., $\theta_1 = 45° - 15° = 30°$ and $\theta_2 = 45° + 15° = 60°$. Assume the ball is caught at the same elevation from which it is thrown.

*12–80. The baseball player A hits the baseball at $v_A = 40$ ft/s and $\theta_A = 60°$ from the horizontal. When the ball is directly overhead of player B he begins to run under it. Determine the constant speed at which B must run and the distance d in order to make the catch at the same elevation at which the ball was hit.

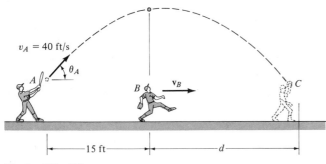

Prob. 12–80

12–81. The football player at A throws the ball in the y-z plane with a speed of $v_A = 50$ ft/s and an angle of $\theta_A = 60°$ with the horizontal. At the instant the ball is thrown, the player at B is running with constant speed along the line BC in order to catch it. Determine this speed, v_B, so that he makes the catch at the same elevation from which the ball was thrown.

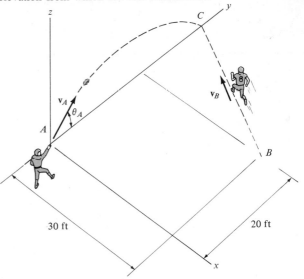

Probs. 12–81/12–82

12–82. The football player at A throws the ball in the y-z plane with a speed of $v_A = 50$ ft/s and an angle of $\theta_A = 60°$ with the horizontal. The player at B reacts within 0.1 s after seeing the ball thrown. If he can run at a maximum speed of $v_B = 23$ ft/s along the line BC, determine if he can arrive at C, running at constant speed, in order to catch the ball.

12–83. A boy at O throws a ball in the air with a speed v_0 at an angle θ_1. If he then throws another ball at the same speed v_0 at an angle $\theta_2 < \theta_1$, determine the time between the throws so the balls collide in mid air at B.

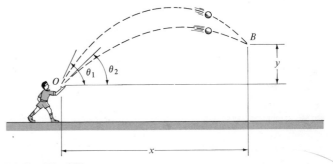

Prob. 12–83

12.6 Curvilinear Motion: Normal and Tangential Components

When the path along which a particle is moving is *known*, it is often convenient to describe the motion using n and t coordinates which act normal and tangent to the path, respectively, and at the instant considered have their *origin located at the particle*.

Planar Motion

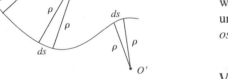

Position

(a)

Consider the particle P shown in Fig. 12–24a, which is moving in a plane such that at a given instant it is located along a fixed curve at position s, measured from point O. We will now consider a coordinate system that has its origin at a *fixed point* on the curve, and at the instant considered this origin happens to *coincide* with the location of the particle. The t axis is *tangent* to the curve at P and is positive in the direction of *increasing s*. We will designate this positive direction with the unit vector $\mathbf{u}_t$. A unique choice for the *normal axis* can be made by considering the fact that geometrically the curve consists of a series of differential arc segments ds. As shown in Fig. 12–24b, each segment ds is constructed from the arc of an associated circle having a *radius of curvature ρ* (rho) and *center of curvature O'*. The normal axis n which will be chosen is perpendicular to the t axis and is directed from P *toward* the center of curvature O', Fig. 12–24a. This positive direction, which is *always* on the concave side of the curve, will be designated by the unit vector $\mathbf{u}_n$. The plane which contains the n and t axes is referred to as the *osculating plane*, and in this case it is fixed in the plane of motion.*

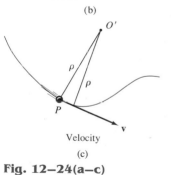

Radius of curvature

(b)

Velocity

(c)

Fig. 12–24(a–c)

Velocity

Since the particle is moving, s is a function of time. As indicated in Sec. 12.3, the particle's velocity $\mathbf{v}$ has a *direction* that is *always tangent to the curve*, Fig. 12–24c, and a *magnitude* that is determined by taking the time derivative of the path function $s = s(t)$, i.e., $v = ds/dt$ (Eq. 12–8). Hence

$$\mathbf{v} = v\mathbf{u}_t \qquad (12\text{–}15)$$

where

$$v = \dot{s} \qquad (12\text{–}16)$$

---

*The osculating plane may also be defined as that plane which has the greatest contact with the curve at a point. It is the limiting position of a plane contacting both the point and the arc segment ds. As noted above, the osculating plane is always coincident with a plane curve; however, each point on a three-dimensional curve has a unique osculating plane.

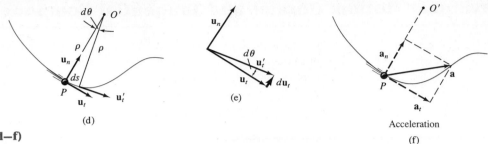

Fig. 12–24(d–f)

(d)

(e)

Acceleration

(f)

Acceleration

The acceleration of the particle is the time rate of change of the velocity. Thus,

$$\mathbf{a} = \dot{\mathbf{v}} = \dot{v}\mathbf{u}_t + v\dot{\mathbf{u}}_t \tag{12–17}$$

The first term, $\dot{v}\mathbf{u}_t$, represents the change in the magnitude of velocity, i.e., $\dot{v}$, whereas the second term, $v\dot{\mathbf{u}}_t$, represents the change in the velocity's direction. In order to compute the time derivative $\dot{\mathbf{u}}_t$, note that as the particle moves along the arc ds in time dt, $\mathbf{u}_t$ preserves its magnitude of unity; however, it changes its *direction*, so that it becomes $\mathbf{u}'_t$, Fig. 12–24d. The change $d\mathbf{u}_t$ in $\mathbf{u}_t$ is shown in Fig. 12–24e. Here $d\mathbf{u}_t$ stretches between the heads of $\mathbf{u}_t$ and $\mathbf{u}'_t$, which lie on an infinitesimal arc of radius $u_t = 1$. Hence, $d\mathbf{u}_t$ has a *magnitude* of $du_t = (1)\, d\theta$ and its *direction* is defined by $\mathbf{u}_n$. Consequently, $d\mathbf{u}_t = d\theta\mathbf{u}_n$ so that the time derivative becomes $\dot{\mathbf{u}}_t = \dot{\theta}\mathbf{u}_n$. Since $ds = \rho\, d\theta$, Fig. 12–24d, then $\dot{\theta} = \dot{s}/\rho$, and consequently

$$\dot{\mathbf{u}}_t = \frac{\dot{s}}{\rho}\mathbf{u}_n = \frac{v}{\rho}\mathbf{u}_n$$

Substituting into Eq. 12–17, $\mathbf{a}$ can be written as the sum of its two components,

$$\mathbf{a} = a_t\mathbf{u}_t + a_n\mathbf{u}_n \tag{12–18}$$

where

$$a_t = \dot{v} \qquad \text{or} \qquad a_t = v\frac{dv}{ds} \tag{12–19}$$

and

$$a_n = \frac{v^2}{\rho} \tag{12–20}$$

These two mutually perpendicular components are shown in Fig. 12–24f, in which case the *magnitude* of acceleration is the positive value of

$$a = \sqrt{a_t^2 + a_n^2} \tag{12–21}$$

Fig. 12–25

To summarize the above concepts consider the following two special cases of motion.

1. If the particle moves along a straight line, then $\rho \to \infty$ and from Eq. 12–20, $a_n = 0$. Thus $a = a_t = \dot{v}$, and we can conclude that the *tangential component of acceleration represents the time rate of change in the magnitude of the velocity*.
2. If the particle moves along a curve with a constant speed, then $a_t = \dot{v} = 0$ and $a = a_n = v^2/\rho$. Therefore, the *normal component of acceleration represents the time rate of change in the direction of the velocity*. Since it *always* acts toward the center of curvature, this component is sometimes referred to as the *centripetal acceleration*.

As a result of these interpretations, a particle moving along the curved path shown in Fig. 12–25 will have accelerations directed as shown in the figure.

Three-Dimensional Motion

If the particle is moving along a space curve, Fig. 12–26, then at a given instant the t axis is uniquely specified; however, an infinite number of straight lines can be constructed normal to the tangent axis at P. As in the case of planar motion, we will choose the positive n axis directed from P toward the path's center of curvature O'. This axis is referred to as the *principal normal* to the curve at P. With the n and t axes so defined, Eqs. 12–15 to 12–21 can be used to determine $\mathbf{v}$ and $\mathbf{a}$. Since $\mathbf{u}_t$ and $\mathbf{u}_n$ are always perpendicular to one another, and lie in the osculating plane, for spatial motion a third unit vector, $\mathbf{u}_b$, defines a *binormal axis* b which is perpendicular to $\mathbf{u}_t$ and $\mathbf{u}_n$, Fig. 12–26. Using the vector cross product, these three vectors are related by the equation $\mathbf{u}_b = \mathbf{u}_t \times \mathbf{u}_n$.

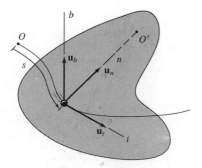

Fig. 12–26

PROCEDURE FOR ANALYSIS

Coordinate System. Provided the *path* of the particle is *known*, we can establish a set of *n* and *t* coordinates having a *fixed origin* which is coincident with the particle at the instant considered. The positive tangent axis acts in the direction of motion and the positive normal axis is directed toward the path's center of curvature. This set of axes is particularly advantageous for studying the velocity and acceleration of the particle, because the *t* and *n* components of **a** have been readily formulated.

Velocity. The particle's *velocity* is tangent to the path. Its magnitude is found from the time derivative of the path function.

$$v = \dot{s}$$

Note that v can be positive or negative depending upon whether it acts in the positive or negative s direction.

Tangential Acceleration. The *tangential component of acceleration is the result of the time rate of change in the magnitude of velocity*. This component acts in the positive s direction if the particle's speed is increasing or in the opposite direction if the speed is decreasing. The magnitudes of **v** and $\mathbf{a}_t$ are related to the path function s and time t by the equations of rectilinear motion, namely,

$$a_t = \dot{v} \qquad a_t \, ds = v \, dv$$

If a_t is *constant*, $a_t = (a_t)_c$, the above equations, when integrated, yield

$$s = s_0 + v_0 t + \tfrac{1}{2}(a_t)_c t^2$$
$$v = v_0 + (a_t)_c t$$
$$v^2 = v_0^2 + 2(a_t)_c(s - s_0)$$

Normal Acceleration. The *normal component of acceleration is the result of the time rate of change in the direction of the particle's velocity*. This component is *always* directed toward the center of curvature of the path, i.e., along the positive n axis. Its magnitude is determined from

$$a_n = \frac{v^2}{\rho}$$

In particular, if the path is expressed as $y = f(x)$, the radius of curvature ρ is computed from the equation*

$$\rho = \left| \frac{[1 + (dy/dx)^2]^{3/2}}{d^2y/dx^2} \right|$$

The following examples numerically illustrate application of these equations.

*The derivation of this result is given in any standard calculus text.

Example 12–14

A skier travels with a constant speed of 6 m/s along the parabolic path $y = \frac{1}{20}x^2$ shown in Fig. 12–27. Determine his velocity and acceleration at the instant he arrives at A. Neglect the size of the skier in the calculation.

SOLUTION

Coordinate System. Although the path has been expressed in terms of its x and y coordinates, we can still establish the origin of the n, t axes at the fixed point A and determine the components of $\mathbf{v}$ and $\mathbf{a}$ along these axes, Fig. 12–27a.

Velocity. By definition, the velocity is always directed tangent to the path. Since $y = \frac{1}{20}x^2$, $dy/dx = \frac{1}{10}x$, then $dy/dx|_{x=10} = 1$. Hence, at A, $\mathbf{v}$ makes an angle of $\theta = \tan^{-1} 1 = 45°$ with the x axis, Fig. 12–27: Therefore,

$$v_A = 6 \text{ m/s} \qquad {}_{45°}\nearrow \mathbf{v}_A \qquad\qquad \textit{Ans.}$$

Acceleration. The acceleration is computed by using $\mathbf{a} = \dot{v}\mathbf{u}_t + (v^2/\rho)\mathbf{u}_n$. However, it is first necessary to determine the radius of curvature of the path at A (10 m, 5 m). Since $d^2y/dx^2 = \frac{1}{10}$, then

$$\rho = \left| \frac{[1 + (dy/dx)^2]^{3/2}}{d^2y/dx^2} \right| = \left| \frac{[1 + (\frac{1}{10}x)^2]^{3/2}}{\frac{1}{10}} \right|_{x=10\text{m}} = 28.3 \text{ m}$$

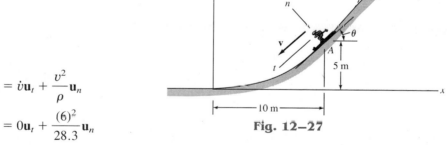

Fig. 12–27

The acceleration becomes

$$\mathbf{a}_A = \dot{v}\mathbf{u}_t + \frac{v^2}{\rho}\mathbf{u}_n$$

$$= 0\mathbf{u}_t + \frac{(6)^2}{28.3}\mathbf{u}_n$$

$$= \{1.27\mathbf{u}_n\} \text{ m/s}^2$$

Since $\mathbf{a}_A$ acts in the direction of the positive n axis, it makes an angle of $\theta + 90° = 135°$ with the positive x axis. Hence,

$$a_A = 1.27 \text{ m/s}^2 \qquad {}^{\mathbf{a}_A}\diagdown{}^{135°} \qquad\qquad \textit{Ans.}$$

Note: By using n, t coordinates, we were able to readily solve this problem since the n and t components account for the *separate* changes in the magnitude and direction of $\mathbf{v}$, and each of these changes have been directly formulated as Eqs. 12–19 and 12–20, respectively.

Example 12–15

A race car C travels around the horizontal circular track that has a radius of 300 ft, Fig. 12–28. If the car increases its speed at a constant rate of 7 ft/s² starting from rest, determine the time needed for it to reach an acceleration of 8 ft/s². What is its speed at this instant?

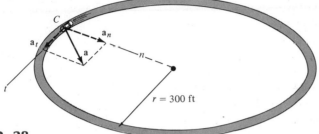

Fig. 12–28

SOLUTION

Coordinate System. The origin of the n and t axes is coincident with the car at the instant considered. The t axis is in the direction of motion, and the positive n axis is directed toward the center of the circle. This coordinate system is selected since the path is known.

Acceleration. The magnitude of acceleration can be related to its components using $a = \sqrt{a_t^2 + a_n^2}$. Here $a_t = 7$ ft/s². Also $a_n = v^2/\rho$, where $\rho = 300$ ft and

$$v = v_0 + (a_t)_c t$$
$$v = 0 + 7t$$

Thus

$$a_n = \frac{v^2}{\rho} = \frac{(7t)^2}{300} = 0.163t^2 \text{ ft/s}^2$$

The time needed for the acceleration to reach 8 ft/s² is therefore

$$a = \sqrt{a_t^2 + a_n^2}$$
$$8 = \sqrt{(7)^2 + (0.163t^2)^2}$$

Solving for the positive value of t yields

$$0.163t^2 = \sqrt{(8)^2 - (7)^2}$$
$$t = 4.87 \text{ s} \qquad \qquad \textit{Ans.}$$

Velocity. The speed at time $t = 4.87$ s is

$$v = 7t = 7(4.87) = 34.1 \text{ ft/s} \qquad \qquad \textit{Ans.}$$

Example 12–16

A car starts from rest at point A and travels along the horizontal track shown in Fig. 12–29a. During the motion, the increase in speed is $a_t = (0.2t)$ m/s^2, where t is in seconds. Determine the magnitude of the car's acceleration when it arrives at point B.

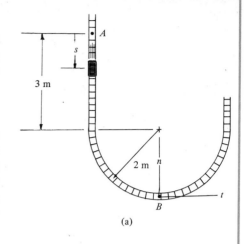

(a)

SOLUTION

Coordinate System. The position of the car at any instant is defined from the fixed point A using the position or path coordinate s, Fig. 12–29a. The car's acceleration is to be determined at B, so the origin of the n, t axes is at this point. Why are n and t coordinates selected to solve this problem?

The acceleration is computed from $\mathbf{a} = \dot{v}\mathbf{u}_t + (v^2/\rho)\mathbf{u}_n$. To apply this equation, however, it is first necessary to formulate v and $\dot{v}$ so that they may be evaluated at B. Since $v_A = 0$ when $t = 0$, then

$$a_t = \dot{v} = 0.2t \qquad (1)$$

$$\int_0^v dv = \int_0^t 0.2t\, dt$$

$$v = 0.1t^2 \qquad (2)$$

The time needed for the car to reach point B can be determined by realizing that the position of B is $s_B = 3 + 2\pi(2)/4 = 6.14$ m, Fig. 12–29a, and since $s_A = 0$ when $t = 0$ we have

$$v = \frac{ds}{dt} = 0.1t^2$$

$$\int_0^{6.14} ds = \int_0^{t_B} 0.1t^2\, dt$$

$$6.14 = 0.0333t_B^3$$

$$t_B = 5.69 \text{ s}$$

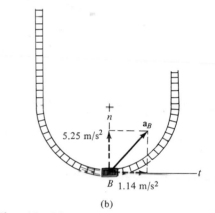

(b)

Fig. 12–29

Substituting into Eqs. (1) and (2) yields

$$(a_t)_B = 0.2(5.69) = 1.14 \text{ m/s}^2$$

$$v_B = 0.1(5.69)^2 = 3.24 \text{ m/s}$$

At B, $\rho_B = 2$ m, so that

$$(a_n)_B = \frac{v_B^2}{\rho_B} = \frac{(3.24)^2}{2} = 5.25 \text{ m/s}^2$$

The magnitude of $\mathbf{a}_B$, Fig. 12–29b, is therefore

$$a_B = \sqrt{(1.14)^2 + (5.25)^2} = 5.37 \text{ m/s}^2 \qquad \textit{Ans.}$$

45

PROBLEMS

*12–84. The car travels along the curve having a radius of 300 m. If its speed is uniformly increased from 15 m/s to 27 m/s in 3 s, determine the magnitude of its acceleration at the instant its speed is 20 m/s.

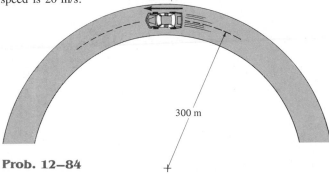

20 m/s

300 m

Prob. 12–84

12–85. At a given instant the jet plane has a speed of 400 ft/s and an acceleration of 70 ft/s² acting in the direction shown. Determine the rate of increase in the plane's speed and the radius of curvature ρ of the path.

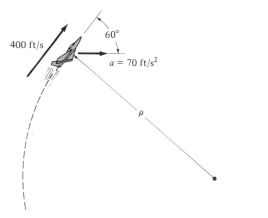

400 ft/s

60°

$a = 70$ ft/s²

ρ

Prob. 12–85

12–86. A particle travels along the path $y = a + bx + cx^2$, where a, b, c are constants. If the speed of the particle is constant, $v = v_0$, determine the x and y components of velocity and the normal component of acceleration when $x = 0$.

12–87. A particle P moves along the curve $y = (x^2 - 9)$ ft with a constant speed of 15 ft/s. Determine the point on the curve where the maximum magnitude of acceleration occurs and compute its value.

*12–88. The truck travels in a circular path having a radius of 50 m at a speed of 4 m/s. For a short distance from $s = 0$, its speed is increased by $\dot{v} = (0.05s)$ m/s², where s is in meters. Determine its speed and the magnitude of its acceleration when it has moved $s = 10$ m.

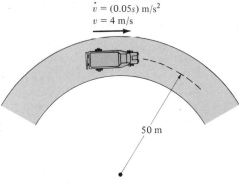

$\dot{v} = (0.05s)$ m/s²
$v = 4$ m/s

50 m

Prob. 12–88

12–89. A rocket follows a path such that its acceleration is defined by $\mathbf{a} = \{16\mathbf{i} + 4t\mathbf{j}\}$ ft/s². If it starts from rest at $\mathbf{r} = \mathbf{0}$, determine the speed of the rocket and the radius of curvature of its path when $t = 10$ s.

12–90. A particle moves along the curve $y = 180/x^2$ such that when it is at $x = 5$ ft it has a speed of 8 ft/s, which is increasing at 12 ft/s². Determine the magnitude of the particle's acceleration at this instant.

12–91. A car moves along a circular track of radius 250 ft such that its speed for a short period of time $0 \leq t \leq 4$ s is $v = 3(t + t^2)$ ft/s, where t is in seconds. Determine the magnitude of its acceleration when $t = 3$ s. How far has it traveled in $t = 3$ s?

*12–92. The plane travels along the vertical parabolic path at a constant speed of 200 m/s. Determine the magnitude of acceleration of the plane when it is at point A.

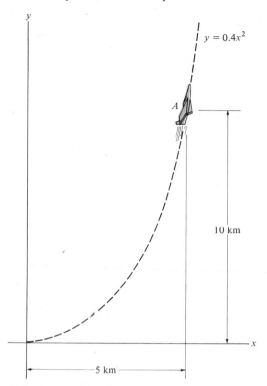

Probs. 12–92/12–93

12–93. The jet plane travels along the vertical parabolic path. When it is at point A it has a speed of 200 m/s, which is increasing at the rate of 0.8 m/s². Determine the magnitude of acceleration of the plane when it is at point A.

12–94. The automobile is originally at rest at $s = 0$. If it then starts to increase its speed according to the equation $\dot{v} = (0.05t^2)$ ft/s², where t is in seconds, determine the magnitudes of its velocity and acceleration when $t = 18$ s.

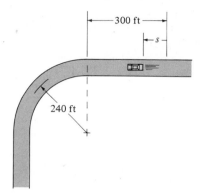

Probs. 12–94/12–95

12–95. The automobile is originally at rest at $s = 0$. If it then starts to increase its speed at $\dot{v} = (0.05t^2)$ ft/s², where t is in seconds, determine the magnitudes of its velocity and acceleration at $s = 550$ ft.

*12–96. The car B turns such that its speed is increased by $\dot{v}_B = (0.5e^t)$ m/s², where t is in seconds. If the car starts from rest when $\theta = 0°$, determine the magnitudes of the velocity and acceleration of the car when the arm AB rotates $\theta = 30°$. Neglect the size of the car.

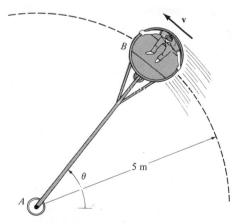

Prob. 12–96

12–97. Starting from rest a bicyclist travels around a horizontal circular path, $\rho = 10$ m, at a speed $v = (0.09t^2 + 0.1t)$ m/s, where t is in seconds. Determine the magnitudes of his velocity and acceleration when he has traveled $s = 3$ m.

12–98. A boat has an initial speed of 16 ft/s. If it increases its speed along a circular path of radius $\rho = 80$ ft at the rate of $\dot{v} = (1.5s)$ ft/s, where s is in feet measured along the path, determine the normal and tangential components of the boat's acceleration at $s = 16$ ft.

12–99. A boat has an initial speed of 16 ft/s. If it increases its speed along a circular path of radius $\rho = 80$ ft at the rate of $\dot{v} = (1.5s)$ ft/s, where s is in feet measured along the path, determine the time needed for the boat to travel 50 ft.

*12–100. The ball is kicked with an initial speed of $v_A = 8$ m/s at an angle of $\theta_A = 40°$ with the horizontal. Find the equation of the path, $y = f(x)$, and then determine the ball's velocity and the normal and tangential components of acceleration when $t = 0.25$ s.

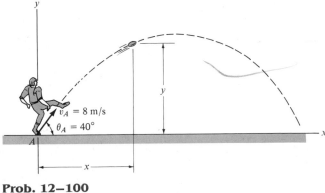

Prob. 12–100

12.7 Curvilinear Motion: Cylindrical Components

In some engineering problems it is often convenient to express the path of motion of a particle in terms of cylindrical coordinates, r, θ, and z. If motion is restricted to the plane, the polar coordinates r and θ are used.

Polar Coordinates

We can specify the location of particle P shown in Fig. 12–30a using both the *radial coordinate r*, which extends outward from the fixed origin O to the particle, and a *transverse coordinate θ*, which is the counterclockwise angle between a fixed reference line and the r axis. The angle is generally measured in degrees or radians, where 1 rad $= 180°/\pi$. The positive directions of the r and θ coordinates are defined by the unit vectors $\mathbf{u}_r$ and $\mathbf{u}_\theta$, respectively. These directions are *perpendicular* to one another. Here $\mathbf{u}_r$ or the radial direction $+r$ extends from P along increasing r, when θ is held fixed, and $\mathbf{u}_\theta$ or $+\theta$ extends from P in a direction that occurs when r is held fixed and θ is increased.

Position

At any instant the position of the particle, Fig. 12–30a, is defined by the position vector

$$\mathbf{r} = r\mathbf{u}_r \qquad (12\text{–}22)$$

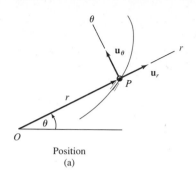

Position
(a)

Velocity

The instantaneous velocity $\mathbf{v}$ is obtained by taking the time derivative of $\mathbf{r}$. Using a dot to represent time differentiation, we have

$$\mathbf{v} = \dot{\mathbf{r}} = \dot{r}\mathbf{u}_r + r\dot{\mathbf{u}}_r$$

In order to evaluate $\dot{\mathbf{u}}_r$, notice that $\mathbf{u}_r$ changes only its direction with respect to time, since by definition the magnitude of this vector is always unity. Hence, when θ changes by an increasing amount $\Delta\theta$ during the time Δt, $\mathbf{u}_r$ becomes unit vector $\mathbf{u}_r'$, Fig. 12–30b. The time change in $\mathbf{u}_r$ is then $\Delta\mathbf{u}_r$. For small angles $\Delta\theta$ this vector has a magnitude $\Delta u_r \approx 1(\Delta\theta)$ and acts in the $\mathbf{u}_\theta$ direction, i.e., $\Delta\mathbf{u}_r = \Delta\theta\mathbf{u}_\theta$. Thus,

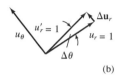

(b)

$$\dot{\mathbf{u}}_r = \lim_{\Delta t \to 0} \frac{\Delta\mathbf{u}_r}{\Delta t} = \left(\lim_{\Delta t \to 0} \frac{\Delta\theta}{\Delta t}\right)\mathbf{u}_\theta$$

$$\dot{\mathbf{u}}_r = \dot{\theta}\mathbf{u}_\theta \qquad (12\text{–}23)$$

Substituting into the above equation, the velocity can be written in component form as

$$\boxed{\mathbf{v} = v_r\mathbf{u}_r + v_\theta\mathbf{u}_\theta} \qquad (12\text{–}24)$$

where

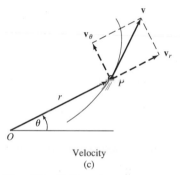

Velocity
(c)

Fig. 12–30(a–c)

$$\boxed{\begin{aligned} v_r &= \dot{r} \\ v_\theta &= r\dot{\theta} \end{aligned}} \qquad (12\text{–}25)$$

These components are shown graphically in Fig. 12–30c. Note that the *radial component* of velocity $\mathbf{v}_r$ is a measure of the rate of increase or decrease in the length of the radial coordinate, i.e., $\dot{r}$, whereas the *transverse component* of velocity, $\mathbf{v}_\theta$, can be interpreted as the rate of motion along the circumference of a circle having a radius r. In particular, the term $\dot{\theta} = d\theta/dt$ is called the *angular velocity* since it provides a measure of the time rate of change of the angle θ. Common units used for this measurement are rad/s.

Since the components are mutually perpendicular, the *magnitude* of velocity is simply the positive value of

$$v = \sqrt{(\dot{r})^2 + (r\dot{\theta})^2} \qquad (12\text{–}26)$$

and the *direction* of $\mathbf{v}$ is, of course, tangent to the path at P, Fig. 12–30c.

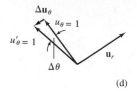

(d)

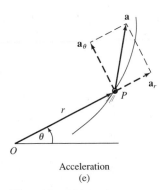

Acceleration
(e)

Fig. 12–30(d, e)

Acceleration

Taking the time derivatives of Eq. 12–24, using Eqs. 12–25, we obtain the particle's instantaneous acceleration,

$$\mathbf{a} = \dot{\mathbf{v}} = \ddot{r}\mathbf{u}_r + \dot{r}\dot{\mathbf{u}}_r + \dot{r}\dot{\theta}\mathbf{u}_\theta + r\ddot{\theta}\mathbf{u}_\theta + r\dot{\theta}\dot{\mathbf{u}}_\theta$$

To evaluate the term involving $\dot{\mathbf{u}}_\theta$, it is necessary to find the change made in the direction of $\mathbf{u}_\theta$. As shown in Fig. 12–30d, when θ changes by an increasing amount $\Delta\theta$, the new or changed position of $\mathbf{u}_\theta$ is designated as $\mathbf{u}_\theta'$. The time change in $\mathbf{u}_\theta$ is thus $\Delta\mathbf{u}_\theta$. For small angles this vector has a magnitude $\Delta u_\theta \approx 1(\Delta\theta)$ and acts in the $-\mathbf{u}_r$ direction, i.e., $\Delta\mathbf{u}_\theta = -\Delta\theta\mathbf{u}_r$. Thus,

$$\dot{\mathbf{u}}_\theta = \lim_{\Delta t \to 0} \frac{\Delta\mathbf{u}_\theta}{\Delta t} = -\left(\lim_{\Delta t \to 0} \frac{\Delta\theta}{\Delta t}\right)\mathbf{u}_r$$

$$\dot{\mathbf{u}}_\theta = -\dot{\theta}\mathbf{u}_r \qquad (12\text{–}27)$$

Substituting this result and Eq. 12–23 into the above equation for $\mathbf{a}$, we can write the acceleration in component form as

$$\boxed{\mathbf{a} = a_r\mathbf{u}_r + a_\theta\mathbf{u}_\theta} \qquad (12\text{–}28)$$

where

$$\boxed{\begin{aligned} a_r &= \ddot{r} - r\dot{\theta}^2 \\ a_\theta &= r\ddot{\theta} + 2\dot{r}\dot{\theta} \end{aligned}} \qquad (12\text{–}29)$$

The term $\ddot{\theta} = d^2\theta/dt^2 = d/dt(d\theta/dt)$ is called the *angular acceleration* since it measures the change made in the rate of change of θ during an instant of time. Common units for this measurement are rad/s^2.

Since $\mathbf{a}_r$ and $\mathbf{a}_\theta$ are always perpendicular, the *magnitude* of acceleration is simply the positive value of

$$a = \sqrt{(\ddot{r} - r\dot{\theta}^2)^2 + (r\ddot{\theta} + 2\dot{r}\dot{\theta})^2} \qquad (12\text{–}30)$$

The *direction* is determined from the vector addition of its two components. In general, $\mathbf{a}$ will *not* be tangent to the path, Fig. 12–30e.

Cylindrical Coordinates

If the particle P moves along a space curve as shown in Fig. 12–31, then its location may be specified by the three *cylindrical coordinates, r, θ, z*. The z coordinate is identical to that used for rectangular coordinates. Since the unit vector defining its direction, $\mathbf{u}_z$, is constant, the time derivatives of this vector are zero and therefore the position, velocity, and acceleration of the particle can be written in terms of its cylindrical coordinates as follows:

$$\mathbf{r}_P = r\mathbf{u}_r + z\mathbf{u}_z$$

$$\mathbf{v} = \dot{r}\mathbf{u}_r + r\dot{\theta}\mathbf{u}_\theta + \dot{z}\mathbf{u}_z \qquad (12\text{–}31)$$

$$\mathbf{a} = (\ddot{r} - r\dot{\theta}^2)\mathbf{u}_r + (r\ddot{\theta} + 2\dot{r}\dot{\theta})\mathbf{u}_\theta + \ddot{z}\mathbf{u}_z \qquad (12\text{–}32)$$

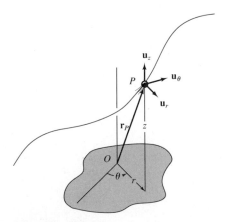

Fig. 12–31

PROCEDURE FOR ANALYSIS

Coordinate System. Polar coordinates are a suitable choice for solving problems for which data regarding the angular motion of the radial coordinate r is given to describe the particle's motion. Also, some paths of motion can conveniently be described in terms of these coordinates. To use cylindrical coordinates, the origin is established at a fixed point, and the radial line r is directed to the particle. Also, the transverse coordinate θ is established, measured counterclockwise from a fixed reference line to the radial line. If necessary, the equation of the path can then be determined by relating r to θ, $r = f(\theta)$.

Time derivatives. Two types of problems generally occur:

1. If the coordinates are specified as time-parametric equations, $r = r(t)$ and $\theta = \theta(t)$, the derivatives are simply $\dot{r} = dr/dt$, $\ddot{r} = d^2r/dt^2$, $\dot{\theta} = d\theta/dt$, and $\ddot{\theta} = d^2\theta/dt^2$ (see Examples 12–17 and 12–18).

2. If the time parametric equations of motion are not given, it will be necessary to specify the path $r = f(\theta)$ and compute the relationship between the time derivatives using the chain rule of calculus. In this case,

$$\dot{r} = \frac{df(\theta)}{d\theta} \dot{\theta} \qquad \text{and} \qquad \ddot{r} = \left(\frac{d^2f(\theta)}{d\theta^2} \dot{\theta} \right) \dot{\theta} + \frac{df(\theta)}{d\theta} \ddot{\theta}$$

Thus, if two of the *four* time derivatives $\dot{r}$, $\ddot{r}$, $\dot{\theta}$, and $\ddot{\theta}$ are *known*, the other two can be obtained from these equations (see Example 12–19). In some problems, however, two of these time derivatives may *not* be known; instead the magnitude of the particle's velocity or acceleration may be specified. If this is the case, $v^2 = \dot{r}^2 + (r\dot{\theta})^2$ and $a^2 = (\ddot{r} - r\dot{\theta}^2)^2 + (r\ddot{\theta} + 2\dot{r}\dot{\theta})^2$ may be used to obtain the necessary relationships involving $\dot{r}$, $\ddot{r}$, $\dot{\theta}$, and $\ddot{\theta}$ (see Example 12–20).

Velocity and Acceleration. Once r and the time derivatives $\dot{r}$, $\ddot{r}$, $\dot{\theta}$, and $\ddot{\theta}$ have been evaluated at the instant considered, their values can be substituted into Eqs. 12–25 and 12–29 to obtain the radial and transverse components of **v** and **a.**

Motion in three dimensions requires a simple extension of the above procedure to include $\dot{z}$ and $\ddot{z}$.

Besides the examples which follow, further examples involving the calculation of a_r and a_θ can be found in the "kinematics" sections of Examples 13–10 to 13–12.

Example 12–17

The ball B shown in Fig. 12–32a is rotating in a horizontal circular path of radius r such that the attached cord has an angular velocity $\dot{\theta}$ and angular acceleration $\ddot{\theta}$. Express the velocity and acceleration of the ball in terms of its polar coordinates.

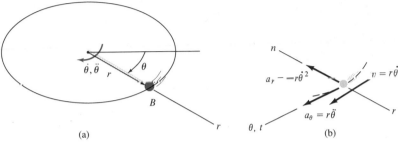

(a)   (b)

Fig. 12–32

SOLUTION

Coordinate System. Since the angular motion of the cord is reported, polar coordinates are chosen for the solution, Fig. 12–32a. Here θ is not related to r, since the radius is constant for all θ.

Time Derivatives. Equations 12–25 and 12–29 will be used for the solution, and so it is first necessary to specify the first and second time derivatives of r and θ. Since r is *constant*, we have

$$r = r, \qquad \dot{r} = 0, \qquad \ddot{r} = 0$$

Velocity
$$v_r = \dot{r} = 0 \qquad\qquad\qquad \textit{Ans.}$$
$$v_\theta = r\dot{\theta} \qquad\qquad\qquad \textit{Ans.}$$

Acceleration
$$a_r = \ddot{r} - r\dot{\theta}^2 = -r\dot{\theta}^2 \qquad\qquad \textit{Ans.}$$
$$a_\theta = r\ddot{\theta} + 2\dot{r}\dot{\theta} = r\ddot{\theta} \qquad\qquad \textit{Ans.}$$

These results are shown in Fig. 12–32b. Also shown are the n, t directions, which in this special case of circular motion happen to be *collinear* with the r and θ directions, respectively. In particular note that $v = v_r = v_t = r\dot{\theta}$. Also,

$$-a_r = a_n = \frac{v^2}{\rho} = \frac{(r\dot{\theta})^2}{r} = r\dot{\theta}^2$$

$$a_\theta = a_t = \frac{dv}{dt} = \frac{d}{dt}(r\dot{\theta}) = \frac{dr}{dt}\dot{\theta} + r\frac{d\dot{\theta}}{dt} = 0 + r\ddot{\theta}$$

Example 12–18

The rod OA, shown in Fig. 12–33a, is rotating in the horizontal x-y plane such that at any instant $\theta = (t^3)$ rad. At the same time, the collar B is sliding outward along OA so that $r = (100t^2)$ mm. If in both cases t is in seconds, determine the velocity and acceleration of the collar when $t = 1$ s.

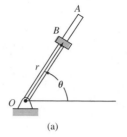

(a)

SOLUTION

Coordinate System. Since time-parametric equations of the path are given, it is not necessary to relate r to θ.

Time Derivatives. Computing the time derivatives, and evaluating when $t = 1$ s, we have

$$r = 100t^2 \Big|_{t=1\,\text{s}} = 100 \text{ mm} \qquad \theta = t^3 \Big|_{t=1\,\text{s}} = 1 \text{ rad} = 57.3°$$

$$\dot{r} = 200t \Big|_{t=1\,\text{s}} = 200 \text{ mm/s} \qquad \dot{\theta} = 3t^2 \Big|_{t=1\,\text{s}} = 3 \text{ rad/s}$$

$$\ddot{r} = 200 \Big|_{t=1\,\text{s}} = 200 \text{ mm/s}^2 \qquad \ddot{\theta} = 6t \Big|_{t=1\,\text{s}} = 6 \text{ rad/s}^2$$

Velocity. Fig. 12–33b

$$\mathbf{v} = \dot{r}\mathbf{u}_r + r\dot{\theta}\mathbf{u}_\theta$$
$$= 200\mathbf{u}_r + 100(3)\mathbf{u}_\theta$$
$$= \{200\mathbf{u}_r + 300\mathbf{u}_\theta\} \text{ mm/s}$$

The magnitude of $\mathbf{v}$ is

$$v = \sqrt{(200)^2 + (300)^2} = 361 \text{ mm/s} \qquad \textit{Ans.}$$

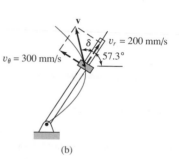

(b)

From Fig. 12–33b,

$$\delta = \tan^{-1}\left(\frac{300}{200}\right) = 56.3° \qquad \delta + 57.3° = 114° \qquad \textit{Ans.}$$

Acceleration. Fig. 12–33c

$$\mathbf{a} = (\ddot{r} - r\dot{\theta}^2)\mathbf{u}_r + (r\ddot{\theta} + 2\dot{r}\dot{\theta})\mathbf{u}_\theta$$
$$= [200 - 100(3)^2]\mathbf{u}_r + [100(6) + 2(200)3]\mathbf{u}_\theta$$
$$= \{-700\mathbf{u}_r + 1800\mathbf{u}_\theta\} \text{ mm/s}^2$$

The magnitude of $\mathbf{a}$ is

$$a = \sqrt{(700)^2 + (1800)^2} = 1930 \text{ mm/s}^2 \qquad \textit{Ans.}$$

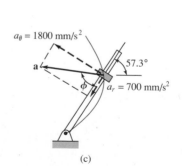

(c)

From Fig. 12–33c,

$$\phi = \tan^{-1}\left(\frac{1800}{700}\right) = 68.7° \qquad \phi - 57.3° = 11.4° \qquad \textit{Ans.} \quad \textbf{Fig. 12–33}$$

Example 12–19

The searchlight shown in Fig. 12–34a casts a spot of light along the face of a wall that is located 100 m from the searchlight. Determine the magnitudes of the velocity and acceleration at which the spot appears to travel across the wall at the instant $\theta = 45°$. The searchlight is rotating about the z axis at a constant rate $\dot{\theta} = 4$ rad/s.

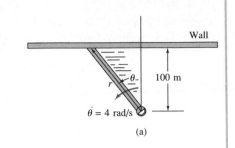

(a)

SOLUTION

Coordinate System. Why should polar coordinates be used to solve this problem? To compute the necessary time derivatives it is first necessary to relate r to θ. From Fig. 12–34a, this relation is

$$r = \frac{100}{\cos \theta} = 100 \sec \theta$$

Time Derivatives. Using the chain rule of calculus, noting that $d(\sec \theta) = \sec \theta \tan \theta \, d\theta$, and $d(\tan \theta) = \sec^2 \theta \, d\theta$, we have

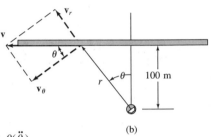

(b)

$$\dot{r} = 100(\sec \theta \tan \theta)\dot{\theta}$$

$$\ddot{r} = 100(\sec \theta \tan \theta)\dot{\theta}(\tan \theta)\dot{\theta} + 100 \sec \theta(\sec^2 \theta)\dot{\theta}(\dot{\theta}) + 100 \sec \theta \tan \theta(\ddot{\theta})$$

$$= 100 \sec \theta \tan^2 \theta(\dot{\theta})^2 + 100 \sec^3 \theta(\dot{\theta})^2 + 100(\sec \theta \tan \theta)\ddot{\theta}$$

Since $\dot{\theta} = 4$ rad/s = constant, then $\ddot{\theta} = 0$, and the above equations, when $\theta = 45°$, become

$$r = 100 \sec 45° = 141.4$$
$$\dot{r} = 400 \sec 45° \tan 45° = 565.7$$
$$\ddot{r} = 1600(\sec 45° \tan^2 45° + \sec^3 45°) = 6788.2$$

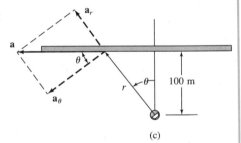

(c)

Fig. 12–34

Velocity   Fig. 12–34b

$$\mathbf{v} = \dot{r}\mathbf{u}_r + r\dot{\theta}\mathbf{u}_\theta$$
$$= 565.7\mathbf{u}_r + 141.4(4)\mathbf{u}_\theta$$
$$= 565.7\mathbf{u}_r + 565.7\mathbf{u}_\theta$$
$$v = \sqrt{v_r^2 + v_\theta^2} = \sqrt{(565.7)^2 + (565.7)^2}$$
$$= 800 \text{ m/s} \qquad \qquad Ans.$$

Acceleration.   Fig. 12–34c

$$\mathbf{a} = (\ddot{r} - r\dot{\theta}^2)\mathbf{u}_r + (r\ddot{\theta} + 2\dot{r}\dot{\theta})\mathbf{u}_\theta$$
$$= [6788.2 - 141.4(4)^2]\mathbf{u}_r + [141.4(0) + 2(565.7)4]\mathbf{u}_\theta$$
$$= 4525.8\mathbf{u}_r + 4525.8\mathbf{u}_\theta$$
$$a = \sqrt{a_r^2 + a_\theta^2} = \sqrt{(4525.8)^2 + (4525.8)^2}$$
$$= 6400 \text{ m/s}^2 \qquad \qquad Ans.$$

Example 12–20

Due to the rotation of the forked rod, the cylindrical peg A in Fig. 12–35a travels around the slotted path, a portion of which is in the shape of a cardioid, $r = 0.5(1 - \cos\theta)$ ft, where θ is in radians. If the peg's velocity is $v = 4$ ft/s and its acceleration is $a = 30$ ft/s^2 at the instant $\theta = 180°$, determine the angular velocity $\dot\theta$ and angular acceleration $\ddot\theta$ of the fork.

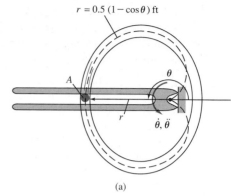

$r = 0.5\,(1 - \cos\theta)\,\text{ft}$

(a)

SOLUTION

Coordinate System. This path is most unusual, and mathematically it is best expressed using polar coordinates rather than rectangular coordinates. Also, $\dot\theta$ and $\ddot\theta$ must be determined so r, θ coordinates are an obvious choice.

Time Derivatives. Computing the time derivatives of r using the chain rule of calculus yields

$$r = 0.5(1 - \cos\theta)$$
$$\dot r = 0.5(\sin\theta)\dot\theta$$
$$\ddot r = 0.5(\cos\theta)\dot\theta(\dot\theta) + 0.5(\sin\theta)\ddot\theta$$

Evaluating these results at $\theta = 180°$, we have

$$r = 1\ \text{ft} \qquad \dot r = 0 \qquad \ddot r = -0.5\dot\theta^2$$

Velocity. Since $v = 4$ ft/s, using Eq. 12–26 to determine $\dot\theta$ yields

$$v = \sqrt{(\dot r)^2 + (r\dot\theta)^2}$$
$$4 = \sqrt{(0)^2 + (1\dot\theta)^2}$$
$$\dot\theta = 4\ \text{rad/s} \qquad\qquad Ans.$$

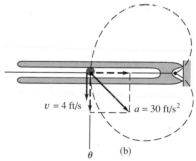

$v = 4$ ft/s

$a = 30$ ft/s^2

(b)

Fig. 12–35

Acceleration. In a similar manner, $\ddot\theta$ can be found using Eq. 12–30.

$$a = \sqrt{(\ddot r - r\dot\theta^2)^2 + (r\ddot\theta + 2\dot r\dot\theta)^2}$$
$$30 = \sqrt{[-0.5(4)^2 - 1(4)^2]^2 + [1\ddot\theta + 2(0)(4)]^2}$$
$$(30)^2 = (-24)^2 + \ddot\theta^2$$
$$\ddot\theta = 18\ \text{rad/s}^2 \qquad\qquad Ans.$$

Vectors **a** and **v** are shown in Fig. 12–35b.

PROBLEMS

12–101. A particle is moving along a circular path having a radius of $r = 0.4$ m. Its angular position as a function of time is given by $\theta = (2t^2)$ rad, where t is in seconds. Determine the magnitude of the particle's acceleration when $\theta = 30°$. The particle starts from rest when $\theta = 0°$.

12–102. A car is traveling along the circular curve having a radius $r = 400$ ft. At the instant shown, its angular rate of rotation is $\dot{\theta} = 0.25$ rad/s, which is decreasing at the rate $\ddot{\theta} = -0.008$ rad/s². Determine the radial and transverse components of the car's velocity and acceleration at this instant and sketch these components on the curve.

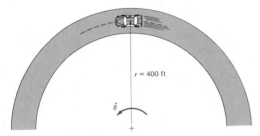

Probs. 12–102/12–103

12–103. A car is traveling along the circular curve of radius $r = 400$ ft with a constant speed $v = 30$ ft/s. Determine the angular rate of rotation $\dot{\theta}$ of the radial line r and the magnitude of the car's acceleration.

*12–104. The arm of the robot has a fixed length so that $r = 3$ ft and its grip A moves along the path $z = (3 \sin 4\theta)$ ft, where θ is in radians. If $\theta = (0.5t)$ rad, where t is in seconds, determine the magnitudes of the grip's velocity and acceleration when $t = 3$ s.

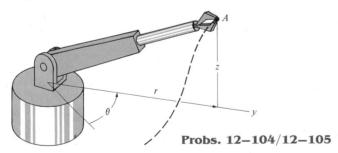

Probs. 12–104/12–105

12–105. For a short time the arm of the robot is extending at a constant rate such that $\dot{r} = 1.5$ ft/s when $r = 3$ ft, $z = (4t^2)$ ft, and $\theta = 0.5t$ rad, where t is in seconds. Determine the magnitudes of the velocity and acceleration of the grip A when $t = 3$ s.

12–106. At the instant shown, the watersprinkler is rotating with an angular speed $\dot{\theta} = 2$ rad/s and an angular acceleration $\ddot{\theta} = 3$ rad/s². If the nozzle lies in the vertical plane and water is flowing through it at a constant rate of 3 m/s, determine the magnitudes of the velocity and acceleration of a water particle as it exits the open end, $r = 0.2$ m.

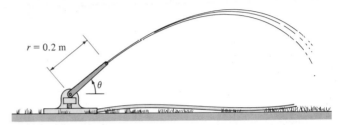

Prob. 12–106

12–107. A particle moves along a spiral path $r = (10/\theta)$ ft, where θ is in radians. If it maintains a constant speed of $v = 20$ ft/s, determine the magnitudes of its components v_r and v_θ as functions of θ and evaluate each at $\theta = 1$ rad. Sketch the path and show the velocity components at $\theta = 1$ rad.

*12–108. The telescopic rod rotates at $\dot{\theta} = 0.8$ rad/s which is decreasing at $\ddot{\theta} = -0.2$ rad/s², and at the same time the rod is moving in and out such that its end A is following the path $r = (0.75\cos\theta + 2)$ ft, where θ is in radians. Determine the radial and transverse components of the velocity and acceleration of A at the instant $\theta = \pi/8$.

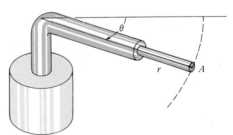

Probs. 12–108/12–109

12–109. The telescopic rod rotates at $\dot{\theta} = 0.8$ rad/s which is decreasing at $\ddot{\theta} = -0.2$ rad/s², and at the same time the rod is moving in and out such that its end A is following the path $r = (1 + 0.75\cos\theta)$ ft, where θ is in radians. Determine the radial and transverse components of the velocity and acceleration of A at the instant $\theta = \pi/8$.

12–110. A particle moves along the spiral path $r = (0.4e^{\theta})$ ft such that $\dot{\theta} = (0.2t)$ rad/s, where t is in seconds. Determine the transverse and radial components of the particle's velocity and acceleration when $t = 1$ s. When $t = 0$, $\theta = \pi/2$ rad.

12–111. A horse on the merry-go-round moves according to the equations $r = 8$ ft, $\theta = (0.6t)$ rad, and $z = (1.5 \sin \theta)$ ft, where t is in seconds. Determine the cylindrical components of the velocity and acceleration of the horse when $t = 4$ s.

Prob. 12–111

*12–112. A particle moves along an Archimedean spiral $r = (8\theta)$ ft, where θ is given in radians. If $\dot{\theta} = 4$ rad/s (constant), compute the radial and transverse components of the particle's velocity and acceleration at the instant $\theta = \pi/2$ rad. Sketch the curve and show the components on the curve.

12–113. Solve Prob. 12–112 if the particle has an angular acceleration $\ddot{\theta} = 5$ rad/s^2 when $\dot{\theta} = 4$ rad/s at $\theta = \pi/2$ rad.

12–114. For a short time the roller coaster travels along the track defined by the equation $r = (200 \cos 2\theta)$ ft, where θ is in radians. If the angular velocity of the radial coordinate line is $\dot{\theta} = (0.008t^2)$ rad/s, where t is in seconds, determine the radial and transverse components of the car's velocity and acceleration at the instant $\theta = 30°$. When $t = 0$, $\theta = 0$.

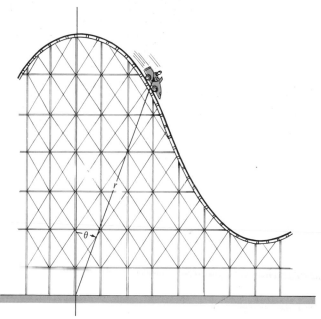

Prob. 12–114

12–115. The car travels around the circular curve with a constant speed of 20 m/s. Determine the car's radial and transverse components of velocity and acceleration at the instant $\theta = \pi/4$ rad.

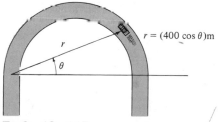

$r = (400 \cos \theta)$m

Prob. 12–115

*12–116. A particle starts from the origin at $t = 0$ and moves along a curve such that it has a velocity of $\mathbf{v} = \{8\mathbf{u}_r + 4t\mathbf{u}_{\theta} + 5t^2\mathbf{u}_z\}$ ft/s, where t is in seconds. Determine the magnitudes of the velocity and acceleration of the particle when $t = 2$ s. At $t = 2$ s, $r = 3$ ft.

12–117. The slotted arm *AB* drives the pin *C* through the spiral groove described by the equation $r = a\theta$. If the angular rate of rotation is constant at $\dot\theta$, determine the radial and transverse components of velocity and acceleration of the pin.

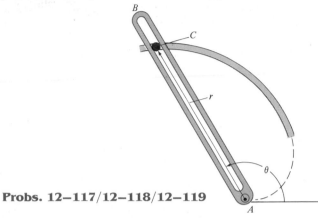

Probs. 12–117/12–118/12–119

12–118. Solve Prob. 12–117 if the spiral path is logarithmic, i.e., $r = a\,e^{k\theta}$.

12–119. The slotted arm *AB* drives the pin *C* through the spiral groove described by the equation $r = (1.5\theta)$ ft, where θ is in radians. If the arm starts from rest when $\theta = 60°$ and is driven at an angular rate of $\dot\theta = (4t)$ rad/s, where *t* is in seconds, determine the radial and transverse components of velocity and acceleration of the pin *C* when $t = 1$ s.

*12–120.** The roller coaster is traveling down along the spiral ramp with a constant speed $v = 6$ m/s. If the track descends a distance of 10 m for every full revolution, $\theta = 2\pi$ rad, determine the magnitude of the roller coaster's acceleration as it moves along the track, $r = 5$ m. *Hint:* For part of the solution, note that the tangent to the ramp at any point is at an angle $\phi = \tan^{-1}[10/2\pi(5)] = 17.66°$ from the horizontal. Use this to determine the velocity components v_θ and v_z, which in turn are used to determine $\dot\theta$ and $\dot z$.

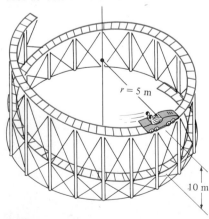

Prob. 12–120

12–121. The rocket is traveling upwards such that when $\theta = 30°$ the radar tracking device is turning at $\dot\theta = 0.6$ rad/s and $\ddot\theta = 0.2$ rad/s². Determine the magnitude of velocity and acceleration of the rocket at this instant.

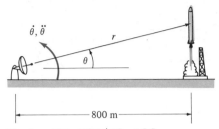

Probs. 12–121/12–122

12–122. As the rocket travels upwards, the radar tracking device has an angular position $\theta = (0.03t^2)$ rad, where *t* is in seconds. Determine the magnitude of velocity and acceleration of the rocket at the instant $t = 3$ s.

12–123. The slotted link is pinned at *O* and as a result of the constant angular velocity $\dot\theta = 3$ rad/s it drives the peg *P* along the path defined by $r = (0.6\sin\theta + 0.3)$ ft, where θ is in radians. Compute the radial and transverse components of the velocity and acceleration of *P* at the instant $\theta = \pi/3$ rad.

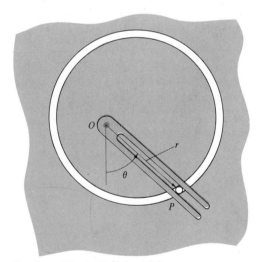

Probs. 12–123/12–124

*12–124.** Solve Prob. 12–123 if the slotted link has an angular acceleration of $\ddot\theta = 4$ rad/s² and $\dot\theta = 3$ rad/s at $\theta = \pi/3$ rad.

12–125. A particle travels along the "four-leaf rose" defined by the equation $r = (5 \cos 2\theta)$ m. If the angular velocity of the radial coordinate line is $\dot{\theta} = (3t^2)$ rad/s, where t is in seconds, determine the radial and transverse components of the particle's velocity and acceleration at the instant $\theta = 30°$. When $t = 0$, $\theta = 0$.

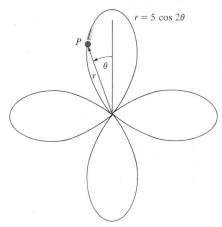

$r = 5 \cos 2\theta$

P

θ

r

Prob. 12–125

12–126. The time rate of change of acceleration is referred to as the *jerk*, which is often used as a means of measuring passenger discomfort. Calculate this vector, $\dot{\mathbf{a}}$, in terms of its cylindrical components, using Eq. 12–32.

■**12–127.** The double collar C is pin-connected together such that one collar slides over a fixed rod and the other slides over a rotating rod AB. If the angular velocity of AB is given as $\dot{\theta} = (e^{0.5t^2})$ rad/s, where t is in seconds, and the path defined by the fixed rod is $r = |(0.4 \sin \theta + 0.2)|$ m, determine the radial and transverse components of the collar's velocity and acceleration when $t = 1$ s. When $t = 0$, $\theta = 0$. Use Simpson's rule to determine θ at $t = 1$ s.

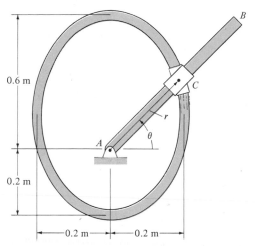

B

0.6 m

C

r

θ

A

0.2 m

0.2 m 0.2 m

Prob. 12–127

*****12–128.** A particle travels along a path such that its position vector is defined by $\mathbf{r} = \{2 \cos (0.1t)\mathbf{i} + 1.5 \sin(0.1t)\mathbf{j} + (2t)\mathbf{k}\}$ m, where t is in seconds and the arguments for the sine and cosine are given in radians. When $t = 10$ s, determine the coordinate direction angles α, β, and γ, which the binormal axis to the osculating plane makes with the x, y, and z axes. Sketch the path in three dimensions. *Hint:* Solve for the velocity $\mathbf{v}$ and acceleration $\mathbf{a}$ of the particle in terms of their $\mathbf{i}$, $\mathbf{j}$, $\mathbf{k}$ components. The binormal is collinear to $\mathbf{v} \times \mathbf{a}$. Why?

12–129. The motion of a particle along a fixed path is defined by the parametric equations $r = 1.5$ m, $\theta = (2t)$ rad, and $z = (t^2)$ m, where t is in seconds. Determine the unit vector that specifies the direction of the binormal axis to the osculating plane with respect to a set of fixed x, y, z coordinate axes when $t = 0.25$ s. *Hint:* Formulate the particle's velocity $\mathbf{v}$ and acceleration $\mathbf{a}$ in terms of their $\mathbf{i}$, $\mathbf{j}$, $\mathbf{k}$ components. Note that $x = r \cos \theta$ and $y = r \sin \theta$. The binormal is collinear to $\mathbf{v} \times \mathbf{a}$. Why?

12.8 Absolute Dependent Motion Analysis of Two Particles

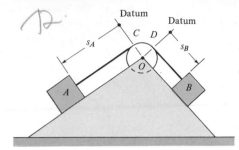

Fig. 12–36

In some types of problems the motion of one particle will *depend* upon the corresponding motion of another particle. This dependency commonly occurs if the particles are interconnected by inextensible cords which are wrapped around pulleys. For example, the movement of block *A* downward along the inclined plane in Fig. 12–36 will cause a corresponding movement of block *B* up the other incline. We can show this mathematically by first specifying the positions of the blocks using the path coordinates s_A and s_B. Note that these coordinates are (1) referenced from a *fixed* point (*O*) or datum, (2) measured along the plane in the direction of motion, and (3) have a positive sense from *C* to *A* and *D* to *B*. Since the total cord length l_T is constant, the coordinates are related by the equation

$$s_A + s_B = l_T - l_{CD} = l'$$

where l_{CD} is the length of cord passing over arc *CD*.* Taking the time derivative of this expression yields a relation between the velocities of the blocks, i.e.,

$$\frac{ds_B}{dt} + \frac{ds_A}{dt} = 0$$

or

$$v_B = -v_A$$

The negative sign indicates that when block *A* has a velocity downward, i.e., in the direction of positive s_A, it causes a corresponding upward velocity of block *B*; i.e., *B* moves in the negative s_B direction.

In a similar manner, time differentiation of the velocities yields the relation between the accelerations, i.e.,

$$a_B = -a_A$$

A more complicated example involving dependent motion of two blocks is shown in Fig. 12–37a. In this case, the position of block *A* is specified by s_A, and the position of the *end* of the cord from which block *B* is suspended is defined by s_B. Here we have chosen coordinates which are (1) referenced from fixed points or datums, (2) measured in the direction of motion of each block, and (3) positive to the right (s_A) and downward (s_B). The *constant* cord length l, which represents the total length of cord *minus* the constant cord segments which are colored, can be related to the position coordinates by the equation

---

*Note that when the blocks move, the position coordinates s_A and s_B measure the lengths of the *changing segments* of the cord. The cord length l_{CD} always remains *constant*.

$$2s_B + h + s_A = l$$

Since l and h are constants, the two time derivatives yield

$$2v_B = -v_A \qquad 2a_B = -a_A$$

Hence, when B moves downward $(+s_B)$, A moves to the left $(-s_A)$ at two times the motion.

This example can also be worked by defining the position of block B from the center of the bottom pulley (a fixed point), Fig. 12–37b. In this case

$$2(h - s_B) + h + s_A = l$$

Time differentiation yields

$$2v_B = v_A \qquad 2a_B = a_A$$

Here the signs are the same. Why?

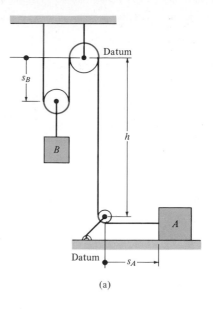

(a)

PROCEDURE FOR ANALYSIS

The above method of relating the dependent motion of one particle to that of another can be performed using algebraic scalars or position coordinates provided each particle moves along a rectilinear path. When this is the case, only the magnitudes of the velocity and acceleration of each particle will change, not their line of direction. The following procedure is required.

Position-Coordinate Equation. Establish position coordinates which have their origin referenced to a *fixed* point or datum and extend from this point to each of the particles. It is *not necessary* that the *origin* be the *same* for both of these coordinates; however, it is *important* that each coordinate axis selected be directed along the *path of motion* of the particle.

Using geometry or trigonometry, relate the coordinates to the total length of the cord, l_T, or to that portion of cord, l, which *excludes* the segments that do not change length as the particles move — such as the arc segments wrapped over pulleys.

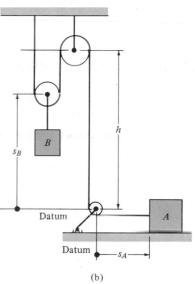

Datum

(b)

Fig. 12–37

Time Derivatives. The two successive time derivatives of the position-coordinate equation yield the required velocity and acceleration equations which relate the motions of the two particles. The signs of the terms in these equations will be consistent with those that specify the positive and negative sense of the position coordinates.

If a problem involves a *system* of two or more cords wrapped around pulleys, then the motion of a point on one cord must be related to the motion of a point on another cord using the above procedure. Separate equations are written for each cord of the system and the positions of the two particles are then related by these equations (see Examples 12–22 and 12–23).

Example 12–21

Determine the speed of block A in Fig. 12–38. Block B has an upward speed of 6 ft/s.

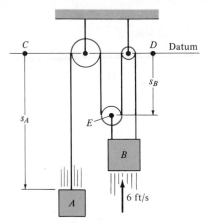

Fig. 12–38

SOLUTION

Position-Coordinate Equation. There is *one* cord in this system having segments which are changing length. Position coordinates s_A and s_B will be used since each is measured from a fixed point (C or D) and extends along the block's *path of motion*. In particular, s_B is directed to point E since motion of B and E is the *same*.

The colored segments of the cord in Fig. 12–38 remain at a constant length and do not have to be considered as the blocks move. The remaining length of cord, l, is also constant and is related to the changing position coordinates s_A and s_B by the equation

$$s_A + 3s_B = l$$

Time Derivative. Taking the time derivative yields

$$v_A + 3v_B = 0$$

so that when $v_B = -6$ ft/s (upward), then

$$v_A = 18 \text{ ft/s} \downarrow \qquad\qquad Ans.$$

Example 12–22

Determine the speed of block A in Fig. 12–39. Block B has an upward speed of 6 ft/s.

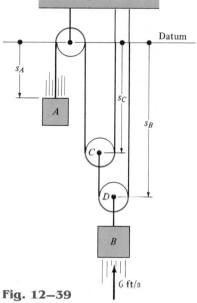

Fig. 12–39

SOLUTION

Position-Coordinate Equation. This system has two cords which have segments that change length. Motion of the end of the cord attached to A can be related to the motion of the end of the cord attached to the pulley at C using coordinates s_A and s_C. Likewise, the motion of the end of the cord at C can be related to the motion of the end of the cord at D using position coordinates s_C and s_B.

The colored segments of the cords in Fig. 12–39 do not have to be considered in the analysis. Why? For the remaining cord lengths, say l_1 and l_2, we have

$$s_A + 2s_C = l_1 \qquad s_B + (s_B - s_C) = l_2$$

Eliminating s_C yields an equation defining the positions of both blocks, i.e.,

$$s_A + 4s_B = 2l_2 + l_1$$

Time Derivative. The time derivative gives

$$v_A + 4v_B = 0$$

so that when $v_B = -6$ ft/s (upward), then

$$v_A = 24 \text{ ft/s} \downarrow \qquad\qquad Ans.$$

Example 12–23

Determine the speed with which block B rises in Fig. 12–40 if the end of the cord at A is pulled down with a speed of 2 m/s.

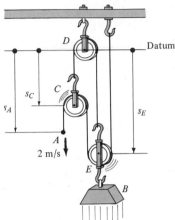

Fig. 12–40

SOLUTION

Position-Coordinate Equation. By inspection this system has two cords. Three position coordinates have been selected. These are all measured from the horizontal datum passing through the fixed pin at pulley D. The position of the end of the cord at A is defined by s_A, the position of the end of the cord at C is defined by s_C, and the position of the end of the cord at E (block B) is defined by s_E.

Excluding the colored segments of the cords in Fig. 12–40, the remaining constant cord lengths l_1 and l_2 (with the hook dimensions) can be expressed in terms of the three position coordinates. We have

$$(s_A - s_C) + (s_E - s_C) + s_E = l_1$$
$$s_C + s_E = l_2$$

Eliminating s_C yields

$$s_A + 4s_E = l_1 + 2l_2$$

This equation relates the position s_E of block B to the position s_A of point A.

Time Derivative. The time derivative gives

$$v_A + 4v_E = 0$$

so that when $v_A = 2$ m/s (downward), then

$$v_E = -0.5 \text{ m/s} = 0.5 \text{ m/s} \uparrow \qquad \qquad \textit{Ans.}$$

Example 12–24

A man at A is hoisting a safe S as shown in Fig. 12–41 by walking to the right with a constant velocity $v_A = 0.5$ m/s. Determine the velocity and acceleration of the safe when it reaches the window elevation at E. The rope is 30 m long and passes over a small pulley at D.

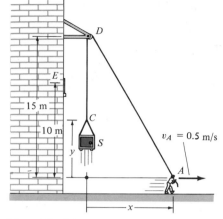

Fig. 12–41

SOLUTION

Position-Coordinate Equation. This problem is unlike the previous examples, since rope segment DA changes both direction and magnitude. However, the ends of the rope, which define the positions of S and A, are specified by means of the x and y coordinates measured from fixed points and *directed along the paths of motion* of the ends of the rope.

The x and y coordinates may be related since the rope has a fixed length $l = 30$ m, which at all times is equal to the length of segment DA plus CD. Using the Pythagorean theorem to determine l_{DA}, we have $l_{DA} = \sqrt{(15)^2 + x^2}$; also, $l_{CD} = 15 - y$. Hence,

$$l = l_{DA} + l_{CD}$$
$$30 = \sqrt{(15)^2 + x^2} + (15 - y)$$
$$y = \sqrt{225 + x^2} - 15 \qquad (1)$$

Time Derivatives. Taking the time derivative, where $v_S = dy/dt$ and $v_A = dx/dt$, yields

$$v_S = \frac{dy}{dt} = \frac{1}{2} \frac{2x}{\sqrt{225 + x^2}} \frac{dx}{dt}$$

$$= \frac{x}{\sqrt{225 + x^2}} v_A \qquad (2)$$

At $y = 10$ m, x is determined from Eq. (1), i.e., $x = 20$ m. Hence, from Eq. (2) with $v_A = 0.5$ m/s,

$$v_S = \frac{20}{\sqrt{225 + (20)^2}} (0.5) = 0.4 \text{ m/s} = 400 \text{ mm/s} \uparrow \qquad \textit{Ans.}$$

The acceleration is determined by taking the time derivative of Eq. (2). Since v_A is constant, then $a_A = dv_A/dt = 0$, and we have

$$a_S = \frac{d^2y}{dt^2} = \left[\frac{-x^2}{(225 + x^2)^{3/2}} \frac{dx}{dt} + \frac{1}{(225 + x^2)^{1/2}} \frac{dx}{dt} \right] v_A$$

$$= \frac{225 v_A^2}{(225 + x^2)^{3/2}}$$

At $x = 20$ m, with $v_A = 0.5$ m/s, the acceleration becomes

$$a_S = \frac{225(0.5)^2}{(225 + (20)^2)^{3/2}} = 0.00360 \text{ m/s}^2 = 3.60 \text{ mm/s}^2 \uparrow \qquad \textit{Ans.}$$

12.9 Relative-Motion Analysis of Two Particles Using Translating Axes

Throughout this chapter the absolute motion of a particle has been determined using a single fixed reference frame for measurement. There are many cases, however, where the path of motion for a particle is complicated, so that it may be feasible to analyze the motion in parts by using two or more frames of reference. For example, the motion of a particle located at the tip of an airplane propeller, while the plane is in flight, is more easily described if one observes first the motion of the airplane from a fixed reference and then superimposes (vectorially) the circular motion of the particle measured from a reference attached to the airplane. Any type of coordinates—rectangular, cylindrical, etc.—may be chosen to describe these two different motions.

In this section only *translating frames of reference* will be considered for the analysis. Relative-motion analysis of particles using rotating frames of reference will be treated in Secs. 16.8 and 20.4, since such an analysis depends upon prior knowledge of the kinematics of line segments.

Position

Consider particles A and B, which move along the arbitrary paths aa and bb, respectively, as shown in Fig. 12–42a. The *absolute position* of each particle, $\mathbf{r}_A$ and $\mathbf{r}_B$, is measured from the common origin O of the *fixed x, y, z* reference frame. The origin of a second frame of reference x', y', z' is attached to and moves with particle A. The axes of this frame do not rotate; rather they are *only permitted to translate* relative to the fixed frame. The *relative position* of "*B* with respect to *A*" is observed from this moving frame and is designated by $\mathbf{r}_{B/A}$, called a *relative-position vector*. Using vector addition, the three vectors shown in Fig. 12–42a can be related by the equation*

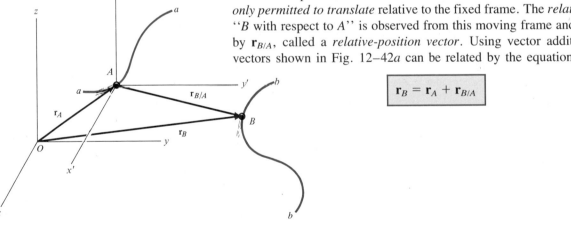

(a)

Fig. 12–42a

$$\boxed{\mathbf{r}_B = \mathbf{r}_A + \mathbf{r}_{B/A}} \qquad (12\text{–}33)$$

Velocity

An equation that relates the velocities of the particles can be determined by taking the time derivative of Eq. 12–33, i.e.,

*An easy way to remember the setup of this equation, and others like it, is to note the "cancellation" of the subscript A between the two terms, i.e., $\mathbf{r}_B = \mathbf{r}_{\not A} + \mathbf{r}_{B/\not A}$.

$$\boxed{\mathbf{v}_B = \mathbf{v}_A + \mathbf{v}_{B/A}} \qquad (12\text{–}34)$$

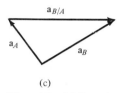

(b)

Here $\mathbf{v}_B = d\mathbf{r}_B/dt$ and $\mathbf{v}_A = d\mathbf{r}_A/dt$ refer to *absolute velocities*, since they are measured from the fixed frame of reference; whereas the *relative velocity* $\mathbf{v}_{B/A} = d\mathbf{r}_{B/A}/dt$ is observed from the translating reference frame. It is important to note that since the x', y', z' axes translate, the *components* of $\mathbf{r}_{B/A}$ will *not* change direction and therefore the time derivative of this vector will only have to account for changes in the vector's magnitude. The above equation therefore states that the velocity of B is equal to the velocity of A plus (vectorially) the relative velocity of "B with respect to A," as measured by a *translating observer* fixed in the x', y', z' reference, Fig. 12–42b.

Acceleration

The time derivative of Eq. 12–34 yields a similar vector relationship between the *absolute* and *relative accelerations* of particles A and B,

(c)

Fig. 12–42(b, c)

$$\boxed{\mathbf{a}_B = \mathbf{a}_A + \mathbf{a}_{B/A}} \qquad (12\text{–}35)$$

Here $\mathbf{a}_{B/A}$ is the acceleration of B as seen by an observer located at A and translating with the x', y', z' reference frame. The vector addition is shown in Fig. 12–42c.

PROCEDURE FOR ANALYSIS

When applying the relative-position equation, $\mathbf{r}_B = \mathbf{r}_A + \mathbf{r}_{B/A}$, it is first necessary to specify the locations of the fixed x, y, z and translating x', y', z' axes. Usually, the origin A of the translating axes is located at a point having a *known position*, $\mathbf{r}_A$, Fig. 12–42a. A graphical representation of the vector addition $\mathbf{r}_B = \mathbf{r}_A + \mathbf{r}_{B/A}$ can be shown, and both the known and unknown quantities labeled on this sketch. Since vector addition forms a triangle there can be at most *two unknowns*, represented by the magnitudes and/or directions of the vector quantities. These unknowns can be solved for either graphically, using trigonometry (law of sines, law of cosines), or by resolving each of the three vectors $\mathbf{r}_B$, $\mathbf{r}_A$, and $\mathbf{r}_{B/A}$ into rectangular or Cartesian components, thereby generating two scalar equations. The latter method is illustrated in the example problems which follow.

The relative-motion equations $\mathbf{v}_B = \mathbf{v}_A + \mathbf{v}_{B/A}$ and $\mathbf{a}_B = \mathbf{a}_A + \mathbf{a}_{B/A}$ are applied in the same manner as explained above, except in this case the origin O of the fixed x, y, z axes does not have to be specified, Figs. 12–42b and 12–42c.

Example 12–25

Water drips from a faucet at the rate of five drops per second as shown in Fig. 12–43. Determine the vertical separation between two consecutive drops after the lower drop has attained a velocity of 3 m/s.

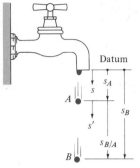

Fig. 12–43

SOLUTION

If the first and second drops of water are denoted as B and A, respectively, Fig. 12–43, the separation $s_{B/A}$ (position of B with respect to A) can be determined by the equation

$$(+\downarrow) \qquad\qquad s_{B/A} = s_B - s_A \qquad\qquad (1)$$

Here the origin of the fixed frame of reference is located at the head of the faucet (datum) with positive s downward. The origin of the moving frame s' is at drop A. Since each drop is subjected to rectilinear motion, having a *constant* downward acceleration $a_c = g = 9.81 \text{ m/s}^2$, Eq. 12–4 may be applied with the initial condition $v_0 = 0$ to determine the time needed for drop B to attain a velocity of 3 m/s.

$$(+\downarrow) \qquad\qquad v = v_0 + a_c t$$
$$3 = 0 + 9.81 t_B$$
$$t_B = 0.306 \text{ s}$$

Since one drop falls every fifth of a second (0.2 s), drop A falls for a time

$$t_A = 0.306 \text{ s} - 0.2 \text{ s} = 0.106 \text{ s}$$

before drop B attains a velocity of 3 m/s.

The position of each drop can be found by means of Eq. 12–5, with the condition $s_0 = 0$, $v_0 = 0$.

$$(+\downarrow) \qquad\qquad s = s_0 + v_0 t + \tfrac{1}{2} a_c t^2$$
$$s_B = 0 + 0 + \tfrac{1}{2}(9.81)(0.306)^2$$
$$= 0.459 \text{ m}$$
$$s_A = 0 + 0 + \tfrac{1}{2}(9.81)(0.106)^2$$
$$= 0.0549 \text{ m}$$

Applying Eq. (1) yields

$$s_{B/A} = s_B - s_A = 0.459 - 0.0549 = 0.404 \text{ m}$$
$$= 404 \text{ mm} \downarrow \qquad\qquad \textit{Ans.}$$

Example 12–26

A train, traveling at a constant speed of 60 mi/h, crosses over a road as shown in Fig. 12–44a. If an automobile A is traveling at 45 mi/h along the road, determine the relative velocity of the train with respect to the automobile.

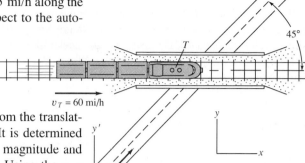

SOLUTION I

Vector Analysis. The relative velocity $\mathbf{v}_{T/A}$ is measured from the translating x', y' axes attached to the automobile, Fig. 12–44a. It is determined from $\mathbf{v}_T = \mathbf{v}_A + \mathbf{v}_{T/A}$. Since $\mathbf{v}_T$ and $\mathbf{v}_A$ are known in *both* magnitude and direction, the unknowns become the components of $\mathbf{v}_{T/A}$. Using the x, y axes in Fig. 12–44a and a Cartesian vector analysis, we have

$$\mathbf{v}_T = \mathbf{v}_A + \mathbf{v}_{T/A}$$
$$60\mathbf{i} = (45 \cos 45° \, \mathbf{i} + 45 \sin 45° \, \mathbf{j}) + \mathbf{v}_{T/A}$$
$$\mathbf{v}_{T/A} = \{28.2\mathbf{i} - 31.8\mathbf{j}\} \text{ mi/h} \qquad \textit{Ans.}$$

The magnitude of $\mathbf{v}_{T/A}$ is thus

$$v_{T/A} = \sqrt{(28.2)^2 + (-31.8)^2} = 42.5 \text{ mi/h} \qquad \textit{Ans.}$$

From the direction of each component, the direction of $\mathbf{v}_{T/A}$ defined from the x axis, is

$$\tan \theta = \frac{(v_{T/A})_y}{(v_{T/A})_x} = \frac{31.8}{28.2}$$
$$\theta = 48.4° \qquad \textit{Ans.}$$

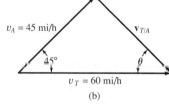

Fig. 12–44

Note that the vector addition shown in Fig. 12–44b indicates the correct sense for $\mathbf{v}_{T/A}$. This figure anticipates the answer and can be used to check it.

SOLUTION II

Scalar Analysis. The components of $\mathbf{v}_{T/A}$ can also be determined by applying a scalar analysis. We will assume these components act in the *positive* x and y directions. Thus,

$$\mathbf{v}_T = \mathbf{v}_A + \mathbf{v}_{T/A}$$
$$\begin{bmatrix} 60 \text{ mi/h} \\ \rightarrow \end{bmatrix} = \begin{bmatrix} 45 \text{ mi/h} \\ \angle 45° \end{bmatrix} + \begin{bmatrix} (v_{T/A})_x \\ \rightarrow \end{bmatrix} + \begin{bmatrix} (v_{T/A})_y \\ \uparrow \end{bmatrix}$$

Resolving each vector into its horizontal and vertical components yields

$(\xrightarrow{+})$ $\qquad\qquad 60 = 45 \cos 45° + (v_{T/A})_x + 0$

$(+\uparrow)$ $\qquad\qquad 0 = 45 \sin 45° + 0 + (v_{T/A})_y$

Solving, we obtain the previous results,

$$(v_{T/A})_x = 28.2 \text{ mi/h} = 28.2 \text{ mi/h} \rightarrow$$
$$(v_{T/A})_y = -31.8 \text{ mi/h} = 31.8 \text{ mi/h} \downarrow$$

Example 12–27

Two jet planes are flying horizontally at the same elevation, as shown in Fig. 12–45a. Plane A is flying along a straight-line path, and at the instant shown it has a speed of 700 km/h and an acceleration of 50 km/h². Plane B is flying along a circular path having a radius of curvature of $\rho_B = 400$ km. Its speed is 600 km/h, which is decreasing at the rate of 100 km/h². Determine the velocity and acceleration of B as measured by the pilot in A.

SOLUTION

Velocity. The fixed x, y axes are located at an arbitrary point. Plane A is traveling with rectilinear motion and a *translating frame of reference* x', y' will be attached to it. Applying the relative-velocity equation in scalar form since the velocity vectors of both planes are parallel at the instant shown, we have

$$(+\uparrow) \qquad v_B = v_A + v_{B/A}$$

$$600 = 700 + v_{B/A}$$

$$v_{B/A} = -100 \text{ km/h} = 100 \text{ km/h} \downarrow \qquad \textbf{\textit{Ans.}}$$

The vector addition is shown in Fig. 12–45b.

Acceleration. Plane B has both tangential and normal components of acceleration, since it is flying along a *curved path*. From Eq. 12–20, the magnitude of the normal component is

$$(a_B)_n = \frac{v_B^2}{\rho} = \frac{(600 \text{ km/h})^2}{400 \text{ km}} = 900 \text{ km/h}^2$$

Applying the relative-acceleration equation, we have

$$\mathbf{a}_B = \mathbf{a}_A + \mathbf{a}_{B/A}$$

$$(900\mathbf{i} - 100\mathbf{j}) = 50\mathbf{j} + \mathbf{a}_{B/A}$$

Thus,

$$\mathbf{a}_{B/A} = \{900\mathbf{i} - 150\mathbf{j}\} \text{ km/h}^2$$

The magnitude and direction of $\mathbf{a}_{B/A}$ are therefore

$$a_{B/A} = 912 \text{ km/h}^2 \qquad \theta = \tan^{-1}\frac{150}{900} = 9.46° \searrow_\theta \qquad \textbf{\textit{Ans.}}$$

The vector addition is shown in Fig. 12–45c.

Notice that the solution to this problem is possible using a translating frame of reference, since the pilot in plane A is "translating." Observation of plane A with respect to plane B, however, must be obtained using a *rotating* set of axes attached to plane B. (This assumes, of course, that the pilot of B does not turn his eyes to follow the motion of A.) The analysis for this case is given in Example 16–19.

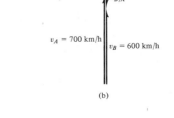

(b)

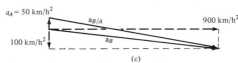

(c)

Fig. 12–45

PROBLEMS

12–130. The cable and pulley system is used to lift the load A. Determine the displacement of the load if the end C of the cable is pulled 4 ft upward.

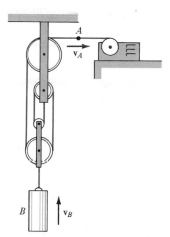

Prob. 12–130

12–131. Determine the constant speed at which the cable at A must be drawn in by the motor in order to hoist the load at B 15 ft in 5 s.

Probs. 12–131/12–132

*12–132. If the motor draws in the cable such that point A is subjected to an acceleration of $a_A = (3t^2)$ ft/s^2, where t is in seconds, determine how high the load at B travels in $t = 1.5$ s starting from rest.

12–133. The pulley system supports the three blocks. If A is moving downward with a speed of 5 ft/s while C is moving upward at 2 ft/s, determine the speed of block B.

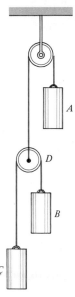

Probs. 12–133/12–134

12–134. If block A of the pulley system is moving downward at 11 ft/s while block C is moving downward at 18 ft/s, determine the relative velocity of block B with respect to C.

12–135. The pipe P is being lifted using the cable and pulley system shown. If point A on the cable is being drawn toward the drum with a speed of 2 m/s, determine the speed of the pipe.

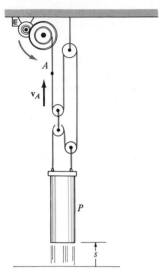

Probs. 12–135/12–136

*12–136. The pipe P is being lifted using the cable and pulley system shown. Starting from rest, point A on the cable is being drawn in with an acceleration $a_A = (5t^2)$ m/s^2, where t is in seconds. If $s = 0$ when $t = 0$, determine the height s of the pipe when $t = 2$ s.

12–137. The motor draws in the cable at C with a constant velocity of $v_C = 4$ m/s. The motor draws in the cable at D with a constant acceleration of $a_D = 8$ m/s^2. If $v_D = 0$ and $h = 3$ m when $t = 0$, determine (a) the time needed for $h = 0$, and (b) the relative velocity of block A with respect to block B when this occurs.

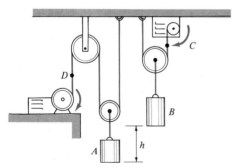

Prob. 12–137

12–138. The hoist is used to lift the load at D. If the end A of the chain is traveling downward at $v_A = 5$ ft/s and the end B is traveling upward at $v_B = 2$ ft/s, determine the velocity of the load at D.

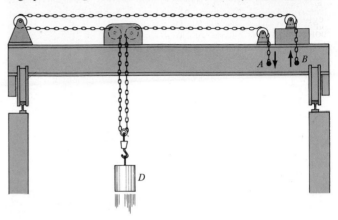

Prob. 12–138

12–139. If the point A on the cable is moving upward at $v_A = 14$ m/s, determine the speed of block B.

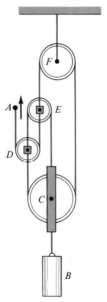

Prob. 12–139

*12–140. The crate C is being lifted by moving the roller at A downward with a constant speed of $v_A = 2$ m/s along the guide. Determine the velocity and acceleration of the crate at the instant $s = 1$ m. When the roller is at B, the crate rests on the ground. Neglect the size of the pulley in the calculation. *Hint:* Relate the coordinates x_C and x_A using the problem geometry, then take the first and second time derivatives.

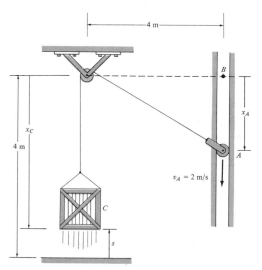

Prob. 12–140

12–141. The cable at A is drawn in at a constant rate $v_A = 6$ m/s. Determine the speed at which the load at B rises as a function of its vertical position y.

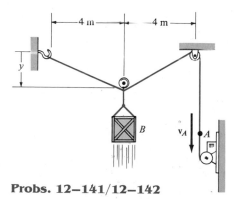

Probs. 12–141/12–142

12–142. If the cable at A is drawn in with an acceleration $a_A = 2$ m/s^2 when it has a speed $v_A = 6$ m/s, determine the velocity and acceleration of the load at B at this instant, which occurs when $y = 3$ m.

12–143. The roller at A is moving upward with a velocity of $v_A = 3$ ft/s and has an acceleration of $a_A = 4$ ft/s^2, when $s_A = 4$ ft. Determine the velocity and acceleration of block B at this instant.

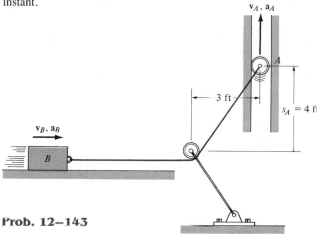

Prob. 12–143

*12–144. The truck at B is used to hoist the cabinet up to the fourth floor of a building using the rope and pulley arrangement shown. If the truck is moving forward at a constant speed of $v_B = 2$ ft/s, determine the speed at which the cabinet rises at the instant $s_A = 40$ ft. Neglect the size of the pulley. When $s_B = 0$, $s_A = 0$, so that points A and B are coincident, i.e., the rope is 100 ft long.

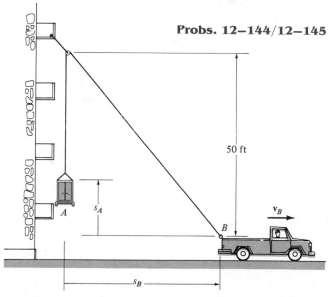

Probs. 12–144/12–145

12–145. Determine the acceleration of the cabinet when $s_A = 40$ ft in Prob. 12–144 if the truck is accelerating at $a_B = 0.6$ ft/s^2 and it has a speed of $v_B = 2$ ft/s when $s_A = 40$ ft.

12–146. Planes A and B are flying at the same altitude. If their velocities are $v_A = 300$ mi/h and $v_B = 250$ mi/h when the angle between their straight-line courses is $35°$ as shown, determine the velocity of plane A with respect to plane B.

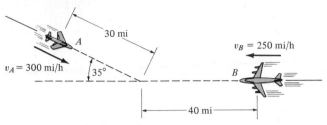

Probs. 12–146/12–147

12–147. At a given instant the planes A and B are located as shown. If they maintain their straight-line courses and speeds, determine the distance between them in $t = 5$ min. Plane A is traveling at $v_A = 300$ mi/h and plane B is traveling at $v_B = 250$ mi/h.

12–148. Two planes, A and B, are flying at the same altitude. If their velocities are $v_A = 600$ km/h and $v_B = 500$ km/h and the angle between their straight-line courses is $\theta = 75°$, determine the velocity of plane B with respect to plane A.

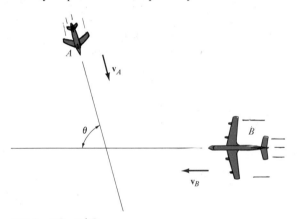

Prob. 12–148

12–149. Two boats leave the shore at the same time and travel in the directions shown. If $v_A = 20$ ft/s and $v_B = 15$ ft/s, determine the speed of boat A with respect to boat B. How long after leaving the shore will the boats be 800 ft apart?

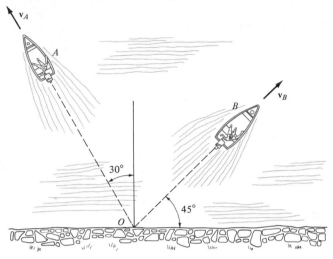

Probs. 12–149/12–150

12–150. Each boat begins from rest at point O and heads in the direction shown at the same instant. If A has an acceleration of $a_A = 2$ ft/s², and B has an acceleration of $a_B = 3$ ft/s², determine the speed of boat A with respect to boat B at the instant they become 800 ft apart. How long will this take?

12–151. At the instant shown, the bicyclist at A is traveling at 7 m/s around the curve on the race track while increasing his speed at 0.5 m/s². The bicyclist at B is traveling at 8.5 m/s along the straight-a-way and increasing his speed at 0.7 m/s². Determine the relative velocity and relative acceleration of A with respect to B at this instant.

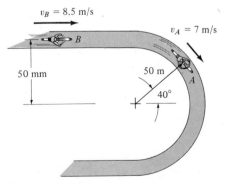

Prob. 12–151

*12–152. A passenger in an automobile observes that raindrops make an angle of 30° with the horizontal as the auto travels forward with a speed of 60 km/h. Compute the terminal (constant) velocity $\mathbf{v}_r$ of the rain if it is assumed to fall vertically.

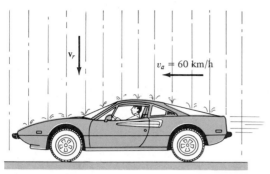

Prob. 12–152

12–153. The boat can travel with a speed of 16 km/h in still water. The point of destination is located along the dashed line. If the water is moving at 4 km/h, determine the bearing angle θ at which the boat must travel to stay on course.

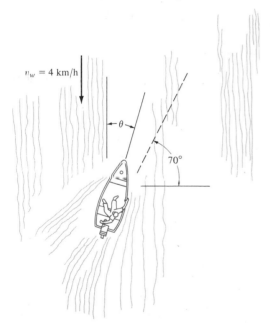

Prob. 12–153

12–154. Cars A and B are traveling around the circular race track. At the instant shown, A has a speed of 90 ft/s and is increasing its speed at the rate of 15 ft/s², whereas B has a speed of 105 ft/s and is decreasing its speed at 25 ft/s². Determine the relative velocity and relative acceleration of car A with respect to car B at this instant.

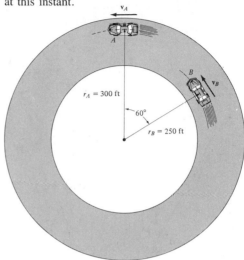

Prob. 12–154

12–155. At a given instant the football player at A throws a football C with a velocity of 20 m/s in the direction shown. Determine the constant speed at which the player at B must run so that he can catch the football at the same elevation at which it was thrown. Also calculate the relative velocity and relative acceleration of the football with respect to B at the instant the catch is made. Player B is 15 m away from A when A starts to throw the football.

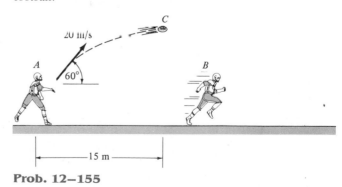

Prob. 12–155

75

13

Kinetics of a Particle: Force and Acceleration

In Chapter 12 we developed the methods needed to formulate the acceleration of a particle in terms of its velocity and position. In this chapter we will use these concepts when applying Newton's second law of motion, $\mathbf{F} = m\mathbf{a}$, to study the effects caused by an unbalanced force acting on a particle. Depending upon the geometry of the path, the analysis of problems will be performed using rectangular, normal and tangential, or cylindrical coordinates. In the last part of the chapter, Newton's second law of motion will be used to study problems in space mechanics.

13.1 Newton's Laws of Motion

Many of the earlier notions about dynamics were dispelled after 1590 when Galileo performed experiments to study the motions of pendulums and falling bodies. The conclusions drawn from these experiments gave some insight as to the effects of forces acting on bodies in motion. The general laws of motion of a body subjected to forces had not been known, however, until 1687, when Isaac Newton first presented three basic laws governing the motion of a particle. In a slightly reworded form, Newton's three laws of motion can be stated as follows:

First Law. A particle originally at rest, or moving in a straight line with a constant velocity, will remain in this state provided the particle is not subjected to an unbalanced force.

Second Law. A particle acted upon by an unbalanced force **F** experiences an acceleration **a** that has the same direction as the force and a magnitude that is directly proportional to the force.*

Third Law. The mutual forces of action and reaction between two particles are equal, opposite, and collinear.

The first and third laws were used extensively in developing the concepts of statics. Although these laws are also considered in dynamics, Newton's second law of motion forms the basis for most of this study, since this law relates the accelerated motion of a particle to the forces that act on it. It should be noted that statics is a special case of dynamics, since Newton's second law yields the results of his first law when the resultant force is equal to zero; namely, no acceleration occurs, and therefore the particle's velocity is constant.

Measurements of force and acceleration can be recorded in a laboratory so that in accordance with the second law, if a known unbalanced force F_1 is applied to a particle, the acceleration a_1 of the particle may be measured. Since the force and acceleration are directly proportional, the constant of proportionality, m, may be determined from the ratio $m = F_1/a_1$. Provided the units of measurement are consistent, a different unbalanced force F_2 applied to the particle will create an acceleration a_2, such that $F_2/a_2 = m$.† In both cases the ratio will be the same and the acceleration and the force, both being vector quantities, will have the same direction. The scalar m is called the *mass* of the particle. Being constant during any acceleration, m provides a quantitative measure of the resistance of the particle to a change in its velocity.

If the mass of the particle is m, Newton's second law of motion may be written in mathematical form as

$$\mathbf{F} = m\mathbf{a}$$

This equation, which is referred to as the *equation of motion,* is one of the most important formulations in mechanics.‡ As stated above, its validity has been based solely upon *experimental evidence.* In 1905, however, Albert

*Stated another way, the unbalanced force acting on the particle is proportional to the time rate of change of the particle's linear momentum. See footnote ‡ below.

†Throughout this discussion, the units of force, mass, length, and time used to measure F, m, and a must be chosen such that the units for one of these quantities are defined in terms of all the others. Such is the case for the SI and FPS units, where it may be recalled that $N = kg \cdot m/s^2$ and $slug = lb \cdot s^2/ft$ (see Sec. 1.3 of *Statics*). Note, however, that if units of force, mass, length, and time are *all* selected arbitrarily, then $F = kma$ and k (a dimensionless constant) would have to be experimentally determined in order to preserve the equality.

‡Since m is constant, we can also write $\mathbf{F} = \dfrac{d}{dt}(m\mathbf{v})$, where $m\mathbf{v}$ is the particle's linear momentum.

Einstein developed the theory of relativity and placed limitations on the use of Newton's second law for describing general particle motion. Through experiments it was proven that *time* is not an absolute quantity as assumed by Newton; and as a result, the equation of motion fails to predict the exact behavior of a particle, especially when the particle's speed approaches the speed of light [0.3 Gm/s]. Developments of the theory of quantum mechanics by Schrödinger and others indicate further that conclusions drawn from using this equation are also invalid when particles move within an atomic distance of one another. For the most part, however, these requirements regarding particle speed and size are not encountered in engineering problems, so that their effects will not be considered in this book.

Newton's Law of Gravitational Attraction

Shortly after formulating his three laws of motion, Newton postulated a law governing the mutual attraction between any two particles. In mathematical form this law can be expressed as

$$F = G\frac{m_1 m_2}{r^2} \qquad (13-1)$$

where F = force of attraction between the two particles

G = universal constant of gravitation; according to experimental evidence $G = 66.73(10^{-12})$ m^3/(kg · s^2)

m_1, m_2 = mass of each of the two particles

r = distance between the centers of the two particles

Any two particles or bodies have a mutually attractive gravitational force acting between them. In the case of a particle located at or near the surface of the earth, however, the only gravitational force having any sizable magnitude is that between the earth and the particle. This force is termed the ''weight'' and, for our purpose, it will be the only gravitational force considered.

Mass and Weight

Mass is a property of matter by which we can compare the action of one body with that of another. As indicated above, this property manifests itself as a gravitational attraction between two bodies and provides a quantitative measure of the resistance of matter to a change in velocity. It is an *absolute* quantity since the measurement of mass can be made at any location. The weight of a body, however, is *not absolute* since it is measured in a gravitational field, and hence its magnitude depends upon where the measurement is made. From Eq. 13–1, we can develop a general expression for finding the weight W of a particle having a mass $m = m_1$. Let m_2 be the mass of the earth and r the distance between the earth's center and the particle. Then if $g = Gm_2/r^2$, we have

$$W = mg$$

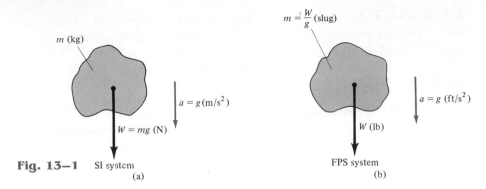

Fig. 13—1   SI system
(a)

FPS system
(b)

By comparison with $F = ma$, we term g the acceleration due to gravity. For most engineering calculations we measure g at a point on the surface of the earth at sea level, and at a latitude of 45°, considered the "standard location."

The mass and weight of a body are measured differently in the SI and FPS systems of units, and the method of defining these units should be thoroughly understood.

SI System of Units. In the SI system the mass of the body is specified in kilograms and the weight must be calculated using the equation of motion, $F = ma$. Hence, if a body has a mass m (kg) and is located at a point where the acceleration due to gravity is g (m/s²), then the weight is expressed in *newtons* as $W = mg$ (N), Fig. 13–1a. In particular, if the body is located at the "standard location," the acceleration due to gravity is $g = 9.806\ 65$ m/s². For calculations, the value $g = 9.81$ m/s² will be used, so that

$$W = mg \text{ (N)} \quad (g = 9.81 \text{ m/s}^2) \tag{13–2}$$

Therefore, a body of mass 1 kg has a weight of 9.81 N; a 2-kg body weighs 19.62 N; and so on.

FPS System of Units. In the FPS system the weight of the body is specified in pounds and the mass must be calculated from $F = ma$. Hence, if a body has a weight W (lb) and is located at a point where the acceleration due to gravity is g (ft/s²), then the mass is expressed in *slugs* as $m = W/g$ (slug), Fig. 13–1b. Since the acceleration of gravity at the standard location is approximately 32.2 ft/s²($= 9.81$ m/s²), the mass of the body measured in slugs is

$$m = \frac{W}{g} \text{(slug)} \quad (g = 32.2 \text{ ft/s}^2) \tag{13–3}$$

Therefore, a body weighing 32.2 lb has a mass of 1 slug; a 64.4-lb body has a mass of 2 slugs; and so on.

13.2 The Equation of Motion

When more than one force acts on a particle, the resultant force is determined by a vector summation of all the forces; i.e., $\mathbf{F}_R = \Sigma\mathbf{F}$. For this more general case, the equation of motion may be written as

$$\Sigma\mathbf{F} = m\mathbf{a} \qquad\qquad (13\text{--}4)$$

To illustrate application of this equation, consider the particle P shown in Fig. 13–2a, which has a mass m and is subjected to the action of two forces, $\mathbf{F}_1$ and $\mathbf{F}_2$. We can graphically account for the magnitude and direction of each force acting on the particle by drawing the particle's *free-body diagram,* Fig. 13–2b. Since the *resultant* of these forces *produces* the vector $m\mathbf{a}$, its magnitude and direction can be represented graphically on the *kinetic diagram,* shown in Fig. 13–2c. The equal sign written between the diagrams symbolizes the *graphical* equivalency between the free-body diagram and the kinetic diagram; i.e., $\Sigma\mathbf{F} = m\mathbf{a}$.* In particular, note that if $\mathbf{F}_R = \Sigma\mathbf{F} = \mathbf{0}$, then the acceleration is also zero, so that the particle will either remain at *rest* or move along a straight-line path with *constant velocity*. Such are the conditions of *static equilibrium,* Newton's first law of motion.

Inertial Frame of Reference

Whenever the equation of motion is applied, it is required that measurements of the acceleration be made from a *Newtonian* or *inertial frame of reference*. Such a *coordinate system does not rotate and is either fixed or translates in a given direction with a constant velocity (zero acceleration)*. This definition ensures that the particle's *acceleration* measured by observers in two different inertial frames of reference will always be the *same*. For example, consider the particle P moving with an acceleration $\mathbf{a}_P$ along a straight path as shown in

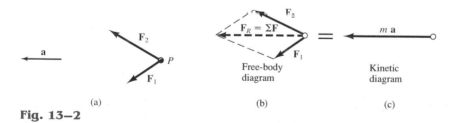

(a)

Free-body diagram (b)

Kinetic diagram (c)

Fig. 13–2

---

*The equation of motion can also be rewritten in the form $\Sigma\mathbf{F} - m\mathbf{a} = \mathbf{0}$. The vector $-m\mathbf{a}$ is referred to as the *inertia force vector*. If it is treated in the same way as a "force vector," then the state of "equilibrium" created is referred to as *dynamic equilibrium*. This method for application of the equation of motion is often referred to as the *D'Alembert principle,* named after the French mathematician Jean le Rond d'Alembert (1717–1783).

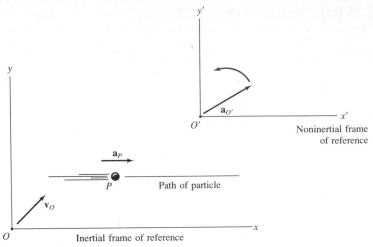

Fig. 13–3

Fig. 13–3. If the observer is *fixed* in the inertial x, y frame of reference, this acceleration, $\mathbf{a}_P$, will be measured by the observer regardless of the direction and magnitude of the velocity $\mathbf{v}_O$ of the frame of reference. On the other hand, if the observer is *fixed* in the noninertial x', y' frame of reference, Fig. 13–3, the observer will not measure the particle's acceleration as $\mathbf{a}_P$. Instead, if the frame is *accelerating* at $\mathbf{a}_{O'}$ the particle will appear to have an acceleration of $\mathbf{a}_{P/O'} = \mathbf{a}_P - \mathbf{a}_{O'}$. Also, if the frame is *rotating*, as indicated by the curl, then the particle will appear to move along a *curved path*, in which case it must have other components of acceleration (see Sec. 16–8). In any case, the measured acceleration cannot be used in Newton's law of motion to determine the forces acting on the particle.

When studying the motions of rockets and satellites it is justifiable to consider the inertial reference frame as fixed to the stars, whereas dynamics problems concerned with motions on or near the surface of the earth may be solved by using an inertial frame which is assumed fixed to the earth. Even though the earth rotates about its own axis and revolves about the sun, the acceleration created by these rotations can be neglected in most computations.

13.3 Equation of Motion for a System of Particles

The equation of motion will now be extended to include a system of n particles isolated within an enclosed region in space, as shown in Fig. 13–4a. In particular, there is no restriction in the way the particles are connected, and as a result the following analysis will apply equally well to the motion of a solid, liquid, or gas system. At the instant considered, the arbitrary ith particle, having a mass m_i, is subjected to a set of internal forces and a resultant external force. The *resultant internal force*, represented symbolically as $\mathbf{f}_i = \displaystyle\sum_{j=1(j\neq i)}^{n} \mathbf{f}_{ij}$, is determined from the forces which the other particles exert on the

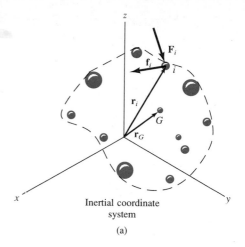

Inertial coordinate
system

(a)

ith particle. Although only one of these particles is shown in Fig. 13–4a, the summation extends over all n particles within the dashed boundary. Note that it is meaningless for $j = i$ since the ith particle cannot exert a force on itself. The *resultant external force* $\mathbf{F}_i$ represents, for example, the effect of gravitational, electrical, magnetic, or contact forces between adjacent bodies or particles *not* included within the system.

The free-body and kinetic diagrams for the ith particle are shown in Fig. 13–4b. Applying the equation of motion yields

$$\Sigma\mathbf{F} = m\mathbf{a}; \qquad \mathbf{F}_i + \mathbf{f}_i = m_i\mathbf{a}_i$$

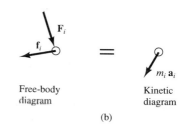

Free-body
diagram

Kinetic
diagram

(b)

Fig. 13–4

When the equation of motion is applied to each of the other particles of the system, similar equations will result. If all these equations are added together *vectorially,* we obtain

$$\Sigma\mathbf{F}_i + \Sigma\mathbf{f}_i = \Sigma m_i\mathbf{a}_i$$

The summation of the internal forces, if carried out, will equal zero since the internal forces between particles all occur in equal, but opposite collinear pairs. Consequently, only the sum of the external forces will remain and therefore the equation of motion, written for the system of particles, becomes

$$\Sigma\mathbf{F}_i = \Sigma m_i\mathbf{a}_i \qquad (13\text{–}5)$$

If $\mathbf{r}_G$ is a position vector which locates the *center of mass G* of the particles, Fig. 13–4a, then by definition of the center of mass $m\mathbf{r}_G = \Sigma m_i\mathbf{r}_i$, where $m = \Sigma m_i$ is the total mass of all the particles. Differentiating this equation twice with respect to time, assuming no mass is entering or leaving the system,* yields

$$m\mathbf{a}_G = \Sigma m_i\mathbf{a}_i$$

Substituting this result into the above equation, we obtain

$$\boxed{\Sigma\mathbf{F} = m\mathbf{a}_G} \qquad (13\text{–}6)$$

This equation states that the sum of the external forces acting on the system of particles is equal to the mass $m = \Sigma m_i$ of a single "fictitious" particle times its acceleration. This fictitious particle is located at the center of mass G of all the particles.

Since, in reality, all particles must have finite size to possess mass, Eq. 13–6 justifies application of the equation of motion to a *body* that is represented as a single particle.

*A case in which m is a function of time (variable mass) is discussed in Sec. 15.9.

13.4 Equations of Motion: Rectangular Coordinates

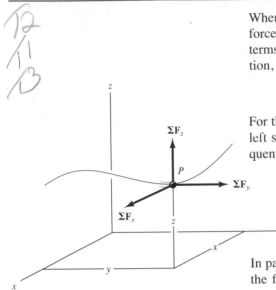

Fig. 13–5

When a particle is moving relative to an inertial x, y, z frame of reference, the forces acting on the particle, as well as its acceleration, may be expressed in terms of their $\mathbf{i}, \mathbf{j}, \mathbf{k}$ components, Fig. 13–5. Applying the equation of motion, we have

$$\Sigma\mathbf{F} = m\mathbf{a}$$

$$\Sigma F_x\mathbf{i} + \Sigma F_y\mathbf{j} + \Sigma F_z\mathbf{k} = m(a_x\mathbf{i} + a_y\mathbf{j} + a_z\mathbf{k})$$

For this equation to be satisfied, the respective $\mathbf{i}, \mathbf{j}$, and $\mathbf{k}$ components on the left side must equal the corresponding components on the right side. Consequently, we may write the following three scalar equations:

$$\begin{aligned} \Sigma F_x &= ma_x \\ \Sigma F_y &= ma_y \\ \Sigma F_z &= ma_z \end{aligned} \qquad (13\text{–}7)$$

In particular, if the particle is constrained to move in the x-y plane, then only the first two of these equations are used to specify the motion.

PROCEDURE FOR ANALYSIS

The equations of motion are used to solve problems which require a relationship between the forces and the accelerated motion they cause. Whenever they are applied, the unknown force and acceleration components should be identified and an equivalent number of equations should be written. If further equations are necessary for the solution, kinematics may be considered. Recall that these equations only relate the geometric properties of the motion, and therefore they become useful if the particle's position or velocity is to be related to its acceleration. Whatever the situation, the following procedure provides a general method for solving problems in kinetics.

Free-Body Diagram. Select the inertial coordinate system. Most often, rectangular or x, y, z coordinates are chosen to analyze problems for which the particle has *rectilinear motion*. If this occurs, one of the axes should extend in the direction of motion. Once the coordinates are established, draw the particle's free-body diagram. Drawing the free-body diagram is *very important* since it provides a graphical representation of *all the forces* ($\Sigma\mathbf{F}$) which act on the particle and thereby makes it possible to resolve these forces into their x, y, z components. The direction and sense of the particle's acceleration $\mathbf{a}$ should also be established. If the sense of its components is unknown, for mathematical convenience assume that they are in the same *direction* as the *positive* inertial coordinate axes. The acceleration may be sketched on the x, y, z coordinate system, *but not on* the free-body diagram, or it may be represented as the $m\mathbf{a}$ vector on the kinetic diagram.* Once the free-body diagram has been constructed, identify the unknowns under consideration.

Equations of Motion.   If the forces can be resolved directly from the free-body diagram, apply the equations of motion in their scalar component form. If the geometry of the problem appears complicated, which often occurs in three dimensions, Cartesian vector analysis can be used for the solution.

Friction If the particle contacts a rough surface, it may be necessary to use the *frictional equation,* which relates the coefficient of kinetic friction μ_k to the magnitudes of the frictional and normal forces $\mathbf{F}_f$ and $\mathbf{N}$ acting at the surfaces of contact, i.e., $F_f = \mu_k N$.

Spring If the particle is connected to an *elastic spring* having negligible mass, the spring force F_s can be related to the deformation of the spring by the equation $F_s = ks$. Here k is the spring's stiffness measured as a force per unit length, and s is the stretch or compression defined as the difference between the deformed length l and the undeformed length l_0, i.e., $s = l - l_0$.

Kinematics.   As stated above, if a complete solution cannot be obtained strictly from the equation of motion, the equations of kinematics may be considered. For example, if the velocity or position of the particle is to be found, it will be necessary to apply the proper kinematic equations once the particle's acceleration is determined from $\Sigma\mathbf{F} = m\mathbf{a}.$

If the *acceleration is a function of time,* use $a = dv/dt$ and $v = ds/dt$ which, when integrated, yield the particle's velocity and position.

If the *acceleration is a function of displacement,* integrate $a\ ds = v\ dv$ to obtain the velocity as a function of position.

If the *acceleration is constant,* use

$$v = v_0 + a_c t \qquad s = s_0 + v_0 t + \tfrac{1}{2}a_c t^2 \qquad v^2 = v_0^2 + 2a_c(s - s_0)$$

to determine the position or velocity of the particle.

---

*It is a convention in this text to always use the kinetic diagram as a graphical aid when developing the proofs and theory.

If the problem involves the dependent motion of several particles, use the method outlined in Sec. 12.8 to relate their accelerations.

In all cases, make sure the positive coordinate directions used for writing the kinematic equations are the same as those used for writing the equations of motion; otherwise, simultaneous solution of the equations will result in errors. Also, if the solution for an unknown vector component yields a negative scalar, it indicates that the component acts in the direction opposite to that which was assumed.

The following examples numerically illustrate application of this procedure.

Example 13–1

The 50-kg crate shown in Fig. 13–6a rests on a horizontal plane for which the coefficient of kinetic friction is $\mu_k = 0.3$. If the crate is subjected to a 400-N towing force as shown, determine the velocity of the crate in 5 s starting from rest.

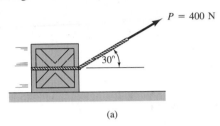

$P = 400$ N

30°

(a)

SOLUTION

This problem requires application of the equations of motion, since the crate's acceleration can be related to the force causing the motion. The crate's velocity can then be determined using kinematics.

Free-Body Diagram. The weight of the crate is $W = mg = 50$ kg $(9.81$ m/s$^2) = 490.5$ N. As shown in Fig. 13–6b, the frictional force has a magnitude $F = \mu_k N_C$ and acts to the left, since it opposes the motion of the crate. The acceleration **a** is assumed to act horizontal, in the positive x direction. There are two unknowns, namely N_C and a.

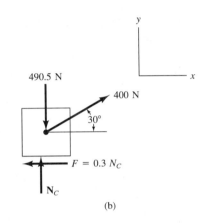

490.5 N

400 N

30°

$F = 0.3\ N_C$

N_C

(b)

Equations of Motion. Using the data shown on the free-body diagram, we have

$$\xrightarrow{+}\Sigma F_x = ma_x; \qquad 400\cos 30° - 0.3\ N_C = 50a \qquad (1)$$
$$+\uparrow \Sigma F_y = ma_y; \qquad N_C - 490.5 + 400\sin 30° = 0 \qquad (2)$$

Solving Eq. (2) for N_C, substituting the result into Eq. (1), and solving for a yields

$$N_C = 290.5 \text{ N}$$
$$a = 5.19 \text{ m/s}^2$$

Kinematics. Since the acceleration is *constant*, and the initial velocity is zero, the velocity of the crate in 5 s is

$(\xrightarrow{+})$

$$v = v_0 + a_c t$$
$$= 0 + 5.19(5)$$
$$= 26.0 \text{ m/s} \rightarrow \qquad\qquad \textit{Ans.}$$

Note the alternative procedure of drawing the crate's free-body *and* kinetic diagrams, Fig. 13–6c, prior to applying the equations of motion.

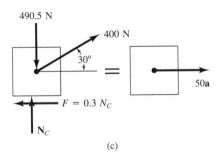

490.5 N

400 N

30°

$F = 0.3\ N_C$

N_C

$=$

50a

(c)

Fig. 13–6

Example 13–2

A 10-kg projectile is fired vertically upward from the ground, with an initial velocity of 50 m/s, Fig. 13–7a. Determine the maximum height to which it will travel if (a) atmospheric resistance is neglected; and (b) atmospheric resistance is measured as $F_D = (0.01v^2)$ N, where v is the speed at any instant, measured in m/s.

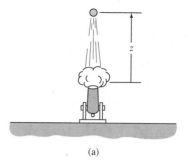

(a)

SOLUTION

In both cases the known force on the projectile can be related to its acceleration using the equation of motion. Kinematics can then be used to relate the projectile's acceleration to its position.

Part (a) Free-Body Diagram. As shown in Fig. 13–7b, the projectile's weight is $W = mg = 10(9.81) = 98.1$ N. We assume the unknown acceleration **a** acts upward in the *positive z* direction.

Equation of Motion

$+\uparrow \Sigma F_z = ma_z;$ $-98.1 = 10a,\ a = -9.81$ m/s²

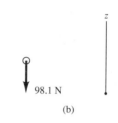

98.1 N

(b)

The result indicates that the projectile, like every object having free-flight motion near the earth's surface, is subjected to a constant downward acceleration of 9.81 m/s².

Kinematics. Initially, $z_0 = 0$ and $v_0 = 50$ m/s, and at the maximum height $z = h$, $v = 0$. Since the acceleration is *constant*, then

$(+\uparrow)$ $v^2 = v_0^2 + 2a_c(z - z_0)$

$0 = (50)^2 + 2(-9.81)(h - 0)$

$h - 127$ m   *Ans.*

Part (b) Free-Body Diagram. Since the force $F_D = (0.01v^2)$ N tends to retard the upward motion of the projectile, it acts downward as shown on the free-body diagram, Fig. 13–7c.

Equation of Motion*

$|\uparrow \Sigma F_z = ma_z;$ $-0.01v^2 - 98.1 = 10a$

$a = -0.001v^2 - 9.81$

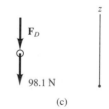

F_D

98.1 N

(c)

Fig. 13–7

Kinematics. Here the acceleration is *not constant;* however, it can be related to the velocity and displacement by using $a\ dz = v\ dv$.

$(+\uparrow)\ a\ dz = v\ dv;$ $(-0.001v^2 - 9.81)dz = v\ dv$

Separating the variables and integrating, realizing that initially $z_0 = 0$, $v_0 = 50$ m/s (positive upward), and at $z = h$, $v = 0$, we have

$$\int_0^h dz = -\int_{50}^0 \frac{v\ dv}{0.001v^2 + 9.81}$$

$$h = -500 \ln (v^2 + 9810)\Big|_{50}^0 = 113 \text{ m} \qquad Ans.$$

The answer indicates a lower elevation than that obtained in Part (a). Why?

---

*Note that if the projectile were fired downward, with z positive downward, the equation of motion would then be $-0.01v^2 + 98.1 = 10a$.

Example 13–3

The crate shown in Fig. 13–8a has a weight of 50 lb and is acted upon by a force having a variable magnitude $P = 20t$, where P is in pounds and t is in seconds. Compute the crate's velocity 2 s after $\mathbf{P}$ has been applied. The crate's initial velocity is $v_0 = 3$ ft/s down the plane, and the coefficient of kinetic friction between the crate and the plane is $\mu_k = 0.3$.

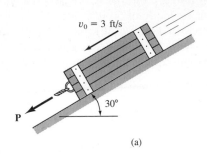

(a)

SOLUTION

Before reading further, do you see *why* it is necessary to apply both the equations of motion *and* kinematics to obtain a solution of this problem?

Free-Body Diagram. As shown in Fig. 13–8b, the frictional force is directed opposite to the crate's sliding motion and has a magnitude $F = \mu_k N_C = 0.3 N_C$. The mass is $m = W/g = 50/32.2 = 1.55$ slug. At this point, there are two unknowns, namely N_C and a.

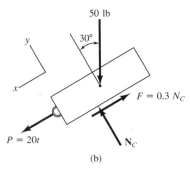

(b)

Fig. 13–8

Equations of Motion. Using the data shown on the free-body diagram, we have

$$+\swarrow \Sigma F_x = ma_x; \quad 20t - 0.3N_C + 50 \sin 30° = 1.55a \qquad (1)$$
$$+\nwarrow \Sigma F_y = ma_y; \quad\quad N_C - 50 \cos 30° = 0 \qquad\qquad (2)$$

Solving for N_C in Eq. (2) ($N_C = 43.3$ lb), substituting into Eq. (1), and simplifying yields

$$a = 12.88t + 7.73 \qquad (3)$$

Kinematics. Since the acceleration is a function of time, the velocity of the crate is obtained by using $a = dv/dt$ with the initial condition that $v_0 = 3$ ft/s at $t = 0$. We have

$$(+\swarrow) \qquad\qquad dv = a\, dt$$
$$\int_3^v dv = \int_0^t (12.88t + 7.73)dt$$
$$v = 6.44t^2 + 7.73t + 3$$

When $t = 2$ s,

$$v = 44.2 \text{ ft/s} \swarrow \qquad\qquad \textbf{\textit{Ans.}}$$

Example 13–4

A smooth 2-kg collar C, shown in Fig. 13–9a, is attached to a spring having a stiffness $k = 3$ N/m and an unstretched length of 0.75 m. If the collar is released from rest at A, determine its acceleration and the normal force of the rod on the collar at the instant $y = 1$ m.

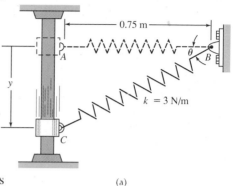

(a)

SOLUTION

Free-Body Diagram. The free-body diagram of the collar when it is located at the arbitrary position y is shown in Fig. 13–9b. Note that the weight is $W = 2(9.81) = 19.62$ N. Furthermore, the collar is *assumed* to be accelerating so that '**a**' acts downward in the *positive* y direction. There are four unknowns, namely N_C, F_s, a, and θ.

Equations of Motion. Using the data in Fig. 13–9b,

$$\xrightarrow{+}\Sigma F_x = ma_x; \qquad -N_C + F_s \cos \theta = 0 \qquad (1)$$
$$+\downarrow \Sigma F_y = ma_y; \qquad 19.62 - F_s \sin \theta - 2a \qquad (2)$$

From Eq. (2) it is seen that the acceleration is not constant; rather, it depends upon the magnitude and direction of the spring force. Solution is possible once F_s and θ are known.

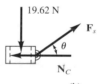

(b)

Fig. 13–9

The magnitude of the spring force is a function of the stretch s of the spring; i.e., $F_s = ks$. Here the unstretched length is $AB = 0.75$ m, Fig. 13–9a; therefore, $s = CB - AB = \sqrt{y^2 + (0.75)^2} - 0.75$. Since $k = 3$ N/m, then

$$F_s = ks = 3(\sqrt{y^2 + (0.75)^2} - 0.75) \qquad (3)$$

From Fig. 13–9a, the angle θ is related to y by trigonometry.

$$\sin \theta = \frac{y}{\sqrt{y^2 + (0.75)^2}} \qquad (4)$$

Substituting $y = 1$ m into Eqs. (3) and (4) yields $F_s = 1.50$ N and $\theta = 53.1°$. Substituting these results into Eqs. (1) and (2), we obtain

$$N_C = 0.900 \text{ N} \qquad\qquad Ans.$$
$$a = 9.21 \text{ m/s}^2 \downarrow \qquad\qquad Ans.$$

Example 13–5

The 100-kg block A shown in Fig. 13–10a is released from rest. If the mass of the pulleys and the cord is neglected, determine the speed of the 20-kg block B in 2 s.

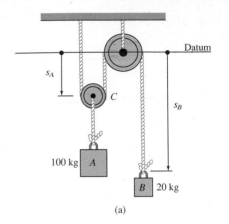

(a)

SOLUTION
Motion of blocks A and B will be analyzed separately.

Free-Body Diagram. Since the mass of the pulleys is neglected, the equilibrium condition for pulley C is shown in Fig. 13–10b. The free-body diagrams for blocks A and B are shown in Fig. 13–10c and d, respectively. Here we *assume* both blocks accelerate downward, in the direction of $+s_A$ and $+s_B$. What are the three unknowns?

Equations of Motion
Block A (Fig. 13–10c):

$$+ \downarrow \Sigma F_y = ma_y; \qquad 981 - 2T = 100a_A \qquad (1)$$

Block B (Fig. 13–10d):

$$+ \downarrow \Sigma F_y = ma_y; \qquad 196.2 - T = 20a_B \qquad (2)$$

(b)

Kinematics. The necessary third equation is obtained by studying the kinematics of the pulley arrangement in order to relate a_A to a_B. Using the technique developed in Sec. 12.8, the coordinates s_A and s_B measure the positions of A and B from the fixed datum, Fig. 13–10a. It is seen that

$$2s_A + s_B = l$$

where l is constant and represents the total vertical length of cord. Differentiating this expression twice with respect to time yields

$$2a_A = -a_B \qquad (3)$$

Hence when block A accelerates *downward,* block B accelerates *upward* at twice the amount. Notice, however, that in writing Eqs. (1) to (3), the *positive direction was always assumed downward.* It is important to be *consistent* in this assumption since we are seeking a simultaneous solution of equations. The solution yields

$$T = 327.0 \text{ N}$$
$$a_A = 3.27 \text{ m/s}^2$$
$$a_B = -6.54 \text{ m/s}^2$$

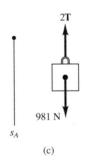

(c)

Since a_B is constant, the velocity of block B in 2 s is thus

$$(+ \downarrow) \qquad v = v_0 + a_B t$$
$$= 0 + (-6.54)(2)$$
$$= -13.1 \text{ m/s} \qquad Ans.$$

The negative sign indicates that block B is moving upward. Why?

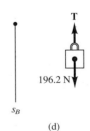

(d)

Fig. 13–10

PROBLEMS

13–1. Venus and Earth have diameters of 12,400 km and 12,760 km, respectively. The mass of Venus is 0.815 times that of the Earth. If a body has a mass of 35 kg on the surface of the Earth, what would its weight be on Venus? Also, what is the acceleration of gravity on Venus?

13–2. The moon has a mass of $73.5(10^{21})$ kg and the earth's mass is $5.98(10^{24})$ kg. If their centers are 384 Mm apart, determine the gravitational attractive force between the two bodies.

13–3. Determine the gravitational attraction between the earth, which has a mass of $5.98(10^{24})$ kg, and the sun, which has a mass of $1.99(10^{30})$ kg. The distance between their mass centers is $150(10^6)$ km.

*13–4. The baggage truck A has a mass of 800 kg and is used to pull each of the 300-kg cars. Determine the tension force in the couplings at B and C if the tractive force $\mathbf{F}$ on the truck is $F = 480$ N. What is the speed of the truck when $t = 2$ s, starting from rest? The car wheels are free to roll. Neglect the mass of the wheels.

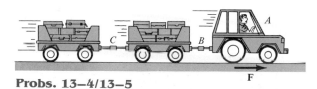

Probs. 13–4/13–5

13–5. The baggage truck A has a mass of 800 kg and is used to pull each of the 300-kg cars. If the tractive force $\mathbf{F}$ on the truck is $F = 480$ N, determine the initial acceleration of the truck. What is the acceleration of the truck if the coupling at C suddenly fails? The car wheels are free to roll. Neglect the mass of the wheels.

13–6. Two men stand on ice and hold the ends of a rope. Man A weighs 150 lb and man B weighs 200 lb. If A pulls horizontally on the rope at 30 lb, determine the force which the rope exerts on man B and the acceleration of each man. Neglect friction.

13–7. The elevator E has a mass of 500 kg and the counterweight at A has a mass of 150 kg. If the motor supplies a constant force of 5 kN on the cable at B, determine the speed of the elevator in $t = 3$ s starting from rest. Neglect the mass of the pulleys and cable.

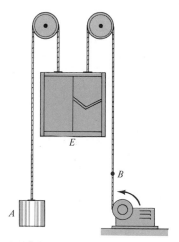

Probs. 13–7/13–8

*13–8. The elevator E has a mass of 500 kg and the counterweight at A has a mass of 150 kg. If the elevator attains a speed of 10 m/s after it rises 40 m, determine the force developed in the cable at B. Neglect the mass of the pulleys and cable.

13–9. Block B has a mass m and is released from rest when it is on top of cart A, which has a mass of $3\,m$. Determine the tension in cord CD needed to hold the cart from moving while B is sliding down A. Neglect friction.

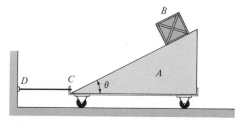

Probs. 13–9/13–10

13–10. Block B has a mass m and is released from rest when it is on top of cart A, which has a mass of $3\,m$. Determine the tension in cord CD needed to hold the cart from moving while B is sliding down A. The coefficient of kinetic friction between A and B is μ.

13–11. At a given instant the 5-lb block A is moving downward with a speed of 4 ft/s. Determine its speed 3 s later. Block B has a weight of 6 lb, and the coefficient of kinetic friction between it and the horizontal plane is $\mu_k = 0.3$. Neglect the mass of the pulleys and cord.

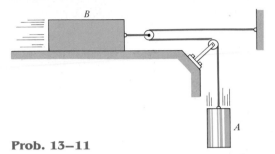

Prob. 13–11

*13–12. The 200-kg crate is suspended from the cable of a crane. Determine the force in the cable in $t = 2$ s, if the crate is moving upward with (a) a constant velocity of 2 m/s, and (b) a speed of $v = (0.2t^2 + 2)$ m/s, where t is in seconds.

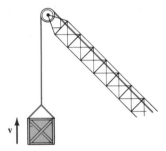

Prob. 13–12

13–13. The man weighs 180 lb and supports the barbells which have a weight of 100 lb. If he lifts them 2 ft in the air in 1.5 s starting from rest, determine the reaction of *both* of his feet on the ground during the lift. Assume the motion is with uniform acceleration.

Prob. 13–13

13–14. The 400-lb cylinder at A is hoisted using the motor and the pulley system shown. If the speed of point B on the cable is increased at a constant rate from zero to $v_B = 10$ ft/s in $t = 5$ s, determine the tension in the cable at B to cause the motion.

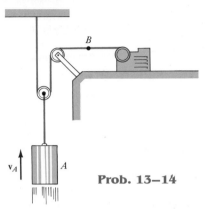

Prob. 13–14

13–15. The speed of the 3500-lb sports car is plotted over the 30-s time period. Plot the variation of the traction force $\mathbf{F}$ needed to cause the motion.

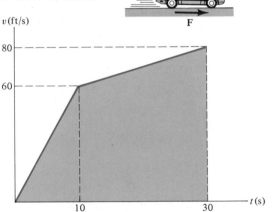

Prob. 13–15

*13–16. Block A has a mass of 20 kg and block B has a mass of 50 kg. If the spring is stretched 0.2 m at the instant shown, determine the acceleration of block B at this instant if it is (a) originally at rest, (b) moving downward with a speed of 3 m/s. Neglect friction.

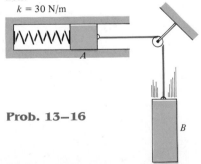

Prob. 13–16

13–17. A force of $F = 15$ lb is applied to the cord. Determine how high the 30-lb block A rises in 2 s starting from rest. Neglect the weight of the pulleys and cord.

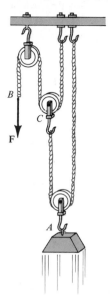

Probs. 13–17/13–18

13–18. Determine the constant force **F** which must be applied to the cord in order to cause the 30-lb block A to have a speed of 12 ft/s when it has been displaced 3 ft upward starting from rest. Neglect the weight of the pulleys and cord.

13–19. A freight elevator, including its load, has a mass of 500 kg. It is prevented from rotating by using the track and wheels mounted along its sides. If the motor M develops a constant tension $T = 1.50$ kN in its attached cable, determine the velocity of the elevator when it has moved upward 3 m starting from rest. Neglect the mass of the pulleys and cables.

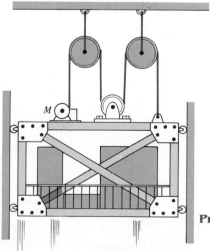

Probs. 13–19/13–20

*13–20. A freight elevator, including its load, has a mass of 500 kg. It is prevented from rotating by using the track and wheels mounted along its sides. Starting from rest, in $t = 2$ s, the motor M draws in the cable with a speed of 6 m/s, *measured relative to the elevator*. Determine the constant acceleration of the elevator and the tension in the cable. Neglect the mass of the pulleys and cables.

13–21. Determine the acceleration of the 5-kg cylinder A. Neglect the mass of the pulleys and cords. The block at B has a mass of 10 kg. Assume the surface at B is smooth.

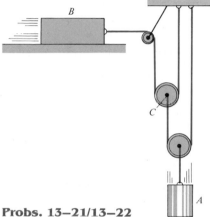

Probs. 13–21/13–22

13–22. Determine the acceleration of the 5-kg cylinder A. Neglect the mass of the pulleys and cords. The block at B has a mass of 10 kg. The coefficient of kinetic friction between block B and the surface is $\mu_k = 0.1$.

13–23. The 40-lb suitcase slides from rest 20 ft down the smooth ramp. Determine the point where it strikes the ground at C. How long does it take to go from A to C?

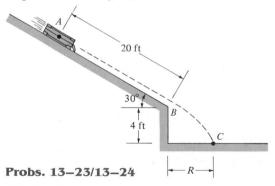

Probs. 13–23/13–24

*13–24. Solve Prob. 13–23 if the suitcase has an initial velocity down the ramp of $v_A = 10$ ft/s and the coefficient of kinetic friction along AB is $\mu_k = 0.2$.

13–25. Block A has a weight of 8 lb and block B has a weight of 6 lb. They rest on a surface for which the coefficient of kinetic friction is $\mu_k = 0.2$. If the spring has a stiffness of $k = 20$ lb/ft, and it is compressed 0.2 ft, determine the acceleration of each block just after they are released.

Prob. 13–25

13–26. The 100-kg crate is hoisted up the incline using the cable and motor M. For a short time, the force in the cable is $F = 800t^2$ N, where t is in seconds. If the crate has an initial velocity $v_1 = 2$ m/s when $t = 0$, determine its velocity when $t = 2$ s. The coefficient of kinetic friction between the crate and the incline is $\mu_k = 0.3$.

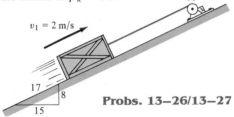

Probs. 13–26/13–27

13–27. The 100-kg crate is hoisted up the incline using the cable and motor M. For a short time, the force in the cable is $F = 800t^2$ N, where t is in seconds. If the crate has an initial velocity $v_1 = 2$ m/s at $s = 0$ and $t = 0$, determine the distance the crate moves up the plane when $t = 2$ s. The coefficient of kinetic friction between the crate and the incline is $\mu_k = 0.3$.

*13–28. The crate has a mass of 80 kg and is being towed by a chain which is always directed at 20° from the horizontal as shown. If the magnitude of $\mathbf{T}$ is increased until the crate begins to slide, determine the crate's initial acceleration if the coefficient of static friction is $\mu_s = 0.5$ and the coefficient of kinetic friction is $\mu_k = 0.3$.

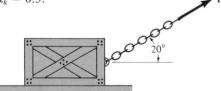

Probs. 13–28/13–29

13–29. The crate has a mass of 80 kg and is being towed by a chain which is always directed at 20° from the horizontal as shown. Determine the crate's acceleration in $t = 2$ s if the coefficient of static friction is $\mu_s = 0.4$ and the coefficient of kinetic friction is $\mu_k = 0.3$ and the towing force is $T = (90t^2)$ N, where t is in seconds.

13–30. A parachutist having a mass m is falling at v_0 when he opens his parachute at a very high altitude. If the atmospheric drag resistance is $F_D = kv^2$, where k is a constant, determine his velocity when he has fallen a distance h. What is his velocity when he lands on the ground? This velocity is referred to as the *terminal velocity*, which is found by letting the distance of fall $y \to \infty$.

Prob. 13–31

13–31. A parachutist having a mass m opens his parachute from an at-rest position at a very high altitude. If the atmospheric drag resistance is $F_D = kv^2$, where k is a constant, determine his velocity when he has fallen for a time t. What is his velocity when he lands on the ground? This velocity is referred to as the *terminal velocity*, which is found by letting the time of fall $t \to \infty$.

*13–32. A car of mass m, is traveling at a slow velocity of v_0. If it is subjected to the drag resistance of the wind, which is proportional to its velocity, i.e., $F_D = kv$, determine the distance and the time the car will travel before its velocity becomes $0.5 \, v_0$. Assume no other frictional forces act on the car.

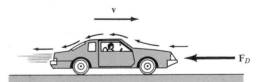

Prob. 13–32

13–33. The device shown is used as a crash sensor in an automobile to activate the inflation of an air bag in the event of a sudden deceleration. During operation, the 0.8-kg smooth "sensing" ball B moves forward and pushes on the rod at A. As the rod rotates about C from the vertical position, the firing pin D is triggered and stabs a primer P, which then inflates the bag. If the spring on the firing pin is compressed 20 mm, determine the required compression of the lever bias spring E so that the mechanism is triggered when the car decelerates at 30 m/s². Neglect friction and the mass of the lever and the small deformation of the springs due to the slight rotation of the lever before firing.

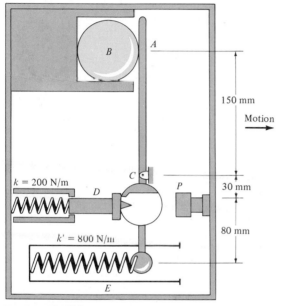

Prob. 13–33

13–34. Each of the two blocks has a mass m. The coefficient of kinetic friction at all surfaces of contact is μ. If a horizontal force P is applied to the bottom block in each case, determine the acceleration of the bottom block.

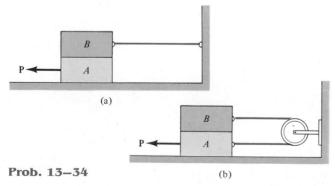

Prob. 13–34

13–35. Each of the three plates has a mass of 10 kg. If the coefficients of static and kinetic friction at each surface of contact are $\mu_s = 0.3$ and $\mu_k = 0.2$, respectively, determine the acceleration of each plate at the instant the three horizontal forces are applied.

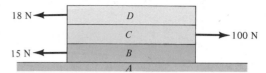

Prob. 13–35

*13–36. Blocks A and B each have a mass m. Determine the largest horizontal force P which can be applied to B so that A will not move relative to B. All surfaces are smooth.

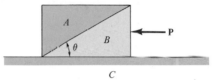

Prob. 13–36

13–37. Blocks A and B each have a mass m. Determine the largest horizontal force P which can be applied to B so that A will not slip up B. The coefficient of static friction between A and B is μ. Neglect any friction between B and C.

■**13–38.** The 10-kg crate rests on the cart for which the coefficient of static friction is $\mu_s = 0.3$ between the crate and cart. Determine the largest angle θ of the plane so that the crate does not slide on the cart when the cart is given an acceleration of $a = 6$ m/s².

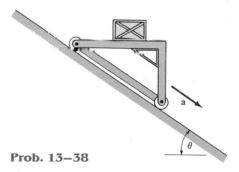

Prob. 13–38

95

13–39. The conveyor belt C is moving downward at 2 m/s. If the coefficient of static friction between the conveyor and the 10-kg box B is $\mu_s = 0.8$, determine the shortest time the conveyor can stop so that the box does not shift or move on the belt.

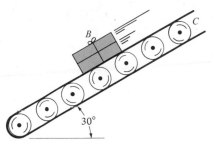

Probs. 13–39/13–40

*13–40. The coefficients of static and kinetic friction between the conveyor and the 10-kg box B are $\mu_s = 0.8$ and $\mu_k = 0.7$, respectively. Starting from rest a point on the conveyor belt moves downward 2 m in 3 s. Determine the acceleration of the box and the conveyor.

13–41. Cylinder B has a mass m and is hoisted using the cord and pulley system shown. Determine the magnitude of force $\mathbf{F}$ as a function of the block's vertical position y so that when $\mathbf{F}$ is applied to the end of the cord the block rises with a constant acceleration $\mathbf{a}_B$. Neglect the mass of the cord and pulleys.

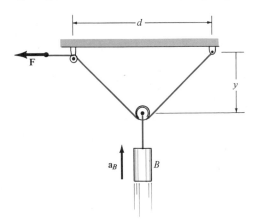

Prob. 13–41

13–42. The spring mechanism is used as a shock absorber for railroad cars. Determine the maximum compression of spring HI if the fixed bumper R of a 5-Mg railroad car, rolling freely at 1.5 m/s, strikes the plate P. Bar AB slides along the guide paths CE and DF. The ends of all springs are attached to their respective members and the springs are originally unstretched. Assume that P and R remain in contact during collision.

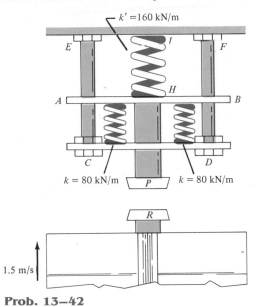

Prob. 13–42

13–43. In the cathode-ray tube, electrons having a mass m are emitted from a source point S and begin to travel horizontally with an initial velocity $\mathbf{v}_0$. While passing between the grid plates a distance l, they are subjected to a vertical force having a magnitude eV/w, where e is the charge of an electron, V the applied voltage acting across the plates, and w the distance between the plates. After passing clear of the plates, the electrons then travel in straight lines and strike the screen at A. Determine the deflection s of the electrons in terms of the dimensions of the voltage plate and tube. Neglect gravity and the slight vertical deflection which occurs between the plates.

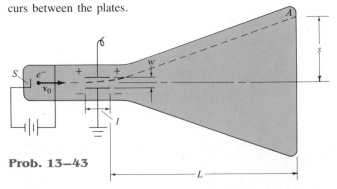

Prob. 13–43

*13–44. Block *A* has a mass of 15 kg and block *B* has a mass of 20 kg. If the horizontal forces of 100 N and 300 N are applied, determine the acceleration of each block. Assume the surfaces of contact are smooth.

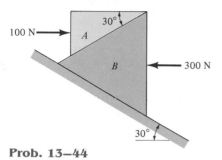

Prob. 13–44

13–45. The smooth block *B* of negligible size has a mass *m* and rests on the horizontal plane. If the board *AC* pushes on the block at an angle θ with a constant acceleration of $\mathbf{a}_0$, determine the velocity of the block along the board and the distance *s* the block moves along the board as a function of time *t*. The block starts from rest when $s = 0$, $t = 0$.

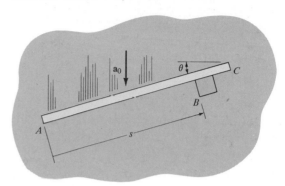

Prob. 13–45

13.5 Equations of Motion: Normal and Tangential Coordinates

When a particle moves over a curved path which is known, the equation of motion for the particle may be written in the normal and tangential directions. We have

$$\Sigma \mathbf{F} = m\mathbf{a}$$
$$\Sigma F_t \mathbf{u}_t + \Sigma F_n \mathbf{u}_n + \Sigma F_b \mathbf{u}_b = m\mathbf{a}_t + m\mathbf{a}_n$$

Here ΣF_n, ΣF_t, and ΣF_b represent the sums of all the force components acting on the particle in the normal, tangential, and binormal directions, respectively, Fig. 13–11. Note that there is no motion of the particle in the binormal direction, since the particle is constrained to move along the path. The above equation is satisfied provided

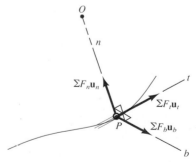

Fig. 13–11

$$\Sigma F_t = ma_t$$
$$\Sigma F_n = ma_n \qquad\qquad (13\text{--}8)$$
$$\Sigma F_b = 0$$

Recall that $a_t (= dv/dt)$ represents the time rate of change in the magnitude of velocity. Consequently, if $\Sigma \mathbf{F}_t$ acts in the direction of motion, the particle's speed will increase, whereas if it acts in the opposite direction, the particle will slow down. Likewise, $a_n \ (= v^2/\rho)$ represents the time rate of change in the velocity's direction. Since this vector *always* acts in the positive n direction, i.e., towards the path's center of curvature, then $\Sigma \mathbf{F}_n$, which causes $\mathbf{a}_n$, also acts in this direction. In particular, when the particle is constrained to travel in a circular path with a constant speed, there is a normal force exerted on the particle by the constraint. Since this force is always directed toward the center of the path, it is often referred to as the *centripetal force*.

PROCEDURE FOR ANALYSIS

When a problem involves the motion of a particle along a *known curved path,* normal and tangential coordinates should be considered for the analysis since the acceleration components can be readily formulated. The method for applying the equations of motion, which relate the forces to the acceleration, has been outlined in the procedure given in Sec. 13.4. Specifically, for n, t, b coordinates it may be stated as follows:

Free-Body Diagram. Establish the inertial n, t, b coordinate system at the particle and draw the particle's free-body diagram. The particle's normal acceleration $\mathbf{a}_n$ *always* acts in the positive n direction. If the tangential acceleration $\mathbf{a}_t$ is unkown, assume it acts in the positive t direction. From the diagram identify the unknowns in the problem.

Equations of Motion. Apply the equations of motion, Eqs. 13–8.

Kinematics. Formulate the tangential and normal components of acceleration; i.e., $a_t = dv/dt$ or $a_t = v \, dv/ds$ and $a_n = v^2/\rho$. If the path is defined as $y = f(x)$, the radius of curvature can be obtained from $\rho = |[1 + (dy/dx)^2]^{3/2}/(d^2y/dx^2)|$.

The following examples numerically illustrate application of this procedure.

Example 13–6

Determine the banking angle θ of the circular track so that the wheels of the sports car shown in Fig. 13–12a will not have to depend upon friction to prevent the car from sliding either up or down the curve. The car travels at a constant speed of 100 ft/s. The radius of the track is 600 ft.

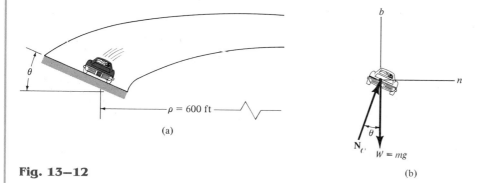

(a)

(b)

Fig. 13–12

SOLUTION

Before looking at the solution to this problem, give some thought as to why it should be solved using n, t, b coordinates.

Free-Body Diagram. As shown in Fig. 13–12b, the car is assumed to have a mass m. As stated in the problem, no frictional force acts on the car. Here N_C represents the *resultant* of all the wheels on the ground. Since a_n can be calculated, the unknowns are N_C and θ.

Equations of Motion. Using the n, b axes shown,

$$+\uparrow \Sigma F_b = 0; \qquad N_C \cos \theta - mg = 0 \qquad (1)$$

$$\xrightarrow{+} \Sigma F_n = ma_n; \qquad N_C \sin \theta = m \frac{v^2}{\rho} \qquad (2)$$

Eliminating N_C and m from these equations by dividing Eq. (2) by Eq. (1), we obtain

$$\tan \theta = \frac{v^2}{g\rho} = \frac{(100)^2}{32.2(600)}$$

$$\theta = \tan^{-1}(0.518)$$

$$= 27.4° \qquad\qquad Ans.$$

A force summation in the tangential direction of motion is of no consequence to the solution. If it were considered, note that $a_t = dv/dt = 0$, since the car moves with *constant speed*.

Example 13–7

The 3-kg disk D is attached to the end of a cord as shown in Fig. 13–13a. The other end of the cord is attached to a ball-and-socket joint located at the center of a platform. If the platform is rotating rapidly, and the disk is placed on it and released from rest as shown, determine the time it takes for the disk to reach a speed great enough to break the cord. The maximum tension the cord can sustain is 100 N, and the coefficient of kinetic friction between the disk and the platform is $\mu_k = 0.1$.

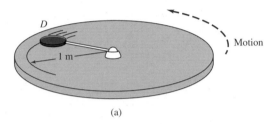

(a)

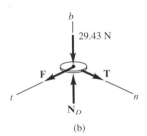

(b)

Fig. 13–13

SOLUTION

Free-Body Diagram. As shown in Fig. 13–13b, the disk has *both* normal and tangential components of acceleration as a result of the unbalanced forces **T** and **F**. Since sliding occurs, the frictional force has a magnitude $F = \mu_k N_D = 0.1 N_D$ and a direction that opposes the *relative motion* of the disk with respect to the platform. The weight of the disk is $W = 3(9.81) = 29.43$ N. Since a_n can be related to v, the unknowns are N_D, a_t and v.

Equations of Motion

$$\Sigma F_b = 0; \qquad\qquad N_D - 29.43 = 0 \qquad\qquad (1)$$

$$\Sigma F_t = ma_t; \qquad\qquad 0.1N_D = 3a_t \qquad\qquad (2)$$

$$\Sigma F_n = ma_n; \qquad\qquad T = 3\left(\frac{v^2}{1}\right) \qquad\qquad (3)$$

Since the maximum tension sustained by the cord is $T = 100$ N, Eq. (3) can be solved for the critical speed v_{cr} of the disk needed to break the cord. Solving all the equations, we obtain

$$N_D = 29.43 \text{ N}$$
$$a_t = 0.981 \text{ m/s}^2$$
$$v_{cr} = 5.77 \text{ m/s}$$

Kinematics. Since a_t is *constant*, the time needed to break the cord is

$$v_{cr} = v_0 + a_t t$$
$$5.77 = 0 + (0.981)t$$
$$t = 5.89 \text{ s} \qquad\qquad \textit{Ans.}$$

Example 13–8

The skier in Fig. 13–14a descends the smooth slope, which may be approximated by a parabola. If she has a weight of 120 lb, determine the normal force she exerts on the ground at the instant she arrives at point A, where her velocity is 30 ft/s. Also compute her acceleration at A.

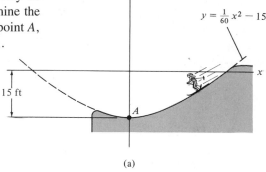

$y = \frac{1}{60}x^2 - 15$

15 ft

(a)

SOLUTION

Why consider using n, t coordinates to solve this problem?

Free-Body Diagram. The free-body diagram for the skier when she is at A is shown in Fig. 13–14b. Since the path is *curved,* there are two components of acceleration, a_n and a_t. Since a_n can be calculated, the unknowns are a_t and N_A.

Equations of Motion

$$+\uparrow \Sigma F_n = ma_n; \qquad N_A - 120 = \frac{120}{32.2}\left(\frac{(30)^2}{\rho}\right) \qquad (1)$$

$$\xleftarrow{+} \Sigma F_t = ma_t; \qquad 0 = \frac{120}{32.2}a_t \qquad (2)$$

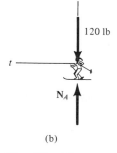

n

120 lb

t

N_A

(b)

Fig. 13–14

The radius of curvature ρ for the path must be computed at point $A(0, -15$ ft$)$. Since $y = \frac{1}{60}x^2 - 15$, $dy/dx = \frac{1}{30}x$, $d^2y/dx^2 = \frac{1}{30}$, then at $x = 0$,

$$\rho = \left|\frac{[1 + (dy/dx)^2]^{3/2}}{d^2y/dx^2}\right|_{x=0} = \left|\frac{[1 + (0)^2]^{3/2}}{\frac{1}{30}}\right| = 30 \text{ ft}$$

Substituting into Eq. (1) and solving for N_A, we have

$$N_A = 232 \text{ lb} \qquad\qquad Ans.$$

Kinematics. From Eq. (2),

$$a_t = 0$$

Thus,

$$a_n = \frac{v^2}{\rho} = \frac{(30)^2}{30} = 30 \text{ ft/s}^2$$

$$a_A = a_n = 30 \text{ ft/s}^2 \uparrow \qquad\qquad Ans.$$

101

Example 13–9

A block having a mass of 2 kg is given an initial velocity of 1 m/s when it is at the top surface of the smooth cylinder shown in Fig. 13–15a. If the block slides along a path of radius 0.5 m, determine the angle $\theta = \theta_{max}$ at which it begins to leave the cylinder's surface.

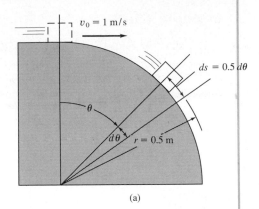

(a)

SOLUTION

Free-Body Diagram. The free-body diagram for the block, when the block is located at the *general position* θ, is shown in Fig. 13–15b. The block must have a tangential acceleration $\mathbf{a}_t$, since its *speed* is always *increasing* as it slides downward. The weight is $W = 2(9.81) = 19.62$ N. Specify the three unknowns.

Equations of Motion

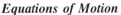

$$+\swarrow \Sigma F_n = ma_n; \qquad -N_B + 19.62 \cos\theta = 2\frac{v^2}{0.5} \qquad (1)$$

$$+\searrow \Sigma F_t = ma_t; \qquad 19.62 \sin\theta = 2a_t \qquad (2)$$

These two equations contain four unknowns, N_B, v, a_t, and θ. At the instant $\theta = \theta_{max}$, however, the block leaves the surface of the cylinder so that $N_B = 0$.

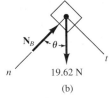

(b)

Fig. 13–15

Kinematics. A third equation for the solution may be obtained by noting that the magnitude of tangential acceleration a_t may be related to the speed of the block v and the angle θ. Since $a_t \, ds = v \, dv$ and $ds = r \, d\theta = 0.5 \, d\theta$, Fig. 13–15a, we have

$$a_t = \frac{v \, dv}{0.5 \, d\theta} \qquad (3)$$

Substituting Eq. (3) into Eq. (2) and separating the variables, we have

$$v \, dv = 4.905 \sin\theta \, d\theta$$

Integrating both sides, realizing that when $\theta = 0°$, $v_0 = 1$ m/s, yields

$$\int_1^v v \, dv = 4.905 \int_{0°}^\theta \sin\theta \, d\theta$$

$$\left.\frac{v^2}{2}\right|_1^v = -4.905 \cos\theta \Big|_{0°}^\theta$$

$$v^2 = 9.81(1 - \cos\theta) + 1$$

Substituting into Eq. (1) with $N_B = 0$ and solving for $\cos\theta_{max}$ gives

$$19.62 \cos\theta_{max} = \frac{2}{0.5}[9.81(1 - \cos\theta_{max}) + 1]$$

$$\cos\theta_{max} = \frac{43.24}{58.86}; \qquad \theta_{max} = 42.7° \qquad \textit{Ans.}$$

PROBLEMS

13–46. Compute the mass of the sun, knowing that the distance from the earth to the sun is 149.6(10⁶) km. *Hint:* Use Eq. 13–1 to represent the force of gravity acting on the earth.

13–47. If the crest of the hill has a radius of curvature of $\rho = 200$ ft, determine the maximum constant speed at which the car can travel over it without leaving the surface of the road. Neglect the size of the car in the calculation. The car has a weight of 3500 lb.

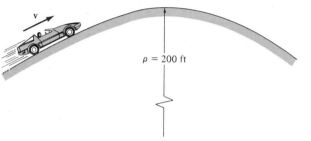

Prob. 13–47

*13–48. Determine the constant speed of the planes on the amusement park ride if it is observed that the supporting cables are directed at $\theta = 30°$ from the vertical. Each plane including its passenger has a mass of 225 kg. Also, determine the normal force that a 50-kg passenger exerts on the seat of the plane during the motion. Neglect friction and the size of the passenger and plane.

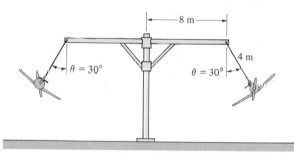

Prob. 13–48

13–49. The 150-lb man lies against the cushion for which the coefficient of static friction is $\mu_s = 0.5$. Determine the resultant normal and frictional forces the cushion exerts on him if, due to rotation about the z axis, he has a constant speed $v = 20$ ft/s. Neglect the size of the man.

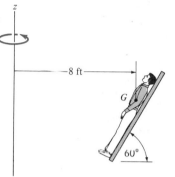

Probs. 13–49/13–50

13–50. The 150-lb man lies against the cushion for which the coefficient of static friction is $\mu_s = 0.3$. Determine the smallest speed v he can have, due to a rotation about the z axis, so that he does not slide downward. Neglect the size of the man.

13–51. Prove that if the block is released from rest at the top point B of a smooth path of *arbitrary shape*, the speed it attains when it reaches point A is equal to the speed it attains when it falls freely through a distance h; i.e., $v = \sqrt{2gh}$.

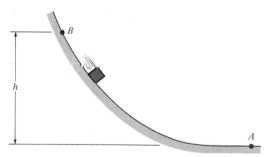

Prob. 13–51

103

*13–52. A skier starts from rest at $A(10 \text{ m}, 0)$ and descends the smooth slope, which may be approximated by a parabola. If she has a mass of 52 kg, determine the normal force she exerts on the ground at the instant she arrives at point B. Neglect the size of the skier. *Hint:* Use the result of Prob. 13–51.

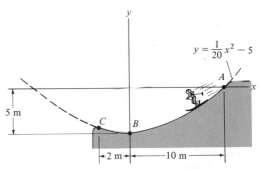

$$y = \frac{1}{20}x^2 - 5$$

Probs. 13–52/13–53

13–53. A skier starts from rest at $A(10 \text{ m}, 0)$ and descends the smooth slope, which may be approximated by a parabola. If she has a mass of 52 kg, determine the normal force she exerts on the ground at the instant she arrives at point C. Neglect the size of the skier. *Hint:* Use the result of Prob. 13–51.

13–54. The pendulum bob has a mass m and is released from rest when $\theta = 0°$. Determine the tension in the cord as a function of the angle of descent θ. Neglect the size of the bob.

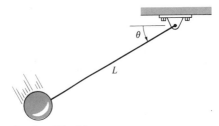

Prob. 13–54

13–55. At the instant $\theta = 60°$ the boy's mass center G has a speed of $v_G = 15$ ft/s. Determine the rate of increase in his speed and the tension in each of the two supporting cords at this instant. The boy has a weight of 60 lb. Neglect his size and the mass of the seat and cords.

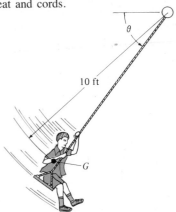

Probs. 13–55/13–56

*13–56. At the instant $\theta = 60°$ the boy's mass center G is at rest. Determine the rate of increase in his speed and the tension in each of the two supporting cords when $\theta = 90°$. The boy has a weight of 60 lb. Neglect his size and the mass of the seat and cords.

13–57. A girl having a mass of 25 kg sits at the edge of the merry-go-round so her center of mass G is at a distance of 1.5 m from the axis of rotation. If the angular motion of the platform is *slowly* increased so that the girl's tangential component of acceleration can be neglected, determine the maximum speed which she can have before she begins to slip off the merry-go-round. The coefficient of static friction between the girl and the merry-go-round is $\mu_s = 0.3$.

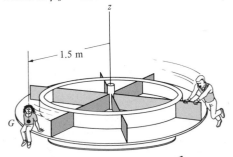

Probs. 13–57/13–58

13–58. Solve Prob. 13–57 assuming the platform starts rotating from rest so that the girl's speed is increased uniformly at $\dot{v} = 0.5$ m/s².

13–59. The cylindrical plug has a weight of 2 lb and it is free to move within the confines of the smooth pipe. The spring has a stiffness of $k = 14$ lb/ft and when no motion occurs the distance $d = 0.5$ ft. Determine the force of the spring on the plug when the plug is at rest with respect to the pipe. The plug is traveling with a constant speed of 15 ft/s which is caused by the rotation of the pipe about the vertical axis.

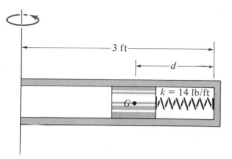

Prob. 13–59

*13–60.** If the ball has a mass of 30 kg and a speed $v = 4$ m/s at the instant it is at its lowest point, $\theta = 0°$, determine the tension in the cord at this instant. Also, determine the angle θ to which the ball swings at the instant it stops. Neglect the mass of the cord.

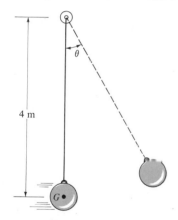

Probs. 13–60/13–61

13–61. The ball has a mass of 30 kg and a speed $v = 4$ m/s at the instant it is at its lowest point, $\theta = 0°$. Determine the tension in the cord and the rate at which the ball's speed is decreasing when $\theta = 20°$. Neglect the mass of the cord.

13–62. The 200-kg snowmobile with passenger is traveling down the hill at a constant speed of 6 m/s. Determine the resultant normal force and the resultant frictional force exerted on the tracks at the instant it reaches point A. Neglect the size of the snowmobile.

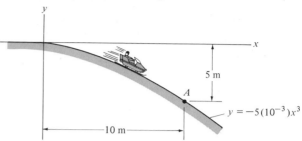

Probs. 13–62/13–63

13–63. The 200-kg snowmobile with passenger is traveling down the hill such that when it is at point A, it is traveling at 4 m/s and increasing its speed at 2 m/s². Determine the resultant normal force and the resultant frictional force exerted on the tracks at this instant. Neglect the size of the snowmobile.

*13–64.** The block has a weight of 5 lb. The attached spring has an unstretched length of 3 ft. If at the instant $\theta = 30°$ the block has a speed $v = 4$ ft/s, determine the normal force on the block and the magnitude of the block's acceleration. Neglect friction.

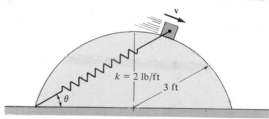

Probs. 13–64/13–65

13–65. Solve Prob. 13–64 if the coefficient of kinetic friction between the block and the surface is $\mu_k = 0.1$.

13–66. The sled is freely coasting down the hill defined by the curve $y = 3(1 - e^{-0.5x})$. If it has a speed of 4 m/s when $x = 4$ m, determine the normal reaction acting on it and the rate of increase in its speed at this instant. The sled and rider have a total mass of 80 kg. Neglect friction and the size of the sled.

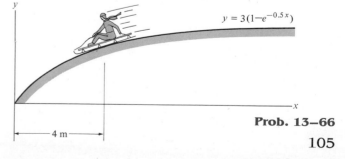

Prob. 13–66

13–67. The 2-kg spool S fits loosely on the inclined rod for which the coefficient of static friction is $\mu_s = 0.2$. If the spool is located 0.25 m from A, determine the minimum constant speed the spool can have so that it does not slip down the rod.

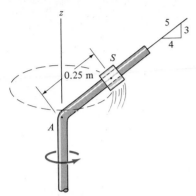

Probs. 13–67/13–68

*13–68. The 2-kg spool S fits loosely on the inclined rod for which the coefficient of static friction is $\mu_s = 0.2$. If the spool is located 0.25 m from A, determine the maximum constant speed the spool can have so that it does not slip up the rod.

13–69. If the bicycle and rider have a total weight of 180 lb, determine the resultant normal force acting on the bicycle when it is at point A while it is freely coasting at $v = 6$ ft/s. Also, compute the increase in the bicyclist's speed at this point. Neglect the resistance due to the wind.

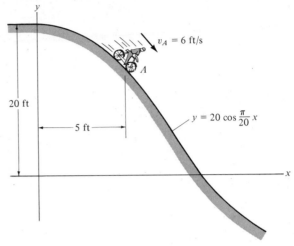

Prob. 13–69

13–70. The 2-kg block B is given a velocity of $v_A = 2$ m/s when it reaches point A. Determine the speed v of the block and the normal force N_B of the plane on the block as a function of θ. Plot these results as v vs. θ and N_B vs. θ and specify the angle at which the normal force is maximum. Neglect friction and the size of the block in the calculation.

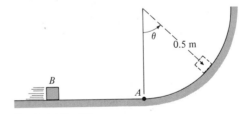

Prob. 13–70

13.6 Equations of Motion: Cylindrical Coordinates

When all the forces acting on a particle are resolved into components along the unit-vector directions $\mathbf{u}_r$, $\mathbf{u}_\theta$, and $\mathbf{u}_z$, Fig. 13–16, the equation of motion may be expressed as

$$\Sigma \mathbf{F} = m\mathbf{a}$$

$$\Sigma F_r\mathbf{u}_r + \Sigma F_\theta\mathbf{u}_\theta + \Sigma F_z\mathbf{u}_z = ma_r\mathbf{u}_r + ma_\theta\mathbf{u}_\theta + ma_z\mathbf{u}_z$$

To satisfy this equation, the respective $\mathbf{u}_r$, $\mathbf{u}_\theta$, and $\mathbf{u}_z$ components on the left side must equal the corresponding components on the right side. Consequently, we may write the following three scalar equations:

$$\boxed{\begin{aligned} \Sigma F_r &= ma_r \\ \Sigma F_\theta &= ma_\theta \\ \Sigma F_z &= ma_z \end{aligned}} \tag{13–9}$$

If the particle is constrained to move only in the r-θ plane, then only the first two of Eqs. 13–9 are used to specify the motion.

Tangential and Normal Forces

The most straightforward type of problem involving cylindrical coordinates requires the determination of the resultant force components ΣF_r, ΣF_θ, and ΣF_z causing a particle to move with a *known* acceleration. If, however, the particle's accelerated motion is not completely specified at the given instant, then some information regarding the directions or magnitudes of the forces acting on the particle must be known or computed in order to solve Eqs.

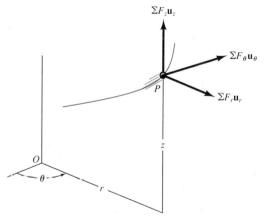

Fig. 13–16

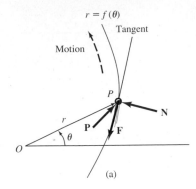

(a)

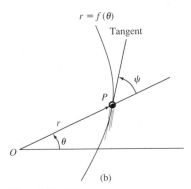

(b)

Fig. 13–17

13–9. For example, the force **P** causes the particle in Fig. 13–17a to move along a path defined in polar coordinates; $r = f(\theta)$. The *normal force* **N** which the path exerts on the particle is always *perpendicular to the tangent of the path*, whereas the frictional force **F** always acts along the tangent in the opposite direction of motion. The *directions* of **N** and **F** can be specified relative to the radial coordinate by computing the angle ψ (psi), Fig. 13–17b, which is defined between the *extended* radial line $r = OP$ and the tangent to the curve. This angle is determined from the equation*

$$\tan \psi = \frac{r}{dr/d\theta} \tag{13–10}$$

If it is calculated as a positive quantity, it is measured from the extended radial line to the tangent in a counterclockwise sense or in the positive direction of θ. If it is negative, it is measured in the opposite direction to positive θ. Application is illustrated numerically in Example 13–12.

PROCEDURE FOR ANALYSIS

Cylindrical or polar coordinates are a suitable choice for the analysis of a problem for which data regarding the angular motion of the radial line r is given, or in cases where the path can be conveniently expressed in terms of these coordinates. Once these coordinates have been established, the equations of motion can be applied in order to relate the forces acting on the particle to its acceleration components. The method for doing this has been outlined in the procedure for analysis given in Sec. 13.4. The following is a summary of this procedure.

Free-Body Diagram. Establish the r, θ, z inertial coordinate system and draw the particle's free-body diagram. Assume that $\mathbf{a}_r$, $\mathbf{a}_\theta$, and $\mathbf{a}_z$ act in the *positive directions* of r, θ, and z if they are unknown. Before applying the equations of motion, identify all the unknowns.

Equations of Motion. Apply the equations of motion, Eqs. 13–9.

Kinematics. Use the methods of Sec. 12.7 to determine r and the time derivatives $\dot{r}$, $\ddot{r}$, $\dot{\theta}$, $\ddot{\theta}$, and $\ddot{z}$, and evaluate the acceleration components $a_r = \ddot{r} - r\dot{\theta}^2$, $a_\theta = r\ddot{\theta} + 2\dot{r}\dot{\theta}$, and $a_z = \ddot{z}$. If any of these three components are computed as negative quantities, it indicates that they act in their negative coordinate directions.

The following examples numerically illustrate application of this procedure. Further application of Eqs. 13–9 is given in Sec. 13.7, where problems involving rockets, satellites, and planetary bodies subjected to gravitational forces are considered.

*The derivation is given in any standard calculus text.

Example 13–10

The 2-lb block in Fig. 13–18a moves on the smooth horizontal plane, such that its path is specified in polar coordinates by the parametric equations $r = (10t^2)$ ft and $\theta = (0.5t)$ rad, where t is in seconds. Determine the magnitude of the *resultant* force **F** causing the motion at the instant $t = 1$ s.

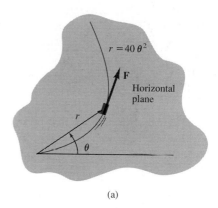

(a)

SOLUTION

Free-Body Diagram. Note that if the parameter t is eliminated between r and θ, the equation of the path becomes $r = 40\theta^2$. The block's free-body diagram when it is located at an arbitrary point along the path is shown in Fig. 13–18b. Here the r and θ components of the resultant force are represented. They, and $\mathbf{a}_r$ and $\mathbf{a}_\theta$, are assumed to act in the *positive* coordinate directions. The four unknowns are F_r, F_θ, a_r, and a_θ.

Equations of Motion

$$\Sigma F_r = ma_r; \qquad F_r = \frac{2}{32.2}(\ddot{r} - r\dot{\theta}^2) \qquad (1)$$

$$\Sigma F_\theta = ma_\theta; \qquad F_\theta = \frac{2}{32.2}(r\ddot{\theta} + 2\dot{r}\dot{\theta}) \qquad (2)$$

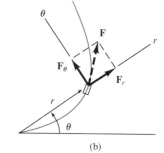

(b)

Fig. 13–18

Kinematics. Since the motion is specified, the coordinates and the required time derivatives can be computed and evaluated at $t = 1$ s.

$$r = 10t^2 \Big|_{t=1\text{s}} = 10 \text{ ft} \qquad \theta = 0.5t \Big|_{t=1\text{s}} = 0.5 \text{ rad}$$

$$\dot{r} = 20t \Big|_{t=1\text{s}} = 20 \text{ ft/s} \qquad \dot{\theta} = 0.5 \text{ rad/s}$$

$$\ddot{r} = 20 \text{ ft/s}^2 \qquad \ddot{\theta} = 0$$

Substituting these results into Eqs. (1) and (2) yields

$$F_r = \frac{2}{32.2}(20 - 10(0.5)^2) = 1.087 \text{ lb}$$

$$F_\theta = \frac{2}{32.2}(0 + 2(20)(0.5)) = 1.242 \text{ lb}$$

Hence, the resultant force has a magnitude of

$$F = \sqrt{(1.087)^2 + (1.242)^2}$$
$$= 1.65 \text{ lb} \qquad\qquad \textbf{\textit{Ans.}}$$

Example 13–11

The smooth 2-kg cylinder C in Fig. 13–19a has a peg P through its center which passes through the slot in arm OA. If the arm rotates in the *vertical plane* at a constant rate $\dot{\theta} = 0.5$ rad/s, determine the force that the arm exerts on the peg at the instant $\theta = 60°$.

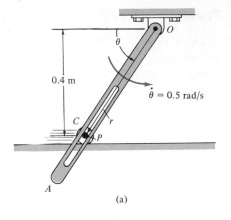

(a)

SOLUTION

Why is it a good idea to use polar coordinates to solve this problem?

Free-Body Diagram. The free-body diagram for the cylinder is shown in Fig. 13–19b. The force of the peg, $\mathbf{F}_P$, acts perpendicular to the slot in the arm. As usual, $\mathbf{a}_r$ and $\mathbf{a}_\theta$ are assumed to act in the directions of *positive r* and θ, respectively. Identify the four unknowns.

Equations of Motion. Using the data in Fig. 13–19b, we have

$$+\swarrow \Sigma F_r = ma_r; \qquad 19.62 \sin \theta - N_C \sin \theta = 2a_r \qquad (1)$$
$$+\searrow \Sigma F_\theta = ma_\theta; \quad 19.62 \cos \theta + F_P - N_C \cos \theta = 2a_\theta \qquad (2)$$

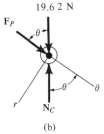

19.62 N

(b)

Fig. 13–19

Kinematics. From Fig. 13–19a, r can be related to θ by the equation

$$r = \frac{0.4}{\sin \theta} = 0.4 \csc \theta$$

Since $d(\csc \theta) = -(\csc \theta \cot \theta)d\theta$ and $d(\cot \theta) = -(\csc^2 \theta)\, d\theta$, then r and the necessary time derivatives become

$$\dot{\theta} = 0.5 \qquad r = 0.4 \csc \theta$$
$$\ddot{\theta} = 0 \qquad \dot{r} = -0.4(\csc \theta \cot \theta)\dot{\theta}$$
$$= -0.2 \csc \theta \cot \theta$$
$$\ddot{r} = -0.2(-\csc \theta \cot \theta)(\dot{\theta}) \cot \theta - 0.2 \csc \theta(-\csc^2 \theta)\dot{\theta}$$
$$= 0.1 \csc \theta(\cot^2 \theta + \csc^2 \theta)$$

Evaluating these formulas at $\theta = 60°$, we get

$$\dot{\theta} = 0.5 \qquad r = 0.462$$
$$\ddot{\theta} = 0 \qquad \dot{r} = -0.133$$
$$\ddot{r} = 0.192$$

$$a_r = \ddot{r} - r\dot{\theta}^2 = 0.192 - 0.462(0.5)^2 = 0.0765$$
$$a_\theta = r\ddot{\theta} + 2\dot{r}\dot{\theta} = 0 + 2(-0.133)(0.5) = -0.133$$

Substituting these results into Eqs. (1) and (2) with $\theta = 60°$ and solving yields
$$N_C = 19.4 \text{ N} \qquad F_P = -0.355 \text{ N} \qquad \qquad \textit{Ans.}$$

The negative sign indicates that $\mathbf{F}_P$ acts opposite to that shown in Fig. 13–19b.

Example 13–12

A can C, having a mass of 0.5 kg, moves along a grooved horizontal slot shown in Fig. 13–20a. The slot is in the form of a spiral, which is defined by the equation $r = (0.1\theta)$ m, where θ is in radians. If the arm OA is rotating at a constant rate $\dot{\theta} = 4$ rad/s in the horizontal plane, determine the force it exerts on the can at the instant $\theta = \pi$ rad. Neglect friction and the size of the can.

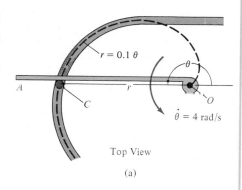

Top View

(a)

SOLUTION

Free-Body Diagram. The driving force $\mathbf{F}_C$ acts perpendicular to the arm OA, whereas the normal force of the wall of the slot on the can, $\mathbf{N}_C$, acts perpendicular to the tangent to the curve at $\theta = \pi$ rad, Fig. 13–20b. As usual, $\mathbf{a}_r$ and $\mathbf{a}_\theta$ act in the *positive directions* of r and θ, respectively. Since the path is specified, the angle ψ which the extended radial line r makes with the tangent, Fig. 13–20c, can be determined from Eq. 13–10. We have $r = 0.1\theta$, so that $dr/d\theta = 0.1$, and therefore

$$\tan \psi = \frac{r}{dr/d\theta} = \frac{0.1\theta}{0.1} = \theta$$

When $\theta = \pi$, $\psi = \tan^{-1} \pi = 72.3°$, so that $\phi = 90° - \psi = 17.7°$, as shown in Fig. 13–20c. Identify the four unknowns in Fig. 13–20b.

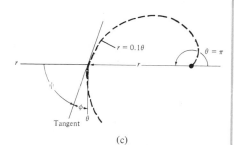

(b)

Equations of Motion. Using the data shown in Fig. 13–20b, we have

$$\xleftrightarrow{\pm}\Sigma F_r = ma_r; \qquad N_C \cos 17.7° = 0.5a_r \qquad (1)$$
$$+\downarrow\Sigma F_\theta = ma_\theta; \qquad F_C - N_C \sin 17.7° = 0.5a_\theta \qquad (2)$$

Kinematics. Computing the time derivatives of r and θ gives

$$\dot{\theta} = 4 \text{ rad/s} \qquad r = 0.1\theta$$
$$\ddot{\theta} = 0 \qquad \dot{r} = 0.1\dot{\theta} = 0.1(4) = 0.4 \text{ m/s}$$
$$\ddot{r} = 0.1\ddot{\theta} = 0$$

At the instant $\theta = \pi$ rad,

$$a_r = \ddot{r} - r\dot{\theta}^2 = 0 - 0.1(\pi)(4)^2 = -5.03 \text{ m/s}^2$$
$$a_\theta = r\ddot{\theta} + 2\dot{r}\dot{\theta} = 0 + 2(0.4)(4) = 3.20 \text{ m/s}^2$$

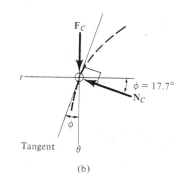

(c)

Fig. 13–20

Substituting these results into Eqs. (1) and (2) and solving yields

$$N_C = -2.64 \text{ N}$$
$$F_C = 0.797 \text{ N} \qquad\qquad\qquad \textit{Ans.}$$

What does the negative sign for N_C indicate?

111

PROBLEMS

13–71. A particle, having a mass of 2 kg, moves along a three-dimensional path defined by the equations $r = (10t^2 + 3t)$ m, $\theta = (0.1t^3)$ rad, and $z = (4t^2 + 15t - 6)$ m, where t is in seconds. Determine the r, θ, and z components of force which the path exerts on the particle when $t = 2$ s.

*13–72. The girl has a mass of 50 kg. She is seated on the horse of the merry-go-round which undergoes constant rotational motion $\dot\theta = 1.5$ rad/s. If the path of the horse is defined by $r = 4$ m, $z = (0.5 \sin \theta)$ m, determine the maximum and minimum force F_z the horse exerts on her during the motion.

Prob. 13–72

13–73. For a short time the 250-kg roller coaster car is traveling along the spiral track at a constant speed of $v = 8$ m/s. If the track descends $d = 12$ m for every full revolution, $\theta = 2\pi$ rad, determine the magnitudes of the components of force $\mathbf{F}_r$, $\mathbf{F}_\theta$, and $\mathbf{F}_z$ which the track exerts on the car. Neglect the size of the car.

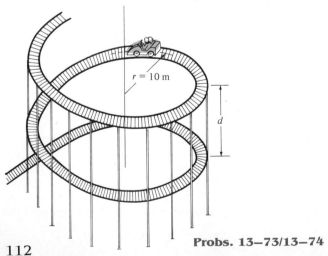

Probs. 13–73/13–74

13–74. For a short time the 250-kg roller coaster car is traveling along the spiral track at a constant speed such that its position measured from the top of the track has components $r = 10$ m, $\theta = (0.2t)$ rad, and $z = (-0.3t)$ m, where t is in seconds. Determine the magnitudes of the components of force $\mathbf{F}_r$, $\mathbf{F}_\theta$, and $\mathbf{F}_z$ which the track exerts on the car at the instant $t = 2$ s. Neglect the size of the car.

13–75. The collar, which has a weight of 3 lb, slides along the smooth rod lying in the *horizontal plane* and having the shape of a parabola $r = 4/(1 - \cos \theta)$, where θ is in radians and r is in feet. If the collar's angular rate is constant and equals $\dot\theta = 4$ rad/s, determine the tangential retarding force P needed to cause the motion and the normal force that the collar exerts on the rod at the instant $\theta = 90°$.

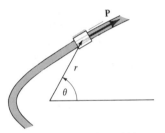

Probs. 13–75/13–76

*13–76. Solve Prob. 13–75 if the parabolic path (rod) lies in the *vertical plane*.

13–77. The 0.5-lb particle is guided along the vertical circular path using the arm OA. If the arm has an angular velocity $\dot\theta = 0.4$ rad/s and an angular acceleration $\ddot\theta = 0.8$ rad/s² at the instant $\theta = 30°$, determine the force of the arm on the particle. Set $r_c = 0.4$ ft.

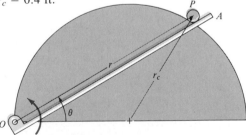

Probs. 13–77/13–78

13–78. The particle of mass m is guided along the vertical circular path of radius r_c using the arm OA. If the arm has a constant angular velocity $\dot\theta_0$, determine the angle θ at which the particle starts to leave the surface of the semicylinder.

13–79. Using air pressure, the 0.5-kg ball is forced to move through the smooth tube lying in the *horizontal plane* and having the shape of a logarithmic spiral. If the tangential force exerted on the ball due to the air is 6 N, determine the magnitude and direction of the rate of increase in the ball's speed at the instant $\theta = \pi/2$. Neglect the size of the ball.

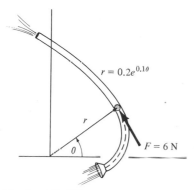

Prob. 13–79

*13–80. Using a forked rod, a smooth cylinder C having a mass of 0.5 kg is forced to move along the *horizontal slotted* path $r = (0.5\theta)$ m, where θ is in radians. If the angular position of the arm is $\theta = (0.5t^2)$ rad, where t is in seconds, determine the force of the forked rod on the cylinder and the normal force of the slot on the cylinder at the instant $t = 2$ s. The cylinder is in contact with only *one edge* of the rod and slot at any instant.

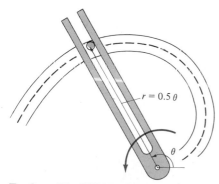

Probs. 13–80/13–81

13–81. Using a forked rod, a smooth cylinder C having a mass of 0.5 kg is forced to move along the *vertical slotted* path $r = (0.5\theta)$ m, where θ is in radians. If the angular position of the arm is $\theta = (0.5t^2)$ rad, where t is in seconds, determine the force of the forked rod on the cylinder and the normal force of the slot on the cylinder at the instant $t = 2$ s. The cylinder is in contact with only *one edge* of the rod and slot at any instant.

13–82. Rod OA rotates counterclockwise in the horizontal plane with a constant angular rate $\dot\theta = 4$ rad/s. The double collar B is pin-connected together such that one collar slides over the rotating rod and the other collar slides over the circular rod described by the equation $r = (1.6 \cos \theta)$ m. If *both* collars have a total mass of 0.5 kg, determine the force which the circular rod exerts on one of the collars and the force that OA exerts on the other collar at the instant $\theta = 45°$.

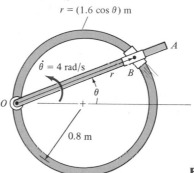

Probs. 13–82/13–83

13–83. Solve Prob. 13–82 if the path is *vertical*.

*13–84. The forked rod is used to move the smooth 2-lb cylinder around the horizontal path in the shape of a limaçon, $r = (2 + \cos \theta)$ ft. If at all times $\dot\theta = 0.5$ rad/s, determine the force which the rod exerts on the cylinder at the instant $\theta = 90°$. The fork and path contact the cylinder on only one side.

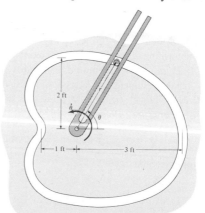

Prob. 13–84/13–85/13–86

13–85. Solve Prob. 13–84 at the instant $\theta = 60°$.

13–86. The forked rod is used to move the smooth 2-lb cylinder around the horizontal path in the shape of a limaçon, $r = (2 + \cos \theta)$ ft. If $\theta = (0.5t^2)$ rad, where t is in seconds, determine the force which the rod exerts on the cylinder at the instant $t = 1$ s. The fork and path contact the cylinder on only one side.

113

13–87. A car of a roller coaster travels along a track which for a short distance is defined by a conical spiral, $r = 0.75z$, $\theta = -1.5z$, where r and z are in meters and θ is in radians. If the angular motion $\dot{\theta} = 1$ rad/s is always maintained, determine the r, θ, z components of reaction exerted on the car by the tracks at the instant $z = 6$ m. The car and passenger have a total mass of 300 kg. Neglect the mass of the wheels and the size of the car.

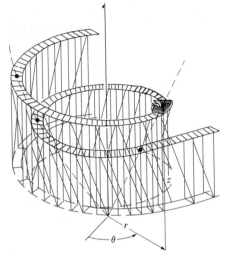

Prob. 13–87

*13–88.** The ball has a mass of 2 kg and a negligible size. It is originally traveling around the horizontal circular path of radius $r_0 = 0.5$ m such that the angular rate of rotation is $\dot{\theta}_0 = 1$ rad/s. If the attached cord ABC is drawn down through the hole at a constant speed of 0.2 m/s, determine the tension the cord exerts on the ball at the instant $r = 0.25$ m. Also, compute the angular velocity of the ball at this instant. Neglect the effects of friction between the ball and horizontal plane. *Hint:* First show that the equation of motion in the θ direction yields $a_\theta = r\ddot{\theta} + 2\dot{r}\dot{\theta} = \dfrac{1}{r}\dfrac{d(r^2\dot{\theta})}{dt} = 0$. When integrated, $r^2\dot{\theta} = C$, where the constant C is determined from the problem data.

Prob. 13–88

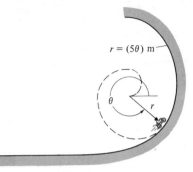

13–89. Determine the normal and frictional driving forces that the spiral track exerts on the 200-kg motorcycle at the instant $\theta = \frac{5}{3}\pi$ rad, $\dot{\theta} = 0.4$ rad/s, and $\ddot{\theta} = 0.8$ rad/s².

$r = (5\theta)$ m

Prob. 13–89

13–90. The pilot of an airplane executes a vertical loop which in part follows the path of a cardioid, $r = 600(1 + \cos\theta)$ ft, where θ is in radians. If his speed at A ($\theta = 0$) is a constant $v_P = 80$ ft/s, determine the vertical force the belt of his seat must exert on him to hold him to his seat when the plane is upside down at A. He weighs 150 lb. *Hint:* To determine the time derivatives necessary to compute the plane's acceleration components a_r and a_θ, take the first and second time derivatives of $r = 600(1 + \cos\theta)$ ft. Then, for further information, use Eq. 12–26 to determine $\dot{r}$ and $\dot{\theta}$. Also, take the time derivative of Eq. 12–26, noting that $\dot{v}_P = 0$, to determine $\ddot{r}$ and $\ddot{\theta}$.

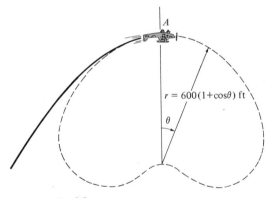

$r = 600(1+\cos\theta)$ ft

Prob. 13–90

13–91. A truck T has a weight of 8000 lb and is traveling along a portion of a road defined by the lemniscate $r^2 = 0.2(10^6) \cos 2\theta$, where r is measured in feet and θ is in radians. If the truck maintains a constant speed of $v_T = 4$ ft/s, determine the magnitude of the resultant frictional force which must be exerted by all the wheels to maintain the motion when $\theta = 0°$. *Hint:* To determine the time derivatives necessary to compute the truck's acceleration components a_r and a_θ, take the first and second time derivatives of $r^2 = 0.2(10^6) \cos 2\theta$. Then, for further information, use Eq. 12–19 to determine $\dot{r}$ and $\dot{\theta}$. Also take the time derivative of Eq. 12–19, noting that $\dot{v}_T = 0$, to determine $\ddot{r}$ and $\ddot{\theta}$.

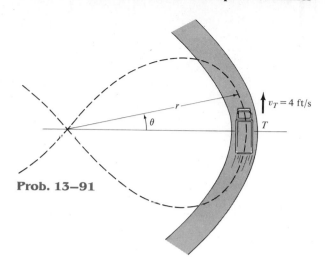

Prob. 13–91

*13.7 Central-Force Motion and Space Mechanics

If a particle is moving only under the influence of a force having a line of action which is always directed toward a fixed point, the motion is called *central-force motion*. This type of motion is commonly caused by electrostatic and gravitational forces.

In Fig. 13–21a, the particle P having a mass m is acted upon only by the central force **F**. The free-body diagram for the particle is shown in Fig. 13–21b. Using polar coordinates (r, θ), the equations of motion, Eqs. 13–9, become

$$-F = m\left[\frac{d^2 r}{dt^2} - r\left(\frac{d\theta}{dt} \right)^2 \right]$$

$$0 = m\left(r\frac{d^2\theta}{dt^2} + 2\frac{dr}{dt}\frac{d\theta}{dt} \right) \tag{13–11}$$

The second of these equations may be written in the form

$$\frac{1}{r}\left[\frac{d}{dt}\left(r^2\frac{d\theta}{dt} \right) \right] = 0$$

so that integrating yields

$$r^2\frac{d\theta}{dt} = h \tag{13–12}$$

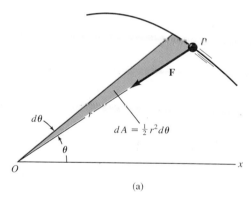

(a)

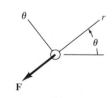

(b)

Fig. 13–21

where h is a constant of integration. From Fig. 13–21a it can be seen that the shaded area described by the radius r, as r moves through an angle $d\theta$, is $dA = \frac{1}{2}r^2\, d\theta$. If the *areal velocity* is defined as

$$\frac{dA}{dt} = \frac{1}{2}r^2\frac{d\theta}{dt} = \frac{h}{2} \qquad (13\text{–}13)$$

then, by comparison with Eq. 13–12, it is seen that the areal velocity for a particle subjected to central-force motion is *constant*. In other words, the particle will sweep out equal segments of area per unit time as it travels along the path. To obtain the *path of motion*, $r = f(\theta)$, the independent variable t must be eliminated from Eqs. 13–11. Using the chain rule of calculus and Eq. 13–12, the time derivatives of Eqs. 13–11 may be replaced by

$$\frac{dr}{dt} = \frac{dr}{d\theta}\frac{d\theta}{dt} = \frac{h}{r^2}\frac{dr}{d\theta}$$

$$\frac{d^2r}{dt^2} = \frac{d}{dt}\left(\frac{h}{r^2}\frac{dr}{d\theta}\right) = \frac{d}{d\theta}\left(\frac{h}{r^2}\frac{dr}{d\theta}\right)\frac{d\theta}{dt} = \left[\frac{d}{d\theta}\left(\frac{h}{r^2}\frac{dr}{d\theta}\right)\right]\frac{h}{r^2}$$

Substituting a new dependent variable (xi) $\xi = 1/r$ into the second equation, it is seen that

$$\frac{d^2r}{dt^2} = -h^2\xi^2\frac{d^2\xi}{d\theta^2}$$

Also, the square of Eq. 13–12 becomes

$$\left(\frac{d\theta}{dt}\right)^2 = h^2\xi^4$$

Substituting these last two equations into the first of Eqs. 13–11, we have

$$-h^2\xi^2\frac{d^2\xi}{d\theta^2} - h^2\xi^3 = -\frac{F}{m}$$

or

$$\frac{d^2\xi}{d\theta^2} + \xi = \frac{F}{mh^2\xi^2} \qquad (13\text{–}14)$$

This differential equation defines the path over which the particle travels when it is subjected to the central force* **F**.

For application, the force of gravitational attraction will be considered. Some common examples of central-force systems which depend upon gravitation include the motion of the moon and artificial satellites about the earth, and the motion of the planets about the sun. As a typical problem in space mechanics, consider the trajectory of a space satellite or space vehicle launched into orbit with an initial velocity v_0, Fig. 13–22. It will be assumed that this velocity is initially *parallel* to the tangent at the surface of the earth, as shown in the figure.† Just after the satellite is released into free flight the

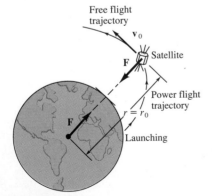

Fig. 13–22

---

*In the derivation, **F** is considered positive when it is directed toward point O. If **F** is oppositely directed, the right side of Eq. 13–14 should be negative.

†The case where v_0 acts at some initial angle θ to the tangent is best described using the conservation of angular momentum (see Prob. 15–73).

only force acting upon it is the gravitational force of the earth. (Gravitational attractions involving other bodies such as the moon or sun will be neglected, since for orbits close to the earth their effect is small in comparison with the gravitational force of the earth.) According to Newton's law of gravitation, force **F** will always act between the mass centers of the earth and the satellite, Fig. 13–22. From Eq. 13–1, this force of attraction has a magnitude of

$$F = G \frac{M_e m}{r^2}$$

where M_e and m represent the mass of the earth and the satellite, respectively, G is the gravitational constant, and r is the distance between the mass centers. Setting $\xi = 1/r$ in the above equation and substituting the result into Eq. 13–14, we obtain

$$\frac{d^2\xi}{d\theta^2} + \xi = \frac{GM_e}{h^2} \qquad (13\text{–}15)$$

This second-order ordinary differential equation has constant coefficients and is nonhomogeneous. The solution is represented as the sum of the complementary and particular solutions. The complementary solution is obtained when the term on the right is equal to zero. It is,

$$\xi_c = C \cos(\theta - \phi)$$

where C and ϕ are constants of integration. The particular solution is

$$\xi_p = \frac{GM_e}{h^2}$$

Thus, the complete solution to Eq. 13–15 is

$$\xi = \xi_c + \xi_p$$
$$= \frac{1}{r} = C \cos(\theta - \phi) + \frac{GM_e}{h^2} \qquad (13\text{–}16)$$

The validity of this result may be checked by substitution into Eq. 13–15.

Equation 13–16 represents the *free-flight trajectory* of the satellite. It is the equation of a conic section expressed in terms of polar coordinates. As shown in Fig. 13–23, a *conic section* is defined as the locus of point P which moves in a plane in such a way that the ratio of its distance from a fixed point F to its distance from a fixed line is constant. The fixed point is called the *focus*, and the fixed line DD is called the *directrix*. The constant ratio is called the *eccentricity* of the conic section and is denoted by e. Thus,

$$e = \frac{FP}{PA}$$

which may be written in the form

$$FP = r = e(PA) = e[p - r \cos(\theta - \phi)]$$

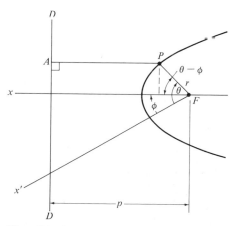

Fig. 13–23

117

or

$$\frac{1}{r} = \frac{1}{p} \cos (\theta - \phi) + \frac{1}{ep}$$

Comparing this equation with Eq. 13–16, it is seen that the eccentricity of the conic section for the trajectory is

$$e = \frac{Ch^2}{GM_e} \qquad (13\text{--}17)$$

and the fixed distance from the focus to the directrix is

$$p = \frac{1}{C} \qquad (13\text{--}18)$$

Provided the polar angle θ is measured from the x axis (an axis of symmetry since it is perpendicular to the directrix DD), the angle ϕ is zero, Fig. 13–23, and therefore Eq. 13–16 reduces to

$$\frac{1}{r} = C \cos \theta + \frac{GM_e}{h^2} \qquad (13\text{--}19)$$

The constants h and C are determined from the data obtained for the position and velocity of the satellite at the end of the *power-flight trajectory*. For example, if the initial height or distance to the space vehicle is r_0 (measured from the center of the earth) and its initial speed is v_0 at the beginning of its free flight, Fig. 13–24, then the constant h may be obtained from Eq. 13–12.

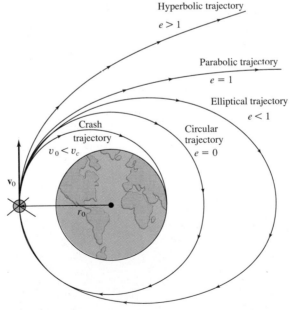

Fig. 13–24

When $\theta = \phi = 0°$, the velocity $\mathbf{v}_0$ has no radial component; therefore, from Eq. 12–25, $v_0 = r_0(d\theta/dt)$, so that

$$h = r_0^2 \frac{d\theta}{dt}$$

or

$$h = r_0 v_0 \tag{13–20}$$

To determine C, use Eq. 13–19 with $\theta = 0°$, $r = r_0$, and substitute Eq. 13–20 for h:

$$C = \frac{1}{r_0}\left(1 - \frac{GM_e}{r_0 v_0^2}\right) \tag{13–21}$$

The equation for the free-flight trajectory therefore becomes

$$\frac{1}{r} = \frac{1}{r_0}\left(1 - \frac{GM_e}{r_0 v_0^2}\right) \cos \theta + \frac{GM_e}{r_0^2 v_0^2} \tag{13–22}$$

The type of path taken by the satellite is determined from the value of the eccentricity of the conic section as given by Eq. 13–17. If

$$
\begin{array}{ll}
e = 0 & \text{free-flight trajectory is a circle} \\
e = 1 & \text{free-flight trajectory is a parabola} \\
e < 1 & \text{free-flight trajectory is an ellipse} \\
e > 1 & \text{free-flight trajectory is a hyperbola}
\end{array}
\tag{13–23}
$$

Each of these trajectories is shown in Fig. 13–24. From the curves it is seen that when the satellite follows a parabolic path, it is "on the border" of not returning to its initial starting point. The initial velocity at launching, $\mathbf{v}_0$, required for the satellite to follow a parabolic path is called the *escape velocity*. The speed, v_e, can be determined by using the second of Eqs. 13–23 with Eqs. 13–17, 13–20, and 13–21. It is left as an exercise to show that

$$v_e = \sqrt{\frac{2GM_e}{r_0}} \tag{13–24}$$

The speed v_c required to launch a satellite into a *circular orbit* can be found using the first of Eqs. 13–23. Since e is related to h and C, Eq. 13–17, C must be zero to satisfy this equation (from Eq. 13–20, h cannot be zero); and therefore, using Eq. 13–21, we have

$$v_c = \sqrt{\frac{GM_e}{r_0}} \tag{13–25}$$

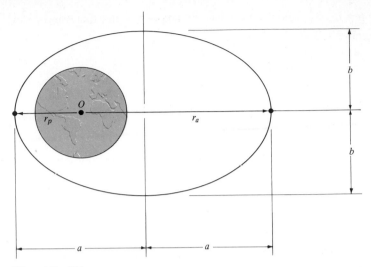

Fig. 13–25

Provided r_0 represents a minimum height for launching, in which frictional resistance from the atmosphere is neglected, speeds at launch which are less than v_c will cause the satellite to reenter the earth's atmosphere and either burn up or crash, Fig. 13–24.

All the trajectories attained by planets and most satellites are elliptical, Fig. 13–25. For an orbit about the earth, the *minimum distance* from the orbit to the center of the earth O (which is located at one of the foci of the ellipse) is r_p and can be found using Eq. 13–22 with $\theta = 0°$. Therefore,

$$r_p = r_0 \qquad (13\text{–}26)$$

This minimum distance is called the *perigee* of the orbit. The *apogee* or maximum distance r_a can be found using Eq. 13–22 with $\theta = 180°$.*
Thus,

$$r_a = \frac{r_0}{\dfrac{2GM_e}{r_0 v_0^2} - 1} \qquad (13\text{–}27)$$

With reference to Fig. 13–25, the semimajor axis a of the ellipse is

$$a = \frac{r_p + r_a}{2} \qquad (13\text{–}28)$$

---

*Actually, the terminology perigee and apogee pertains only to orbits about the *earth*. If any other body is located at the focus of an elliptical orbit, the minimum and maximum distances are referred to respectively as the *periapsis* and *apoapsis* of the orbit.

Using analytical geometry it can be shown that the minor axis b is determined from the equation

$$b = \sqrt{r_p r_a} \tag{13–29}$$

Furthermore, by direct integration, the area of an ellipse is

$$A = \pi a b = \frac{\pi}{2}(r_p + r_a)\sqrt{r_p r_a} \tag{13–30}$$

The areal velocity has been defined by Eq. 13–13, $dA/dt = h/2$. Integrating yields $A = hT/2$, where T is the *period* of time required to make one orbital revolution. From Eq. 13–30, the period is

$$\boxed{T = \frac{\pi}{h}(r_p + r_a)\sqrt{r_p r_a}} \tag{13–31}$$

In addition to predicting the orbital trajectory of earth satellites, the theory developed in this section is valid, to a surprisingly close approximation, at predicting the actual motion of the planets traveling around the sun. In this case the mass of the sun, M_s, should be substituted for M_e when the appropriate formulas are used.

The fact that the planets do indeed follow elliptic orbits about the sun was discovered by the German astronomer Johannes Kepler (1571–1630). His discovery was made *before* Newton had developed the laws of motion and the law of gravitation. Kepler's laws, stipulated after 20 years of planetary observation, are summarized by the following three statements:

1. Every planet moves in its orbit such that the line joining it to the sun sweeps over equal areas in equal intervals of time, whatever the line's length.
2. The orbit of every planet is an ellipse with the sun placed at one of its foci.
3. The square of the period of any planet is directly proportional to the cube of the minor axis of its orbit.

A mathematical statement of the first and second laws is given by Eqs. 13–13 and 13–22. The third law can be shown from Eq. 13–31 (see Prob. 13–103).

Example 13–13

A satellite is launched 600 km from the surface of the earth, with an initial velocity of 30 Mm/h acting parallel to the tangent at the surface of the earth, Fig. 13–26. Assuming that the radius of the earth is 6378 km and that its mass is $5.976(10^{24})$ kg, determine (a) the eccentricity of the orbital path, and (b) the velocity of the satellite at apogee.

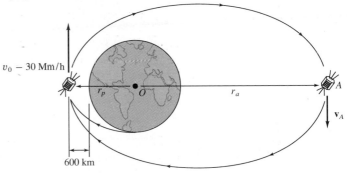

Fig. 13–26

SOLUTION

Part (a). The eccentricity of the orbit is obtained using Eq. 13–17. The constants h and C are first determined from Eqs. 13–20 and 13–21. Since

$$r_p = 6378 \text{ km} + 600 \text{ km} = 6.978(10^6) \text{ m}$$
$$v_0 = 30 \text{ Mm/h} = 8333.3 \text{ m/s}$$

then

$$h = r_p v_0 = 6.978(10^6)(8333.3) = 5.815(10^{10}) \text{ m}^2/\text{s}$$

$$C = \frac{1}{r_p}\left(1 - \frac{GM_e}{r_p v_0^2}\right)$$

$$= \frac{1}{6.978(10^6)}\left(1 - \frac{6.673(10^{-11})[5.976(10^{24})]}{6.978(10^6)(8333.3)^2}\right) = 2.54(10^{-8}) \text{ m}^{-1}$$

Hence,

$$e = \frac{Ch^2}{GM_e} = \frac{2.54(10^{-8})[5.815(10^{10})]^2}{6.673(10^{-11})[5.976(10^{24})]} = 0.215 < 1 \qquad Ans.$$

From Eq. 13–23, observe that the orbit is an *ellipse*.

Part (b). If the satellite were launched at the apogee A shown in Fig. 13–26, with a velocity $\mathbf{v}_A$, the same orbit would be maintained provided

$$h = r_p v_0 = r_a v_A = 5.815(10^{10}) \text{ m}^2/\text{s}$$

Using Eq. 13–27, we have

$$r_a = \frac{r_p}{\dfrac{2GM_e}{r_p v_0^2} - 1} = \frac{6.978(10^6)}{\left(\dfrac{2[6.673(10^{-11})][5.976(10^{24})]}{6.978(10^6)(8333.3)^2} - 1\right)} = 10.804(10^6)$$

Thus,

$$v_A = \frac{5.815(10^{10})}{10.804(10^6)} = 5382.3 \text{ m/s} = 19.4 \text{ Mm/h} \qquad Ans.$$

PROBLEMS

In the following problems, assume that the radius of the earth is 6378 km, the earth's mass is $5.976(10^{24})$ kg, the mass of the sun is $1.99(10^{30})$ kg, and the gravitational constant is $G = 66.73(10^{-12})$ m^3/(kg · s^2).

*13–92. A satellite is placed into orbit at a velocity of 6 km/s, parallel to the surface of the earth. Determine the proper altitude of the satellite above the earth's surface such that its orbit remains circular. What will happen to the satellite if its initial velocity is 5 km/s when placed tangentially into orbit at the calculated altitude?

13–93. A communications satellite is to be placed into an equatorial circular orbit around the earth so that it always remains directly over a point on the earth's surface. If this requires the period to be 24 hours (approximately), determine the radius of the orbit and the satellite's velocity.

13–94. A satellite is launched with an initial velocity of $v_0 = 2500$ mi/h, parallel to the surface of the earth. Determine the required altitude (or range of altitudes) above the earth's surface for launching if the free-flight trajectory is to be (a) circular, (b) parabolic, (c) elliptical, and (d) hyperbolic. Take $G = 34.4(10^{-9})$(lb · ft^2)/slug^2, $M_e = 409(10^{21})$ slug, the earth's radius $r_e = 3960$ mi, and 1 mi = 5280 ft.

13–95. The rocket is in a free-flight elliptical orbit about the earth such that the eccentricity of its orbit is e and its perigee is r_0. Determine the minimum increment of speed it should have in order to escape the earth's gravitational field. Where along the path should this occur?

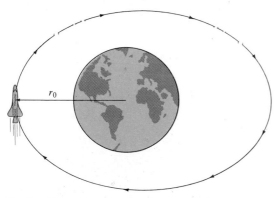

Prob. 13–95

*13–96. The rocket shown is originally in a circular orbit 6 Mm above the surface of the earth. It is required that it travel in another circular orbit having an altitude of 14 Mm. To do this, the rocket is given a short pulse of power at A so that it travels in free flight along the dashed elliptical path from the first orbit to the second orbit. Determine the necessary speed it must have at A just after the power pulse, and the time required to get to the outer orbit along the path AA'. What adjustment in speed must be made at A' to maintain the second circular orbit?

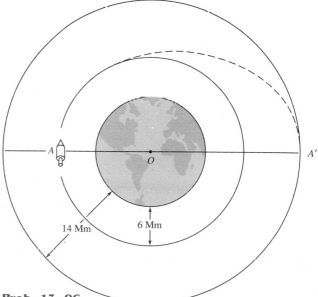

Prob. 13–96

13–97. A rocket is in circular orbit about the earth at an altitude of $h = 4$ Mm. Determine the minimum increment in speed it must have in order to escape the earth's gravitational field.

13–98. A satellite is in an elliptic orbit around the earth such that $e = 0.156$. If its perigee is 5 Mm, determine its velocity at this point and also the distance OB when it is at point B, located 135° away as shown.

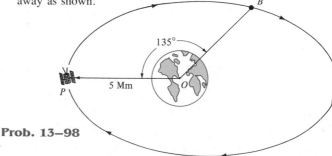

Prob. 13–98

123

13–99. A satellite is to be placed into an elliptical orbit about the earth such that at the perigee of its orbit it has an *altitude* of 800 km and at apogee its *altitude* is 2400 km. Determine its required launch velocity tangent to the earth's surface at perigee and the period of its orbit.

*13–100. An asteroid is in an elliptical orbit about the sun such that its periapsis is $9.30(10^9)$ km. If the eccentricity of the orbit is $e = 0.073$, determine the apoapsis of the orbit.

13–101. The rocket is traveling in a free-flight elliptical orbit about the earth such that $e = 0.76$ and its perigee is 9 Mm as shown. Determine its velocity when it is at point B. Also determine the sudden decrease in speed the rocket must experience at A in order to travel in a circular orbit about the earth.

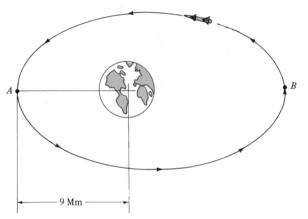

Prob. 13–101

13–102. The planet Jupiter travels around the sun in an elliptical orbit such that the eccentricity is $e = 0.048$. If the periapsis between Jupiter and the sun is $r_0 = 440(10^6)$ mi, determine (a) Jupiter's speed at periapsis and (b) the apoapsis of the orbit. Take $G = 34.4(10^{-9})$ lb · ft^2/slug^2, $M_s = 197(10^{27})$ slug, and 1 mi = 5280 ft.

13–103. Prove Kepler's third law of motion. *Hint:* Use Eqs. 13–19, 13–28, 13–29, and 13–31.

*13–104. Show that the velocity of a satellite launched into a circular orbit about the earth is given by Eq. 13–25. Determine the velocity of a satellite launched parallel to the surface of the earth so that it travels in a circular orbit 800 km from the earth's surface.

13–105. The elliptical path of a satellite has an eccentricity $e = 0.130$. If it has a speed of 15 Mm/h when it is at perigee, P, determine its speed when it arrives at apogee, A. Also, how far is it from the earth's surface when it is at A?

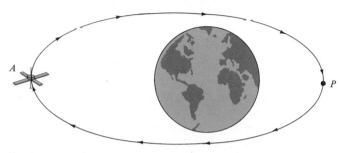

Prob. 13–105

13–106. A satellite S travels in a circular orbit around the earth. A rocket is located at the apogee of its elliptical orbit for which $e = 0.58$. Determine the sudden change in velocity that must occur at A so that the rocket can enter the satellite's orbit while in free flight along the dashed elliptical trajectory. When it arrives at B, determine the sudden adjustment in speed that must be given to the rocket in order to maintain the circular orbit.

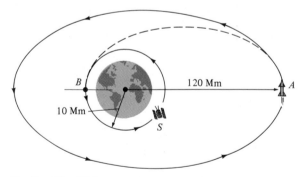

Prob. 13–106

14

Kinetics of a Particle: Work and Energy

In this chapter we will integrate the equation of motion with respect to displacement and thereby obtain the principle of work and energy. This principle is useful for solving problems which involve force, velocity, and displacement. Later in the chapter the concept of power is discussed and a method is presented for solving kinetic problems using the theorem of conservation of energy. Before presenting these topics, however, It is first necessary to define the work done by various types of forces.

14.1 The Work of a Force

In mechanics a force **F** does *work* only when it undergoes a *displacement in the direction of the force*. For example, consider the force **F** having a location on the path s which is specified by the position vector **r**, Fig. 14–1. If the force moves along the path to a new position **r**′, the displacement is then $d\mathbf{r} = \mathbf{r}' - \mathbf{r}$. The magnitude of $d\mathbf{r}$ is represented by ds, the differential segment along the path. If the angle between the tails of $d\mathbf{r}$ and **F** is θ, Fig. 14–1, then the work dU which is done by **F** is a *scalar quantity,* defined by

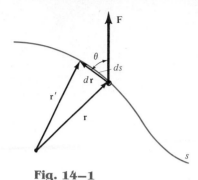

Fig. 14–1

$$dU = F \, ds \cos \theta$$

By definition of the dot product (see Eq. C–14) the above equation may also be written as

$$dU = \mathbf{F} \cdot d\mathbf{r}$$

Work as expressed by the above equation may be interpreted in one of two ways: either as the product of F and the component of displacement in the direction of the force, i.e., $ds \cos \theta$, or as the product of ds and the component of force in the direction of displacement, i.e., $F \cos \theta$. Note that if $0° \leq \theta \leq 90°$, then the force component and the displacement have the *same sense* so that the work is *positive;* whereas if $90° < \theta \leq 180°$, these vectors have an *opposite sense* and therefore the work is *negative*. Also, $dU = 0$ if the force is *perpendicular* to displacement, since $\cos 90° = 0$, or if the force is applied at a *fixed point,* in which case the displacement is zero.

The basic unit for work in the SI system is called a joule (J). This unit combines the units of force and displacement. Specifically, 1 *joule* of work is done when a force of 1 newton moves 1 meter along its line of action (1 J = 1 N · m). The moment of a force has this same combination of units (N · m); however, the concepts of moment and work are in no way related. A moment is a vector quantity, whereas work is a scalar. In the FPS system work is generally defined by writing the units as ft · lb, which is distinguished from the units for a moment written as lb · ft.

Work of a Variable Force

If a force undergoes a finite displacement along its path from $\mathbf{r}_1$ to $\mathbf{r}_2$ or s_1 to s_2, Fig. 14–2a, the work is determined by integration. If $\mathbf{F}$ is expressed as a function of position, $F = F(s)$, we have

$$U_{1-2} = \int_{\mathbf{r}_1}^{\mathbf{r}_2} \mathbf{F} \cdot d\mathbf{r} = \int_{s_1}^{s_2} F \cos \theta \, ds \qquad (14\text{–}1)$$

If the working component of the force, $F \cos \theta$, is plotted versus s, Fig. 14–2b, the integral represented in this equation can be interpreted as the *area under the curve* from position s_1 to position s_2.

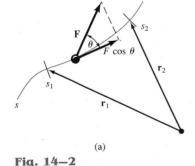

(a)

Fig. 14–2

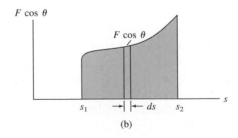

(b)

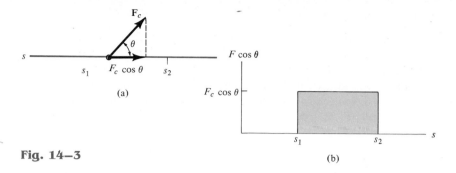

Fig. 14–3

(a)

(b)

Work of a Constant Force Moving Along a Straight Line

If the force $\mathbf{F}_c$ has a constant magnitude and acts at a constant angle θ from its straight-line path, Fig. 14–3a, then the component of $\mathbf{F}_c$ in the direction of displacement is $F_c \cos \theta$. The work done by $\mathbf{F}_c$ when it is displaced from s_1 to s_2 is determined by Eq. 14–1, in which case

$$U_{1-2} = F_c \cos \theta \int_{s_1}^{s_2} ds$$

or

$$\boxed{U_{1-2} = F_c \cos \theta \, (s_2 - s_1)} \qquad (14\text{–}2)$$

Here the work of $\mathbf{F}_c$ represents the *area under the rectangle* in Fig. 14–3b.

Work of a Weight

Consider a particle which moves up along the path s shown in Fig. 14–4 from position s_1 to position s_2. At an intermediate point, the displacement $d\mathbf{r} = dx\,\mathbf{i} + dy\,\mathbf{j} + dz\,\mathbf{k}$. Since $\mathbf{W} = -W\mathbf{j}$, applying Eq. 14–1 yields

$$U_{1-2} = \int \mathbf{F} \cdot d\mathbf{r} = \int_{\mathbf{r}_1}^{\mathbf{r}_2} (-W\mathbf{j}) \cdot (dx\,\mathbf{i} + dy\,\mathbf{j} + dz\,\mathbf{k})$$

$$= \int_{y_1}^{y_2} -W \, dy = -W(y_2 - y_1)$$

or

$$\boxed{U_{1-2} = -W\Delta y} \qquad (14\text{–}3)$$

Thus, the work done is equal to the magnitude of the particle's weight times its vertical displacement. In the case shown in Fig. 14–4 the work is *negative*, since W is downward and Δy is upward. Note, however, that if the particle is displaced *downward* $(-\Delta y)$ the work of the weight is *positive*. Why?

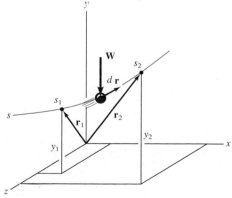

Fig. 14–4

127

Work of a Spring Force

The magnitude of force developed in a linear elastic spring when the spring is displaced a distance s from its unstretched position is $F_s = ks$, where k is the spring stiffness. If the spring is elongated or compressed from a position s_1 to a further position s_2, Fig. 14–5a, the work done *on the spring* by $\mathbf{F}_s$ is *positive*, since in each case the force and displacement are in the *same direction*. We require

$$U_{1-2} = \int_{s_1}^{s_2} F_s \, ds = \int_{s_1}^{s_2} ks \, ds$$
$$= \tfrac{1}{2}ks_2^2 - \tfrac{1}{2}ks_1^2$$

This equation represents the trapezoidal area under the line $F_s = ks$ versus s, Fig. 14–5b.

If a particle (or body) is attached to a spring, then the force $\mathbf{F}_s$ exerted on the particle is *opposite* to that exerted on the spring, Fig. 14–5c. Consequently, the force will do *negative work* on the particle when the particle is moving so as to further elongate (or compress) the spring. Hence, the above equation becomes

$$U_{1-2} = -(\tfrac{1}{2}ks_2^2 - \tfrac{1}{2}ks_1^2) \tag{14–4}$$

When this equation is used, a mistake in sign can be eliminated if one simply notes the direction of the spring force acting on the particle and compares it with the direction of displacement of the particle—if both are in the *same direction, positive work* results; if they are *opposite* to one another, the *work is negative*.

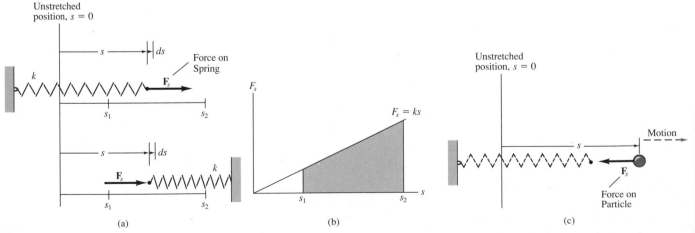

Fig. 14–5

Example 14–1

The 10-kg block shown in Fig. 14–6a rests on the smooth incline. If the spring is originally unstretched, determine the total work done by all the forces acting on the block when a horizontal force $P = 400$ N pushes the block up the plane $s = 2$ m.

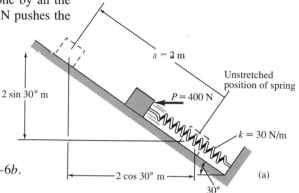

Unstretched position of spring

$s = 2$ m

$2 \sin 30°$ m

$P = 400$ N

$k = 30$ N/m

$2 \cos 30°$ m

$30°$

(a)

SOLUTION

The free-body diagram of the block is shown in Fig. 14–6b.

98.1 N $30°$

$P = 400$ N $30°$

N_B F_s

(b)

Fig. 14–6

Horizontal Force **P.** Since the force is *constant*, the work is computed using Eq. 14–2. The result can be calculated as the force times the component of displacement in the direction of the force, i.e.,

$$U_P = 400(2 \cos 30°) = 692.8 \text{ J}$$

or, the displacement times the component of force in the direction of displacement, i.e.,

$$U_P = 400 \cos 30°(2) = 692.8 \text{ J}$$

Spring Force **F$_s$.** Since the spring is originally unstretched and in the final position is stretched 2 m, the work of **F$_s$** is

$$U_s = -\tfrac{1}{2}(30)(2)^2 = -60 \text{ J}$$

Why is the work negative?

Weight **W.** Since the weight acts in the opposite direction to its vertical displacement the work is negative; i.e.,

$$U_W = -98.1(2 \sin 30°) = -98.1 \text{ J}$$

Note that it is also possible to consider the component of weight in the direction of displacement; i.e.,

$$U_W = -(98.1 \sin 30°)2 = -98.1 \text{ J}$$

Normal Force **N$_B$.** This force does *no work* since it is *always* perpendicular to the displacement.

The work of all the forces when the block is displaced 2 m is thus

$$U_T = 692.8 - 60 - 98.1 = 535 \text{ J} \qquad \textit{Ans.}$$

14.2 Principle of Work and Energy

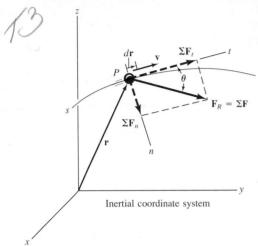

Fig. 14-7

Consider a particle P in Fig. 14–7, which at the instant considered is located at position $\mathbf{r}$ on the path as measured from an inertial coordinate system. If the particle has a mass m and is subjected to a system of external forces represented by the resultant $\mathbf{F}_R = \Sigma\mathbf{F}$, then the equation of motion for the particle is $\Sigma\mathbf{F} = m\mathbf{a}$. When the particle undergoes a displacement $d\mathbf{r}$ along the path, the work done by the forces is

$$\Sigma\mathbf{F} \cdot d\mathbf{r} = m\mathbf{a} \cdot d\mathbf{r}$$

If we establish the n and t axes at the particle, then $\Sigma\mathbf{F}$ can be resolved into its normal and tangential components, Fig. 14–7. Realizing that the magnitude of $d\mathbf{r}$ is ds, it can be seen that $\Sigma\mathbf{F} \cdot d\mathbf{r} = \Sigma F \, ds \cos\theta = \Sigma F_t \, ds$. In other words, the work of $\Sigma\mathbf{F}$ is computed *only from its tangential components*. The normal components $\Sigma F_n = \Sigma F \sin\theta$ do *no work* since they cannot move in the normal direction. Since $\mathbf{a} \cdot d\mathbf{r} = a_t \, ds$, the above equation can also be written as

$$\Sigma\mathbf{F} \cdot d\mathbf{r} = \Sigma F_t \, ds = ma_t \, ds$$

Using the kinematic equation $a_t \, ds = v \, dv$, and integrating both sides, assuming initially that the particle has a position $\mathbf{r} = \mathbf{r}_1$ and a speed $v = v_1$, and later $\mathbf{r} = \mathbf{r}_2$, $v = v_2$, yields

$$\Sigma \int_{\mathbf{r}_1}^{\mathbf{r}_2} \mathbf{F} \cdot d\mathbf{r} = \int_{v_1}^{v_2} mv \, dv$$

or

$$\Sigma \int_{\mathbf{r}_1}^{\mathbf{r}_2} \mathbf{F} \cdot d\mathbf{r} = \tfrac{1}{2}mv_2^2 - \tfrac{1}{2}mv_1^2 \tag{14–5}$$

Using Eq. 14–1, the final result may be written as

$$\Sigma U_{1-2} = \tfrac{1}{2}mv_2^2 - \tfrac{1}{2}mv_1^2 \tag{14–6}$$

This equation represents the *principle of work and energy* for the particle. The term on the left is the sum of the work done by *all* the forces acting on the particle as the particle moves from point 1 to point 2. The two terms on the right side, which are of the form $T = \tfrac{1}{2}mv^2$, define the particle's final and initial *kinetic energy*, respectively. These terms are *positive* scalar quantities since they do not depend on the direction of the particle's velocity. Furthermore, Eq. 14–6 is dimensionally homogeneous so that the kinetic energy has the same units as work, e.g., joules (J) or ft · lb.

When Eq. 14–6 is applied, it is often symbolized in the form

$$\boxed{T_1 + \Sigma U_{1-2} = T_2} \tag{14–7}$$

which states that the particle's initial kinetic energy plus the work done by all the forces acting on the particle as it moves from its initial to its final position is equal to the particle's final kinetic energy.

As noted from the derivation, the principle of work and energy represents an integrated form of $\Sigma F_t = ma_t$, acquired by using the kinematic equation $a_t = v \, dv/ds$. As a result, this principle will provide a convenient substitution for $\Sigma F_t = ma_t$ when solving those types of kinetic problems which involve force, velocity, and displacement, since these variables are involved in the terms of Eq. 14-7. For example, if a particle's initial speed is known, and the work of all the forces acting on the particle can be computed, then Eq. 14-7 provides a *direct means* of obtaining the final speed v_2 of a particle after it undergoes a specified displacement. If instead v_2 is determined by means of the equation of motion, a two-step process is necessary; i.e., apply $\Sigma F_t = ma_t$ to obtain a_t, then integrate $a_t = v \, dv/ds$ to obtain v_2. Note that the principle of work and energy cannot be used, for example, to determine forces directed *normal* to the path of motion, since these forces do no work on the particle. Instead $\Sigma F_n = ma_n$ must be applied. For curved paths, however, the magnitude of the normal force is a function of velocity. Hence, it is generally easier to obtain the velocity using the principle of work and energy, and then substitute this quantity into the equation of motion $\Sigma F_n = mv^2/\rho$ to obtain the normal force.

PROCEDURE FOR ANALYSIS

As stated above, the principle of work and energy is used to solve kinetic problems that involve *velocity, force,* and *displacement,* since these terms are involved in the equation. For application it is suggested that the following procedure be used.

Work (Free-Body Diagram). Establish the inertial coordinate system and draw a free-body diagram of the particle in order to account for all the forces that do work on the particle as it moves along its path.

Principle of Work and Energy. Apply the principle of work and energy, $T_1 + \Sigma U_{1-2} = T_2$. The kinetic energy at the initial and final points is always positive, since it involves the speed squared ($T = \frac{1}{2}mv^2$). The work done by each force shown on the free-body diagram is computed by using the appropriate equations developed in Sec. 14.1. Since *algebraic addition* of the work terms is required, it is important that the proper sign of each term be specified. Specifically, work is positive when the force component is in the same direction as its displacement, otherwise it is negative.

Numerical application of this procedure is illustrated in the examples following Sec. 14.3.

14.3 Principle of Work and Energy for a System of Particles

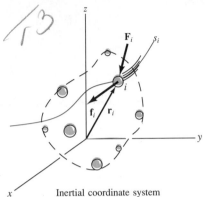

Inertial coordinate system

Fig. 14–8

The principle of work and energy can be extended to include a system of n particles isolated within an enclosed region of space as shown in Fig. 14–8. Here the arbitrary ith particle, having a mass m_i, is subjected to a resultant external force $\mathbf{F}_i$ and a resultant internal force $\mathbf{f}_i = \sum_{j=1(j \neq i)}^{n} \mathbf{f}_{ij}$ which each of the other particles exerts on the ith particle. Using Eq. 14–5, the principle of work and energy written for the ith particle is thus

$$\tfrac{1}{2} m_i v_{i1}^2 + \int_{\mathbf{r}_{i1}}^{\mathbf{r}_{i2}} \mathbf{F}_i \cdot d\mathbf{r}_i + \int_{\mathbf{r}_{i1}}^{\mathbf{r}_{i2}} \mathbf{f}_i \cdot d\mathbf{r}_i = \tfrac{1}{2} m_i v_{i2}^2$$

Similar equations result if the principle of work and energy is applied to each of the other particles of the system. Since both work and kinetic energy are scalars, the results may be added together algebraically, so that

$$\Sigma \tfrac{1}{2} m_i v_{i1}^2 + \Sigma \int_{\mathbf{r}_{i1}}^{\mathbf{r}_{i2}} \mathbf{F}_i \cdot d\mathbf{r}_i + \Sigma \int_{\mathbf{r}_{i1}}^{\mathbf{r}_{i2}} \mathbf{f}_i \cdot d\mathbf{r}_i = \Sigma \tfrac{1}{2} m_i v_{i2}^2$$

We can write this equation symbolically as

$$\Sigma T_1 + \Sigma U_{1-2} = \Sigma T_2 \tag{14–8}$$

Hence, the system's initial kinetic energy (ΣT_1) plus the work done by all the external and internal forces acting on the particles of the system (ΣU_{1-2}) is equal to the system's final kinetic energy (ΣT_2). To maintain this balance of energy, strict accountability of the work done by all the forces must be made. In this regard, note that although the internal forces on adjacent particles occur in equal but opposite collinear pairs, the total work done by each of these forces will, in general, *not cancel* out since the paths over which corresponding particles travel will be *different*. There are, however, two important exceptions to this rule which often occur in practice. If the particles are contained within the boundary of a *translating rigid body*, the internal forces all undergo displacement by an equal amount, and therefore the internal work will be zero. Also, particles connected by inextensible cables make up a system which has internal forces which are displaced by an equal amount. In this case, adjacent particles exert equal but opposite internal forces which have components that undergo the same displacement, and therefore the work of these forces cancels. On the other hand, note that if the body is assumed to be *nonrigid*, the particles of the body are displaced along *different paths*, and some of the energy due to force interactions would be given off and lost as heat or stored in the body if permanent deformations occur. We will briefly discuss these effects at the end of this section and in Sec. 15–4. Throughout this text, however, the principle of work and energy will be applied to the solution of problems only when direct accountability of these energy losses does not have to be made.

The procedure for analysis outlined in Sec. 14.2 provides a method for applying Eq. 14–8; however, only one equation applies for the entire system. If the particles are connected by cords, other equations can generally be obtained by using the kinematic principles outlined in Sec. 12.8 in order to relate the particles' speeds. When doing this, assume that the particles are displaced in the *positive coordinate directions* when writing *both* the kinematic equation and the principle of work and energy (see Example 14–6).

Work of Friction Caused by Sliding

A special class of problems will now be investigated which requires a careful application of Eq. 14–8. These problems all involve cases where a body is sliding over the surface of another body in the presence of friction. Consider, for example, a block which is translating a distance s over a rough surface as shown in Fig. 14–9. If the applied force **P** just balances the *resultant* frictional force $\mu_k N$, then due to equilibrium a constant velocity **v** is maintained and one would expect Eq. 14–8 to be applied as follows:

$$\tfrac{1}{2}mv^2 + Ps - \mu_k Ns = \tfrac{1}{2}mv^2$$

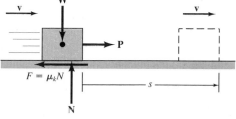

Fig. 14–9

Indeed this equation is satisfied if $P = \mu_k N$; however, as one realizes from experience, the sliding motion will *generate heat,* a form of energy which seems not to be accounted for in the work-energy equation. In order to explain this paradox and thereby more accurately represent the nature of friction, we should actually consider a model of the surfaces of contact which are *deformable* (nonrigid).* Recall that the rough portions at the bottom of the block act as "teeth," and when the block slides these teeth *deform slightly* and either break off or vibrate due to interlocking effects and pulling away from "teeth" at the contacting surface. As a result, frictional forces that act at these points are diminished; however, they are replaced by other frictional forces as other points of contact are made. At any instant, the *resultant* of the frictional forces remains essentially constant, i.e., $\mu_k N$; however, due to the localized deformations, the actual displacement s' of $\mu_k N$ is *not* the same displacement s as the applied force **P.** Instead, s' will be *less* than s $(s' < s)$, and therefore the *external work* done by the resultant frictional force will be $\mu_k Ns'$. The remaining amount of work, $\mu_k N(s - s')$, manifests itself as an increase in *internal energy,* which in fact causes the block's temperature to rise.

In summary then, Eq. 14–8 can be applied to problems involving sliding friction; however, *it should be fully realized that the work of the resultant frictional force is not represented by $\mu_k Ns$; instead, this term represents both the external work of friction ($\mu_k Ns'$) and internal work [$\mu_k N(s - s')$] which is converted into various forms of internal energy.†*

*See Chapter 9 of *Engineering Mechanics: Statics.*

†See B. A. Sherwood and W. H. Bernard, "Work and Heat Transfer in the Presence of Sliding Friction," *Am. J. Phys.* 52, 1001 (1984).

Example 14–2

The 3500-lb automobile shown in Fig. 14–10a is traveling down the 10° inclined road at a speed of 20 ft/s. If the driver wishes to stop his car, determine how far s his tires skid on the road if he jams on the brakes, causing his wheels to lock. The coefficient of kinetic friction between the wheels and the road is $\mu_k = 0.5$.

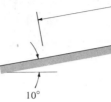

SOLUTION

This problem can be solved using the principle of work and energy since it involves force, velocity, and displacement.

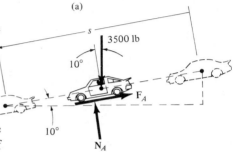

Fig. 14–10

Work (Free-Body Diagram). As shown in Fig. 14–10b, the normal force $\mathbf{N}_A$ does no work since it never undergoes displacement along its line of action. The weight, 3500 lb, is displaced $s \sin 10°$ and does positive work, whereas the frictional force $\mathbf{F}_A$ does both external and internal work when it is *thought* to undergo a displacement s. Applying the equation of equilibrium normal to the road, we have

$$+\nwarrow \Sigma F_n = 0; \quad N_A - 3500 \cos 10° = 0 \quad N_A = 3446.8 \text{ lb}$$

Thus,

$$F_A = 0.5 N_A = 1723.4 \text{ lb}$$

Principle of Work and Energy

$$\{T_1\} + \{\Sigma U_{1-2}\} - \{T_2\}$$

$$\left\{\frac{1}{2}\left(\frac{3500}{32.2}\right)(20)^2\right\} + \{3500(s \sin 10°) - 1723.4s\} = \{0\}$$

Solving for s yields
$$s = 19.5 \text{ ft} \qquad \qquad \textit{Ans.}$$

Note that if this problem is solved by using the equation of motion, *two steps* are involved. First, from the free-body diagram, Fig. 14–10b, the equation of motion is applied along the incline. This yields

$$+\swarrow \Sigma F_s = ma_s; \quad 3500 \sin 10° - 1723.4 = \frac{3500}{32.2}a$$

$$a = -10.3 \text{ ft/s}^2$$

Then, using the integrated form of $a\, ds = v\, dv$ (kinematics) since a is constant, we have

$$(+\swarrow) \quad v^2 = v_0^2 + 2a_c(s - s_0); \quad (0)^2 = (-20)^2 + 2(-10.3)(s - 0)$$

$$s = 19.5 \text{ ft} \qquad \textit{Ans.}$$

Example 14–3

A 10-kg block rests on the horizontal surface shown in Fig. 14–11a. The spring, which is not attached to the block, has a stiffness $k = 500$ N/m and is initially compressed 0.2 m from C to A. After the block is released from rest at A, determine its velocity when it passes point D. The coefficient of kinetic friction between the block and the plane is $\mu_k = 0.2$.

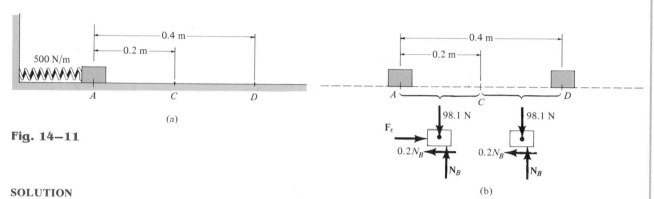

(a)

Fig. 14–11

(b)

SOLUTION

Why can this problem be solved using the principle of work and energy?

Work (Free-Body Diagrams). Two free-body diagrams for the block are shown in Fig. 14–11b. The block moves under the influence of the spring force $\mathbf{F}_s$ along the 0.2-m-long path AC, after which it continues to slide along the plane to point D. With reference to either free-body diagram, $\Sigma F_y = 0$; hence, $N_B = 98.1$ N. Only the spring and friction forces do work during the displacement—the spring force does positive work from A to C, whereas the frictional force does negative work. Why?

Principle of Work and Energy

$$\{T_A\} + \{\Sigma U_{A-D}\} = \{T_D\}$$

$$\{\tfrac{1}{2}m(v_A)^2\} + \{\tfrac{1}{2}ks_{AC}^2 - (0.2N_B)s_{AD}\} = \{\tfrac{1}{2}m(v_D)^2\}$$

$$\{0\} + \left\{\frac{1}{2}(500)(0.2)^2 - 0.2(98.1)(0.4)\right\} = \left\{\frac{1}{2}(10)(v_D)^2\right\}$$

Solving for v_D, we get

$$v_D = 0.656 \text{ m/s} \rightarrow \qquad \qquad \textit{Ans.}$$

Example 14–4

The platform P, shown in Fig. 14–12a, has negligible mass and is tied down so that the 0.4-m-long cords keep the spring compressed 0.6 m when *nothing* is on the platform. If a 2-kg block is placed on the platform and released from rest after the platform is pushed down 0.1 m, determine the maximum height h the block rises in the air, measured from the ground.

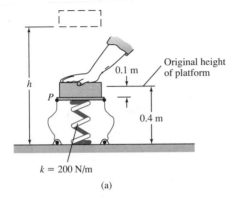

Original height of platform

0.1 m

0.4 m

P

h

$k = 200$ N/m

(a)

SOLUTION

Work (Free-Body Diagram). Since the block is released from rest and later reaches its maximum height, the initial and final velocities are zero. The free-body diagram of the block when it is still in contact with the platform is shown in Fig. 14–12b. Note that the weight does negative work and the spring force does positive work. Why? In particular, the *initial compression* in the spring is $s_1 = 0.6$ m $+ 0.1$ m $= 0.7$ m. Due to the cords, the spring's *final compression* is $s_2 = 0.6$ m (after the block leaves the platform). The bottom of the block rises from a height of $(0.4$ m $- 0.1$ m$) = 0.3$ m to a final height h.

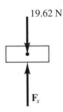

19.62 N

F_s

(b)

Fig. 14–12

Principle of Work and Energy

$$\{T_1\} + \{\Sigma U_{1-2}\} = \{T_2\}$$
$$\{\tfrac{1}{2}mv_1^2\} + \{-(\tfrac{1}{2}ks_2^2 - \tfrac{1}{2}ks_1^2) - W\Delta y\} = \{\tfrac{1}{2}mv_2^2\}$$

Note that here $s_2 < s_1$, so that the work of the spring will indeed be positive. Thus,

$$\{0\} + \left\{-\left(\frac{1}{2}(200)(0.6)^2 - \frac{1}{2}(200)(0.7)^2\right) - (19.62)[h - (0.3)]\right\} = \{0\}$$

Solving yields

$$h = 0.963 \text{ m} \qquad\qquad Ans.$$

Example 14–5

A block having a mass of 2 kg is given an initial velocity of $v_0 = 1$ m/s when it is at the top surface of the smooth cylinder shown in Fig. 14–13a. If the block travels along a path of 0.5-m radius, determine the angle $\theta = \theta_{max}$ at which it begins to leave the cylinder's surface.

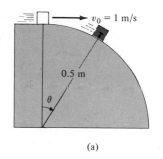

(a)

SOLUTION

Work (Free–Body Diagram). Inspection of Fig. 14–13b reveals that only the weight $W = 2(9.81) = 19.62$ N does work during the displacement. If the block is assumed to leave the surface when $\theta = \theta_{max}$ then the weight moves through a vertical displacement of $0.5(1 - \cos \theta_{max})$ m, as shown in the figure.

Principle of Work and Energy

$$\{T_1\} + \{\Sigma U_{1-2}\} = \{T_2\}$$

$$\left\{\frac{1}{2}(2)(1)^2\right\} + \{19.62(0.5)(1 - \cos \theta_{max})\} = \left\{\frac{1}{2}(2)v_2^2\right\}$$

$$v_2^2 = 9.81(1 - \cos \theta_{max}) + 1 \qquad (1)$$

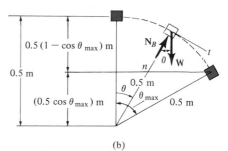

(b)

Fig. 14–13

Equation of Motion. There are two unknowns in Eq. (1), θ_{max} and v_2. A second equation relating these two variables may be obtained by applying the equation of motion in the *normal direction* to the forces on the free-body diagram. Thus,

$$+\swarrow \Sigma F_n = ma_n; \qquad -N_B + 19.62 \cos \theta = 2\left(\frac{v^2}{0.5}\right)$$

When the block leaves the surface of the cylinder at $\theta = \theta_{max}$, $N_B = 0$ and $v = v_2$; hence,

$$\cos \theta_{max} = \frac{v_2^2}{4.905} \qquad (2)$$

Eliminating the unknown v_2^2 between Eqs. (1) and (2) gives

$$4.905 \cos \theta_{max} = 9.81(1 - \cos \theta_{max}) + 1$$

Solving, we have

$$\cos \theta_{max} = 0.735$$

$$\theta_{max} = 42.7° \qquad \qquad Ans.$$

This problem has also been solved in Example 13–9. If the two methods of solution are compared it will be apparent that a work-energy approach yields a more direct solution.

Example 14–6

The blocks A and B shown in Fig. 14–14a have a mass of 10 kg and 100 kg, respectively. Determine the distance B travels from the point where it is released from rest to the point where its speed becomes 2 m/s.

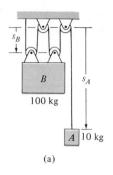

(a)

SOLUTION

This problem may be solved by considering the blocks separately and applying the principle of work and energy to each block. However, the work of the (unknown) cable tension can be eliminated from the analysis by considering blocks A and B together as a *system*. The solution will require simultaneous solution of the equations of work and energy *and* kinematics. To be consistent with our sign convention, we will assume both blocks move in the positive *downward* direction.

Work (Free-Body Diagram). As shown on the free-body diagram of the system, Fig. 14–14b, the cable force $\mathbf{T}$ and reactions $\mathbf{R}_1$ and $\mathbf{R}_2$ do *no work*, since these forces represent the reactions at the supports and consequently do not move while the blocks are being displaced.

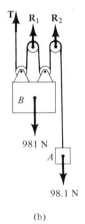

Principle of Work and Energy. Realizing the blocks are released from rest, and as stated previously, assuming *both* move *downward*, we have

$$\{\Sigma T_1\} + \{\Sigma U_{1-2}\} = \{\Sigma T_2\}$$

$$\{\tfrac{1}{2}m_A(v_A)_1^2 + \tfrac{1}{2}m_B(v_B)_1^2\} + \{W_B\Delta s_B + W_A\Delta s_A\} = \{\tfrac{1}{2}m_A(v_A)_2^2 + \tfrac{1}{2}m_B(v_B)_2^2\}$$

$$\{0 + 0\} + \{981(\Delta s_B) + 98.1(\Delta s_A)\} = \left\{\frac{1}{2}(10)(v_A)_2^2 + \frac{1}{2}(100)(2)^2\right\} \qquad (1)$$

(b)

Fig. 14–14

Kinematics. Using the methods of kinematics discussed in Sec. 12.8, it may be seen from Fig. 14–14a that at any given instant the total length l of all the vertical segments of cable may be expressed in terms of the position coordinates s_A and s_B as

$$s_A + 4s_B = l$$

Hence, a change in position yields the displacement equation

$$\Delta s_A + 4\,\Delta s_B = 0$$
$$\Delta s_A = -4\,\Delta s_B \qquad (2)$$

Taking the time derivative yields

$$v_A = -4v_B = -4(2) = -8 \text{ m/s}$$

Retaining the negative sign in Eq. (2) and substituting into Eq. (1) yields

$$\Delta s_B = 0.883 \text{ m} \downarrow \qquad\qquad \textit{Ans.}$$

PROBLEMS

14–1. A woman having a mass of 70 kg stands in an elevator which has a downward acceleration of 4 m/s^2 starting from rest. Determine the work done by her weight and the work of the normal force which the floor exerts on her when the elevator descends 6 m. Explain why the work of these forces is different.

14–2. The car having a mass of 2 Mg is towed along an inclined road. If the car starts from rest and attains a speed of 5 m/s after traveling a distance of 150 m, determine the constant towing force F applied to the car. Neglect friction and the mass of the wheels.

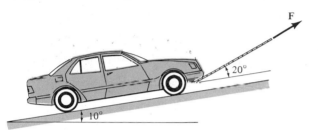

Prob. 14–2

14–3. The "air spring" A is used to protect the support structure B and prevent damage to the conveyor belt tensioning weight C in the event of a belt failure D. The force developed by the spring as a function of its deflection is shown by the graph. Determine the suspended height d of the 50-lb counterweight C so that if the conveyor belt fails the counterweight will deform the spring $s_{max} = 0.2$ ft. Neglect the mass of the pulley and belt.

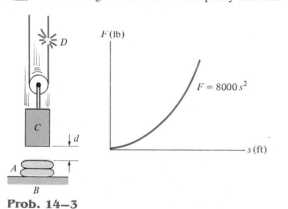

Prob. 14–3

***14–4.** Marbles having a mass of 5 g fall from rest at A through the glass tube and accumulate in the can at C. Determine the placement R of the can from the end of the tube and the speed at which the marbles fall into the can. Neglect the size of the can.

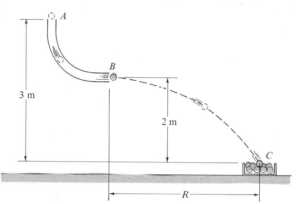

Prob. 14–4

14–5. Packages having a weight of 15 lb are transferred horizontally from one conveyor to the next using a ramp for which the coefficient of kinetic friction is $\mu_k = 0.15$. The top conveyor is moving at 6 ft/s and the packages are spaced 3 ft apart. Determine the required speed of the bottom conveyor so no sliding occurs when the packages come horizontally in contact with it. What is the spacing s between the packages on the bottom conveyor?

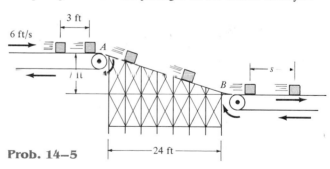

Prob. 14–5

139

14–6. The motion of a 6500-lb boat is arrested using a bumper which provides a resistance as shown in the graph. Determine the distance the boat dents the bumper if it is coasting at 3 ft/s when it approaches it.

F(lb)

$v = 3$ ft/s

s

$F = 3(10^3)s^3$

s(ft)

Prob. 14–6

14–7. The conveyor belt delivers each 12-kg crate to the ramp at A such that the crate's velocity is $v_A = 2.5$ m/s, directed down *along* the ramp. If the coefficient of kinetic friction between each crate and the ramp is $\mu_k = 0.3$, determine the speed at which each crate slides off the ramp at B. Assume that no tipping occurs.

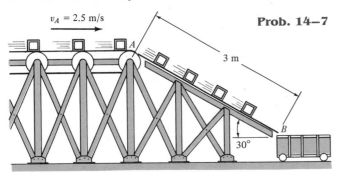

$v_A = 2.5$ m/s

Prob. 14–7

3 m

B

$30°$

*14–8. Cylinder A has a mass of $m_A = 3$ kg and cylinder B has a mass of $m_B = 8$ kg. Determine the speed of A after it moves upward 2 m starting from rest. Neglect the mass of the cord and pulleys.

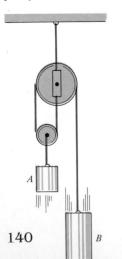

A

Prob. 14–8

B

14–9. The smooth plug has a weight of 20 lb and is pushed against a series of Belleville spring washers so that the compression in the spring is $s = 0.05$ ft. If the force of the spring on the plug is $F = (3s^{1/3})$ lb, where s is given in feet, determine the speed of the plug after it moves away from the spring. Neglect friction.

Prob. 14–9

14–10. Block A has a weight of $W_A = 60$ lb and block B has a weight of $W_B = 10$ lb. Determine the speed of A just after it moves downward 3 ft starting from rest. Neglect the mass of the cord and pulleys.

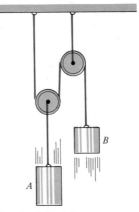

B

A

Prob. 14–10

14–11. The 50-kg block has a speed of $v_A = 8$ m/s when it reaches point A. Determine the normal force it exerts on the incline when it reaches point B. Neglect friction and the block's size.

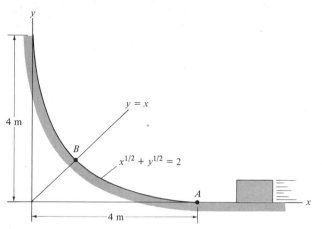

4 m

$y = x$

B

$x^{1/2} + y^{1/2} = 2$

A

4 m

Prob. 14–11

*14–12. The spring has a stiffness $k = 50$ lb/ft and an *unstretched length* of 2 ft. As shown, it is confined by the plate and wall using cables so that its length is 1.5 ft. Determine the speed v_A of the 4-lb block when it is at A, so that it slides along the horizontal plane and strikes the plate and pushes it forward 0.25 ft before stopping. The coefficient of kinetic friction between the block and plane is $\mu_k = 0.2$. Neglect the mass of the plate and spring and the energy loss between the plate and block during the collision.

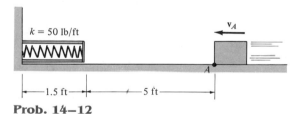

$k = 50$ lb/ft

v_A

1.5 ft — 5 ft

A

Prob. 14–12

14–13. The 120-lb man acts as a human cannon ball by being "fired" from the spring-loaded cannon shown. If the greatest acceleration he can experience is $a = 10g = 322$ ft/s^2, determine the required stiffness of the spring if it is compressed 2 ft at the moment of firing. With what velocity will he exit the cannon barrel when the cannon is fired? When the spring is compressed $s = 2$ ft then $d = 8$ ft. Neglect friction and assume the man holds himself in a rigid position throughout the motion.

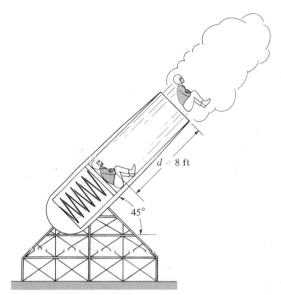

$d = 8$ ft

45°

Prob. 14–13

14–14. Packages having a weight of 50 lb are delivered to the chute at $v_A = 3$ ft/s using a conveyor belt. Determine their speed when they reach points B, C, and D. Also calculate the normal force of the chute on the packages at B and C. Neglect friction and the size of the packages.

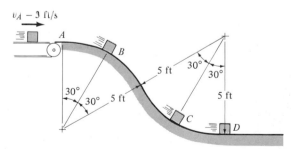

$v_A = 3$ ft/s

A

B

5 ft 30° 30°

30° 30° 5 ft

5 ft C 5 ft

D

Prob. 14–14

141

14–15. The 0.5-kg ball of negligible size is fired up the vertical circular track using the spring plunger. The plunger keeps the spring compressed 0.08 m when $s = 0$. Determine how far s it must be pulled back and released so that the ball will begin to leave the track when $\theta = 135°$.

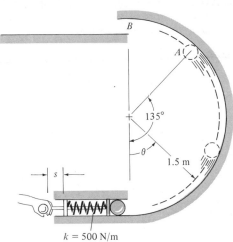

Probs. 14–15/14–16

*14–16. The 0.5-kg ball of negligible size is fired up the vertical circular track using the spring plunger. The plunger keeps the spring compressed 0.08 m when $s = 0$. Determine how far s it must be pulled back and released so that the ball will just make it around the loop and land on the platform at B.

14–17. The smooth block has a mass m and is attached to a spring having an unstretched length when $y = d$. Determine the work done by the horizontal force **P** needed to raise the block from an at-rest position at $y = d$ to an at-rest position when the elevation is y.

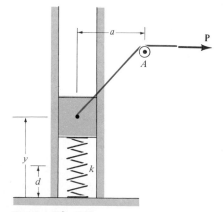

Prob. 14–17

14–18. Block A has a weight of $W_A = 3$ lb and rests on a surface for which the coefficient of kinetic friction is $\mu_k = 0.3$. Block B has a weight of $W_B = 8$ lb. Determine the distance B must descend so that A has a speed of $v_A = 5$ ft/s starting from rest.

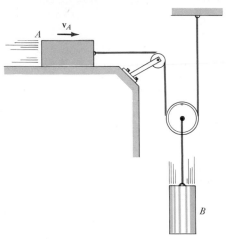

Prob. 14–18

14–19. The 200-kg roller coaster car is accelerated uniformly from rest at A until it reaches a maximum speed at B in $t = 3.5$ s, at which point it begins to travel freely along the spiral loop. Determine this maximum speed at B so that the car just makes the loop without falling off the track. Also, compute the constant horizontal force F needed to give the car the necessary acceleration from A to B. The radius of curvature at C is $\rho_C = 25$ m.

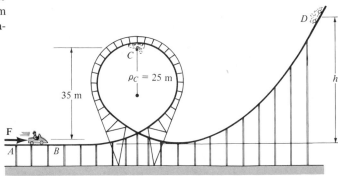

Probs. 14–19/14–20

*14–20. Determine the height h to the top of the incline D to which the 200-kg roller coaster car will reach, if it is launched at B with a speed just sufficient for it to round the top of the loop at C without leaving the tracks. The radius of curvature at C is $\rho_C = 25$ m.

14–21. The collar has a mass of 5 kg and is moving at 8 m/s when $x = 0$. A force of $F = 60$ N is applied to it as shown. The direction θ of this force varies such that $\theta = 10x$, where x is in meters and θ is clockwise, measured in degrees. Determine the speed of the collar when $x = 3$ m. The coefficient of kinetic friction between the collar and the rod is $\mu_k = 0.3$.

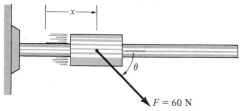

Prob. 14–21

■**14–22.** The force **F** acting in a constant direction on the 20-kg block has a magnitude which varies with the position s of the block. Determine how far the block slides before its velocity becomes 5 m/s. When $s = 0$ the block is moving to the right at 2 m/s. The coefficient of kinetic friction between the block and surface is $\mu_k = 0.3$.

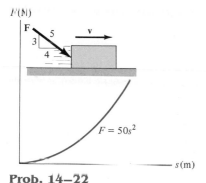

Prob. 14–22

14–23. As indicated by the derivation, the principle of work and energy is valid for observers in *any* inertial reference frame. Show that this is so, by considering the 10-kg block which rests on the smooth surface and is subjected to a horizontal force of 6 N. If observer A is in a *fixed* frame x, determine the final speed of the block if it has an initial speed of 5 m/s and travels 10 m, both directed to the right and measured from the fixed frame. Compare the result with that obtained by an observer B, attached to the x' axis and moving at a constant velocity of 2 m/s relative to A. *Hint:* The distance the block travels will first have to be computed for observer B before applying the principle of work and energy.

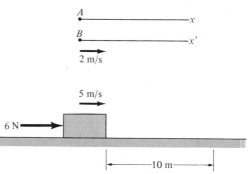

Prob. 14–23

*****14–24.** The 2-kg block is subjected to a force having a constant direction and a magnitude $F = (300/(1 + s))$ N, where s is measured in meters. When $s = 4$ m, the block is moving to the left with a speed of 8 m/s. Determine its speed when $s = 12$ m. The coefficient of kinetic friction between the block and the ground is $\mu_k = 0.25$.

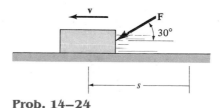

Prob. 14–24

14–25. The steel ingot has a mass of 1800 kg. It travels along the conveyor at a speed of $v = 0.5$ m/s when it collides with the "nested" spring assembly. Determine the maximum deflection in each spring needed to stop the motion of the ingot. Neglect the mass of the springs.

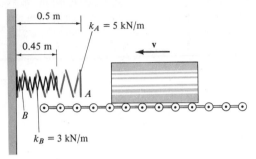

Prob. 14–25

14–26. The collar has a mass of 20 kg and is supported on the smooth rod. The attached springs are undeformed when $d = 0.5$ m. Determine the speed of the collar after the applied force $F = 100$ N causes it to be displaced so that $d = 0.3$ m. When $d = 0.5$ m the collar is at rest.

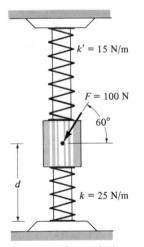

Probs. 14–26/14–27

14–27. The collar has a mass of 20 kg and is supported on the smooth rod. The attached springs are both compressed 0.4 m when $d = 0.5$ m. Determine the speed of the collar after the applied force $F = 100$ N causes it to be displaced so that $d = 0.3$ m. When $d = 0.5$ m the collar is at rest.

*14–28. A rocket of mass m is fired vertically from the surface of the earth, i.e., at $r = r_1$. Assuming that no mass is lost as it travels upward, determine the work it must do against gravity to reach a distance r_2. The force of gravity is $F = GM_e m/r^2$ (Eq. 13–1), where M_e is the mass of the earth and r the distance between the rocket and the center of the earth.

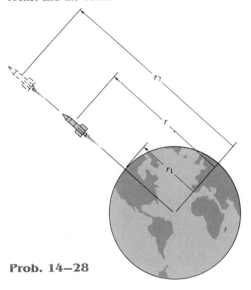

Prob. 14–28

14–29. The 5-lb cylinder is dropped from A with an initial speed of $v_A = 10$ ft/s onto the platform. Determine the maximum displacement of the platform caused by the collision. The spring has an unstretched length of 1.75 ft and is originally kept in compression by the 1-ft long cables attached to the platform. Neglect the mass of the platform and spring and any energy lost during the collision.

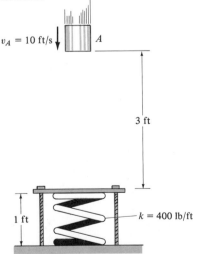

Prob. 14–29

14.4 Power and Efficiency

Power

Power is defined as the amount of work performed per unit of time. Hence, the *power* generated by a machine or engine that performs an amount of work dU within the time interval dt is

$$P = \frac{dU}{dt}$$

(14–9)

Provided the work dU is expressed by $dU = \mathbf{F} \cdot d\mathbf{r}$, then it is also possible to write

$$P = \frac{dU}{dt} = \frac{\mathbf{F} \cdot d\mathbf{r}}{dt}$$

or

$$P = \mathbf{F} \cdot \mathbf{v}$$

(14–10)

Hence, power is a *scalar,* where in the formulation $\mathbf{v}$ represents the velocity of the point which is acted upon by the force $\mathbf{F}$.

The basic units of power used in the SI and FPS systems are the watt (W) and horsepower (hp), respectively. These units are defined as

$$1 \text{ W} = 1 \text{ J/s} = 1 \text{ N} \cdot \text{m/s}$$

$$1 \text{ hp} = 550 \text{ ft} \cdot \text{lb/s}$$

For conversion between the two systems of units, 1 hp = 746 W.

The term "power" provides a useful basis for determining the type of motor or machine which is required to do a certain amount of work in a given time. For example, two pumps may each be able to empty a reservoir if given enough time; however, the pump having the larger power will complete the job sooner.

Efficiency

The *mechanical efficiency* of a machine is defined as the ratio of the output of useful power produced by the machine to the input of power supplied to the machine. Hence,

$$\varepsilon = \frac{\text{power output}}{\text{power input}} \qquad (14\text{–}11)$$

If energy applied to the machine occurs during the *same time interval* at which it is removed, then the efficiency may also be expressed in terms of the ratio of output energy to input energy; i.e.

$$\varepsilon = \frac{\text{energy output}}{\text{energy input}} \qquad (14\text{–}12)$$

Since machines consist of a series of moving parts, frictional forces will always be developed within the machine, and as a result, extra energy or power is needed to overcome these forces. Consequently, *the efficiency of a machine is always less than 1*.

PROCEDURE FOR ANALYSIS

When computing the power supplied to a body, one must first determine the unbalanced external force **F** acting on the body which causes the motion. This force is usually developed by a machine or engine placed either within or external to the body. If the body is accelerating, it may be necessary to draw its free-body diagram and apply the equation of motion ($\Sigma\mathbf{F} = m\mathbf{a}$) to determine **F**. Once **F** and the velocity **v** of the point where **F** is applied have been computed, the power is determined by multiplying the force magnitude by the component of velocity acting in the direction of **F**, i.e., $P = \mathbf{F} \cdot \mathbf{v} = Fv \cos\theta$.

In some problems the power may also be computed by calculating the work done per unit of time ($P_{\text{avg}} = \Delta U/\Delta t$, or $P = dU/dt$). Depending upon the problem, this work can be done either by the external or internal force of a machine or engine, by the weight of the body, or by an elastic spring force acting on the body.

Example 14-7

The motor M of the hoist shown in Fig. 14–15a operates with an efficiency of 0.85. Determine the power that must be supplied to the motor to lift the crate C having a weight of 75 lb. Point P on the cable is being drawn in with an acceleration of 4 ft/s², and at the instant shown its speed is 2 ft/s. Neglect the mass of the pulley and cable.

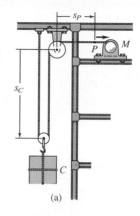

(a)

SOLUTION

In order to compute the power output of the motor, it is first necessary to determine the tension in the cable since this force is developed by the motor.

From the free-body diagram, Fig. 14–15b, we have

$$+\uparrow \Sigma F_y = ma_y; \qquad 2T - 75 = -\frac{75}{32.2}a_C \qquad (1)$$

The acceleration of the crate can be obtained by using kinematics to relate the motion of the crate to the known motion of point P, Fig. 14–15a. Hence, by the methods of Sec. 12.8, the coordinates s_C and s_P in Fig. 14–15a can be related to a constant portion of cable length l which is changing in the vertical and horizontal directions. We have $2s_C + s_P = l$. Taking the second time derivative of this equation yields

$$2a_C = -a_P \qquad (2)$$

(b)

Fig. 14-15

Since $a_P = +4$ ft/s², then $a_C = (-4$ ft/s²$)/2 = -2$ ft/s². Substituting this result into Eq. (1) and *retaining* the negative sign since the acceleration in *both* Eqs. (1) and (2) is considered positive downward, we have

$$2T - 75 = -\frac{75}{32.2}(-2)$$

$$T = 39.8 \text{ lb}$$

The power output, measured in units of horsepower, required to draw the cable in at a rate of 2 ft/s is therefore

$$P = \mathbf{T} \cdot \mathbf{v} = (39.8 \text{ lb})(2 \text{ ft/s})(1 \text{ hp/550 lb} \cdot \text{ft/s})$$

$$= 0.145 \text{ hp}$$

This *power output* requires that the motor provide a *power input* of

$$\text{power input} = \frac{1}{\varepsilon}(\text{power output})$$

$$= \frac{1}{0.85}(0.145) = 0.170 \text{ hp} \qquad \textbf{\textit{Ans.}}$$

Since the velocity of the crate is constantly changing, notice that this power requirement is *instantaneous*.

Example 14–8

The car shown in Fig. 14–16a has a mass of 2 Mg and an engine running efficiency of $\varepsilon = 0.63$. As it moves forward, the wind creates a drag resistance on the car of $F_D = 1.2v^2$ N, where v is the velocity in m/s. If the car is traveling at a speed of 50 m/s, determine the maximum power supplied by the engine.

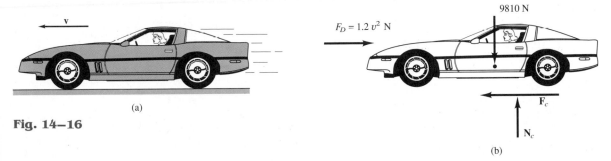

(a)

Fig. 14–16

(b)

SOLUTION

As shown on the free-body diagram, Fig. 14–16b, the normal force N_C and frictional force $\mathbf{F}_C$ represent the *resultant forces* of all four wheels. In particular, the unbalanced frictional force drives or pushes the car *forward*. This effect is, of course, created by the rotating motion of the rear wheels on the pavement and is developed by the power of the engine.

Applying the equation of motion in the x direction, we have

$$\xrightarrow{+} \Sigma F_x = ma_x; \qquad F_C - 1.2v^2 = 2000\ \frac{dv}{dt}$$

Since the car is traveling with *constant velocity*, $dv/dt = 0$. Hence, with $v = 50$ m/s,

$$F_C = 1.2(50)^2 = 3000\ \text{N}$$

The power output of the car is manifested by the driving (frictional) force $\mathbf{F}_C$. Thus

$$P = \mathbf{F}_C \cdot \mathbf{v} = (3000\ \text{N})(50\ \text{m/s}) = 150\ \text{kW}$$

The power supplied by the engine (power input) is therefore

$$\text{power input} = \frac{1}{\varepsilon}\ (\text{power output}) = \frac{1}{0.63}\ (150) = 238\ \text{kW} \quad \textit{Ans.}$$

PROBLEMS

14–30. A boy having a weight of 130 lb can run with constant speed up a staircase that is 20 ft high in 8 s. Determine the average power he produces.

14–31. A train having a weight of $4(10^6)$ lb is traveling at 20 ft/s. If it is brought to a stop, determine how many days a 75-W light bulb must burn to expend the same amount of energy.

*14–32. Determine the power input for a motor necessary to lift 300 lb at a constant rate of 5 ft/s. The efficiency of the motor is $\varepsilon = 0.65$.

14–33. The jeep has a weight of 2500 lb and an engine which transmits a power of 100 hp to *all* the wheels. Assuming the wheels do not slip on the ground, determine the angle θ of the largest incline the jeep can climb at a constant speed $v = 30$ ft/s.

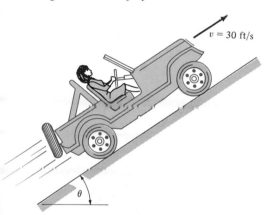

$v = 30$ ft/s

Prob. 14–33

14–34. A motor for a mine hoist lifts a 5000-lb load from a speed of 2 ft/s to a speed of 10 ft/s in $t = 5$ s with constant acceleration. Determine the power developed by the motor when $t = 5$ s.

14–35. The locomotive of a train exerts a constant pull of 400 kN on the cars which have a total mass of 2 Gg. The cars have a total frictional resistance of 6.5 kN and begin traveling at 2 m/s up a slope of 1°. Determine the speed of the cars after they travel 2 km. What is the power output of the locomotive when it has reached this point?

*14–36. The motor M is used to hoist the loaded 500-kg elevator upward with a constant velocity $v_E = 8$ m/s. If the motor draws 60 kW of electrical power, determine the motor's efficiency. Neglect the mass of the pulleys and cable.

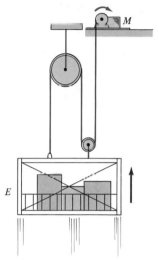

M

E

Probs. 14–36/14–37

14–37. The 500-kg elevator starts from rest and travels upward with a constant acceleration $a_c = 2$ m/s². Determine the power output of the motor M when $t = 3$ s. Neglect the mass of the pulleys and cable.

14–38. The 10-lb collar starts from rest at A and is lifted by applying a constant horizontal force of $F = 25$ lb to the cord. If the rod is smooth, determine the power developed by the force at the instant $\theta = 60°$.

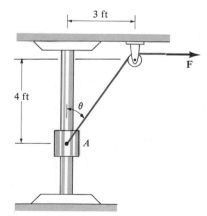

3 ft

F

4 ft

θ

A

Prob. 14–38

14–39. A car has a mass m and accelerates along a horizontal straight road from rest such that the power is always a constant amount P. Determine how far it must travel to reach a speed of v.

*14–40. The 50-lb load is hoisted by the pulley system and motor M. If the crate starts from rest and by constant acceleration attains a speed of 15 ft/s after rising $s = 6$ ft, determine the power that must be supplied to the motor when $s = 6$ ft. The motor has an efficiency of $\varepsilon = 0.76$. Neglect the mass of the pulleys and cable.

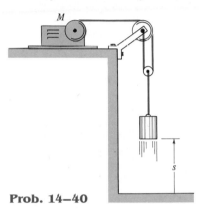

Prob. 14–40

14–41. The crate has a mass of 150 kg and rests on a surface for which the coefficients of static and kinetic friction are $\mu_s = 0.3$ and $\mu_k = 0.2$, respectively. If the motor M supplies a cable force of $F = (8t^2 + 20)$ N, where t is in seconds, determine the power output developed by the motor when $t = 5$ s.

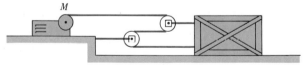

Prob. 14–41

14–42. The rocket has a mass of 3 Mg and is fired from rest. If the engine provides a constant vertical thrust $T = 150$ kN, determine the power output of the engine as a function of time. Neglect the change in weight, loss of fuel mass, and drag due to air resistance, i.e., $F_D = 0$.

Probs. 14–42/14–43

14–43. Solve Prob. 14–42 if the drag due to air resistance is $F_D = (0.06v^2)$ kN, where v is in m/s.

*14–44. The 150-lb crate rests on the rough surface for which the coefficients of static and kinetic friction are $\mu_s = 0.3$ and $\mu_k = 0.25$, respectively. A towing force $F = (50 + 0.3s^2)$ lb, where s is the horizontal displacement in ft, acts on the crate in the direction shown. If the crate is originally at rest, determine the power developed by the force when the crate has moved $s = 3$ ft.

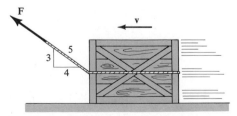

Prob. 14–44

14–45. The block has a weight of 80 lb and rests on the floor for which the coefficient of kinetic friction is $\mu_k = 0.4$. If the motor draws in the cable at a constant rate of 6 ft/s, determine the power output of the motor at the instant $\theta = 30°$. Neglect the mass of the cable and pulleys.

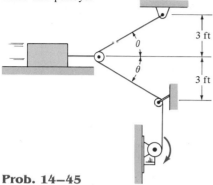

Prob. 14–45

14.5 Conservative Forces and Potential Energy

Conservative Force

A particularly simple type of force acting on a particle is one that depends *only* on the particle's position and is independent of the particle's velocity and acceleration. Furthermore, if the work done by this force in moving the particle from one point to another is *independent of the path* followed by the particle, this force is called a *conservative force*. The weight of a particle and the force of an elastic spring are two examples of conservative forces often encountered in mechanics.

Weight. The work done by the weight of a particle is *independent of the path;* rather, it depends only on the particle's *vertical displacement*. If this displacement is Δy (positive upward), then from Eq. 14–3,

$$U = -W(\Delta y)$$

Elastic Spring. The work done by a spring force *acting on a particle* is *independent of the path* of the particle, but depends only on the extension or compression s of the spring. If the spring is elongated or compressed from position s_1 to a further position s_2, then from Eq. 14–4,

$$U = -(\tfrac{1}{2}ks_2^2 - \tfrac{1}{2}ks_1^2)$$

Friction. In contrast to a conservative force, consider the force of friction exerted *on a moving object* by a fixed surface. The work done by the frictional force *depends upon the path*—the longer the path, the greater the work. Consequently, *frictional forces are nonconservative*. The work is dissipated from the body in the form of heat.

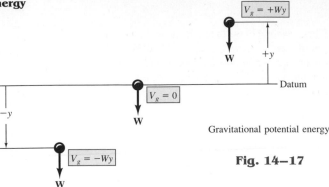

Gravitational potential energy

Fig. 14–17

Potential Energy

Energy may be defined as the capacity for doing work. When energy comes from the *motion* of the particle, it is referred to as *kinetic energy*. When it comes from the *position* of the particle, measured from a fixed datum or reference plane, it is called *potential energy*. Thus, potential energy is a measure of the amount of work a force will do when it moves from a given position to the datum. In mechanics, the potential energy due to both gravity (weight) and an elastic spring is important.

    Gravitational Potential Energy. If a particle is located a distance y *above* an arbitrarily selected datum, as shown in Fig. 14–17, the particle's weight **W** has positive *gravitational potential energy*, V_g, since **W** has the capacity of doing positive work when the particle is moved back down to the datum. Likewise, if the particle is located a distance y *below* the datum, V_g is negative since the weight does negative work when the particle is moved back up to the datum. At the datum $V_g = 0$.

    In general, if y is *positive upward,* the gravitational potential energy of the particle of weight W is thus*

$$V_g = Wy \tag{14–13}$$

    Elastic Potential Energy. When an elastic spring is elongated or compressed a distance s from its unstretched position, the elastic potential energy V_e due to the spring's configuration can be expressed as

$$V_e = +\tfrac{1}{2}ks^2 \tag{14–14}$$

Here V_e is *always positive* since, in the deformed position, the force of the spring has the *capacity* for always doing positive work on the particle when the spring is returned to its unstretched position, Fig. 14–18.

---

*Here the weight is assumed to be *constant*. This assumption is suitable for small differences in elevation Δy. If the elevation change is significant, however, a variation of weight with elevation must be taken into account (see Prob. 14–59).

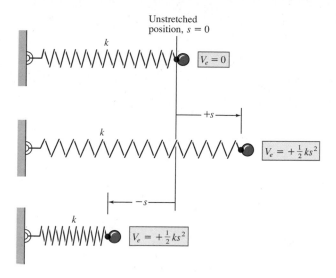

Elastic potential energy

Fig. 14–18

Potential Function

In the general case, if a particle is subjected to both gravitational and elastic forces, the particle's potential energy can be expressed as a *potential function*, which is the algebraic sum

$$V = V_g + V_e \qquad (14\text{--}15)$$

Measurement of V depends upon the location of the particle with respect to a selected datum in accordance with Eqs. 14–13 and 14–14. If the particle is located at an arbitrary point (x, y, z) in space, this potential function is then $V = V(x, y, z)$. The work done by a conservative force in moving the particle from point (x_1, y_1, z_1) to point (x_2, y_2, z_2) is measured by the *difference* of this function, i.e.,

$$U_{1\text{--}2} = V_1 - V_2 \qquad (14\text{--}16)$$

For example, the potential function for a particle of weight W suspended from a spring can be expressed in terms of its position, s, measured from a datum located at the unstretched length of the spring, Fig. 14–19. We have

$$V = V_g + V_e$$
$$= -Ws + \tfrac{1}{2}ks^2$$

If the particle moves from s_1 to a lower position s_2, then applying Eq. 14–16 it can be seen that the work of $\mathbf{W}$ and $\mathbf{F}_s$ is

$$U_{1\text{--}2} = V_1 - V_2 = (-Ws_1 + \tfrac{1}{2}ks_1^2) - (-Ws_2 + \tfrac{1}{2}ks_2^2)$$
$$= W(s_2 - s_1) - (\tfrac{1}{2}ks_2^2 - \tfrac{1}{2}ks_1^2)$$

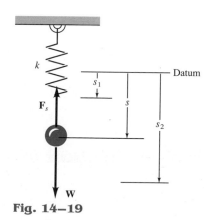

Fig. 14–19

153

When the displacement path is infinitesimal, i.e., from point (x, y, z) to $(x + dx, y + dy, z + dz)$, Eq. 14–16 becomes

$$dU = V(x, y, z) - V(x + dx, y + dy, z + dz)$$
$$= -dV(x, y, z) \qquad (14\text{–}17)$$

Provided both the force and displacement are defined using rectangular coordinates, then the work can also be expressed as

$$dU = \mathbf{F} \cdot d\mathbf{r} = (F_x\mathbf{i} + F_y\mathbf{j} + F_z\mathbf{k}) \cdot (dx\mathbf{i} + dy\mathbf{j} + dz\mathbf{k})$$
$$= F_x dx + F_y dy + F_z dz$$

Substituting this result into Eq. 14–17 and expressing the differential $dV(x, y, z)$ in terms of partial derivatives yields

$$F_x dx + F_y dy + F_z dz = -\left(\frac{\partial V}{\partial x}dx + \frac{\partial V}{\partial y}dy + \frac{\partial V}{\partial z}dz\right)$$

Since changes in x, y, and z are all independent of one another, this equation is satisfied provided

$$F_x = -\frac{\partial V}{\partial x}, \qquad F_y = -\frac{\partial V}{\partial y}, \qquad F_z = -\frac{\partial V}{\partial z} \qquad (14\text{–}18)$$

Thus,

$$\mathbf{F} = -\frac{\partial V}{\partial x}\mathbf{i} - \frac{\partial V}{\partial y}\mathbf{j} - \frac{\partial V}{\partial z}\mathbf{k}$$
$$= -\left(\frac{\partial}{\partial x}\mathbf{i} + \frac{\partial}{\partial y}\mathbf{j} + \frac{\partial}{\partial z}\mathbf{k}\right) V$$

or

$$\boxed{\mathbf{F} = -\nabla V} \qquad (14\text{–}19)$$

where ∇ (del) represents the vector operator $\nabla = (\partial/\partial x)\mathbf{i} + (\partial/\partial y)\mathbf{j} + (\partial/\partial z)\mathbf{k}$.

Equation 14–19 relates a force $\mathbf{F}$ to its potential function V and thereby provides a mathematical criterion for proving that $\mathbf{F}$ is conservative. For example, the gravitational potential function for a weight located a distance y above a datum is $V_g = Wy$, Eq. 14–13. To prove that $\mathbf{W}$ is conservative, it is necessary to show that it satisfies Eq. 14–19 (or Eq. 14–18), in which case

$$F_y = -\frac{\partial V}{\partial y}; \qquad\qquad F = -\frac{\partial}{\partial y}(Wy) = -W$$

The negative sign indicates that $\mathbf{W}$ acts downward, opposite to positive y, which is upward.

14.6 Conservation of Energy

When a particle is acted upon by a system of *both* conservative and nonconservative forces, the portion of the work done by the *conservative forces* can be written in terms of the difference in their potential energies using Eq.

14–16, i.e., $(\Sigma U_{1-2})_{\text{cons.}} = V_1 - V_2$. As a result, the principle of work and energy can be written as

$$T_1 + V_1 + (\Sigma U_{1-2})_{\text{noncons.}} = T_2 + V_2 \qquad (14\text{–}20)$$

Here $(\Sigma U_{1-2})_{\text{noncons.}}$ represents the work of the nonconservative forces acting on the particle. If *only conservative forces* are applied to the body, this term is zero and then we have

$$\boxed{T_1 + V_1 = T_2 + V_2} \qquad (14\text{–}21)$$

This equation is referred to as the *conservation of mechanical energy* or simply the *conservation of energy*. It states that during the motion the sum of the particle's kinetic and potential energies remains *constant*. For this to occur, kinetic energy must be transformed into potential energy, and vice versa. For example, if a ball of weight **W** is dropped from a height h above the ground (datum), Fig. 14–20, the potential energy of the ball is maximum before it is dropped, at which time its kinetic energy is zero. The total mechanical energy of the ball in its initial position is thus

$$E = T_1 + V_1 = 0 + Wh = Wh$$

When the ball has fallen a distance $h/2$, its speed can be determined by using $v^2 = v_0^2 + 2a_c(y - y_0)$, which yields $v = \sqrt{2g(h/2)} = \sqrt{gh}$. The energy of the ball at the mid-height position is therefore

$$E = V_2 + T_2 = W\frac{h}{2} + \frac{1}{2}\frac{W}{g}(\sqrt{gh})^2 = Wh$$

Just before the ball strikes the ground, its potential energy is zero and its speed is $v = \sqrt{2gh}$. Here, again, the total energy of the ball is

$$E = V_3 + T_3 = 0 + \frac{1}{2}\frac{W}{g}(\sqrt{2gh})^2 = Wh$$

Note that when the ball comes in contact with the ground, it deforms somewhat, and provided the ground is hard enough, the ball will rebound off the surface, reaching a new height h', which will be less than the height h from which it was first released. Neglecting air friction, the difference in height accounts for an energy loss, $E_l = W(h - h')$, occurring at the moment of collision. Portions of this loss produce noise, localized deformation of the ball and ground, and heat.

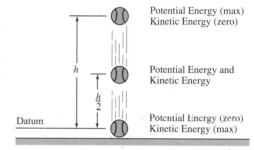

Fig. 14–20

System of Particles

If a system of particles is *subjected only to conservative forces,* then an equation similar to Eq. 14–21 can be written for the particles. Applying the ideas of the preceding discussion, Eq. 14–8 ($\Sigma T_1 + \Sigma U_{1-2} = \Sigma T_2$) becomes

$$\boxed{\Sigma T_1 + \Sigma V_1 = \Sigma T_2 + \Sigma V_2} \qquad\qquad (14\text{--}22)$$

Here, the sum of the system's initial kinetic and potential energy is equal to the sum of the system's final kinetic and potential energy. In other words, the system's kinetic energy and potential energy, caused by *both* the internal and external forces, remain constant; i.e., $\Sigma T + \Sigma V = $ const.

PROCEDURE FOR ANALYSIS

The conservation of energy equation is used to solve problems involving *velocity, displacement,* and *conservative force systems*. It is generally *easier to apply* than the principle of work and energy. This is because the energy equation just requires specifying the particle's kinetic and potential energy at only *two points* along the path, rather than determining the work when the particle moves through a *displacement*. For application it is suggested that the following procedure be used.

Potential Energy.   Draw two diagrams showing the particle located at its initial and final points along the path. If the particle is subjected to a vertical displacement, establish the fixed horizontal datum from which to measure the particle's gravitational potential energy V_g. Although this position can be selected arbitrarily, it is best to locate the datum either at the initial or final point of the path, since at the datum $V_g = 0$. Data pertaining to the elevation y of the particle from the datum and the extension or compression s of any connecting springs can be determined from the geometry associated with the two diagrams. Recall that the potential energy $V = V_g + V_e$. Here $V_g = Wy$, where y is positive upward from the datum, and $V_e = \frac{1}{2}ks^2$, which is always positive.

Conservation of Energy.   Apply the equation $T_1 + V_1 = T_2 + V_2$. When computing the kinetic energy, $T = \frac{1}{2}mv^2$, the particle's speed v must be measured from an inertial reference frame.

It is important to remember that only problems involving conservative force systems may be solved by using the conservation of energy theorem. As stated previously, friction or other drag-resistant forces, which depend upon velocity or acceleration, are nonconservative. A portion of the work done by such forces is transformed into thermal energy used to heat up the surfaces of contact, and consequently this energy dissipates into the surroundings and may not be recovered. Therefore, problems involving frictional forces can either be solved by using the principle of work and energy written in the form of Eq. 14–20, if it applies, or the equation of motion.

The following example problems numerically illustrate application of the procedure described above.

Example 14–9

The boy and his bicycle shown in Fig. 14–21a have a total weight of 125 lb and a center of mass at G. If he is coasting, i.e., not pedaling, with a speed of 10 ft/s at the top of the hill A, determine the normal force exerted on both wheels of the bicycle when he arrives at B, where the radius of curvature of the road is $\rho = 50$ ft. Neglect friction.

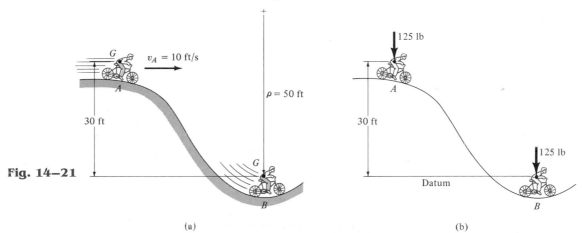

Fig. 14–21

(a)   (b)

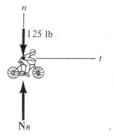

(c)

SOLUTION

Since the normal force does *no work,* it must be obtained using the equation of motion, $\Sigma F_n = m(v^2/\rho)$. We can, however, determine the bicycle's speed at B using the conservation of energy equation. Why?

Potential Energy. Figure 14–21b shows the bicycle at points A and B. For convenience, the potential-energy datum has been established through the center of mass when the bicycle is located at B.

Conservation of Energy

$$\{T_A\} + \{V_A\} = \{T_B\} + \{V_B\}$$

$$\left\{\frac{1}{2}\left(\frac{125}{32.2}\right)(10)^2\right\} + \{(125)(30)\} = \left\{\frac{1}{2}\left(\frac{125}{32.2}\right)(v_B)^2\right\} + \{0\}$$

$$v_B = 45.1 \text{ ft/s}$$

Equation of Motion. Using the data tabulated on the free-body diagram when the bicycle is at B, Fig. 14–21c, we have

$$+\uparrow \Sigma F_n = ma_n; \qquad N_B - 125 = \frac{125}{32.2}\frac{(45.1)^2}{50}$$

$$N_B = 283 \text{ lb} \qquad\qquad \textbf{\textit{Ans.}}$$

Example 14–10

The ram R shown in Fig. 14–22a has a mass of 100 kg and is released from rest 0.75 m from the top of a spring A that has a stiffness $k_A = $ 12 kN/m. If a second spring B, having a stiffness $k_B = 15$ kN/m, is "nested" in A, determine the maximum deflection of A needed to stop the downward motion of the ram. The unstretched length of each spring is indicated in the figure. Neglect the mass of the springs.

SOLUTION

Potential Energy. We will *assume* that the ram compresses *both* springs at the instant it comes to rest. The datum is located through the center of gravity of the ram at its initial position, Fig. 14–22b. When the kinetic energy is reduced to zero ($v_2 = 0$), A is compressed a distance s_A so that B compresses $s_B = s_A - 0.1$ m.

Conservation of Energy

$$T_1 + V_1 = T_3 + V_3$$

$$\{0\} + \{0\} = \{0\} + \{\tfrac{1}{2}k_A s_A^2 + \tfrac{1}{2}k_B(s_A - 0.1)^2 - Wh\}$$

$$\{0\} + \{0\} = \{0\} + \left\{\frac{1}{2}(12000)s_A^2 + \frac{1}{2}(15000)(s_A - 0.1)^2 - 981(0.75 + s_A)\right\}$$

Rearranging the terms,

$$13500s_A^2 - 2481s_A - 660.75 = 0$$

Using the quadratic formula, and solving for the positive root,* we have

$$s_A = 0.331 \text{ m} \qquad\qquad \textbf{\textit{Ans.}}$$

Since $s_B = 0.331 - 0.1 = 0.231$ m, which is positive, indeed the assumption that *both* springs are compressed by the ram is correct.

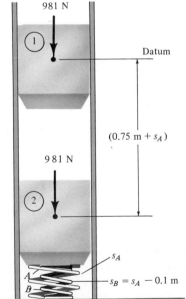

(a)

(b)

Fig. 14–22

---

*The second root, $s_A = -0.148$ m, does not represent the physical situation. Since positive s is measured downward, the negative sign indicates that spring A would have to be "extended" by an amount of 0.148 m to stop the ram.

Example 14–11

A smooth 2-kg collar *C*, shown in Fig. 14–23*a*, fits loosely on the vertical shaft. If the spring is unstretched when the collar is in the dashed position *A*, determine the speed at which the collar is moving when *y* = 1 m, if (a) it is released from rest at *A*, and (b) it is released at *A* with an *upward* velocity v_A = 2 m/s.

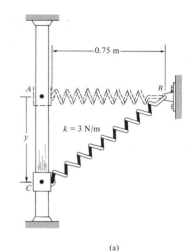

(a)

SOLUTION
Part (a)
Potential Energy. For convenience, the datum is established through *AB*, Fig. 14–23*b*. When the collar is at *C*, the gravitational potential energy is $-(mg)y$, since the collar is *below* the datum, and the elastic potential energy is $\frac{1}{2}ks_{CB}^2$. Here s_{CB} = 0.5 m, which represents the *stretch* in the spring as computed in the figure.

Conservation of Energy

$$\{T_A\} + \{V_A\} = \{T_C\} + \{V_C\}$$

$$\{0\} + \{0\} = \{\tfrac{1}{2}mv_C^2\} + \{\tfrac{1}{2}ks_{CB}^2 - mgy\}$$

$$\{0\} + \{0\} = \left\{\frac{1}{2}(2)v_C^2\right\} + \left\{\frac{1}{2}(3)(0.5)^2 - 2(9.81)(1)\right\}$$

$$v_C = 4.39 \text{ m/s} \downarrow \qquad\qquad \textit{Ans.}$$

This problem can also be solved by using the equation of motion or the principle of work and energy. Note that in *both* of these methods the variation of the magnitude and direction of the spring force must be taken into account (see Example 13–4). Here, however, the above method of solution is clearly advantageous since the calculations depend *only* on data calculated at the initial and final points of the path.

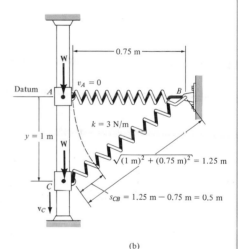

(b)

Fig. 14–23

Part (b)
Conservation of Energy. If v_A = 2 m/s, using the data in Fig. 14–23*b*, we have

$$\{T_A\} + \{V_A\} = \{T_C\} + \{V_C\}$$

$$\{\tfrac{1}{2}mv_A^2\} + \{0\} = \{\tfrac{1}{2}mv_C^2\} + \{\tfrac{1}{2}ks_{CB}^2 - mgy\}$$

$$\left\{\frac{1}{2}(2)(2)^2\right\} + \{0\} = \left\{\frac{1}{2}(2)v_C^2\right\} + \left\{\frac{1}{2}(3)(0.5)^2 - 2(9.81)(1)\right\}$$

$$v_C = 4.82 \text{ m/s} \downarrow \qquad\qquad \textit{Ans.}$$

Note that the kinetic energy of the collar depends only on the *magnitude* of velocity, and therefore it is immaterial if the collar is moving up or down at 2 m/s when released at *A*.

PROBLEMS

14–46. Solve Prob. 14–8 using the conservation of energy equation.

14–47. Solve Prob. 14–10 using the conservation of energy equation.

*14–48. Solve Prob. 14–20 using the conservation of energy equation.

14–49. Solve Prob. 14–14 using the conservation of energy equation.

14–50. Solve Prob. 14–15 using the conservation of energy equation.

14–51. Solve Prob. 14–25 using the conservation of energy equation.

*14–52. The 20-lb collar is constrained to move on the smooth rod. It is attached to the three springs which are unstretched when $s = 0$. If the collar is displaced $s = 0.5$ ft and released from rest, determine its speed when $s = 0$.

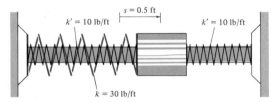

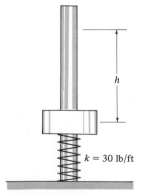

Prob. 14–52

14–53. The collar has a weight of 8 lb. If it is pushed down so as to compress the spring 2 ft and then released from rest ($h = 0$), determine its speed when it is displaced $h = 4.5$ ft. The spring is not attached to the collar. Neglect friction.

Prob. 14–53

14–54. The ball has a weight of 3 lb and is released from rest when $\theta = 0°$. Determine its speed just after the cord strikes the peg at B. Also, what is the speed of the ball and the tension in the cord when the ball reaches its lowest point C as shown?

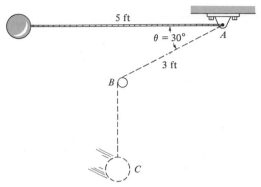

Prob. 14–54

14–55. Each of the two elastic rubberbands of the slingshot has an unstretched length of 200 mm. If it is pulled back to the position shown and released from rest, determine the speed of the 25-g pellet just after the rubberbands become unstretched. Neglect the mass of the rubberbands and the change in elevation of the pellet while it is constrained by the rubberbands. Each rubberband has a stiffness of $k = 50$ N/m.

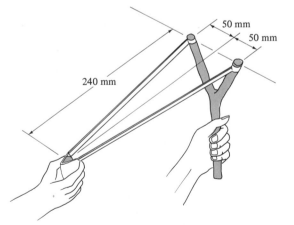

Prob. 14–55

*14–56. The ball has a weight of 15 lb and is fixed to a rod having a negligible mass. If it is released from rest when $\theta = 0°$, determine the angle θ at which the compressive force in the rod becomes zero.

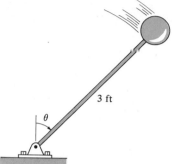

3 ft

θ

Probs. 14–56/14–57

14–57. The ball has a weight of 15 lb and is fixed to a rod having a negligible mass. If it is given a speed of 2 ft/s when $\theta = 0°$, determine the angle θ when the compressive force in the rod becomes 4 lb.

14–58. The cylinder has a mass of 20 kg and is released from rest when $h = 0$. Determine its speed when $h = 3$ m. The springs each have an unstretched length of 2 m.

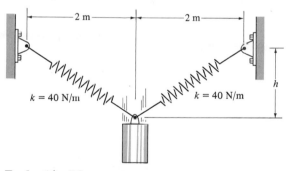

2 m   2 m

$k = 40$ N/m $k = 40$ N/m

h

Prob. 14–58

14–59. If the mass of the earth is M_e, show that the gravitational potential energy of a body of mass m located a distance r from the center of the earth is $V_g = -GM_em/r$. Recall that the gravitational force acting between the earth and the body is $F = G(M_em/r^2)$, Eq. 13–1. For the calculation, locate the datum at $r \rightarrow \infty$. Also, prove that $\mathbf{F}$ is a conservative force.

*14–60. A 60-kg satellite is traveling in free flight along an elliptical orbit such that at A, where $r_A = 20$ Mm, it has a speed $v_A = 40$ Mm/h. What is the speed of the satellite when it reaches point B, where $r_B = 80$ Mm? *Hint:* See Prob. 14–59, where $M_e = 5.976(10^{24})$ kg and $G = 66.73(10^{-12})$ m³/(kg · s²).

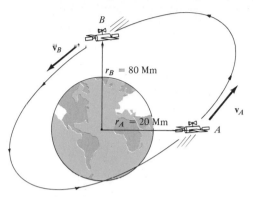

B

v_B

$r_B = 80$ Mm

v_A

$r_A = 20$ Mm A

Prob. 14–60

14–61. The assembly consists of two blocks A and B which have a mass of 20 kg and 30 kg, respectively. Determine the speed of each block when B descends 1.5 m. The blocks are released from rest. Neglect the mass of the pulleys and cords.

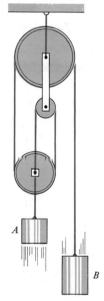

A

B

Prob. 14–61

14–62. Four inelastic cables C are attached to a plate P and hold the 1-ft-long spring 0.25 ft in compression when *no weight* is on the plate. There is also an undeformed spring nested within this compressed spring. If the block, having a weight of 10 lb, is moving downward at $v = 4$ ft/s, when it is 2 ft above the plate, determine the maximum compression in each spring after it strikes the plate. Neglect the mass of the plate and spring and any energy lost in the collision.

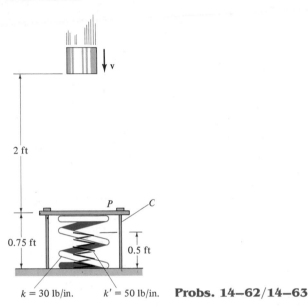

$k = 30$ lb/in. $k' = 50$ lb/in.   **Probs. 14–62/14–63**

14–63. Four inelastic cables C are attached to a plate P and hold the 1-ft-long spring 0.25 ft in compression when *no weight* is on the plate. There is also an undeformed spring nested within this compressed spring. Determine the speed v of the 10-lb block when it is 2 ft above the plate, so that after it strikes the plate, it compresses the spring having a stiffness of 50 lb/in. an amount of 0.20 ft. Neglect the mass of the plate and spring and any energy lost in the collision.

*14–64. The ride at an amusement park consists of a gondola which is lifted to a height of 120 ft at A. If it is released from rest and falls along the parabolic track, determine its speed at the instant $y = 20$ ft. Also determine the normal reaction of the tracks on the gondola at this instant. The gondola and passenger have a total weight of 500 lb. Neglect the effects of friction, the mass of the wheels and the size of the gondola.

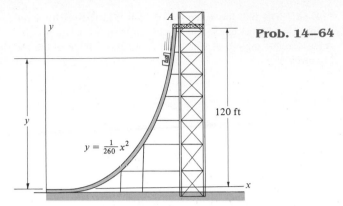

Prob. 14–64

$$y = \frac{1}{260}x^2$$

120 ft

14–65. The spring has a stiffness $k = 3$ lb/ft and an unstretched length of 2 ft. If it is attached to the 5-lb smooth collar and the collar is released from rest at A, determine the speed of the collar just before it strikes the end of the rod at B. Neglect the size of the collar.

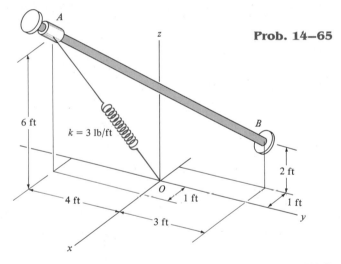

Prob. 14–65

$k = 3$ lb/ft

6 ft

4 ft

3 ft

1 ft

1 ft

2 ft

14–66. The double-spring bumper is used to stop the 1500-lb steel billet in the rolling mill. Determine the maximum deflection of the plate A caused by the billet if it strikes the plate with a speed of 8 ft/s. Neglect the mass of the rollers, the springs, and the plates A and B, and the energy lost during the collision.

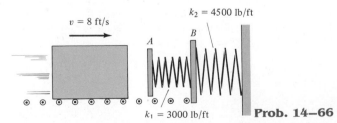

$v = 8$ ft/s

$k_2 = 4500$ lb/ft

$k_1 = 3000$ lb/ft

Prob. 14–66

15

Kinetics of a Particle: Impulse and Momentum

In this chapter we will integrate the equation of motion with respect to time and thereby obtain the principle of impulse and momentum. This principle is useful for solving problems involving force, velocity, and time. It will be shown that impulse and momentum principles also provide the necessary means for analyzing problems of impact, steady fluid flow, and systems which gain or lose mass.

15.1 Principle of Linear Impulse and Momentum

The equation of motion for a particle of mass m can be written as

$$\Sigma \mathbf{F} = m\mathbf{a} = m\frac{d\mathbf{v}}{dt} \qquad (15\text{--}1)$$

where $\mathbf{a}$ and $\mathbf{v}$ are both measured from an inertial frame of reference. Rearranging the terms and integrating between the limits $\mathbf{v} = \mathbf{v}_1$ at $t = t_1$ and $\mathbf{v} = \mathbf{v}_2$ at $t = t_2$, we have

$$\Sigma \int_{t_1}^{t_2} \mathbf{F} \, dt = m \int_{\mathbf{v}_1}^{\mathbf{v}_2} d\mathbf{v}$$

or

$$\Sigma \int_{t_1}^{t_2} \mathbf{F} \, dt = m\mathbf{v}_2 - m\mathbf{v}_1 \qquad (15\text{--}2)$$

This equation is referred to as the *principle of linear impulse and momentum*. From the derivation it can be seen that it is simply a time integration of the equation of motion. It provides a *direct means* of obtaining the particle's final velocity $\mathbf{v}_2$ after a specified time period when the particle's initial velocity is known and the forces acting on the particle are either constant or can be expressed as functions of time. Notice that if $\mathbf{v}_2$ is determined using the equation of motion, a two-step process is necessary; i.e., apply $\Sigma\mathbf{F} = m\mathbf{a}$ to obtain $\mathbf{a}$, then integrate $\mathbf{a} = d\mathbf{v}/dt$ to obtain $\mathbf{v}_2$.

Linear Impulse

The integral $\mathbf{I} = \int \mathbf{F} \, dt$ in Eq. 15–2 is defined as the *linear impulse*. This term is a vector quantity which measures the effect of a force during the time the force acts. The impulse vector acts in the same direction as the force, and its magnitude has units of force–time, e.g., N · s or lb · s. If the force is expressed as a function of time, the impulse may be determined by direct evaluation of the integral. In particular, if $\mathbf{F}$ acts in a *constant direction* during the time period t_1 to t_2, the magnitude of the impulse $\mathbf{I} = \int_{t_1}^{t_2} \mathbf{F} \, dt$ can be represented experimentally by the shaded area under the curve of force versus time, Fig. 15–1. However, if the force is constant in magnitude and direction, the resulting impulse becomes $\mathbf{I} = \int_{t_1}^{t_2} \mathbf{F}_c \, dt = \mathbf{F}_c(t_2 - t_1)$, which represents the shaded rectangular area shown in Fig. 15–2.

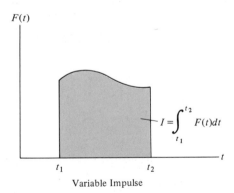

Variable Impulse

Fig. 15–1

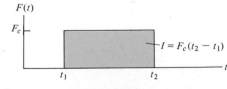

Fig. 15–2   Constant Impulse

Linear Momentum

Each of the two vectors of the form $\mathbf{L} = m\mathbf{v}$ in Eq. 15–2 is defined as the *linear momentum* of the particle. Since m is a scalar, the linear-momentum vector has the same direction as $\mathbf{v}$, and its magnitude mv has units of mass–velocity, e.g., kg · m/s, slug · ft/s.*

Principle of Linear Impulse and Momentum

For problem solving, Eq. 15–2 will be rewritten in the form

$$mv_1 + \Sigma \int_{t_1}^{t_2} F\, dt = mv_2 \tag{15–3}$$

which states that the initial momentum of the particle at t_1 plus the vector sum of all the impulses applied to the particle during the time interval t_1 to t_2 is equivalent to the final momentum of the particle at t_2. These three terms are illustrated graphically on the *impulse and momentum diagrams* shown in Fig. 15–3. Each of these diagrams graphically accounts for all the vectors in the equation $m\mathbf{v}_1 + \Sigma \int_{t_1}^{t_2} \mathbf{F}\, dt = m\mathbf{v}_2$. The two *momentum diagrams* are simply outlined shapes of the particle which indicate the direction and magnitude of the particle's initial and final momenta, $m\mathbf{v}_1$ and $m\mathbf{v}_2$, respectively, Fig. 15–3. Similar to the free-body diagram, the *impulse diagram* is an outlined shape of the particle showing all the impulses that act on the particle when it is located at some intermediate point along its path. In general, whenever the magnitude or direction of a force *varies*, the impulse of the force is determined by integration and represented on the impulse diagram as $\mathbf{I} = \int_{t_1}^{t_2} \mathbf{F}\, dt$. If the force is *constant* for the time interval $(t_2 - t_1)$, the impulse applied to the particle is $\mathbf{I} = \mathbf{F}_c(t_2 - t_1)$, acting in the same direction as $\mathbf{F}_c$.

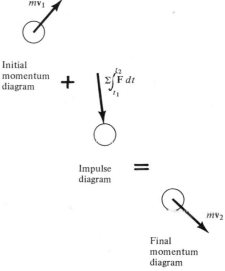

Fig. 15–3

Scalar Equations

If each of the vectors in Eq. 15–3 or on Fig. 15–3 is resolved into its x, y, and z components, we can write symbolically the following three scalar equations:

$$m(v_x)_1 + \Sigma \int_{t_1}^{t_2} F_x\, dt = m(v_x)_2$$

$$m(v_y)_1 + \Sigma \int_{t_1}^{t_2} F_y\, dt = m(v_y)_2 \tag{15–4}$$

$$m(v_z)_1 + \Sigma \int_{t_1}^{t_2} F_z\, dt = m(v_z)_2$$

These equations represent the principle of linear impulse and momentum for the particle in the x, y, and z directions, respectively.

*Although the units for impulse and momentum are defined differently, show that Eq. 15–2 is dimensionally homogeneous.

PROCEDURE FOR ANALYSIS

The principle of linear impulse and momentum is used to solve problems involving *force, time,* and *velocity,* since these terms are involved in the formulation. For application it is suggested that the following procedure be used.

Free-Body Diagram. Establish the x, y, z inertial frame of reference and draw the particle's free-body diagram in order to account for all the forces that produce impulses on the particle. The direction and sense of the particle's initial and final velocities should also be established. If any of these are unknown, assume that the sense of its components is in the direction of the positive inertial coordinate(s).

As an alternative procedure draw the impulse and momentum diagrams for the particle as discussed in reference to Fig. 15–3.*

Principle of Impulse and Momentum. Apply the principle of linear impulse and momentum, $m\mathbf{v}_1 + \Sigma \int_{t_1}^{t_2} \mathbf{F}\, dt = m\mathbf{v}_2$. If motion occurs in the x-y plane, the two scalar component equations can be formulated by either resolving the vector components of $\mathbf{F}$ from the free-body diagram, or by using the data on the impulse and momentum diagrams.

If the problem involves the dependent motion of several particles, use the method outlined in Sec. 12.8 to relate their velocities. Make sure the positive coordinate directions used for writing these kinematic equations are the *same* as those used for writing the equations of impulse and momentum.

The following examples numerically illustrate application of this procedure.

*This procedure will be followed when developing the proofs and theory in the text.

Example 15–1

The 100-kg crate shown in Fig. 15–4a is originally at rest on the smooth horizontal surface. If a force of 200 N, acting at an angle of 45°, is applied to the crate for 10 s, determine the final velocity of the crate and the normal force which the surface exerts on the crate during the time interval.

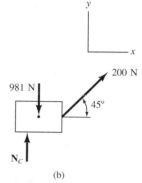

(a)

SOLUTION

This problem can be solved using the principle of impulse and momentum since it involves force, velocity, and time.

Free-Body Diagram. This diagram is shown in Fig. 15–4b. Here it has been assumed that during the motion the crate remains on the surface and after 10 s the crate moves to the right with a velocity v_2. Since all the forces acting on the crate are *constant*, the respective impulses are simply the product of the force magnitude and 10 s ($\mathbf{I} = \mathbf{F}_c(t_2 - t_1)$).

Principle of Impulse and Momentum. Resolving the vectors in Fig. 15–4b along the x,y axes, and applying Eqs. 15–4, yields

$(\xrightarrow{+})$

$$m(v_x)_1 + \Sigma \int_{t_1}^{t_2} F_x\, dt = m(v_x)_2$$

$$0 + 200(10)\cos 45° = 100v_2$$

$$v_2 = 14.1 \quad \text{m/s} \rightarrow \qquad \textit{Ans.}$$

$(+\uparrow)$

$$m(v_y)_1 + \Sigma \int_{t_1}^{t_2} F_y\, dt = m(v_y)_2$$

$$0 + N_C(10) - 981(10) + 200(10)\sin 45° = 0$$

$$N_C = 840 \text{ N} \qquad \textit{Ans.}$$

Since no motion occurs in the y direction, direct application of $\Sigma F_y = 0$ gives the same result for N_C.

Note the alternative procedure of drawing the crate's impulse and momentum diagrams, Fig. 15–4c, prior to applying Eqs. 15–4.

Fig. 15–4

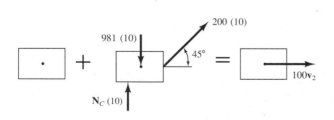

Example 15–2

The crate shown in Fig. 15–5a has a weight of 50 lb and is acted upon by a force having a variable magnitude $P = (20t)$ lb, where t is in seconds. Compute the crate's velocity 2 s after **P** has been applied. The crate has an initial velocity $v_1 = 3$ ft/s down the plane, and the coefficient of kinetic friction between the crate and the plane is $\mu_k = 0.3$.

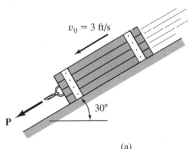

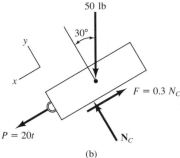

Fig. 15–5

(a)   (b)

SOLUTION

Free-Body Diagram. Fig. 15–5b. Since the magnitude of force $P = 20t$ *varies* with time, the impulse it creates must be determined by *integrating* over the 2-s time interval. The weight, normal force, and frictional force (which acts opposite to the direction of motion) are all *constant*, so that the impulse created by each of these forces is simply the magnitude of the force times 2 s.

Principle of Impulse and Momentum. Applying Eqs. 15–4 in the x direction, we have

$$(+\swarrow) \qquad m(v_x)_1 + \Sigma \int_{t_1}^{t_2} F_x \, dt = m(v_x)_2$$

$$\frac{50}{32.2}(3) + \int_0^2 20t \, dt - 0.3N_C(2) + (50)(2) \sin 30° = \frac{50}{32.2}v_2$$

$$4.66 + 40 - 0.6N_C + 50 = 1.55v_2$$

The equation of equilibrium can be applied in the y direction. Why?

$$+\nwarrow \Sigma F_y = 0; \qquad N_C - 50 \cos 30° = 0$$

Solving,

$$N_C = 43.3 \text{ lb} \qquad v_2 = 44.2 \text{ ft/s} \swarrow \qquad \textbf{\textit{Ans.}}$$

This problem has also been solved using the equation of motion in Example 13–3. The two methods of solution should be compared. Since *force, velocity,* and *time* are involved in the problem, application of the principle of impulse and momentum eliminates the need for using kinematics ($a = dv/dt$) and thereby yields an easier method for solution.

Example 15-3

Blocks A and B shown in Fig. 15–6a have a mass of 3 kg and 5 kg, respectively. If the system is released from rest, determine the velocity of block B in 6 s. Neglect the mass of the pulleys and cord.

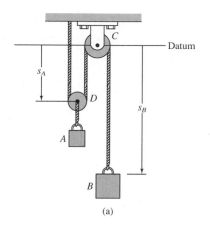

(a)

SOLUTION

Free-Body Diagram. Fig. 15–6b. Since the impulses of the blocks' weights are constant, the impulses of the cord tensions will also be constant. Furthermore, since the mass of pulley D is neglected, the cord tension $T_A = 2T_B$, Fig. 15–6c.

Principle of Impulse and Momentum
Block A:

$$(+\downarrow) \qquad m(v_A)_1 + \Sigma \int_{t_1}^{t_2} F_y \, dt = m(v_A)_2$$

$$0 - 2T_B(6) + 3(9.81)(6) = 3(v_A)_2 \qquad (1)$$

Block B:

$$(+\downarrow) \qquad m(v_B)_1 + \Sigma \int_{t_1}^{t_2} F_y \, dt = m(v_B)_2$$

$$0 + 5(9.81)(6) - T_B(6) = 5(v_B)_2 \qquad (2)$$

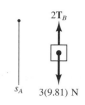

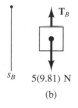

(b)

Kinematics. Since the blocks are subjected to dependent motion, the velocity of A may be related to that of B by using the kinematic analysis discussed in Sec. 12.8. A horizontal datum is established through the fixed point at C, Fig. 15–6a, and the changing positions of the blocks, s_A and s_B, are related to the constant total length l of the vertical segments of the cord by the equation

$$2s_A + s_B = l$$

Taking the time derivative yields

$$2v_A = -v_B \qquad (3)$$

As indicated by the negative sign, when B moves downward A moves upward.* Substituting this result into Eq. (1) and solving Eqs. (1) and (2) yields

$$(v_B)_2 = 35.8 \text{ m/s} \downarrow \qquad \textbf{\textit{Ans.}}$$

$$T_B = 19.2 \text{ N}$$

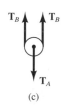

(c)

Fig. 15–6

*Note that the *positive* (downward) directions for $(\mathbf{v}_A)_2$ and $(\mathbf{v}_B)_2$ are *consistent* in Fig. 15–6a and b and in Eqs. (1) to (3). Why is this important?

PROBLEMS

15–1. A ball has a mass of 30 kg and is thrown upwards with a speed of 15 m/s. Determine how long it takes for it to stop. Also, how high does it rise before stopping? Use the principle of impulse and momentum for the solution.

15–2. A 2-lb ball is thrown in the direction shown with an initial speed of $v_A = 18$ ft/s. Determine the time needed for it to reach its highest point B and the speed at which it is traveling at B. Use the principle of impulse and momentum for the solution.

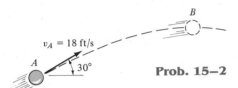

$v_A = 18$ ft/s

B

A

$30°$

Prob. 15–2

15–3. A 40-g golf ball is struck in 3 ms by a driver such that it is given a velocity of $v_1 = 35$ m/s directed 30° from the horizontal. Determine the average force exerted on the ball and the momentum of the ball in $t = 1$ s after it leaves the ground.

*15–4. The 5-kg block is moving downward at $v_1 = 2$ m/s when it is 8 m from the sandy surface. Determine the impulse of the sand on the block necessary to stop its motion. Neglect the distance the block dents into the sand and assume the block does not rebound. Take the weight of the block to be nonimpulsive during the impact.

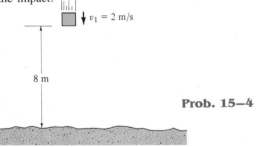

$v_1 = 2$ m/s

8 m

Prob. 15–4

15–5. A tankcar has a mass of 20 Mg and is freely rolling to the right with a speed of 0.75 m/s. If it strikes the barrier, determine the horizontal impulse needed to stop the car if the spring in the bumper B has a stiffness (a) $k \to \infty$ (bumper is rigid), and (b) $k = 15$ kN/m.

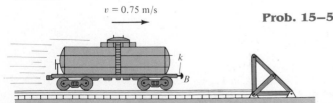

$v = 0.75$ m/s

k

B

Prob. 15–5

15–6. A man kicks the 200-g ball such that it leaves the ground at an angle of 30° with the horizontal and strikes the ground at the same elevation a distance of 15 m away. Determine the impulse of his foot F on the ball. Neglect the impulse caused by the ball's weight while it is being kicked.

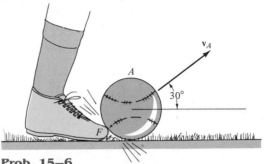

v_A

A

$30°$

F

Prob. 15–6

15–7. If it takes 35 s for the 50-Mg towboat to increase its speed uniformly to 25 km/h, starting from rest, determine the force developed on the rope between the towboat and the barge. The propeller provides the propulsion force F which gives the towboat forward motion, whereas the barge moves freely. Also, determine F acting on the towboat. The barge has a mass of 75 Mg.

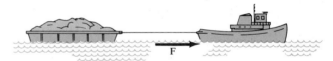

F

Prob. 15–7

*15–8. The tennis ball has a horizontal speed of 15 m/s when it is struck by the racket. If it then travels away at an angle of 25° from the horizontal and reaches a maximum altitude of 10 m, measured from the height of the racket, determine the magnitude of the net impulse of the racket on the ball. The ball has a mass of 180 g. Neglect the weight of the ball during the time the racket strikes the ball.

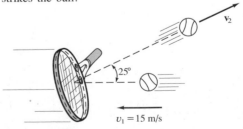

v_2

$25°$

$v_1 = 15$ m/s

Prob. 15–8

15–9. As indicated by the derivation, the principle of impulse and momentum is valid for observers in *any* inertial reference frame. Show that this is so, by considering the 10-kg block which rests on the smooth surface and is subjected to a horizontal force of 6 N. If observer A is in a *fixed* frame x, determine the final speed of the block in 4 s if it has an initial speed of 5 m/s measured from the fixed frame. Compare the result with that obtained by an observer B, attached to the x' axis and moving at a constant velocity of 2 m/s relative to A.

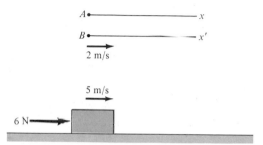

Prob. 15–9

15–10. The rocket sled has a mass of 3 Mg and starts from rest when t = 0. If the engines provide a horizontal thrust T which varies as shown in the graph, determine the sled's velocity in t = 4 s. Neglect air resistance, friction, and the loss of fuel during the motion.

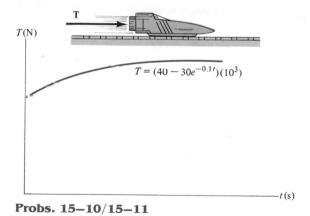

Probs. 15–10/15–11

15–11. The rocket sled has a mass of 3 Mg and starts from rest when t = 0. If the engines provide a horizontal thrust T, which varies as shown in the graph, determine the time needed for the sled to obtain a speed of 18 m/s. Neglect air resistance, friction, and the loss of fuel during the motion.

*15–12. The uniform beam has a weight of 5000 lb. Determine the average tension in each of the two cables AB and AC if the beam is given an upward speed of 8 ft/s in 1.5 s starting from rest. Neglect the mass of the cables.

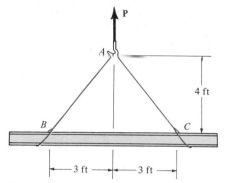

Prob. 15–12

15–13. Block A weighs 10 lb and block B weighs 3 lb. If B is moving downward with a velocity of $(v_B)_1 = 3$ ft/s at t = 0, determine the velocity of A when t = 1 s. Assume that the horizontal plane is smooth. Neglect the mass of the pulleys and cord.

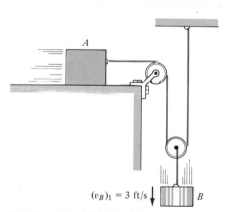

Probs. 15–13/15–14

15–14. Solve Prob. 15–13 if the coefficient of kinetic friction between the horizontal plane and block A is $\mu_A = 0.15$.

171

15–15. The block B has a mass of 70 kg. The winch delivers a horizontal towing force **F** to its cable at A which varies as shown in the graph. Determine the speed of the block when $t = 18$ s. Originally the block is moving upwards at $v_1 = 3$ m/s.

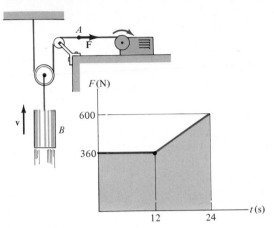

Prob. 15–15

15–18. The motor pulls on the cable at A with a force $F = (30 + t^2)$ lb, where t is in seconds. If the 34-lb crate is originally at rest at $t = 0$, determine its speed in $t = 4$ s. Neglect the mass of the cable and pulleys. *Hint:* First find the time needed to begin lifting the crate.

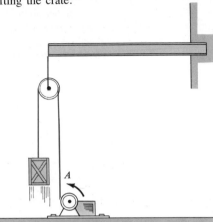

Prob. 15–18

*15–16. The force acting on the 50-lb crate has a magnitude of $F = (2.4t^2 + 15t)$ lb, where t is in seconds. If the crate starts from rest, determine its speed when $t = 2$ s. Also determine the distance the crate moves in 2 s. Neglect friction.

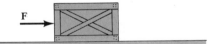

Probs. 15–16/15–17

15–17. The force acting on the 50-lb crate has a magnitude of $F = (2.4t^2)$ lb, where t is in seconds. If the crate starts from rest, determine its speed when $t = 5$ s. The coefficients of static and kinetic friction between the crate and floor are $\mu_s = 0.3$ and $\mu_k = 0.2$, respectively.

15–19. The 50-kg block B is hoisted using the cable and motor arrangement shown. If the block is moving upwards at $v_1 = 2$ m/s when $t = 0$, and the motor develops a tension in the cord of $T = (500 + 120\sqrt{t})$ N, where t is in seconds, determine the velocity of the block when $t = 2$ s. Neglect the mass of the pulleys and cable.

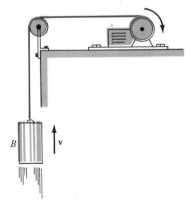

Prob. 15–19

15.2 Principle of Linear Impulse and Momentum for a System of Particles

The principle of linear impulse and momentum for a system of particles moving relative to an inertial reference, Fig. 15–7, may be obtained from the equation of motion $\Sigma \mathbf{F}_i = \Sigma m_i \mathbf{a}_i$, Eq. 13–5, which may be rewritten as

$$\Sigma \mathbf{F}_i = \Sigma m_i \frac{d\mathbf{v}_i}{dt} \qquad (15\text{–}5)$$

The term on the left side represents only the sum of all the *external forces* acting on the system of particles. Recall that the internal forces $\mathbf{f}_i$ between particles do not appear with this summation, since by Newton's third law they occur in equal but opposite collinear pairs and therefore cancel out. Multiplying both sides of Eq. 15–5 by dt and integrating between the limits $t = t_1$, $\mathbf{v}_i = (\mathbf{v}_i)_1$ and $t = t_2$, $\mathbf{v}_i = (\mathbf{v}_i)_2$ yields

$$\Sigma m_i(\mathbf{v}_i)_1 + \Sigma \int_{t_1}^{t_2} \mathbf{F}_i \, dt = \Sigma m_i(\mathbf{v}_i)_2 \qquad (15\text{–}6)$$

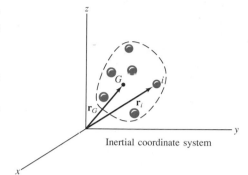

Inertial coordinate system

Fig. 15–7

This equation states that the initial linear momenta of the system added vectorially to the impulses of all the *external forces* acting on the system during the time period t_1 to t_2 are equal to the system's final linear momenta.

By definition, the location of the mass center G of the system is determined from $m\mathbf{r}_G = \Sigma m_i \mathbf{r}_i$, where $m = \Sigma m_i$ is the total mass of all the particles, and $\mathbf{r}_G$ and $\mathbf{r}_i$ are defined in Fig. 15–7. Taking the time derivatives, we have

$$m\mathbf{v}_G = \Sigma m_i \mathbf{v}_i$$

which states that the total linear momentum of the system of particles is equivalent to the linear momentum of a "fictitious" aggregate particle of mass $m = \Sigma m_i$ moving with the velocity of the mass center G of the system. Substituting into Eq. 15–6 yields

$$m(\mathbf{v}_G)_1 + \Sigma \int_{t_1}^{t_2} \mathbf{F} \, dt = m(\mathbf{v}_G)_2 \qquad (15\text{–}7)$$

This equation states that the initial linear momentum of the aggregate particle plus (vectorially) the external impulses acting on the system of particles during the time interval t_1 to t_2 is equal to the aggregate particle's final linear momentum. Since in reality all particles must have finite size to possess mass, the above equation justifies application of the principle of linear impulse and momentum to a rigid body represented as a single particle.

15.3 Conservation of Linear Momentum for a System of Particles

When the sum of the *external impulses* acting on a system of particles is *zero*, Eq. 15–6 reduces to a simplified form, namely,

$$\Sigma m_i(\mathbf{v}_i)_1 = \Sigma m_i(\mathbf{v}_i)_2 \qquad (15\text{–}8)$$

This equation is referred to as the *conservation of linear momentum*. It states that the vector sum of the linear momenta for a system of particles remains constant throughout the time period t_1 to t_2. Since $m\mathbf{v}_G = \Sigma m_i\mathbf{v}_i$, we can also write

$$(\mathbf{v}_G)_1 = (\mathbf{v}_G)_2 \qquad (15\text{–}9)$$

which indicates that the velocity $\mathbf{v}_G$ of the mass center for the system of particles does not change when no external impulses are applied to the system.

The conservation of linear momentum is often applied when particles collide or interact. For application, a careful study of the free-body diagram for the *entire* system of particles should be made. By doing this, one will be able to identify the forces as creating either external or internal impulses and thereby determine in what direction(s) linear momentum is conserved. As stated earlier, for the system the *internal impulses* will always cancel out since they occur in equal but opposite collinear pairs. If the time period over which the motion is studied is *very short*, some of the external impulses may also be neglected or considered approximately equal to zero. The forces causing these negligible impulses are called *nonimpulsive forces*. By comparison, forces which are very large, act for a very short period of time, and yet produce a significant change in momentum are called *impulsive forces*. They, of course, cannot be neglected in the impulse-momentum analysis.

Impulsive forces normally occur due to an explosion or the striking of one body against another, whereas nonimpulsive forces may include the weight of a body, the force imparted by a slightly deformed spring having a relatively small stiffness, or for that matter, any force that is very small compared to other larger (impulsive) forces. When making this distinction between impulsive and nonimpulsive forces, it is important to realize that it only applies during a *specific time period*. To illustrate, consider the effect of striking a baseball with a bat. During the *very short* time of interaction, the force of the bat on the ball (or particle) is impulsive since it drastically changes the ball's momentum. By comparison the ball's weight will have a negligible effect on the change in momentum, and therefore it is nonimpulsive. Consequently, it can be neglected from an impulse-momentum analysis during this time period. It should be pointed out, however, that if an impulse-momentum analysis is considered during the much longer time of flight after the ball-bat interaction, then the impulse of the ball's weight is important since it, along with air resistance, causes the change in the momentum of the ball.

PROCEDURE FOR ANALYSIS

Generally, the principle of linear impulse and momentum or the conservation of linear momentum is applied to a *system of particles* in order to determine the final velocities of the particles *just after* the time period considered. By applying these equations to the entire system, the internal impulses acting within the system, which may be unknown, are *eliminated* from the analysis. For application it is suggested that the following procedure be used.

Free-Body Diagram. Establish the x, y, z inertial frame of reference and draw the free-body diagram for each particle of the system in order to identify the internal and external forces. The conservation of linear momentum applies to the system in a given direction when *no external impulsive forces* act on the system in that direction. Also, establish the direction and sense of the particles' initial and final velocities. If the sense is unknown, assume it is along a positive inertial coordinate.

As an alternative procedure, draw the impulse and momentum diagrams for each particle of the system just before, during, and just after the impulsive forces are applied. Then investigate the impulse diagram in order to clearly distinguish the external impulsive and nonimpulsive forces from the system's internal impulses.

Momentum Equations. Apply the principle of linear impulse and momentum or the conservation of linear momentum in the appropriate directions. If the particles are subjected to dependent motion using cables and pulleys, kinematics as discussed in Sec. 12.8 can be used to relate the velocities.

If it is necessary to determine the *internal impulsive force* acting on only one particle of a system, the particle must be *isolated* (free-body diagram) and the principle of linear impulse and momentum must be applied *to the particle*. After the impulse $\int F \, dt$ is calculated, then, provided the time Δt for which the impulse acts is known, the *average impulsive force* F_{avg} can be determined from $F_{avg} = \int F \, dt/\Delta t$.

The following examples numerically illustrate application of this procedure.

Example 15–4

The 15-Mg boxcar A is coasting freely at 1.5 m/s on the horizontal track when it encounters a tank car B having a mass of 12 Mg and coasting at 0.75 m/s toward it as shown in Fig. 15–8a. If the cars meet and couple together, determine (a) the speed of both cars just after the coupling, and (b) the average force between them if the coupling takes place in 0.8 s.

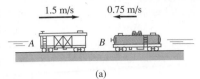
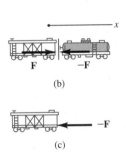

(a)

SOLUTION

Part (a)

Free-Body Diagram.* As shown in Fig. 15–8b, we have considered *both* cars as a single system. By inspection, momentum is conserved in the x direction since the coupling force $\mathbf{F}$ is *internal* to the system and will therefore cancel out. It is assumed both cars, when coupled, move at $\mathbf{v}_2$ in the positive x direction.

(b)

(c)

Fig. 15–8

Conservation of Linear Momentum

$$(\overset{+}{\rightarrow}) \qquad m_A(v_A)_1 + m_B(v_B)_1 = (m_A + m_B)v_2$$

$$15\,000(1.5) - 12\,000(0.75) = 27\,000v_2$$

$$v_2 = 0.5 \text{ m/s} \rightarrow \qquad\qquad Ans.$$

Part (b). The average (impulsive) coupling force, $\mathbf{F}_{avg}$, can be determined by applying the principle of linear momentum to *either one* of the cars.

Free-Body Diagram. As shown in Fig. 15–8c, by isolating the boxcar the coupling force is *external* to the car.

Principle of Impulse and Momentum. Since $\int F\,dt = F_{avg}\,\Delta t = F_{avg}(0.8)$, we have

$$(\overset{+}{\rightarrow}) \qquad m_A(v_A)_1 + \Sigma \int F\,dt = m_A v_2$$

$$15\,000(1.5) - F_{avg}(0.8) = 15\,000(0.5)$$

$$F_{avg} = 18.8 \text{ kN} \qquad\qquad Ans.$$

Solution was possible since the boxcar's final velocity was obtained in Part (a). Try solving for F_{avg} by applying the principle of impulse and momentum to the tank car.

*Only horizontal forces are shown on the free-body diagram.

Example 15—5

The 1200-lb cannon shown in Fig. 15–9a fires an 8-lb projectile with a muzzle velocity of 1500 ft/s. If firing takes place in 0.03 s, determine (a) the recoil velocity of the cannon just after firing, and (b) the average impulsive force acting on the projectile. The cannon support is fixed to the ground and the horizontal recoil of the cannon is absorbed by two springs.

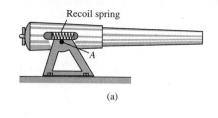

(a)

SOLUTION

Part (a)

Free-Body Diagram.* As shown in Fig. 15–9b, we have considered the projectile and cannon as a single system, since then the impulsive forces, **F,** between the cannon and projectile are *internal* to the system and will therefore cancel from the analysis. Furthermore, during the time $\Delta t =$ 0.03 s, the two recoil springs which are attached to the support each exert a *nonimpulsive force* $\mathbf{F}_s$ on the cannon. This is because Δt is very short, so that during this time the cannon only moves through a very small distance† s. Consequently, $F_s = ks \approx 0$. Hence it may be concluded that momentum for the system is conserved in the *horizontal direction*. Here we will assume that the cannon moves to the left, while the projectile moves to the right after firing.

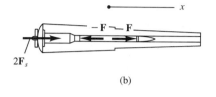

(b)

(c)

Fig. 15—9

Conservation of Linear Momentum.

$$(\xrightarrow{+}) \qquad m_c(v_c)_1 + m_p(v_p)_1 = -m_c(v_c)_2 + m_p(v_p)_2$$

$$0 + 0 = -\frac{1200}{32.2}(v_c)_2 + \frac{8}{32.2}(1500)$$

$$(v_c)_2 = 10 \text{ ft/s} \leftarrow \qquad\qquad Ans.$$

Part (b). The average impulsive force exerted by the cannon on the projectile can be determined by applying the principle of linear impulse and momentum to the projectile (or to the cannon).

Principle of Impulse and Momentum. Using the data on the free-body diagram, Fig. 15–9c, noting that $\int F\, dt = F_{avg}\, \Delta t = F_{avg}(0.03)$, we have

$$(\xrightarrow{+}) \qquad m(v_p)_1 + \Sigma \int F\, dt = m(v_p)_2$$

$$0 + F_{avg}(0.03) = \frac{8}{32.2}(1500)$$

$$F_{avg} = 12.4(10^3) \text{ lb} = 12.4 \text{ kip} \qquad Ans.$$

*Only horizontal forces are shown on the free-body diagram.
†If the cannon is firmly fixed to its support (no springs), the reactive force of the support on the cannon must be considered as an external impulse to the system, since the support would allow no movement of the cannon.

Example 15–6

The 350-Mg tugboat T shown in Fig. 15–10a is used to pull the 50-Mg barge B with a rope R. If the tugboat is coasting freely with an initial velocity of $(v_T)_1 = 3$ m/s while the rope is *slack*, determine the velocity of the tugboat *directly after* the rope becomes taut. Assume the rope does not stretch. Neglect the frictional effects of the water.

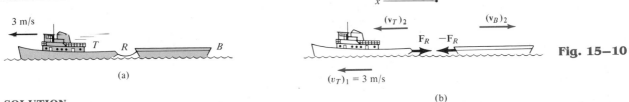

(a)

(b)

Fig. 15–10

SOLUTION

*Free-Body Diagram.** As shown in Fig. 15–10b, we have considered the entire system (tugboat and barge). Hence, the impulsive force created by the rope and the barge is *internal* to the system, and therefore momentum of the system is conserved during the instant of towing.

Conservation of Momentum. Noting that $(v_B)_2 = (v_T)_2$, we have

$$(\xleftarrow{+}) \qquad m_T(v_T)_1 + m_B(v_B)_1 = m_T(v_T)_2 + m_B(v_B)_2$$

$$350(10^3)(3) + 0 = 350(10^3)(v_T)_2 + 50(10^3)(v_T)_2$$

Solving,

$$(v_T)_2 = 2.63 \text{ m/s} \leftarrow \qquad\qquad \textbf{\textit{Ans.}}$$

This value represents the tugboat's velocity *just after* the towing impulse. Use this result and show that the towing impulse is 131 kN · s.

The alternative procedure of drawing the system's impulse and momentum diagrams is shown in Fig. 15–10c. The conservation of momentum is formulated as above by writing the vector terms directly from the diagrams.

*Only horizontal forces are shown on the free-body diagram.

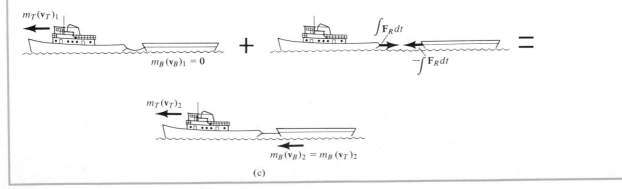

(c)

Example 15–7

A rigid pile P shown in Fig. 15–11a has a mass of 800 kg and is driven into the ground using a hammer H that has a mass of 300 kg. The hammer falls from rest from a height $y_0 = 0.5$ m and strikes the top of the pile. Determine the impulse which the hammer imparts on the pile if the pile is surrounded entirely by loose sand so that after striking the hammer does *not* rebound off the pile.

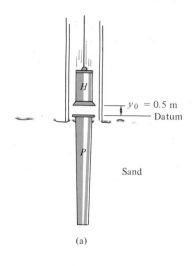

(a)

SOLUTION

Conservation of Energy. The velocity at which the hammer strikes the pile can be determined using the conservation of energy equation applied to the hammer. With the datum at the top of the pile, Fig. 15–11a, we have

$$T_0 + V_0 = T_1 + V_1$$

$$\tfrac{1}{2}m_H(v_H)_0^2 + W_H y_0 = \tfrac{1}{2}m_H(v_H)_1^2 + W_H y_1$$

$$0 + 300(9.81)(0.5) = \frac{1}{2}(300)(v_H)_1^2 + 0$$

$$(v_H)_1 = 3.13 \text{ m/s}$$

Free-Body Diagram. From the physical aspects of the problem, the free-body diagram of the hammer and pile, Fig. 15–11b, indicates that during the *short time* occurring just before to just after the *collision* the weights of the hammer and pile and the resistance force $\mathbf{F}_s$ of the sand are all *nonimpulsive*. Furthermore, the impulsive force $\mathbf{R}$ is internal to the system and therefore cancels. Consequently, momentum is conserved in the vertical direction.

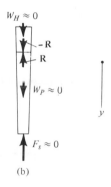

(b)

Conservation of Momentum. Just after collision we will assume $(v_H)_2 = (v_P)_2 = v_2$.

$$(+\downarrow) \qquad m_H(v_H)_1 + m_P(v_P)_1 = m_H v_2 + m_P v_2$$

$$300(3.13) + 0 = 300v_2 + 800v_2$$

$$v_2 = 0.854 \text{ m/s}$$

Principle of Impulse and Momentum. The impulse which the hammer imparts to the pile can now be determined since $\mathbf{v}_2$ is known. From the free-body diagram for the hammer, Fig. 15–11c, we have

(c)

Fig. 15–11

$$(+\downarrow) \qquad m_H(v_H)_1 + \Sigma \int_{t_1}^{t_2} F_y \, dt = m_H v_2$$

$$300(3.13) - \int R \, dt = 300(0.854)$$

$$\int R \, dt = 683 \text{ N} \cdot \text{s} \qquad\qquad \textbf{\textit{Ans.}}$$

Try finding the impulse by applying the principle of impulse and momentum to the pile.

Example 15–8

A boy having a mass of 40 kg stands on the back of a 15-kg toboggan which is originally at rest, Fig. 15–12a. If he walks to the front B and stops, determine the distance the toboggan moves. Neglect friction between the bottom of the toboggan and the ground (ice).

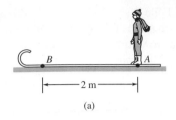

(a)

SOLUTION I

Free-Body Diagram.* The unknown frictional force of the boy's shoes on the bottom of the toboggan can be *excluded* from the analysis if the toboggan and boy on it are considered as a single system. In this way the frictional force **F** becomes internal and the conservation of momentum applies. Since both the initial and final momenta of the system are zero (because the initial and final velocities are zero), the system's momentum must also be zero when the boy is at some intermediate point between A and B, Fig. 15–12b.

Conservation of Momentum

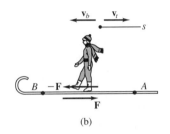

(b)

$$(\pm\rightarrow) \qquad -m_b v_b + m_t v_t = 0 \qquad (1)$$

Here the two unknowns v_b and v_t represent the velocities of the boy and the toboggan measured from a *fixed inertial reference*, s, on the ground. The *positions* of point A on the toboggan and the boy must be determined by integration. Assuming the initial position of point A to be at the origin, then at the intermediate position we have

$$-m_b s_b + m_t s_t = 0$$

If in the final position point A on the toboggan is located $s_t = +s$ to the right (in the same direction as $\mathbf{v}_t$), Fig. 15–12c, then when the boy is at B this point has the position $s_b = (2 - s)$ m to the left (in the same direction as $\mathbf{v}_b$). Hence,

$$-m_b(2 - s) + m_t s = 0 \qquad (2)$$

$$s = \frac{2m_b}{m_b + m_t} = \frac{2(40)}{40 + 15} = 1.45 \text{ m} \qquad Ans.$$

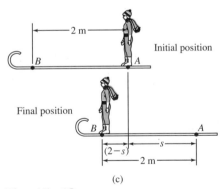

(c)

Fig. 15–12

SOLUTION II

The problem may also be solved by considering the relative motion of the boy with respect to the toboggan, $\mathbf{v}_{b/t}$. This velocity is related to the velocities of the boy and toboggan by the equation $\mathbf{v}_b = \mathbf{v}_t + \mathbf{v}_{b/t}$, Eq. 12–34. Since positive motion is assumed to be to the right in Eq. (1), $\mathbf{v}_b$ and $\mathbf{v}_{b/t}$ are negative, because the boy's motion is to the left. Hence, in scalar form, $-v_b = v_t - v_{b/t}$, and Eq. (1) then becomes $m_b(v_t - v_{b/t}) + m_t v_t = 0$. Integrating gives

$$m_b(s_t - s_{b/t}) + m_t s_t = 0$$

Assuming point A on the toboggan moves a distance s to the right, Fig. 15–12c, realizing $s_{b/t} = 2$ m, we have $m_b(s - 2) + m_t s = 0$, which is the same as Eq. (2).

*Only horizontal forces are shown on the free-body diagram.

PROBLEMS

*15–20. The car A has a weight of 4500 lb and is traveling to the right at 3 ft/s. Meanwhile a 3000-lb car B is traveling at 6 ft/s to the left. If the cars crash head-on and become entangled, determine their common velocity just after the collision. Assume that the brakes are not applied during collision.

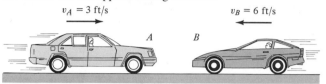

$v_A = 3$ ft/s $v_B = 6$ ft/s

Prob. 15–20

15–21. A railroad car having a mass of 15 Mg is coasting at 1.5 m/s on a horizontal track. At the same time another car having a mass of 12 Mg is coasting at 0.75 m/s in the opposite direction. If the cars meet and couple together, determine the speed of both cars just after the coupling. Compute the difference between the total kinetic energy before and after coupling has occurred, and explain qualitatively what happened to this energy.

15–22. The man M weighs 150 lb and jumps onto the boat B which has a weight of 200 lb. If he has a horizontal component of velocity *relative to the boat* of 3 ft/s, just before he enters the boat, and the boat is traveling at $v_B = 2$ ft/s away from the pier when he makes the jump, determine the resulting velocity of the man and boat.

$v_B = 2$ ft/s

Prob. 15–22

15–23. A 20-g bullet is fired horizontally into the 300-g block which rests on the smooth surface. After the bullet becomes embedded into the block, the block moves to the right 0.3 m before momentarily coming to rest. Determine the speed $(v_B)_1$ of the bullet. The spring has a stiffness of $k = 200$ N/m and is originally unstretched.

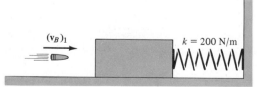

$(v_B)_1$ $k = 200$ N/m

Prob. 15–23

*15–24. The boy and girl are each standing on ice skates and holding on to the ends of the rope. If they begin to pull in the rope, such that their relative speed of approach is $v_{rel} = 2$ m/s, determine their velocities at the instant they meet. If the rope is 5 m long, how long will it take before they embrace? The boy has a mass of 60 kg, and the girl has a mass of 50 kg.

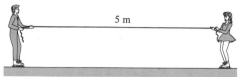

5 m

Prob. 15–24

15–25. Blocks A and B have masses of 40 kg and 60 kg, respectively. They are placed on a smooth surface and the spring connected between them is stretched 2 m. If they are released from rest, determine the speeds of both blocks the instant the spring becomes unstretched.

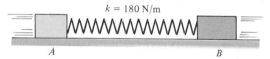

$k = 180$ N/m

A B

Prob. 15–25

15–26. The block has a mass of 50 kg and rests on the surface of the cart having a mass of 75 kg. If the spring which is attached to the cart and not the block is compressed 0.2 m and the system is released from rest, determine the speed of the block after the spring becomes undeformed. Neglect the mass of the cart's wheels and spring in the calculation. Also neglect friction. Take $k = 300$ N/m.

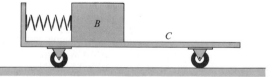

B C

Probs. 15–26/15–27

15–27. The block has a mass of 50 kg and rests on the surface of the cart having a mass of 75 kg. If the spring which is attached to the cart and not the block is compressed 0.2 m and the system is released from rest, determine the speed of the block with respect to the cart after the spring becomes undeformed. Neglect the mass of the wheels and spring in the calculation. Also neglect friction. Take $k = 300$ N/m.

181

*15–28. A boy having a weight of 75 lb stands at the end of the 50-lb canoe. If the canoe is not tied to the pier P and the boy runs at 4 ft/s relative to the canoe, determine how far d the canoe has moved away from the pier at the instant he leaps off at A. Neglect water resistance. Originally $d = 0$.

Prob. 15–28

15–29. The boy jumps off the flat car at A with a velocity of $v' = 4$ ft/s relative to the car as shown. If he lands on the second flat car B, determine the final speed of both cars after the motion. Each car has a weight of 80 lb. The boy's weight is 60 lb. Both cars are originally at rest. Neglect the mass of the car's wheels.

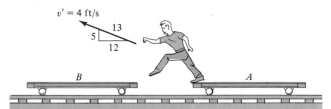

Prob. 15–29

15–30. Two men A and B, each having a weight of 160 lb, stand on the 200-lb cart. Each can run with a speed of 3 ft/s measured relative to the cart. Determine the final speed of the cart if (a) A runs and jumps off, then B runs and jumps off the same end, and (b) both run at the same time and jump off at the same time. Neglect the mass of the wheels. The jumps are made horizontally.

Prob. 15–30

15–31. A boy A having a weight of 80 lb and a girl B having a weight of 65 lb stand motionless at the ends of the toboggan, which has a weight of 20 lb. If they exchange positions, A going to B and then B going to A's original position, determine the final position of the toboggan just after the motion. Neglect friction.

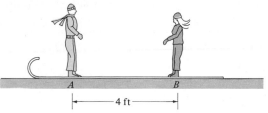

Probs. 15–31/15–32

*15–32. Solve Prob. 15–26 if B first moves to A, then A moves to B's original position.

15–33. The spring-loaded gun rests on the smooth surface and has a mass of 5 kg. It fires a ball having a mass of 1 kg with a velocity of $v' = 6$ m/s relative to the gun in the direction shown. If the gun is originally at rest, determine the horizontal distance d the ball is from the gun at the instant the ball strikes the ground at B. Neglect the height of the gun.

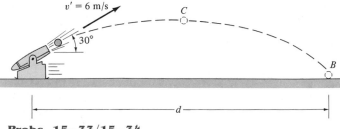

Probs. 15–33/15–34

15–34. The spring-loaded gun rests on the smooth surface and has a mass of 5 kg. It fires a ball having a mass of 1 kg with a velocity of $v' = 6$ m/s relative to the gun in the direction shown. If the gun is originally at rest, determine the horizontal distance the ball is from the gun at the instant the ball reaches its highest elevation C. Neglect the height of the gun.

15–35. The 10-lb projectile is fired from ground level with an initial velocity of $v_A = 80$ ft/s in the direction shown. When it reaches its highest point it explodes into two 5-lb fragments. If one fragment travels vertically upwards at 12 ft/s, determine the distance between the fragments after they strike the ground. Neglect the size of the gun.

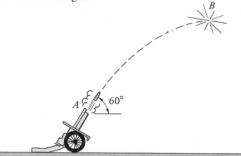

Probs. 15–35/15–36

*15–36. The 10-lb projectile is fired from ground level with an initial velocity of 80 ft/s in the direction shown. When it reaches its highest point it explodes into two 5-lb fragments. If one fragment is seen to travel vertically upwards, and after they fall they are 150 ft apart, determine the speed of each fragment just after the explosion. Neglect the size of the gun.

15–37. The crate has a mass m and rests on the barge which has a mass M. The coefficient of kinetic friction between the crate and barge is μ. The rope is pulled (or jerked) such that it snaps and as a result the initial impulse gives the barge a sudden velocity v_0. Determine the time the crate slides on the barge before coming to rest on the barge. Also, what is the final velocity of the barge? The crate and barge have zero velocity just before the impulse is applied.

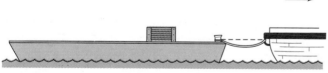

Prob. 15–37

15–38. Using the data in Prob. 15–37, determine the distance the crate slides on the barge before coming to rest.

15–39. The block A has a mass of 5 kg and is placed on the smooth triangular block B having a mass of 30 kg. If the system is released from rest, determine the distance B moves from point O when A reaches the bottom. Neglect the size of block A and friction.

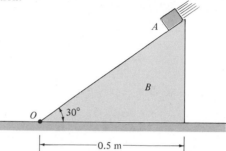

Probs. 15–39/15–40

*15–40. Solve Prob. 15–39 if the coefficient of kinetic friction between A and B is $\mu_k = 0.3$. Neglect friction between block B and the horizontal plane.

15.4 Impact

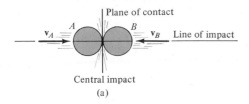

Central impact

(a)

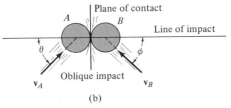

Oblique impact

(b)

Fig. 15–13

Impact occurs when two bodies collide with each other during a very *short* interval of time, causing relatively large (impulsive) forces to be exerted between the bodies. The striking of a hammer and nail, or a golf club and ball, are common examples of impact loadings.

In general, there are two types of impact. *Central impact* occurs when the direction of motion of the mass centers of the two colliding particles is along a line passing through the mass centers of the particles. This line is called the *line of impact,* Fig. 15–13a. When the motion of one or both of the particles is at an angle with the line of impact, Fig. 15–13b, the impact is said to be *oblique impact.*

Central Impact

To illustrate the method for analyzing the mechanics of impact, consider the case involving the central impact of two *smooth* particles A and B shown in Fig. 15–14.

1. The particles have the initial momenta shown in Fig. 15–14a. Provided $(v_A)_1 > (v_B)_1$, collision will eventually occur.
2. During the collision the material of the particles must be thought of as *deformable* or nonrigid. The particles will undergo a *period of deformation* such that they exert an equal but opposite deformation impulse $\int \mathbf{P} \, dt$ on each other, Fig. 15–14b.
3. Only at the instant of *maximum deformation* will both particles move with a common velocity $\mathbf{v}$, Fig. 15–14c.
4. Afterward a *period of restitution* occurs, in which case the material from which the particles are made will either return to its original shape or remain permanently deformed. The equal but opposite *restitution impulse* $\int \mathbf{R} \, dt$ pushes the particles apart from one another, Fig. 15–14d. In reality, the physical properties of any two bodies are such that the deformation impulse is *always greater* than that of restitution, i.e., $\int P \, dt > \int R \, dt$.
5. Just after separation the particles will have the final momenta shown in Fig. 15–14e, where $(v_B)_2 > (v_A)_2$.

In most problems the initial velocities of the particles will be *known* and it will be necessary to determine their final velocities $(v_A)_2$ and $(v_B)_2$. In this regard, *momentum* for the *system of particles* is *conserved* since during collision the internal impulses of deformation and restitution *cancel*. Hence, referring to Fig. 15–14a and e,

$$(\pm) \qquad m_A(v_A)_1 + m_B(v_B)_1 = m_A(v_A)_2 + m_B(v_B)_2 \qquad (15\text{–}10)$$

In order to obtain a second equation, necessary to solve for $(v_A)_2$ and $(v_B)_2$, we must apply the principle of impulse and momentum to *each particle*. For example, during the deformation phase for particle A, Fig. 15–14a, b, and c, we have

$(\overset{+}{\rightarrow})$ $$m_A(v_A)_1 - \int P\,dt = m_A v$$

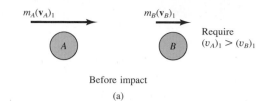

Before impact

(a)

For the restitution phase, Fig. 15–14c, d, and e,

$(\overset{+}{\rightarrow})$ $$m_A v - \int R\,dt = m_A(v_A)_2$$

The ratio of the restitution impulse to the deformation impulse is called the *coefficient of restitution, e*. From the above equations, this value for particle A is

$$e = \frac{\int R\,dt}{\int P\,dt} = \frac{v - (v_A)_2}{(v_A)_1 - v}$$

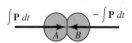

Deformation Impulse

(b)

In a similar manner, we can establish e by considering particle B, Fig. 15–14. This yields

$$e = \frac{\int R\,dt}{\int P\,dt} = \frac{(v_B)_2 - v}{v - (v_B)_1}$$

Maximum deformation

(c)

If the unknown v is eliminated from the above two equations, the coefficient of restitution can be expressed in terms of the particles' initial and final velocities as

$(\overset{+}{\rightarrow})$ $$\boxed{e = \frac{(v_B)_2 - (v_A)_2}{(v_A)_1 - (v_B)_1}}$$ (15–11)

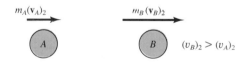

Restitution Impulse

(d)

Provided a value for e is specified, Eqs. 15–10 and 15–11 may be solved simultaneously to obtain $(v_A)_2$ and $(v_B)_2$. In doing so, however, it is important to carefully establish a sign convention for defining the positive directions for both $\mathbf{v}_B$ and $\mathbf{v}_A$, and then use it *consistently* when writing *both* equations. As noted from the application above, and indicated symbolically by the arrow in parentheses, we have defined the positive direction to the right when referring to the motions of both A and B. Consequently, if a negative value results from the solution of either $(v_A)_2$ or $(v_B)_2$, it indicates motion is to the left.

After impact

(e)

Fig. 15–14

Coefficient of Restitution. With reference to Fig. 15–14a and e, it is seen that Eq. 15–11 states that the coefficient of restitution is equal to the ratio of the relative velocity of the particles' separation *just after impact*, $(v_B)_2 - (v_A)_2$, to the relative velocity of the particles' approach *just before impact*, $(v_A)_1 - (v_B)_1$. By measuring these relative velocities experimentally, it has been found that e varies appreciably with impact velocity as well as with the size and shape of the colliding bodies. For these reasons the coefficient of restitution is reliable only when used under conditions which closely approximate those which were known to exist when measurements of it were made. In general e has a value between zero and one, and one should be aware of the physical meaning of these two limits.

Elastic Impact (e = 1). If the collision between the two particles is *perfectly elastic*, the deformation impulse ($\int \mathbf{P}\, dt$) is equal and opposite to the restitution impulse ($\int \mathbf{R}\, dt$). Although in reality this can never be achieved, $e = 1$ for an elastic collision.

Plastic Impact (e = 0). The impact is said to be *inelastic or plastic* when $e = 0$. In this case there is no restitution impulse given to the particles ($\int \mathbf{R}\, dt = \mathbf{0}$), so that after collision both particles couple or stick *together* and move with a common velocity.

From the above derivation it should be evident that the principle of work and energy cannot be used for the analysis of this problem since it is not possible to know how the *internal forces* of deformation and restitution vary or move during the collision. By knowing the particle's velocities before and after collision, however, the energy loss during collision can be calculated on the basis of the difference in the particle's kinetic energy. This energy loss, $\Sigma U_{1-2} = \Sigma T_2 - \Sigma T_1$, occurs because some of the initial kinetic energy of the particle is transformed into thermal energy as well as creating sound and localized deformation of the material when the collision occurs. In particular, if the impact is *perfectly elastic*, no energy would be lost in the collision (see Prob. 15–43); whereas if the collision is *plastic*, the energy lost during collision would be a maximum.

PROCEDURE FOR ANALYSIS (CENTRAL IMPACT)

In most cases the *final velocities* of two smooth particles are to be determined *just after* they are subjected to direct central impact. Provided the coefficient of restitution, the mass of each particle, and each particle's initial velocity *just before* impact are known, the solution to the problem can be obtained using the following two equations:

1. The conservation of momentum applies to the system of particles, $\Sigma m v_1 = \Sigma m v_2$.
2. The coefficient of restitution, $e = ((v_B)_2 - (v_A)_2)/((v_A)_1 - (v_B)_1)$, relates the relative velocities of the particles along the line of impact, just before and just after collision.

When applying these two equations, the sense of an unknown velocity can be assumed. If the solution yields a negative magnitude, the velocity acts in the opposite sense.

Oblique Impact

When oblique impact occurs between two smooth particles, the particles move away from each other with velocities having unknown directions as well as unknown magnitudes. Provided the initial velocities are known, four unknowns are present in the problem. As shown in Fig. 15–15a, these unknowns may be represented as $(v_A)_2$, $(v_B)_2$, θ_2, and ϕ_2.

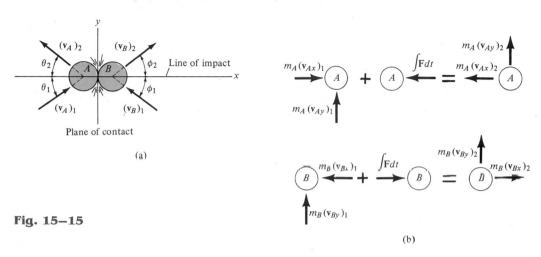

Fig. 15–15

(a)

(b)

PROCEDURE FOR ANALYSIS (OBLIQUE IMPACT)

If the y axis is established within the plane of contact and the x axis along the line of impact, the impulsive forces of deformation and restitution act *only in the x direction,* Fig. 15–15b. Resolving the velocity or momentum vectors into components along the x and y axes, Fig. 15–15b, it is possible to write four independent scalar equations in order to determine $(v_{Ax})_2$, $(v_{Ay})_2$, $(v_{Bx})_2$, and $(v_{By})_2$.

1. Momentum of the system is conserved *along the line of impact, x* axis, so that $\Sigma m(v_x)_1 = \Sigma m(v_x)_2$.
2. The coefficient of restitution, $e = ((v_{Bx})_2 - (v_{Ax})_2)/((v_{Ax})_1 - (v_{Bx})_1)$, relates the relative-velocity *components* of the particles *along the line of impact* (x axis).
3. Momentum of particle A is conserved along the y axis, perpendicular to the line of impact, since no impulse acts on the particle in this direction.
4. Momentum of particle B is conserved along the y axis, perpendicular to the line of impact, since no impulse acts on the particle in this direction.

Application of these four equations is illustrated numerically in Example 15–11.

Example 15–9

The bag A, having a weight of 6 lb, is released from rest at the position $\theta = 0°$, as shown in Fig. 15–16a. It strikes an 18-lb box B when $\theta = 90°$. If the coefficient of restitution between the bag and box is $e = 0.5$, determine the velocities of the bag and box just after impact.

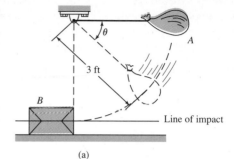

(a)

SOLUTION

This problem involves central impact. Why? Before analyzing the mechanics of the impact, however, it is first necessary to obtain the velocity of the bag *just before* it strikes the box.

Conservation of Energy. With the datum at $\theta = 0°$, Fig. 15–16b, we have

$$T_0 + V_0 = T_1 + V_1$$

$$0 + 0 = \frac{1}{2}\left(\frac{6}{32.2}\right)(v_A)_1^2 - 6(3)$$

$$(v_A)_1 = 13.9 \text{ ft/s}$$

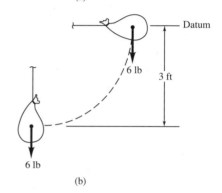

(b)

Conservation of Momentum. After impact we will assume A and B travel to the left. Applying the conservation of momentum to the system, we have

$(\xleftrightarrow{+})$
$$m_B(v_B)_1 + m_A(v_A)_1 = m_B(v_B)_2 + m_A(v_A)_2$$

$$0 + \frac{6}{32.2}(13.9) = \frac{18}{32.2}(v_B)_2 + \frac{6}{32.2}(v_A)_2$$

$$(v_A)_2 = 13.9 - 3(v_B)_2 \tag{1}$$

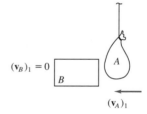

$(v_B)_1 = 0$

$(v_A)_1$

Coefficient of Restitution. Realizing that for separation to occur after collision $(v_B)_2 > (v_A)_2$, Fig. 15–16c, we have

$(\xleftrightarrow{+})$
$$e = \frac{(v_B)_2 - (v_A)_2}{(v_A)_1 - (v_B)_1}$$

$$0.5 = \frac{(v_B)_2 - (v_A)_2}{13.9 - 0}$$

$$(v_A)_2 = (v_B)_2 - 6.95 \tag{2}$$

Solving Eqs. (1) and (2) simultaneously yields

$$(v_A)_2 = -1.74 \text{ ft/s} = 1.74 \text{ ft/s} \rightarrow \quad \text{and} \quad (v_B)_2 = 5.21 \text{ ft/s} \leftarrow \textit{Ans.}$$

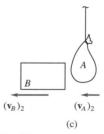

$(v_B)_2$ $(v_A)_2$

(c)

Fig. 15–16

Example 15–10

The ball B shown in Fig. 15–17a has a mass of 1.5 kg and is suspended from the ceiling by a 1-m-long elastic cord. If the cord is *stretched* downward 0.25 m and the ball is released from rest, determine how far the cord stretches after the ball rebounds from the ceiling. The stiffness of the cord is $k = 800$ N/m and the coefficient of restitution between the ball and ceiling is $e = 0.8$. The ball makes a central impact with the ceiling.

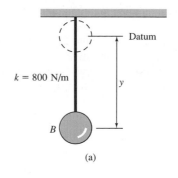

(a)

SOLUTION

It is necessary to first obtain the initial velocity of the ball *just before* it strikes the ceiling.

Conservation of Energy. With the datum located as shown in Fig. 15–17a, realizing that initially $y = y_0 = (1 + 0.25)$ m $= 1.25$ m, we have

$$T_0 + V_0 = T_1 + V_1$$

$$\tfrac{1}{2}m(v_B)_0^2 - W_B y_0 + \tfrac{1}{2}ks^2 = \tfrac{1}{2}m(v_B)_1^2 + 0$$

$$0 - 1.5(9.81)(1.25) + \frac{1}{2}(800)(0.25)^2 = \frac{1}{2}(1.5)(v_B)_1^2$$

$$(v_B)_1 = 2.97 \text{ m/s} \uparrow$$

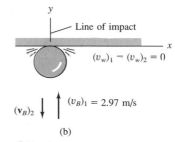

(b)

Fig. 15–17

The interaction of the ball with the ceiling will now be considered using the principles of impact.* Note that the y axis, Fig. 15–17b, represents the line of impact for the ball. Since an unknown portion of the mass of the ceiling is involved in the impact, the conservation of momentum for the ball–ceiling system will not be written. The "velocity" of this portion of ceiling is zero since it remains at rest *both* before and after impact.

Coefficient of Restitution. Since $(\mathbf{v}_B)_2$ is downward, we have

$$(+\uparrow) \quad e = \frac{(v_B)_2 - (v_A)_2}{(v_A)_1 - (v_B)_1}; \quad 0.8 = \frac{-(v_B)_2 - 0}{0 - 2.97}$$

$$(v_B)_2 = 2.37 \text{ m/s} \downarrow$$

Conservation of Energy. The maximum stretch s in the cord may be determined by again applying the conservation of energy equation to the ball just after collision. Assuming that $y = y_3 = (1 + s_3)$ m, Fig. 15–17a, then

$$T_2 + V_2 = T_3 + V_3$$

$$\tfrac{1}{2}m(v_B)_2^2 + 0 = \tfrac{1}{2}m(v_B)_3^2 - W_B y_3 + \tfrac{1}{2}ks_3^2$$

$$\frac{1}{2}(1.5)(2.37)^2 = 0 - 9.81(1.5)(1 + s_3) + \frac{1}{2}(800)s_3^2$$

$$400s_3^2 - 14.72s_3 - 18.94 = 0$$

Solving this quadratic equation for the positive root yields

$$s_3 = 0.237 \text{ m} = 237 \text{ mm} \qquad\qquad Ans.$$

*The weight of the ball is considered a nonimpulsive force.

Example 15–11

Two smooth disks A and B, having a mass of 1 and 2 kg, respectively, collide with initial velocities as shown in Fig. 15–18a. If the coefficient of restitution for the disks is $e = 0.75$, determine the x and y components of the final velocity of each disk after collision. Neglect friction.

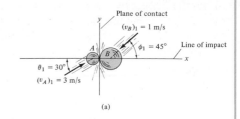

(a)

SOLUTION

The problem involves *oblique impact*. Why? In order to seek a solution, we have established the x and y axes along the line of impact and the plane of contact, respectively, Fig. 15–18a.

Resolving each of the initial velocities into x and y components, we have

$$(v_{Ax})_1 = 3 \cos 30° = 2.60 \text{ m/s}, \qquad (v_{Ay})_1 = 3 \sin 30° = 1.50 \text{ m/s}$$

$$(v_{Bx})_1 = -1 \cos 45° = -0.707 \text{ m/s}, \quad (v_{By})_1 = -1 \sin 45° = -0.707 \text{ m/s}$$

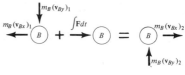

Since the impact occurs only in the x direction (line of impact), Fig. 15–18b, the conservation of momentum for *both* disks can be applied in this direction. Why?

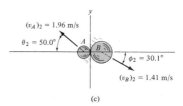

(b)

Conservation of "x" Momentum. In reference to the momentum diagrams, we have

$(\xrightarrow{+})$
$$m_A(v_{Ax})_1 + m_B(v_{Bx})_1 = m_A(v_{Ax})_2 + m_B(v_{Bx})_2$$
$$1(2.60) + 2(-0.707) = 1(v_{Ax})_2 + 2(v_{Bx})_2$$
$$(v_{Ax})_2 + 2(v_{Bx})_2 = 1.18 \tag{1}$$

Coefficient of Restitution (x). Both disks are *assumed* to have components of velocity in the $+x$ direction after collision, Fig. 15–18b.

$(\xrightarrow{+})$
$$e = \frac{(v_{Bx})_2 - (v_{Ax})_2}{(v_{Ax})_1 - (v_{Bx})_1}; \quad 0.75 = \frac{(v_{Bx})_2 - (v_{Ax})_2}{2.60 - (-0.707)}$$
$$(v_{Bx})_2 - (v_{Ax})_2 = 2.48 \tag{2}$$

Solving Eqs. (1) and (2) for $(v_{Ax})_2$ and $(v_{Bx})_2$ yields

$$(v_{Ax})_2 = -1.26 \text{ m/s} = 1.26 \text{ m/s} \leftarrow \qquad (v_{Bx})_2 = 1.22 \text{ m/s} \rightarrow \textit{ Ans.}$$

Conservation of "y" Momentum. The momentum of *each disk* is *conserved* in the y direction (plane of contact), since the disks are smooth and therefore *no impact* occurs in this direction. From Fig. 15–18b,

$(+\uparrow)$
$$m_A(v_{Ay})_1 = m_A(v_{Ay})_2 \qquad (v_{Ay})_2 = 1.50 \text{ m/s} \uparrow \qquad \textit{Ans.}$$

$(+\uparrow)$
$$m_B(v_{By})_1 = m_B(v_{By})_2 \quad (v_{By})_2 = -0.707 \text{ m/s} = 0.707 \text{ m/s} \downarrow \quad \textit{Ans.}$$

Show that when the velocity components are summed, one obtains the results shown in Fig. 15–18c.

Fig. 15–18

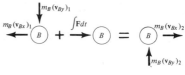

(c)

PROBLEMS

15–41. Disks A and B have a mass of 2 kg and 4 kg, respectively. If they have the velocities shown, and $e = 0.4$, determine their velocities just after direct central impact.

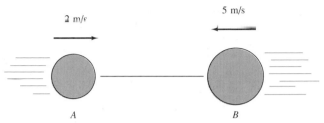

Prob. 15–41

15–42. The cue ball A is given an initial velocity $(v_A)_1 = 5$ m/s. If it makes a direct collision with ball B ($e = 0.8$), determine the velocity of B and the angle θ just after it rebounds from the cushion at C ($e' = 0.6$). Each ball has a mass of 0.4 kg. Neglect the size of each ball.

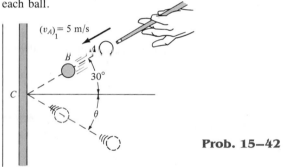

Prob. 15–42

15–43. If two disks A and B have the same mass and are subjected to direct central impact such that the collision is perfectly elastic ($e = 1$), prove that the kinetic energy before collision equals the kinetic energy after collision. The surface upon which they slide is smooth.

*15–44. Block A has a mass of 3 kg and is sliding on a rough horizontal surface with a velocity $(v_A)_1 = 2$ m/s when it makes a direct collision with block B, which has a mass of 2 kg and is originally at rest. If the collision is perfectly elastic ($e = 1$), determine the velocity of each block just after the collision and determine the distance between the blocks when they stop sliding. The coefficient of kinetic friction between the blocks and the plane is $\mu_k = 0.3$.

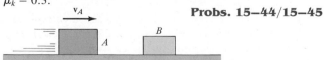

Probs. 15–44/15–45

15–45. Block A has a mass of 3 kg and is sliding on a rough horizontal surface with a velocity $(v_A)_1 = 2$ m/s when it makes a direct collision with block B, which has a mass of 2 kg and is originally at rest. If $e = 0.6$, determine the velocity of each block just after the collision and determine the distance between the blocks when they stop sliding. The coefficient of kinetic friction between the blocks and the plane is $\mu_k = 0.3$.

15–46. Determine the horizontal velocity $\mathbf{v}_A$ at which the girl must throw the ball so that it bounces once on the smooth surface and then lands into the cup at C. Take $e = 0.6$ and neglect the size of the cup.

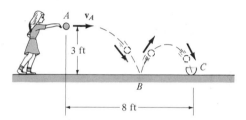

Prob. 15–46

15–47. The 5-lb box B is dropped from rest 5 ft from the top of the 10-lb plate P, which is supported by the spring having a stiffness of $k = 30$ lb/ft. If $e = 0.6$ between the box and plate, determine the maximum compression imparted to the spring. Neglect the mass of the spring.

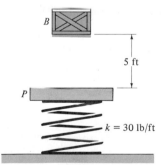

Prob. 15–47

*15–48. The three balls each have the same mass m. If A has a speed v just before a direct collision with B, determine the speed of C just after collision. The coefficient of restitution between each ball is e. Neglect the size of each ball.

Prob. 15–48

15–49. The 20-lb suitcase A is released from rest at C. After it slides down the smooth ramp it strikes the 10-lb suitcase B, which is originally at rest. If the coefficient of restitution between the suitcases is $e = 0.3$ and the coefficient of kinetic friction between the floor DE and each suitcase is $\mu_k = 0.4$, determine (a) the velocity of A just before impact, (b) the velocities of A and B just after impact, and (c) the distance B slides before coming to rest.

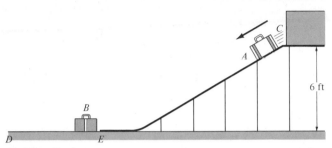

Prob. 15–49

15–50. Block A, having a mass m, is released from rest, falls a distance h and strikes the plate B having a mass 2m. If the coefficient of restitution between A and B is e, determine the velocity of the plate just after collision. The spring has a stiffness k.

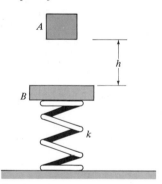

Prob. 15–50

15–51. The 0.2-lb ball bearing travels over the edge A with a velocity of $v_A = 3$ ft/s. Determine the speed at which it rebounds from the smooth inclined plane at B. Take $e = 0.8$.

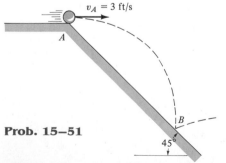

Prob. 15–51

*15–52. The ball is dropped from rest and falls a distance of 4 ft before striking the smooth plane at A. If $e = 0.8$, determine the distance d to where it again strikes the surface at B.

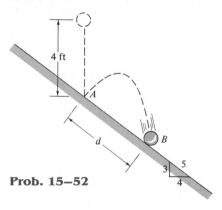

Prob. 15–52

15–53. The slider block B is confined to move within the smooth slot. It is connected to two springs, each of which has a stiffness of $k = 30$ N/m. They are originally stretched 0.5 m when $s = 0$ as shown. Determine the maximum distance, s_{max}, block B moves after it is hit by another block A which is originally traveling at $(v_A)_1 = 8$ m/s. Take $e = 0.4$ and the mass of each block to be $m = 1.5$ kg.

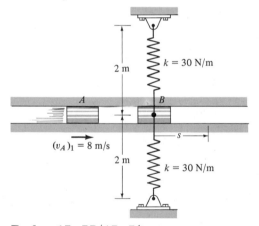

Probs. 15–53/15–54

15–54. In Prob. 15–53 determine the average net force exerted between the two blocks A and B during impact, if the impact occurs in 0.005 s.

15–55. A ball has a mass m and is dropped onto a surface from a height h. If the coefficient of restitution is e between the ball and the surface, determine the time needed for the ball to stop bouncing.

*15–56. A ball of negligible size and mass m is given a velocity of $\mathbf{v}_0$ on the center of the cart which has a mass M and is originally at rest. If the coefficient of restitution between the ball and walls A and B is e, determine the velocity of the ball and the cart just after the ball strikes A. Also, determine the total time needed for the ball to strike A, rebound, then strike B, and rebound and then return to the center of the cart.

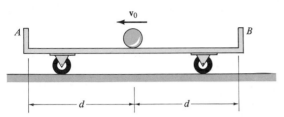

Prob. 15–56

15–57. Two smooth disks A and B each have a mass of 0.5 kg. If both disks are moving with the velocities shown when they collide, determine their final velocities just after collision. The coefficient of restitution is $e = 0.75$.

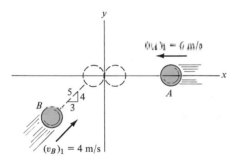

Prob. 15–57

15–58. Ball A strikes ball B with an initial velocity $(v_A)_1$ as shown. If both balls have the same mass and the collision is perfectly elastic, determine the angle θ after collision. Ball B is originally at rest. Neglect the size of each ball.

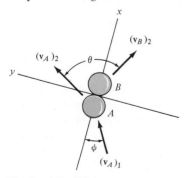

Prob. 15–58

15–59. Two balls A and B each have a mass of 2 kg and the initial velocities shown just before they collide. If the coefficient of restitution is $e = 0.5$, determine their speeds just after impact. Neglect the size of each ball.

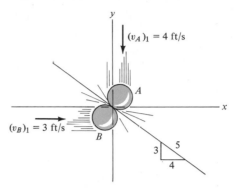

Prob. 15–59

*15–60. Two coins A and B, each having the same mass, slide on a smooth surface with the motion shown. Determine the speed of each coin after collision if they move off along the dashed paths. *Hint:* Since the line of impact has not been defined, apply the conservation of momentum along the x and y axes, respectively.

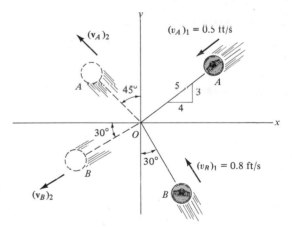

Prob. 15–60

15–61. If disk A is sliding forward along the tangent to disk B and strikes B with a velocity $\mathbf{v}$, determine the velocity of B after the collision and compute the loss of kinetic energy during the collision. Neglect friction. The coefficient of restitution is e, and each disk has the same size and mass m.

15–62. The sphere of mass m falls and strikes the triangular block with a vertical velocity $\mathbf{v}$. If the block rests on a smooth surface and has a mass $3m$, determine its velocity just after the collision. The coefficient of restitution is e.

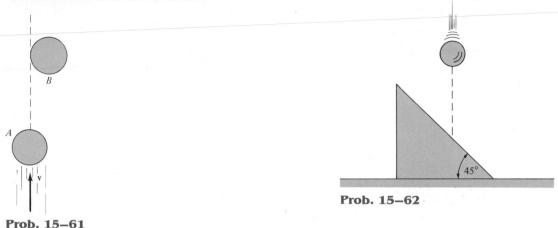

Prob. 15–62

Prob. 15–61

15.5 Angular Momentum

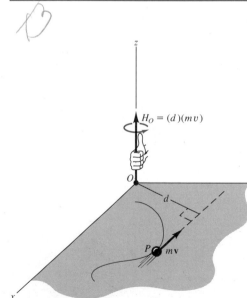

Fig. 15–19

The *angular momentum* of a particle about point O is defined as the "moment" of the particle's linear momentum about O. Since this concept is analogous to finding the moment of a force about a point, the angular momentum, $\mathbf{H}_O$, is sometimes referred to as the *moment of momentum*.

Scalar Formulation

If a particle is moving along a curve lying in the x-y plane, Fig. 15–19, the angular momentum at any instant can be computed about point O (actually the z axis) by using a scalar formulation. The *magnitude* of $\mathbf{H}_O$ is

$$(H_O)_z = (d)(mv) \qquad (15\text{–}12)$$

Here d is the moment arm or perpendicular distance from O to the line of action of $m\mathbf{v}$. Common units for this magnitude are kg $\cdot$ m^2/s or slug $\cdot$ ft^2/s. The *direction* of $\mathbf{H}_O$ is defined by the right-hand rule. As shown in Fig. 15–19, the curl of the fingers of the right hand indicates the sense of rotation of $m\mathbf{v}$ about O, so that in this case the thumb (or $\mathbf{H}_O$) is directed perpendicular to the x-y plane along the $+z$ axis.

Vector Formulation

If the particle is moving along a space curve, Fig. 15–20, the vector cross product can be used to determine the *angular momentum* about O. In this case

$$\mathbf{H}_O = \mathbf{r} \times m\mathbf{v} \qquad (15\text{–}13)$$

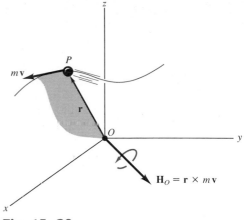

Here $\mathbf{r}$ denotes a position vector drawn from the moment point O to the particle P. As shown in the figure, $\mathbf{H}_O$ is *perpendicular* to the shaded plane containing $\mathbf{r}$ and $m\mathbf{v}$.

In order to evaluate the cross product, $\mathbf{r}$ and $m\mathbf{v}$ should be expressed in terms of their Cartesian components, so that the angular momentum is determined by evaluating the determinant:

$$\mathbf{H}_O = \begin{vmatrix} \mathbf{i} & \mathbf{j} & \mathbf{k} \\ r_x & r_y & r_z \\ mv_x & mv_y & mv_z \end{vmatrix} \qquad (15\text{–}14)$$

Fig. 15–20

15.6 Relation Between Moment of a Force and Angular Momentum

The moments about point O of all the forces acting on the particle may be related to the particle's angular momentum by using the equation of motion. If the mass of the particle is constant, we may write

$$\Sigma \mathbf{F} = m\dot{\mathbf{v}}$$

Performing a cross-product multiplication of each side of this equation by the position vector to obtain the moments of the forces about point O, we have

$$\Sigma \mathbf{M}_O = \mathbf{r} \times \Sigma \mathbf{F} = \mathbf{r} \times m\dot{\mathbf{v}}$$

From Appendix C, the derivative of $\mathbf{r} \times m\mathbf{v}$ can be written as

$$\dot{\mathbf{H}}_O = \frac{d}{dt}(\mathbf{r} \times m\mathbf{v}) = \dot{\mathbf{r}} \times m\mathbf{v} + \mathbf{r} \times m\dot{\mathbf{v}}$$

The first term on the right side, $\dot{\mathbf{r}} \times m\mathbf{v} = m(\dot{\mathbf{r}} \times \dot{\mathbf{r}}) = \mathbf{0}$, since the cross product of a vector with itself is zero. Hence, the above equation becomes

$$\Sigma \mathbf{M}_O = \dot{\mathbf{H}}_O \qquad (15\text{–}15)$$

This equation states that *the resultant moment about point O of all the forces acting on the particle is equal to the time rate of change of the particle's angular momentum about point O.* This result is similar to Eq. 15–1, i.e.,

$$\boxed{\Sigma \mathbf{F} = \dot{\mathbf{L}}} \tag{15-16}$$

Here $\mathbf{L} = m\mathbf{v}$, so that *the resultant force acting on the particle is equal to the time rate of change of the particle's linear momentum.*

From the derivations, it is seen that Eqs. 15–15 and 15–16 are actually another way of stating Newton's second law of motion. In other sections of this book it will be shown that these equations have many practical applications when extended and applied to the solution of problems involving either a system of particles or rigid bodies.

System of Particles

An equation having the same form as Eq. 15–15 may be derived for the system of n particles shown in Fig. 15–21. The forces acting on the arbitrary ith particle of the system consist of a resultant *external force* $\mathbf{F}_i$ and a resultant *internal force* $\mathbf{f}_i = \sum\limits_{j=1(j \neq i)}^{n} \mathbf{f}_{ij}$. Expressing the moments of these forces about point O, using the form of Eq. 15–15, we have

$$(\mathbf{r}_i \times \mathbf{F}_i) + (\mathbf{r}_i \times \mathbf{f}_i) = (\dot{\mathbf{H}}_i)_O$$

Here $\mathbf{r}_i$ represents the position vector drawn from the origin O of an inertial frame of reference to the ith particle, and $(\mathbf{H}_i)_O$ is the angular momentum of the ith particle about O. Similar equations can be written for each of the other particles of the system. When the results are summed vectorially, the result is

$$\Sigma(\mathbf{r}_i \times \mathbf{F}_i) + \Sigma(\mathbf{r}_i \times \mathbf{f}_i) = \Sigma(\dot{\mathbf{H}}_i)_O$$

The second term is zero since the internal forces occur in equal but opposite pairs, and since each pair of forces has the same line of action, the moment of each pair of forces about point O is therefore zero. Hence, dropping the index notation, the above equation can be written in a simplified form as

$$\Sigma \mathbf{M}_O = \dot{\mathbf{H}}_O \tag{15-17}$$

which states that *the sum of the moments about point O of all the external forces acting on a system of particles is equal to the time rate of change of the total angular momentum of the system of particles about point O.* Although point O has been chosen here as the origin of coordinates, it actually can represent any point *fixed* in the inertial frame of reference.

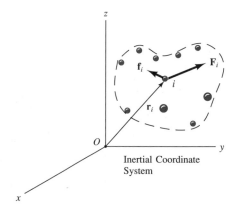

Inertial Coordinate System

Fig. 15–21

Example 15–12

The box shown in Fig. 15–22a has a mass m and is traveling down the smooth circular ramp such that when it is at the angle θ it has a speed v. Compute its angular momentum about point O at this instant and the rate of increase in its speed.

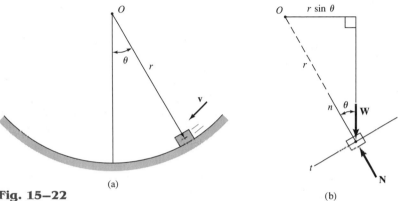

Fig. 15–22

(a)

(b)

SOLUTION

Since **v** is tangent to the path, applying Eq. 15–12 the angular momentum is

$$H_O = rmv \downarrow \qquad\qquad Ans.$$

The rate of increase in its speed (dv/dt) can be found by applying Eq. 15–15. From the free-body diagram of the block, Fig. 15–22b, it is seen that only the weight contributes a moment about point O. We have

$$\curvearrowright + \Sigma M_O = \dot{H}_O; \qquad mg(r \sin \theta) = \frac{d}{dt}(rmv)$$

Since r and m are constant,

$$mgr \sin \theta = rm\frac{dv}{dt}$$

$$\frac{dv}{dt} = g \sin \theta \qquad\qquad Ans.$$

This same result can, of course, be obtained from the equation of motion applied in the tangential direction, Fig. 15–22b, i.e.,

$$+ \swarrow \Sigma F_t = ma_t; \qquad mg \sin \theta = m\left(\frac{dv}{dt}\right)$$

$$\frac{dv}{dt} = g \sin \theta \qquad\qquad Ans.$$

15.7 Angular Impulse and Momentum Principles

Principle of Angular Impulse and Momentum

If Eq. 15–15 is rewritten in the form $\Sigma \mathbf{M}_O \, dt = d\mathbf{H}_O$ and integrated, we have, assuming that at time $t = t_1$, $\mathbf{H}_O = (\mathbf{H}_O)_1$ and at time $t = t_2$, $\mathbf{H}_O = (\mathbf{H}_O)_2$,

$$\Sigma \int_{t_1}^{t_2} \mathbf{M}_O \, dt = (\mathbf{H}_O)_2 - (\mathbf{H}_O)_1$$

or

$$(\mathbf{H}_O)_1 + \Sigma \int_{t_1}^{t_2} \mathbf{M}_O \, dt = (\mathbf{H}_O)_2 \qquad (15\text{–}18)$$

This equation is referred to as the *principle of angular impulse and momentum*. The initial and final angular momenta $(\mathbf{H}_O)_1$ and $(\mathbf{H}_O)_2$ are defined as the moment of the linear momentum of the particle $(\mathbf{H}_O = \mathbf{r} \times m\mathbf{v})$ at the instants t_1 and t_2, respectively. The second term on the left side, $\Sigma \int_{t_1}^{t_2} \mathbf{M}_O \, dt$, is called the *angular impulse*. It is computed on the basis of integrating, with respect to time, the moments of all the forces acting on the particle over the time interval t_1 to t_2. Since the moment of a force about point O is defined as $\mathbf{M}_O = \mathbf{r} \times \mathbf{F}$, the angular impulse may be expressed in vector form as

$$\text{angular impulse} = \int_{t_1}^{t_2} \mathbf{M}_O \, dt = \int_{t_1}^{t_2} (\mathbf{r} \times \mathbf{F}) \, dt \qquad (15\text{–}19)$$

Here $\mathbf{r}$ is a position vector which extends from point O to any point on the line of action of $\mathbf{F}$.

In a similar manner, using Eq. 15–17, the principle of angular impulse and momentum for a system of particles may be written as

$$\Sigma(\mathbf{H}_O)_1 + \Sigma \int_{t_1}^{t_2} \mathbf{M}_O \, dt = \Sigma(\mathbf{H}_O)_2 \qquad (15\text{–}20)$$

Here the first and third terms represent the angular momenta of the system of particles $(\Sigma \mathbf{H}_O = \Sigma(\mathbf{r}_i \times m\mathbf{v}_i))$ at the instants t_1 and t_2. The second term is the vector sum of the angular impulses given to all the particles during the time period t_1 to t_2. Recall that these impulses are created only by the moments of the external forces acting on the system where, for the ith particle, $\mathbf{M}_O = \mathbf{r}_i \times \mathbf{F}_i$.

Vector Formulation. Using impulse and momentum principles, it is therefore possible to write two vector equations which define the particle motion; namely, Eqs. 15–3 and 15–18, restated as

$$m\mathbf{v}_1 + \Sigma \int_{t_1}^{t_2} \mathbf{F}\, dt = m\mathbf{v}_2$$

$$(\mathbf{H}_O)_1 + \Sigma \int_{t_1}^{t_2} \mathbf{M}_O\, dt = (\mathbf{H}_O)_2$$

(15–21)

Scalar Formulation. In general, the above equations may be expressed in x, y, z component form, yielding a total of six independent scalar equations. If the particle is confined to move in the x-y plane, three independent scalar equations may be written to express the motion, namely,

$$m(v_x)_1 + \Sigma \int_{t_1}^{t_2} F_x\, dt = m(v_x)_2$$

$$m(v_y)_1 + \Sigma \int_{t_1}^{t_2} F_y\, dt = m(v_y)_2$$

$$(H_O)_1 + \Sigma \int_{t_1}^{t_2} M_O\, dt = (H_O)_2$$

(15–22)

The first two of these equations represent the principle of linear impulse and momentum in the x and y directions, and the third equation represents the principle of angular impulse and momentum about the z axis.

Conservation of Angular Momentum

When the angular impulses acting on a particle are all zero during the time t_1 to t_2, Eq. 15–18 reduces to the following simplified form,

$$(\mathbf{H}_O)_1 = (\mathbf{H}_O)_2$$

(15–23)

This equation is known as the *conservation of angular momentum*. It states that from t_1 to t_2 the particle's angular momentum remains constant. Obviously, if no external impulse is applied to the particle, both linear and angular momentum will be conserved. In some cases, however, the particle's angular momentum will be conserved and linear momentum may not. An example of this occurs when the particle is subjected *only* to a *central force* (see Sec. 13.7). As shown in Fig. 15–23, the impulsive central force $\mathbf{F}$ is always directed toward point O as the particle moves along the path. Hence, the angular impulse (moment) created by $\mathbf{F}$ about the z axis passing through point O is always zero, and therefore angular momentum of the particle is conserved about this axis.

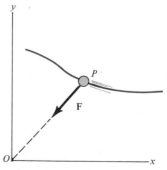

Fig. 15–23

Using Eq. 15–20, we can also write the conservation of angular momentum for a system of particles, namely,

$$\Sigma(\mathbf{H}_O)_1 = \Sigma(\mathbf{H}_O)_2 \qquad (15\text{–}24)$$

In this case the summation must include the angular momenta of all n particles in the system.

PROCEDURE FOR ANALYSIS

When applying the principles of angular impulse and momentum, or the conservation of angular momentum, it is suggested that the following procedure be used.

Free-Body Diagram. Draw the particle's free-body diagram in order to determine any axis about which angular momentum may be conserved. For this to occur, the moments of the forces (or impulses) about the axis must be zero throughout the time period t_1 to t_2. Apart from the free-body diagram, the direction and sense of the particle's initial and final velocities should also be established.

An alternative procedure to that mentioned above requires drawing the impulse and momentum diagrams for the particle.

Momentum Equations. Apply the principle of angular impulse and momentum, $(\mathbf{H}_O)_1 + \Sigma \int_{t_1}^{t_2} \mathbf{M}_O \, dt = (\mathbf{H}_O)_2$, or if appropriate, the conservation of angular momentum, $(\mathbf{H}_O)_1 = (\mathbf{H}_O)_2$.

If other equations are needed for the problem solution, when appropriate, use the principle of linear impulse and momentum, the principle of work and energy, the equations of motion, or kinematics.

The following examples numerically illustrate application of the above procedure.

Example 15–13

The 5-kg block of negligible size rests on the smooth horizontal plane, Fig. 15–24a. It is attached at A to a slender rod of negligible mass. The rod is attached to a ball-and-socket joint at B. If a moment $M = (3t)$ N · m, where t is in seconds, is applied to the rod and a horizontal force $P = 10$ N is applied to the block as shown, determine the speed of the block in 4 s starting from rest.

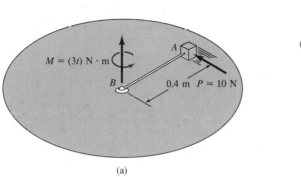

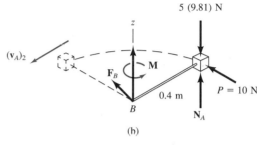

(b)

Fig. 15–24   (a)

SOLUTION

Free-Body Diagram.   If we consider the system of both the rod and block, Fig. 15–24b, then the resultant force reaction $\mathbf{F}_B$ can be eliminated from the analysis by applying the principle of angular impulse and momentum about the z axis. Why? If this is done, the angular impulses created by the weight and normal reaction $\mathbf{N}_A$ are also eliminated, since they act parallel to the z axis and therefore create zero moment about this axis.

Principle of Angular Impulse and Momentum

$$(H_z)_1 + \Sigma \int_{t_1}^{t_2} M_z \, dt = (H_z)_2$$

$$(H_z)_1 + \int_{t_1}^{t_2} M \, dt + r_{BA}P(\Delta t) = (H_z)_2$$

$$0 + \int_{0}^{4} 3t \, dt + (0.4)(10)(4) = 5(v_A)_2(0.4)$$

$$24 + 16 = 2(v_A)_2$$

$$(v_A)_2 = 20 \text{ m/s} \qquad \qquad Ans.$$

Example 15–14

The ball B, shown in Fig. 15–25a, has a weight of 0.8 lb and is attached to a cord which passes through a hole at A in a smooth table. When the ball is $r_1 = 1.75$ ft from the hole, it is rotating around in a circle such that its speed is $v_1 = 4$ ft/s. If by applying a force $\mathbf{F}$ the cord is pulled downward through the hole with a constant speed $v_c = 6$ ft/s, determine (a) the speed of the ball at the instant it is $r_2 = 0.6$ ft from the hole, and (b) the amount of work done by the force $\mathbf{F}$ in shortening the radial distance r. Neglect the size of the ball.

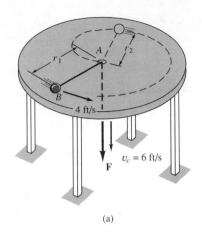

(a)

SOLUTION

Part (a)

Free-Body Diagram. As the ball moves from radial position r_1 to r_2, Fig. 15–25b, the three forces acting on it are all unknown; however, since the cord tension $\mathbf{F}$ passes through the z axis and $\mathbf{W}$ and $\mathbf{N}_B$ are parallel to it, the moments, or angular impulses created by the forces, are all *zero* about this axis. Hence, the conservation of angular momentum applies about the z axis.

Conservation of Angular Momentum. As indicated in Fig. 15–25b, the ball's velocity $\mathbf{v}_2$ is resolved into two components. The radial component, 6 ft/s, is known; however, it produces zero angular momentum about the z axis. Thus,

$$\mathbf{H}_1 = \mathbf{H}_2$$

$$r_1 m_B v_1 = r_2 m_B v_2'$$

$$1.75\left(\frac{0.8}{32.2}\right)4 = 0.6\left(\frac{0.8}{32.2}\right)v_2'$$

$$v_2' = 11.67 \text{ ft/s}$$

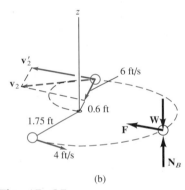

(b)

Fig. 15–25

The speed of the ball is thus

$$v_2 = \sqrt{(11.67)^2 + (6)^2}$$

$$= 13.1 \text{ ft/s} \qquad \textit{Ans.}$$

Part (b). The only force that does work on the ball is $\mathbf{F}$. (The normal force and weight do not move vertically.) The initial and final kinetic energies of the ball can be determined so that from the principle of work and energy we have

$$T_1 + \Sigma U_{1-2} = T_2$$

$$\frac{1}{2}\left(\frac{0.8}{32.2}\right)(4)^2 + U_F = \frac{1}{2}\left(\frac{0.8}{32.2}\right)(13.1)^2$$

$$U_F = 1.94 \text{ ft} \cdot \text{lb} \qquad \textit{Ans.}$$

Example 15–15

The 2-kg block shown in Fig. 15–26a rests on a smooth horizontal surface and is attached to an elastic cord that has a stiffness $k_c = 20$ N/m and is initially unstretched. If the block is given a velocity $(v_B)_1 = 1.5$ m/s, perpendicular to the cord, determine the rate at which the cord is being stretched and the speed of the block at the instant the cord is stretched 0.2 m.

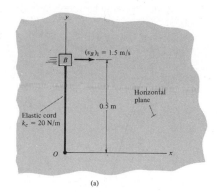

(a)

SOLUTION

Free-Body Diagram.   After the block has been launched, it slides along the dashed path shown in Fig. 15–26b. By inspection, angular momentum about point O (or the z axis) is *conserved*, since the moment of the central force $\mathbf{F}_e$ about O is always zero. Also, when the distance is 0.7 m, only the component of $(\mathbf{v}_B')_2$ is effective in producing angular momentum of the block about O.

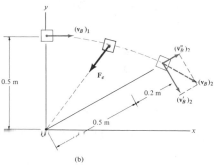

(b)

Conservation of Angular Momentum.   The component $(\mathbf{v}_B')_2$ can be obtained by applying the conservation of angular momentum about O (the z axis), i.e.,

$$(\mathbf{H}_O)_1 = (\mathbf{H}_O)_2$$

$\zeta +$

$$r_1 m_B(v_B)_1 = r_2 m_B(v_B')_2$$
$$0.5(2)(1.5) = 0.7(2)(v_B')_2$$
$$(v_B')_2 = 1.07 \text{ m/s}$$

Fig. 15–26

Conservation of Energy.   The speed of the block may be obtained by applying the conservation of energy equation at the point where the block was launched and at the point where the cord is stretched 0.2 m.

$$T_1 + V_1 = T_2 + V_2$$

$$\frac{1}{2}(2)(1.5)^2 + 0 = \frac{1}{2}(2)(v_B)_2^2 + \frac{1}{2}(20)(0.2)^2$$

Thus,

$$(v_B)_2 = 1.36 \text{ m/s} \qquad \textit{Ans.}$$

Having determined $(v_B)_2$ and its component $(v_B')_2$, the rate of stretch of the cord $(v_B'')_2$ is determined from the Pythagorean theorem,

$$(v_B'')_2 = \sqrt{(v_B)_2^2 - (v_B')_2^2}$$
$$= \sqrt{(1.36)^2 - (1.07)^2}$$
$$= 0.838 \text{ m/s} \qquad \textit{Ans.}$$

PROBLEMS

15–63. Determine the angular momentum $\mathbf{H}_O$ of each of the two particles about point O.

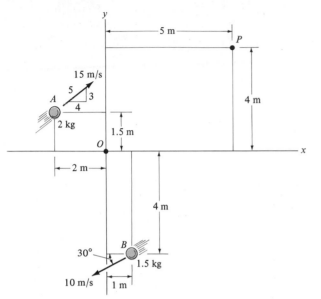

Probs. 15–63/15–64

15–67. A particle has a linear momentum $\mathbf{L} = \{-8\mathbf{i} + 14\mathbf{j} - 12\mathbf{k}\}$ kg · m/s. If it is located at point A (3 m, -4 m, 2 m), determine its angular momentum about point B (-1 m, -2 m, 8 m).

*15–68. The two spheres each have a mass of 3 kg and are rigidly attached to the rod of negligible mass. If a torque of $M = (6e^{0.2t})$ N · m, where t is in seconds, is applied to the rod as shown, determine the speed of each of the spheres in 2 s, starting from rest.

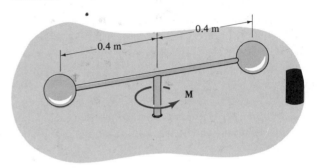

Prob. 15–68

*15–64. Determine the angular momentum $\mathbf{H}_P$ of each of the two particles about point P.

15–65. Determine the angular momentum of the 2-lb particle A about point O.

15–69. The ball B has a mass of 10 kg and is attached to the end of a rod whose mass may be neglected. If the rod is subjected to a torque $M = (3t^2 + 5t + 2)$ N · m, where t is in seconds, determine the speed of the ball when $t = 2$ s. The ball has a speed $v = 2$ m/s when $t = 0$.

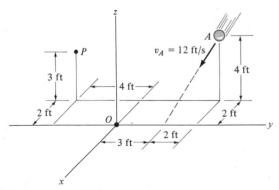

Probs. 15–65/15–66

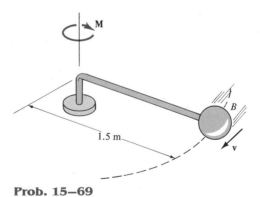

Prob. 15–69

15–66. Determine the angular momentum of the 2-lb particle A about point P.

15–70. A child having a mass of 50 kg holds her legs up as shown as she swings downward from rest at $\theta_1 = 30°$. Her center of mass is located at point G_1. When she is at the bottom position $\theta = 0°$, she *suddenly* lets her legs come down, shifting her center of mass to position G_2. Determine her speed in the upswing due to this sudden movement and the angle θ_2 to which she swings before coming to rest. Treat the child's body as a particle.

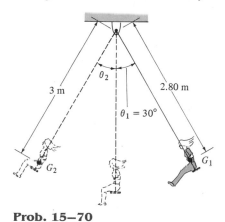

Prob. 15–70

15–71. A 4-lb ball B is traveling around in a circle of radius $r_1 = 3$ ft with a speed of $(v_B)_1 = 6$ ft/s. If the attached cord is pulled down through the hole with a constant speed of $v_r = 2$ ft/s, determine the ball's speed at the instant $r_2 = 2$ ft. How much work has to be done to pull down the cord? Neglect friction and the size of the ball.

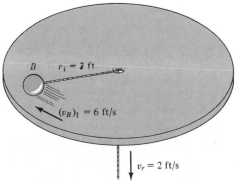

Prob. 15–71

*15–72. The elastic cord has an unstretched length $l_0 = 1.5$ ft and a stiffness $k = 12$ lb/ft. It is attached to a fixed point at A and a block at B, which has a weight of 2 lb. If the block is released from rest from the position shown, determine its speed when it reaches point C after it slides along the smooth guide. After leaving the guide, it is launched onto the smooth *horizontal* plane. Determine its speed at D at the instant the cord becomes unstretched. Also, calculate the angular momentum of the block about point A, at any instant after it passes point C.

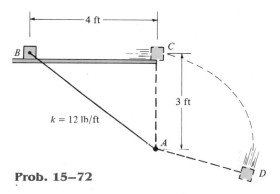

Prob. 15–72

15–73. An earth satellite of mass 700 kg is launched into a free-flight trajectory about the earth with an initial speed $v_A = 10$ km/s when the distance from the center of the earth is $r_A = 15$ Mm. If the launch angle at this position is $\phi_A = 70°$, determine the speed v_B of the satellite and its closest distance r_B from the center of the earth. The earth has a mass $M_e = 5.976(10^{24})$ kg. *Hint:* Under these conditions, the satellite is subjected only to the earth's gravitational force, $F = GM_e m_s/r^2$, Eq. 13–1. For part of the solution, use the conservation of energy (see Prob. 14–59.)

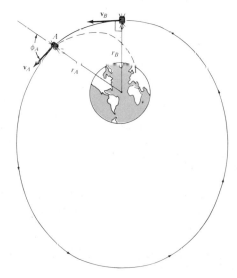

Prob. 15–73

15–74. The 800-lb roller coaster car starts from rest on the track having the shape of a cylindrical helix. If the helix descends 8 ft for every one revolution, determine the speed of the car in $t = 4$ s. Also, how far has the car descended in this time? Neglect friction.

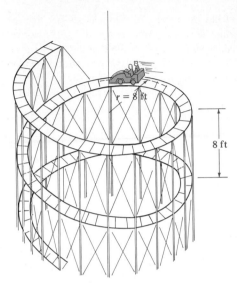

r = 8 ft

8 ft

Prob. 15–74

★15.8 Steady Fluid Streams

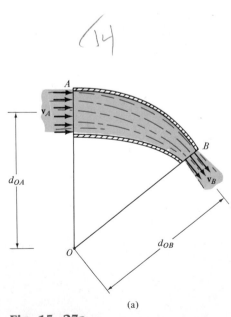

Fig. 15–27a

Knowledge of the forces developed by steadily moving fluid streams is of importance in the design and analysis of turbines, pumps, blades, and fans. To illustrate how the principle of impulse and momentum may be used to determine these forces, consider the diversion of a steady stream of fluid (liquid or gas) by a fixed pipe, Fig. 15–27a. The fluid enters the pipe with a velocity $\mathbf{v}_A$ and exits with a velocity $\mathbf{v}_B$. The impulse and momentum diagrams for the fluid stream are shown in Fig. 15–27b. The force $\Sigma\mathbf{F}$, shown on the impulse diagram, represents the resultant of all the external forces acting on the fluid stream. It is this loading which gives the fluid stream an impulse whereby the original momentum of the fluid is changed in both its magnitude and direction. Since the flow is steady, $\Sigma\mathbf{F}$ will be *constant* during the time interval dt. During this time the fluid stream is in motion, and as a result a small amount of fluid, having a mass dm, is about to enter the pipe with a velocity $\mathbf{v}_A$ at time t. If this element of mass and the mass of fluid in the pipe is considered as a "closed system," then at time $t + dt$ a corresponding element of mass dm must leave the pipe with a velocity $\mathbf{v}_B$. The fluid stream *within* the pipe section has a mass m and an *average velocity* $\mathbf{v}$ which is constant during the time interval dt. Applying the principle of linear impulse and momentum to the fluid stream, we have

$$dm\ \mathbf{v}_A + m\mathbf{v} + \Sigma\mathbf{F}\ dt = dm\ \mathbf{v}_B + m\mathbf{v}$$

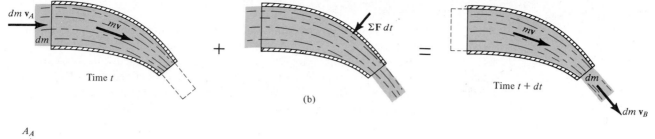

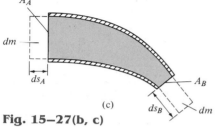

Fig. 15—27(b, c)

Force Resultant

Solving for the resultant force yields

$$\Sigma \mathbf{F} = \frac{dm}{dt}(\mathbf{v}_B - \mathbf{v}_A) \qquad (15\text{--}25)$$

Provided the motion of the fluid can be represented in the x-y plane, it is usually convenient to express this vector equation in the form of two scalar component equations, i.e.,

$$\Sigma F_x = \frac{dm}{dt}(v_{Bx} - v_{Ax})$$
$$\Sigma F_y = \frac{dm}{dt}(v_{By} - v_{Ay}) \qquad (15\text{--}26)$$

The term dm/dt is called the *mass flow* and indicates the constant amount of fluid which flows either into or out of the pipe per unit of time. If the cross-sectional areas and densities of the fluid at the entrance A and exit B are A_A, ρ_A and A_B, ρ_B, respectively, Fig. 15–27c, then *continuity of mass* requires that $dm = \rho\, dV = \rho_A(ds_A\, A_A) = \rho_B(ds_B\, A_B)$. Hence, during the time dt, since $v_A = ds_A/dt$ and $v_B = ds_B/dt$, we have

$$\frac{dm}{dt} = \rho_A v_A A_A = \rho_B v_B A_B = \rho_A Q_A = \rho_B Q_B \qquad (15\text{--}27)$$

Here $Q = vA$ is the volumetric *flow rate,* which measures the volume of fluid flowing per unit of time.

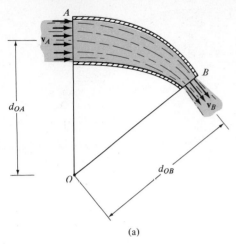

Fig. 15–27a

Moment Resultant

In some cases it is necessary to obtain the support reactions on the fluid-carrying device. If Eq. 15–25 does not provide enough information to do this, the principle of angular impulse and momentum must be used. The formulation of this principle applied to fluid streams can be obtained from Eq. 15–17, $\Sigma \mathbf{M}_O = \dot{\mathbf{H}}_O$, which states that the moment of all the external forces acting on the system about point O is equal to the time rate of change of angular momentum about O. In the case of the pipe shown in Fig. 15–27a, the flow is steady in the x-y plane; hence we have

$(\curvearrowleft +)$

$$\Sigma M_O = \frac{dm}{dt}(d_{OB}\, v_B - d_{OA}\, v_A) \qquad (15\text{–}28)$$

where the moment arms d_{OB} and d_{OA} are directed from O to the *center* of the openings at A and B.

PROCEDURE FOR ANALYSIS

The following procedure provides a method for solving problems involving steady flow.

Kinematic Diagram. In problems where the device is *moving,* a *kinematic diagram* may be helpful for determining the entrance and exit velocities of the fluid flowing onto the device, since a *relative-motion analysis* of velocity will be involved. The *kinematic diagram* in this case is simply a graphical representation of the velocities showing the vector addition of the relative-motion components. Once the velocity of the fluid flowing onto the device is determined, the mass flow is calculated using Eq. 15–27.

Free-Body Diagram. Draw a free-body diagram of the device which is directing the fluid in order to establish the forces $\Sigma \mathbf{F}$ acting on it. These external forces will include the support reactions, the weight of the device and the fluid contained within it, and the static pressure forces of the fluid at the entrance and exit sections of the device.*

Equations of Steady Flow. Apply the equations of steady flow, Eqs. 15–26 and 15–28, using the appropriate components of velocity and force shown on the kinematic and free-body diagrams.

The following examples numerically illustrate application of this procedure.

*In the SI system pressure is measured using the *pascal* (Pa), where 1 Pa = 1 N/m².

Example 15–16

Determine the reaction components which the pipe joint at A exerts on the elbow in Fig. 15–28a, if water flowing through the pipe is subjected to a static pressure of 100 kPa at A. The discharge at B is $Q_B = 0.2 \text{ m}^3/\text{s}$. Water has a density $\rho_w = 1000 \text{ kg/m}^3$, and the water-filled elbow has a mass of 20 kg and center of mass at G.

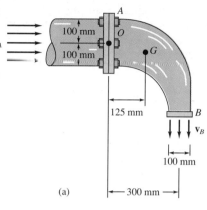

SOLUTION

The absolute velocity of flow at A and B and the mass flow rate can be obtained from Eq. 15–27. Since the density of water is constant, $Q_B = Q_A = Q$. Hence,

$$\frac{dm}{dt} = \rho_w Q = (1000 \text{ kg/m}^3)(0.2 \text{ m}^3/\text{s}) = 200 \text{ kg/s}$$

$$v_B = \frac{Q}{A_B} = \frac{0.2 \text{ m}^3/\text{s}}{\pi(0.05 \text{ m})^2} = 25.46 \text{ m/s} \downarrow$$

$$v_A = \frac{Q}{A_A} = \frac{0.2 \text{ m}^3/\text{s}}{\pi(0.1 \text{ m})^2} = 6.37 \text{ m/s} \rightarrow$$

Free-Body Diagram. As shown on the free-body diagram, Fig. 15–28b, the *fixed* connection at A exerts a resultant couple $\mathbf{M}_O$ and force components $\mathbf{F}_x$ and $\mathbf{F}_y$ on the elbow. Due to the static pressure of water in the pipe, the pressure force acting on the fluid at A is $F_A = p_A A_A$. Since 1 kPa = 1000 N/m²,

$$F_A = p_A A_A; \quad F_A = (100(10^3) \text{ N/m}^2)[\pi(0.1 \text{ m})^2] = 3141.6 \text{ N}$$

There is no static pressure acting at B, since the water is discharged at atmospheric pressure; i.e., the pressure measured by a gauge at B is equal to zero, $p_B = 0$.

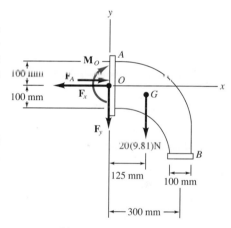

Fig. 15–28

Equations of Steady Flow

$$\xrightarrow{\ \ } \Sigma F_x = \frac{dm}{dt}(v_{Bx} - v_{Ax}), \quad F_x + 3141.6 - 200(0 - 6.37)$$

$$F_x = 4.42 \text{ kN} \qquad\qquad Ans.$$

$$+\uparrow \Sigma F_y = \frac{dm}{dt}(v_{By} - v_{Ay}); \quad -F_y - 20(9.81) = 200(-25.46 - 0)$$

$$F_y = 4.90 \text{ kN} \qquad\qquad Ans.$$

If moments are summed about point O, Fig. 15–28b, $\mathbf{F}_x$, $\mathbf{F}_y$, and the static pressure $\mathbf{F}_A$ are eliminated, as well as the moment of momentum of the water entering at A, Fig. 15–28a. Hence,

$$\zeta+ \Sigma M_O = \frac{dm}{dt}(d_{OB}v_B - d_{OA}v_A)$$

$$M_O + 20(9.81)(0.125) = 200[(0.3)(25.46) - 0]$$

$$M_O = 1.50 \text{ kN} \cdot \text{m} \qquad\qquad Ans.$$

Example 15-17

A 2-in.-diameter water jet having a velocity of 25 ft/s impinges upon a single moving blade, Fig. 15-29a. If the blade is moving at 5 ft/s away from the jet, determine the horizontal and vertical components of force which the blade is exerting on the water. What power does the water generate on the blade? Water has a specific weight of $\gamma_w = 62.4$ lb/ft³.

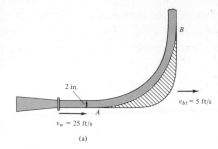

(a)

SOLUTION

Kinematic Diagram. From a fixed coordinate system, Fig. 15-29b, the rate at which water enters the blade is

$$\mathbf{v}_A = \{20\mathbf{i}\} \text{ ft/s}$$

The *relative-flow velocity* of the water onto the blade is $\mathbf{v}_{w/bl} = \mathbf{v}_w - \mathbf{v}_{bl} = 25\mathbf{i} - 5\mathbf{i} = \{20\mathbf{i}\}$ ft/s. Since the blade is moving with a velocity of $\mathbf{v}_{bl} = \{5\mathbf{i}\}$ ft/s, the velocity of flow at B measured from x, y is the vector sum, shown in Fig. 15-29b. Here,

$$\mathbf{v}_B = \mathbf{v}_{bl} + \mathbf{v}_{w/bl}$$
$$= \{5\mathbf{i} + 20\mathbf{j}\} \text{ ft/s}$$

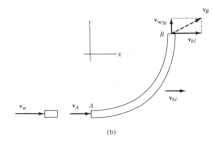

(b)

The mass flow of water *onto* the blade undergoes a momentum change. Thus,

$$\frac{dm}{dt} = \rho_w(v_{w/bl})A_A = \frac{62.4}{32.2}(20)\left[\pi\left(\frac{1}{12}\right)^2\right] = 0.846 \text{ slug/s}$$

Free-Body Diagram. The free-body diagram of a section of water acting on the blade is shown in Fig. 15-29c. The weight of the water will be neglected in the calculation, since this force is small compared to the reactive components $\mathbf{F}_x$ and $\mathbf{F}_y$.

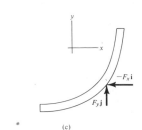

(c)

Fig. 15-29

Equations of Steady Flow

$$\Sigma\mathbf{F} = \frac{dm}{dt}(\mathbf{v}_B - \mathbf{v}_A)$$

$$-F_x\mathbf{i} + F_y\mathbf{j} = 0.846(5\mathbf{i} + 20\mathbf{j} - 25\mathbf{i})$$

Equating the respective $\mathbf{i}$ and $\mathbf{j}$ components gives

$$F_x = 0.846(20) = 16.9 \text{ lb} \leftarrow \qquad Ans.$$
$$F_y = 0.846(20) = 16.9 \text{ lb} \uparrow \qquad Ans.$$

The water exerts equal but opposite forces on the blade.

Since the water force which causes the blade to move forward horizontally with a velocity of 5 ft/s is $F_x = 16.9$ lb, then from Eq. 14-10 the power is

$$P = \mathbf{F} \cdot \mathbf{v}; \qquad P = \frac{16.9(5)}{550} = 0.154 \text{ hp} \qquad Ans.$$

*15.9 Propulsion with Variable Mass

In Sec. 15.8 the case in which a *constant* amount of mass dm entered and left a "closed system" was considered. There are, however, two other important cases involving mass flow, which are represented by a system which is either gaining or losing mass. In this section these cases will be discussed separately.

A System That Loses Mass

Consider a device which at an instant of time has a mass m and a forward velocity **v**, Fig. 15–30a. At this same instant the device is expelling an amount of mass m_e such that the mass flow velocity $\mathbf{v}_e$ is *constant* when the measurement is made a small distance away from the device. For the analysis, consider the "closed system" at an instant to include *both the mass m of the device and the expelled mass* m_e, as shown by the dashed line in the figure. The impulse and momentum diagrams for the system are shown in Fig. 15–30b. During the time interval dt, the velocity of the device is increased from **v** to $\mathbf{v} + d\mathbf{v}$. This increase in forward velocity, however, does not change the velocity $\mathbf{v}_e$ of the expelled mass, since this mass moves at a constant speed once it has been ejected. To increase the velocity of the device during the time dt, an amount of mass dm_e has been ejected and thereby gained in the exhaust. The impulses are created by $\Sigma\mathbf{F}_s$, which represents the resultant of all the external forces which *act on the system* in the direction of motion. This force resultant *does not include* the force which causes the device to move forward, since this force (called a *thrust*) is *internal to the system;* that is, the thrust acts with equal magnitude but opposite direction on the mass m of the device and the expelled exhaust mass m_e.* Applying the principle of impulse and momentum to the system, in reference to Fig. 15–30b, we have

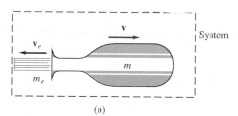

(a)

$$(\overset{+}{\rightarrow}) \quad mv - m_e v_e + \Sigma F_s\, dt = (m - dm_e)(v + dv) - (m_e + dm_e)v_e$$

or

$$\Sigma F_s\, dt = -v\, dm_e + m\, dv - dm_e\, dv - v_e\, dm_e$$

Without loss of accuracy, the third term on the right side of this equation may be neglected since it is a "second-order" differential. Dividing by dt gives

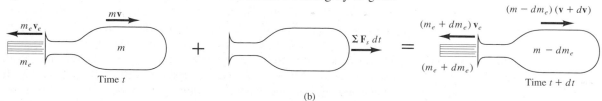

(b)

Fig. 15–30

*$\Sigma\mathbf{F}_s$ represents the external resultant force *acting on the system,* which is different from $\Sigma\mathbf{F}$, the resultant force acting only on the device.

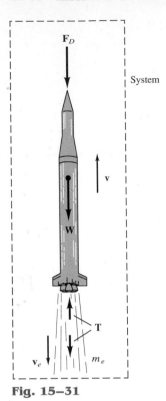

Fig. 15-31

$$\Sigma F_s = m\frac{dv}{dt} - (v + v_e)\frac{dm_e}{dt}$$

Noting that the relative velocity of the device as seen by an observer moving with the particles of the ejected mass is $v_{D/e} = (v + v_e)$, the final result can be written as

$$\Sigma F_s = m\frac{dv}{dt} - v_{D/e}\frac{dm_e}{dt} \qquad (15\text{--}29)$$

Here the term dm_e/dt represents the rate at which mass is being ejected.

To illustrate an application of Eq. 15–29, consider the rocket shown in Fig. 15–31, which has a weight **W** and is moving upward against an atmospheric drag force **F**$_D$. The system to be considered consists of the mass of the rocket and the mass of ejected gas m_e. Applying Eq. 15–29 to this system gives

$$(+\uparrow) \qquad -F_D - W = \frac{W}{g}\frac{dv}{dt} - v_{D/e}\frac{dm_e}{dt}$$

The last term of this equation represents the *thrust* **T** which the engine exhaust exerts on the rocket, Fig. 15–31. Recognizing that $dv/dt = a$, we may therefore write

$$(+\uparrow) \qquad T - F_D - W = \frac{W}{g}a$$

If a free-body diagram of the rocket is drawn, it becomes obvious that this equation represents an application of $\Sigma\mathbf{F} = m\mathbf{a}$ for the rocket.

A System That Gains Mass

A device such as a scoop or a shovel may gain mass as it moves forward. Consider, for example, the device shown in Fig. 15–32a, which at an instant of time has a mass m and is moving forward with a velocity **v.** At the same instant, the device is collecting a particle stream of mass m_i. The flow velocity **v**$_i$ of this injected mass is constant and independent of the velocity **v.** It is required that $v > v_i$. The system to be considered at this instant includes both the mass of the device and the mass of the injected fluid, as shown by the dashed line in the figure. The impulse and momentum diagrams for this system are shown in Fig. 15–32b. With an increase in mass dm_i gained by the device, there is an assumed increase in velocity dv during the time interval dt. This increase is caused by the impulse created by $\Sigma\mathbf{F}_s$, the resultant of all the external forces *acting on the system* in the direction of motion. The force summation does not include the retarding force of the injected mass acting on the device. Why? Applying the principle of impulse and momentum to the system, we have

$$(\xrightarrow{+}) \qquad mv + m_i v_i + \Sigma F_s\, dt = (m + dm_i)(v + dv) + (m_i - dm_i)v_i$$

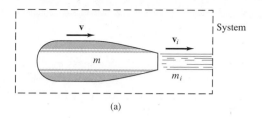

(a)

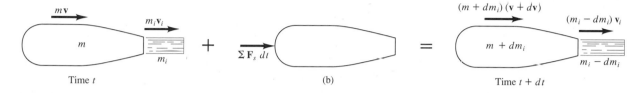

Time t (b)   Time $t + dt$

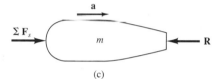

Fig. 15–32   (c)

Using the same procedure as in the previous case, we may write this equation as

$$\Sigma F_s = m\frac{dv}{dt} + (v - v_i)\frac{dm_i}{dt}$$

Since the relative velocity of the device as seen by an observer moving with the particles of the injected mass is $v_{D/i} = (v - v_i)$, the final result can be written as

$$\boxed{\Sigma F_s = m\frac{dv}{dt} + v_{D/i}\frac{dm_i}{dt}} \qquad (15\text{–}30)$$

where dm_i/dt is the rate of mass injected into the device. The last term in this equation represents the magnitude of force **R,** which the injected mass *exerts on the device*. Since $dv/dt = a$, Eq. 15–30 becomes

$$\Sigma F_s - R = ma$$

This is the application of $\Sigma \mathbf{F} = m\mathbf{a}$, Fig. 15–32c.

As in the case of steady flow, problems which are solved using Eqs. 15–29 and 15–30 should be accompanied by the necessary free-body diagram. With this diagram one can then determine ΣF_s *for the system* and isolate the force exerted on the device by the particle stream.

213

Example 15–18

The initial combined mass of a rocket and its fuel is m_0. A total mass m_f of fuel is consumed at a constant rate of $dm_e/dt = c$ and expelled at a constant speed of u relative to the rocket. Determine the velocity of the rocket at the instant the fuel runs out. Neglect the change in the rocket's weight with altitude and the drag resistance of the air. The rocket is fired vertically from rest.

SOLUTION

Since the rocket is losing mass as it moves upward, Eq. 15–29 can be used for the solution. The only *external force* acting on the *system* consisting of the rocket and a portion of the expelled mass is the weight **W**, Fig. 15–33. Hence,

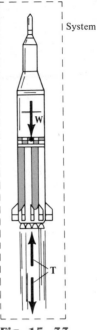

$$+\uparrow \; \Sigma F_s = m\frac{dv}{dt} - v_{D/e}\frac{dm_e}{dt}; \qquad -W = m\frac{dv}{dt} - uc \qquad (1)$$

At any given instant t during the flight, the mass of the rocket can be expressed as $m = m_0 - (dm_e/dt)t = m_0 - ct$. Since $W = mg$, Eq. (1) becomes

$$-(m_0 - ct)g = (m_0 - ct)\frac{dv}{dt} - uc$$

Fig. 15–33

Rearranging the terms and integrating, realizing that $v = 0$ at $t = 0$, we have

$$\int_0^v dv = \int_0^t \left(\frac{uc}{m_0 - ct} - g\right) dt$$

$$v = -u\,\ln(m_0 - ct) - gt \Big|_0^t$$

$$v = u\,\ln\left(\frac{m_0}{m_0 - ct}\right) - gt \qquad (2)$$

This equation gives the speed of the rocket at any instant of time. The time t' needed to consume all the fuel is

$$m_f = (dm_e/dt)t' = ct'$$

Hence,

$$t' = m_f/c$$

Substituting into Eq. (2) yields

$$v = u\,\ln\left(\frac{m_0}{m_0 - m_f}\right) - \frac{gm_f}{c} \qquad \qquad \textit{Ans.}$$

Example 15-19

A chain of length l, Fig. 15–34a, has a mass m. Determine the magnitude of force $\mathbf{F}$ required to (a) raise the chain with a constant speed v_c, starting from rest when $y = 0$; and (b) lower the chain with a constant speed v_c, starting from rest when $y = l$.

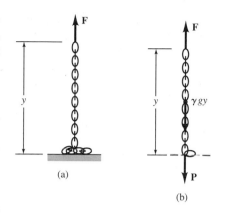

(a)

(b)

SOLUTION

Part (a). As the chain is raised, all the suspended links are given a sudden impulse downward by each added link which is lifted off the ground. Thus, the *suspended portion* of the chain may be considered as a device which is *gaining mass*. A free-body diagram of a portion of the chain which is located at an arbitrary height y above the ground is shown in Fig. 15–34b. The system to be considered is the length of chain y which is suspended by $\mathbf{F}$ at any instant, including the next link which is about to be added but is still at rest. The forces acting on this system *exclude* the internal force $\mathbf{P}$ which the added link exerts on the suspended portion of the chain. Hence, $\Sigma F_s = F - (m/l)gy$.

It now becomes necessary to find the rate at which mass is being added to the system. The velocity $\mathbf{v}_c$ of the chain at a given instant is equivalent to $v_{D/i}$. Why? Since v_c is constant, $dv_c/dt = 0$ and $dy/dt = v_c$. Integrating, using the initial condition that $y = 0$ at $t = 0$, gives $y = v_c t$. Thus, the mass of the system at any instant is $m = (m/l)y = (m/l)v_c t$, and therefore the *rate* at which mass is *added* to the suspended chain is

$$\frac{dm_l}{dt} = (m/l)v_c$$

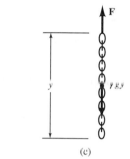

Fig. 15–34

(c)

Applying Eq. 15–30 to the system, using this data, we have

$$+ \uparrow \ \Sigma F_s = m\frac{dv_c}{dt} + v_{D/i}\frac{dm_i}{dt}$$

$$F - (m/l)gy = 0 + v_c(m/l)v_c$$

Hence,
$$F = (m/l)(gy + v_c^2) \qquad \textit{Ans.}$$

Part (b). When the chain is being lowered, the links which are expelled (given zero velocity) *do not* impart an impulse to the *remaining* suspended links. Why? Thus, the system in Part (a) cannot be considered. Instead, the equation of motion will be used to obtain the solution. At time t the portion of chain still off the floor is y. The free-body diagram for a suspended portion of the chain is shown in Fig. 15–34c. Thus,

$$+ \uparrow \Sigma F = ma; \qquad F - (m/l)gy = 0$$

$$F = (m/l)gy \qquad \textit{Ans.}$$

PROBLEMS

15–75. Sand is deposited from a chute onto a conveyor belt which is moving at 0.5 m/s. If the sand is assumed to fall vertically on the belt at A at the rate of 4 kg/s, determine the required belt tension F_B to the right of A. The belt is free to move over the conveyor rollers and its tension to the left of A is $F_C = 400$ N.

Prob. 15–75

*15–76. The fan draws air through a vent with a speed of 12 ft/s. If the cross-sectional area of the vent is 2 ft², determine the thrust on the blade. The specific weight of the air is $\gamma_a = 0.076$ lb/ft³.

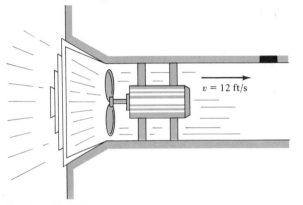

Prob. 15–76

15–77. A jet of water has a cross-sectional area of 5 in². If it strikes the fixed blade with a speed of 60 ft/s, determine the magnitude of force which the water exerts on the blade. $\gamma_w = 62.4$ lb/ft³.

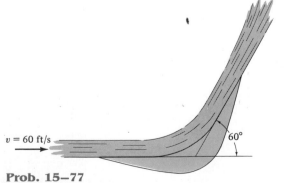

Prob. 15–77

15–78. The nozzle discharges water at a constant flow rate of 2 ft³/s. The cross-sectional area of the nozzle at A is 4 in² and at B the cross-sectional area is 12 in². If the static pressure due to the water at B is 2 lb/in², determine the magnitude of force which must be applied at section B to hold the nozzle in place. $\gamma_w = 62.4$ lb/ft³.

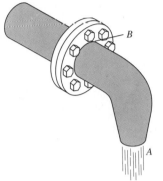

Prob. 15–78

15–79. The static pressure of water at C is 40 lb/in². If water flows out of the pipe at A and B with velocities $v_A = 12$ ft/s and $v_B = 25$ ft/s, determine the horizontal and vertical components of force exerted on the elbow at C necessary to hold the pipe assembly in equilibrium. Neglect the weight of water within the pipe and the weight of the pipe. The pipe has a diameter of 0.75 in. at C, and at A and B the diameter is 0.5 in. $\gamma_w = 62.4$ lb/ft³.

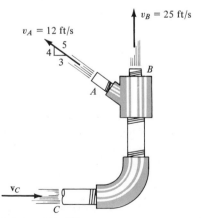

Prob. 15–79

*15–80. The boy at A aims the water hose in the direction shown. If the water is discharged at $30°$ from the horizontal, and the cross-sectional area of the water stream is approximately 2 in^2, determine the force the boy at B must exert on the shield in order to hold it perpendicular to the water. $\gamma_w = 62.4$ lb/ft^3.

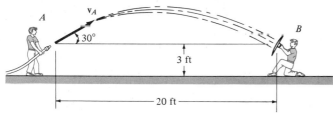

Prob. 15–80

15–81. The vane on a turbine is moving at 80 ft/s when a jet of water having a velocity of $v_A = 150$ ft/s strikes it. If the cross-sectional area of the jet is 1.5 in^2, and it is diverted as shown, determine the power developed by the water on the blade. $\gamma_w = 62.4$ lb/ft^3.

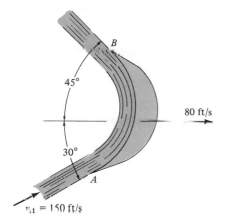

Prob. 15–81

15–82. The hose on the fire boat ejects water in a stream which reaches a height of 6 m as shown. If the water is ejected at 0.08 m^3/s, determine the horizontal thrust T provided by the engines of the boat needed to hold the boat steady. $\rho_w = 1$ Mg/m^3.

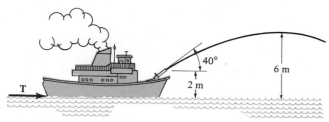

Prob. 15–82

15–83. A rocket burns 2000 lb of fuel in 2 min. If the velocity of the gas is 6500 ft/s with respect to the rocket, determine the thrust of the rocket engine.

*15–84. A plow located on the front of a locomotive scoops up snow at the rate of 10 ft^3/s and stores it in the train. If the locomotive is traveling at a constant speed of 10 mi/h, determine the resistance to motion caused by the shoveling. The specific weight of snow is $\gamma_s = 6$ lb/ft^3.

15–85. A boat weighs 400 lb and is traveling forward on a river with a constant velocity of 45 ft/s. The river is flowing in the opposite direction at 10 ft/s. A scoop is placed in the water in the direction of motion of the boat. If the scoop collects 100 lb of water in the boat in 2 minutes, determine the horizontal thrust T which is required to propel the boat forward. Neglect the drag effect of the water on the boat.

15–86. The second stage of the two-stage rocket weighs 2000 lb (empty) and is launched from the first stage with a velocity of 3000 mi/h. The fuel in the second stage weighs 1000 lb. If it is consumed at the rate of 50 lb/s and ejected with a relative velocity of 8000 ft/s, determine the acceleration of the second stage at the instant the engine is fired. What is the rocket's acceleration after all the fuel is consumed? Neglect the effect of gravitation.

15–87. A rocket has an empty weight of 500 lb and carries 300 lb of fuel. If the fuel is burned at the rate of 15 lb/s and ejected with a relative velocity of 4400 ft/s, determine the maximum speed attained by the rocket starting from rest. Neglect the effect of gravitation on the rocket.

*15–88. The missile weighs 40 kip. The constant thrust provided by the turbojet engine is $T = 15$ kip. Additional thrust is provided by *two* rocket boosters B. There is 600 lb of propellant in each booster which is burned at a constant rate of 150 lb/s, with a relative exhaust velocity of 3000 ft/s. If the mass of the propellant lost by the turbojet engine can be neglected, determine the velocity of the missile after the 4 s burn time of the boosters. The initial velocity of the missile is 300 mi/h.

Prob. 15–88

15–89. The 12-Mg jet airplane has a constant speed of 950 km/h when it is flying along a horizontal straight-line path. Air enters the intake scoops S at the rate of 50 m³/s. If the engine burns fuel at the rate of 0.4 kg/s and the gas (air and fuel) is exhausted relative to the plane with a speed of 450 m/s, determine the resultant drag force exerted on the plane by air resistance. Assume that air has a constant density of 1.22 kg/m³. *Hint:* Since mass both enters and exits the plane, Eqs. 15–29 and 15–30 must be combined to yield

$$\Sigma F_s = m\frac{dv}{dt} - v_{D/e}\frac{dm_e}{dt} + v_{D/i}\frac{dm_i}{dt}$$

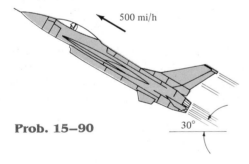

$v = 950$ km/h

Prob. 15–89

15–90. The jet is traveling at a speed of 500 mi/h, 30° with the horizontal. If the fuel is being spent at 3 lb/s, and the engine takes in air at 400 lb/s, whereas the exhaust gas (air and fuel) has a relative speed of 32,800 ft/s, determine the acceleration of the plane at this instant. The drag resistance of the air is $F_D = (0.7v^2)$ lb, where the speed is measured in ft/s. The jet has a weight of 15,000 lb. *Hint:* See Prob. 15–89.

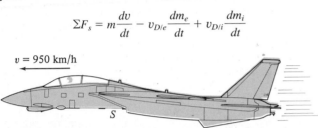

500 mi/h

30°

Prob. 15–90

15–91. The rocket car has a mass of 3 Mg (empty) and carries 150 kg of fuel. If the fuel is consumed at a constant rate of 40 kg/s and ejected from the car with a relative velocity of 250 m/s, determine the maximum speed attained by the car starting from rest. The drag resistance due to the atmosphere is $F_D = (0.6v^2)$ N, where v is the speed measured in m/s.

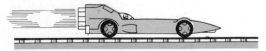

Prob. 15–91

*15–92. The chain has a total length $L < d$ and a mass per unit length of m'. If a portion h of the chain is suspended over the table and released, determine the velocity of its end A as a function of its position y. Neglect friction.

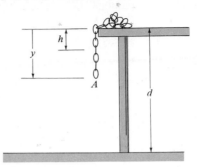

h

y

A

d

Prob. 15–92

15–93. If the chain is lowered at a constant velocity $v = 4$ ft/s, determine the normal reaction exerted on the floor as a function of time. The chain has a weight of 5 lb/ft and a total length of 20 ft.

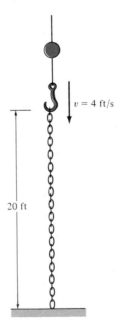

$v = 4$ ft/s

20 ft

Prob. 15–93

15–94. The car has a mass m_0 and is used to tow the smooth chain having a total length l and a mass per unit of length m'. If the chain is originally piled up, determine the tractive force F that must be supplied by the rear wheels of the car to maintain a constant speed v while the chain is being drawn out. Assume the chain is smooth.

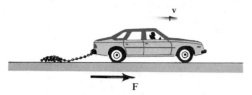

Prob. 15–94

15–95. The chain has a total length of 10 ft and a weight of 2 lb/ft. Determine the magnitude of force **F** as a function of time which must be applied to the end A of the chain to suspend it if the platform upon which it rests is descending at a constant rate of 3 ft/s. Neglect the size of the chain. Initially $y = 0$ at $t = 0$.

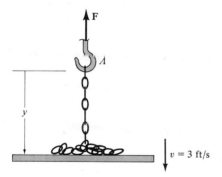

Prob. 15–95

***15–96.** The rocket has an initial mass m_0, including the fuel. For practical reasons desired for the passengers, it is required that it maintain a constant upward acceleration a_0. If the fuel is expelled from the rocket at a relative speed $v_{e/r}$, determine the rate at which the fuel should be consumed to maintain the motion. Neglect air resistance, and assume that the gravitational acceleration is constant.

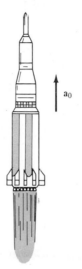

Prob. 15–96

Review 1:
Kinematics and Kinetics of a Particle

The various topics and problems presented in Chapters 12 through 15 have all been *categorized* in order to provide a *clear focus* for learning the various problem-solving principles involved. In engineering practice, however, it is most important to be able to *identify* an appropriate method for the solution of a particular problem. In this regard, one must fully understand the limitations and use of the equations of dynamics, and be able to recognize which equations and principles to use for the problem's solution. For these reasons, we will now summarize the equations and principles of particle dynamics and provide the opportunity for applying them to a variety of problems.

Kinematics

Problems in kinematics require a study only of the geometry of motion, and do not account for the forces causing the motion. When the equations of kinematics are applied, one should clearly establish a fixed origin and select an appropriate coordinate system used to define the position of the particle. Once the positive direction of each coordinate axis is established, then the directions of the components of position, velocity, and acceleration can be determined from the algebraic sign of their numerical quantities.

Rectilinear Motion

Variable Acceleration. If a mathematical (or graphical) relationship is established between *any two* of the *four* variables s, v, a, and t, then a *third* variable can be determined by solving one of the following equations which relates all three variables.

$$v = \frac{ds}{dt} \qquad a = \frac{dv}{dt} \qquad a \, ds = v \, dv$$

Constant Acceleration. Be *absolutely certain* that the acceleration is constant when using the following equations:

$$s = s_0 + v_0 t + \tfrac{1}{2} a_c t^2 \qquad v = v_0 + a_c t \qquad v^2 = v_0^2 + 2a_c(s - s_0)$$

Curvilinear Motion

x, y, z Coordinates. These coordinates are often used when the motion can be resolved into horizontal and vertical components. Also they are useful for studying projectile motion since the acceleration of the projectile is *always* downward.

$$v_x = \dot{x} \qquad a_x = \dot{v}_x$$
$$v_y = \dot{y} \qquad a_y = \dot{v}_y$$
$$v_z = \dot{z} \qquad a_z = \dot{v}_z$$

n, t, b Coordinates. These coordinates are particularly advantageous for studying the particle's *acceleration* along a known path. This is because the *t* and *n* components of **a** represent the separate changes in the magnitude and direction of the velocity, and these components can be readily formulated.

$$v = \dot{s}$$

$$a_t = \dot{v} = v\frac{dv}{ds}$$

$$a_n = \frac{v^2}{\rho}$$

where $\rho = \left| \dfrac{[1 + (dy/dx)^2]^{3/2}}{d^2y/dx^2} \right|$ when the path $y = f(x)$ is given.

r, θ, z Coordinates. These coordinates are used when data regarding the angular motion of the radial coordinate *r* is given to describe the particle's motion. Also, some paths of motion can conveniently be described using these coordinates.

$$v_r = \dot{r} \qquad a_r = \ddot{r} - r\dot{\theta}^2$$
$$v_\theta = r\dot{\theta} \qquad a_\theta = r\ddot{\theta} + 2\dot{r}\dot{\theta}$$
$$v_z = \dot{z} \qquad a_z = \ddot{z}$$

Relative Motion. If the origin of a translating coordinate system is established at particle *A*, then for particle *B*

$$\mathbf{r}_B = \mathbf{r}_A + \mathbf{r}_{B/A}$$
$$\mathbf{v}_B = \mathbf{v}_A + \mathbf{v}_{B/A}$$
$$\mathbf{a}_B = \mathbf{a}_A + \mathbf{a}_{B/A}$$

Kinetics

Problems in kinetics involve the analysis of forces which cause the motion. When applying the equations of kinetics, it is absolutely necessary that measurements of the motion be made from an *inertial coordinate system*, i.e., one that does not rotate and is either fixed or translates with constant velocity. If a problem requires *simultaneous solution* of the equations of kinetics and kinematics, then the coordinate systems selected for writing each of the equations should define the *positive directions* of the axes in the *same* manner.

Equations of Motion. These equations are used to solve for the particle's acceleration or the forces causing the motion. If they are used to determine a particle's position, velocity, or time of motion, then kinematics will also have to be considered in the solution. Before applying the equations of motion, *always draw a free-body diagram* to identify all the forces acting on the particle. Also, establish the direction of the particle's acceleration or its components. (A kinetic diagram may accompany the solution in order to graphically account for the *ma* vector.)

$$\Sigma F_x = ma_x \qquad \Sigma F_n = ma_n \qquad \Sigma F_r = ma_r$$
$$\Sigma F_y = ma_y \qquad \Sigma F_t = ma_t \qquad \Sigma F_\theta = ma_\theta$$
$$\Sigma F_z = ma_z \qquad \Sigma F_b = 0 \qquad \Sigma F_z = ma_z$$

Work and Energy. The equation of work and energy represents an integrated form of the tangential equation of motion, $\Sigma F_t = ma_t$, combined with kinematics ($a_t\, ds = v\, dv$). *It is used to solve problems involving force, velocity, and displacement.* Before applying this equation, *always draw a free-body diagram in order to identify the forces* which do work on the particle.

$$T_1 + \Sigma U_{1-2} = T_2$$

where

$$T = \tfrac{1}{2}mv^2 \quad \text{(kinetic energy)}$$

$$U_F = \int_{s_1}^{s_2} F \cos \theta\, ds \quad \text{(variable force)}$$

$$U_{F_c} = F_c \cos \theta\, (s_2 - s_1) \quad \text{(constant force)}$$

$$U_W = W\, \Delta y \quad \text{(weight)}$$

$$U_s = -(\tfrac{1}{2}ks_2^2 - \tfrac{1}{2}ks_1^2) \quad \text{(elastic spring)}$$

If the forces acting on the particle are *conservative forces*, i.e., those that *do not* cause a dissipation of energy such as friction, then apply the conservation of energy equation. This equation is easier to use than the equation of work and energy since it applies only at *two points* on the path and *does not* require calculation of the work done by a force as the particle moves along the path.

$$T_1 + V_1 = T_2 + V_2$$

where

$$V_g = Wy \quad \text{(gravitational potential energy)}$$

$$V_e = \tfrac{1}{2}ks^2 \quad \text{(elastic potential energy)}$$

If the *power* developed by a force is to be calculated, use

$$P = \frac{dU}{dt} = \mathbf{F} \cdot \mathbf{v}$$

where $\mathbf{v}$ is the velocity of a particle acted upon by the force $\mathbf{F}$.

223

Impulse and Momentum. The equation of *linear impulse and momentum* is an integrated form of the equation of motion, $\Sigma\mathbf{F} = m\mathbf{a}$, combined with kinematics ($\mathbf{a} = d\mathbf{v}/dt$). *It is used to solve problems involving force, velocity, and time.* Before applying this equation, one should *always draw the free-body diagram,* in order to identify all the forces that cause impulses on the particle. From the diagram the impulsive and nonimpulsive forces should be identified. Recall that the nonimpulsive forces can be neglected in the analysis. Also, establish the direction of the particle's velocity just before and just after the impulses are applied. As an alternative procedure, the impulse and momentum diagrams may accompany the solution in order to graphically account for the terms in the equation.

$$m\mathbf{v}_1 + \Sigma \int_{t_1}^{t_2} \mathbf{F}\, dt = m\mathbf{v}_2$$

If several particles are involved in the problem, consider applying the *conservation of momentum* to the system in order to eliminate the internal impulses from the analysis. This can be done in a specified direction, provided no external impulses act on the particles in that direction.

$$\Sigma m\mathbf{v}_1 = \Sigma m\mathbf{v}_2$$

If the problem involves impact and the coefficient of restitution e is given, then apply the following equation.

$$e = \frac{(v_B)_2 - (v_A)_2}{(v_A)_1 - (v_B)_1} \qquad \text{(along line of impact)}$$

The *principle of angular impulse and momentum* and the *conservation of angular momentum* may be applied about an axis in order to *eliminate* some of the unknown impulses acting on the particle during the time period when its motion is studied. Investigation of the particle's free-body diagram (or the impulse diagram) will aid in choosing the axis for application.

$$(\mathbf{H}_O)_1 + \Sigma \int_{t_1}^{t_2} \mathbf{M}_O\, dt = (\mathbf{H}_O)_2$$

$$(\mathbf{H}_O)_1 = (\mathbf{H}_O)_2$$

The following problems involve application of the above concepts. They are presented in *random order* so as to provide practice in identifying the various types of problems and developing the skills necessary for their solution.

PROBLEMS

R1–1. A projectile, initially at the origin, moves along a straight-line path through a fluid medium such that its velocity is defined as $v = 1800(1 - e^{-0.3t})$ mm/s, where t is measured in seconds. Determine the displacement of the projectile during the first 3 s.

R1–2. An electric train car, having a mass of 25 Mg, travels up a 10° incline with a constant speed of 80 km/h. Determine the power required to overcome the force of gravity.

R1–3. A sports car can accelerate at 6 m/s² and decelerate at 8 m/s². If the maximum speed it can attain is 60 m/s, determine the shortest time it takes to travel 900 m starting from rest and then stopping when $s = 900$ m.

*R1–4. The 2-kg collar C is free to slide along the smooth shaft AB. Determine the acceleration of collar C if (a) the shaft is fixed from moving, (b) collar A, which is fixed to shaft AB, moves to the left at a constant velocity of 3 m/s along the horizontal guide, and (c) collar A is subjected to an acceleration of 2 m/s² to the left. In all cases, the shaft moves in the vertical plane.

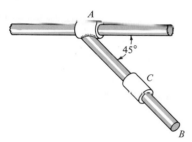

Prob. R1–4

R1–5. The speed of a train during the first minute of its motion has been recorded as follows:

$t(s)$	0	20	40	60
v(m/s)	0	16	21	24

Plot the v-t graph, approximating the curve as straight line segments between the given points. Determine the total distance traveled.

R1–6. A crate has a weight of 1500 lb. If it is pulled along the ground at a constant speed for a distance of 20 ft, and the towing cable makes an angle of 15° with the horizontal, determine the tension in the cable and the work done by the towing force. The coefficient of kinetic friction between the ground and the crate is $\mu_k = 0.55$.

R1–7. If a 150-lb crate is released from rest at A, determine its speed after it slides 30 ft down the plane. The coefficient of kinetic friction between the crate and plane is $\mu_k = 0.3$.

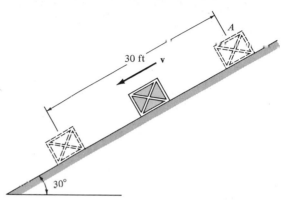

Prob. R1–7

*R1–8. A girl having a weight of 80 lb is skating around in a circle of radius $r_A = 20$ ft with a speed of $(v_A)_1 = 5$ ft/s while holding on to the end of a rope. If her partner starts to pull the rope inward with a constant speed of $v_r = 2$ ft/s, determine the girl's speed at the instant $r_B = 10$ ft. How much work is done by her partner after pulling in the rope? Neglect friction and assume the girl remains in a rigid position.

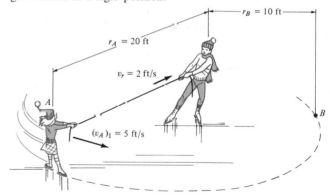

Prob. R1–8

225

R1–9. Determine the tension developed in the two cords and the acceleration of each block. Neglect the mass of the pulleys and cords. *Hint:* Since the system consists of *two* cords, relate the motion of block A to C, and of block B to C. Then, by elimination, relate the motion of A to B.

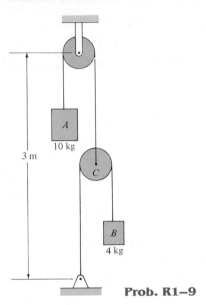

Prob. R1–9

R1–10. An automobile is traveling with a *constant speed* along a horizontal circular curve that has a radius of $\rho = 750$ ft. If the magnitude of acceleration is $a = 8$ ft/s^2, determine the speed at which the automobile is traveling.

R1–11. Disk A weighs 2 lb and is sliding on a smooth horizontal plane with a velocity of 3 ft/s. Disk B weighs 11 lb and is initially at rest. If after the impact A has a velocity of 1 ft/s directed along the positive x axis, determine the velocity of B after impact. How much kinetic energy is lost in the collision?

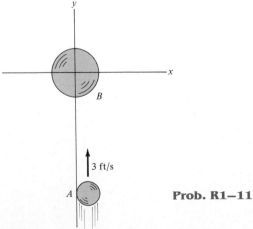

Prob. R1–11

*R1–12. At a given instant the 10-lb block A is moving downward with a speed of 6 ft/s. Determine its speed 2 s later. Block B has a weight of 4 lb, and the coefficient of kinetic friction between it and the horizontal plane is $\mu_k = 0.2$. Neglect the mass of the pulleys and cord.

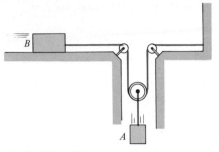

Prob. R1–12

R1–13. A 20-kg block is originally at rest on a horizontal surface for which the coefficient of both static and kinetic friction is $\mu = 0.6$. If a horizontal force F is applied such that it varies with time as shown, determine the speed of the block in 10 s. *Hint:* First determine the time needed to overcome friction and start the block moving.

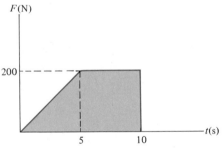

Prob. R1–13

R1–14. Two smooth billiard balls A and B have an equal mass of $m = 200$ g. If A strikes B with a velocity of $(v_A)_1 = 2$ m/s as shown, determine their final velocities just after collision. Ball B is originally at rest and the coefficient of restitution is $e = 0.75$.

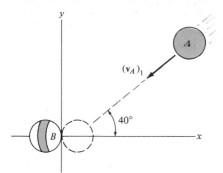

Prob. R1–14

R1–15. The 100-kg crate is subjected to the action of two forces, $F_1 = 800$ N and $F_2 = 1.5$ kN, as shown. If it is originally at rest, determine the distance it slides in order to attain a speed of 6 m/s. The coefficient of kinetic friction between the crate and the surface is $\mu_k = 0.2$.

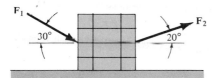

Prob. R1–15

*R1–16. If a horizontal force of $P = 10$ lb is applied to block A, determine the acceleration of block B. Neglect friction. *Hint:* Show that $a_B = a_A \tan 15°$.

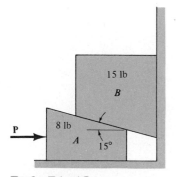

Prob. R1–16

R1–17. The spool, which has a mass of 4 kg, slides along the rotating rod. At the instant shown, the angular rate of rotation of the rod is $\dot\theta = 6$ rad/s and the rotation is increasing at $\ddot\theta = 2$ rad/s². At this same instant, the spool has a velocity of 3 m/s and an acceleration of 1 m/s², both measured relative to the rod and directed away from the center O when $r = 0.5$ m. Determine the radial frictional force and the normal force, both exerted by the rod on the spool at this instant.

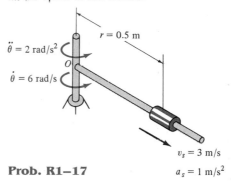

Prob. R1–17

R1–18. A freight train starts from rest at $t = 0$ and travels with a constant acceleration of 0.5 ft/s². After a time t' it maintains a constant speed so that when $t = 160$ s the train has traveled 2000 ft. Determine the time t' and draw the v-t graph for the motion.

R1–19. The block has a mass of $m = 0.5$ kg and moves within the smooth vertical slot. If it starts from rest when the *attached* spring is in the unstretched position at A, determine the *constant* vertical force F which must be applied to the cord so that the block attains a speed of $v_B = 2.5$ m/s when it reaches B; $s_B = 0.15$ m. Neglect the mass of the cord and pulley. *Hint:* The work of $\mathbf{F}$ can be determined by finding the difference Δl in cord lengths AC and BC and using $U_F = F\ \Delta l$.

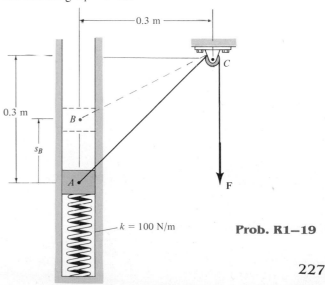

Prob. R1–19

227

Review 1: Kinematics and Kinetics of a Particle

*R1–20. An ivory ball having a mass of 200 g is released from rest at a height of 400 mm above a very large fixed metal surface. If the ball rebounds to a height of 325 mm above the surface, determine the coefficient of restitution between the ball and the surface.

R1–21. If a particle has an initial velocity of $v_0 = 12$ ft/s to the right, determine its position when $t = 10$ s, if $a = 2$ ft/s^2 to the left. Originally $s_0 = 0$.

R1–22. A 3-lb block, initially at rest at point A, slides along the smooth parabolic surface. Determine the normal force acting on the block when it reaches point B. Neglect the size of the block.

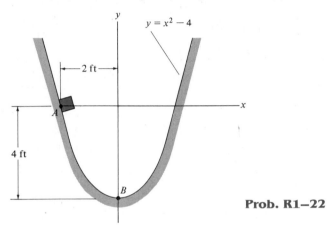

Prob. R1–22

R1–23. The crate, having a weight of 50 lb, is hoisted by the pulley system and motor M. If the crate starts from rest and, by constant acceleration, attains a speed of 12 ft/s after rising $s = 10$ ft, determine the power that must be supplied to the motor at the instant $s = 10$ ft. The motor has an efficiency of $\varepsilon = 0.74$.

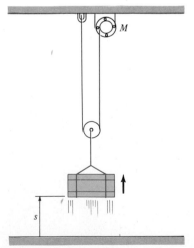

Prob. R1–23

*R1–24. A rifle has a mass of 2.5 kg. If it is loosely gripped and a 1.5-g bullet is fired from it with a muzzle velocity of 1400 m/s, determine the recoil velocity of the rifle just after firing.

R1–25. Two planes, A and B, are flying at the same altitude. If their velocities are $v_A = 450$ mi/h and $v_B = 400$ mi/h such that the angle between their straight-line courses is $\theta = 75°$, determine the velocity of plane B with respect to plane A.

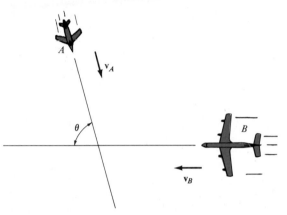

Prob. R1–25

R1–26. The 90-lb force is required to drag the 200-lb block 60 ft up the *rough* inclined plane at constant velocity. If the force is removed when the block reaches point B, and the block is then released from rest, determine the block's velocity when it slides back down the plane and reaches point A.

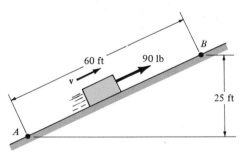

Prob. R1–26

228

R1–27. The boy of mass 40 kg is sliding down the spiral slide at a constant speed such that his position, measured from the top of the chute, has components $r = 2$ m, $\theta = (0.7t)$ rad, and $z = (-0.5t)$ m, where t is in seconds. Determine the components of force $\mathbf{F}_r$, $\mathbf{F}_\theta$, and $\mathbf{F}_z$ which he exerts on the chute at the instant $t = 2$ s.

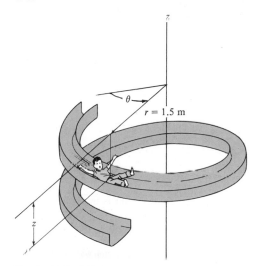

Prob. R1–27

*R1–28. A spring having a stiffness of 5 kN/m is compressed 400 mm. The stored energy in the spring is used to drive a machine which requires 80 W of power. Determine how long the spring can supply energy at the required rate.

R1–29. A tugboat T having a mass of 19 Mg is tied to a barge B having a mass of 75 Mg. If the rope is "elastic" such that it has a stiffness of $k = 600$ kN/m, determine the maximum stretch in the rope during the initial towing. Originally both the tugboat and barge are moving in the same direction with speeds of $(v_T)_1 = 15$ km/h and $(v_B)_1 = 10$ km/h, respectively. Neglect the resistance of the water. When taut, the rope becomes horizontal.

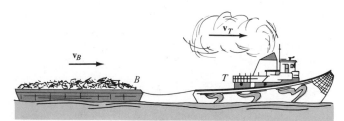

Prob. R1–29

R1–30. The conveyor belt delivers each 15-lb crate to the ramp at A such that the crate's velocity is $v_A = 4$ ft/s, directed down *along* the ramp. If the coefficient of kinetic friction between each crate and the ramp is $\mu_k = 0.3$, determine the speed at which each crate slides off the ramp at B. Assume that no tipping occurs.

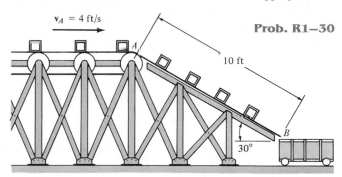

Prob. R1–30

R1–31. The slotted arm AB drives the pin C through the spiral groove described by the equation $r = 1.5\theta$, where θ is in radians and r is in feet. If the arm starts from rest when $\theta = 60°$ and is driven at an angular rate of $\dot{\theta} = 4t$ rad/s, determine the radial and transverse components of velocity and acceleration of the pin when $t = 1$ s.

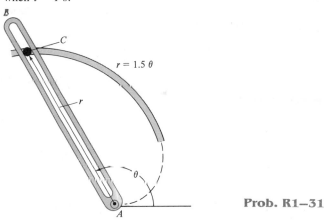

Prob. R1–31

*R1–32. A skier starts from rest at A (30 ft, 0) and descends the smooth slope, which may be approximated by a parabola. If she has a weight of 120 lb, determine the normal force she exerts on the ground at the instant she arrives at point B.

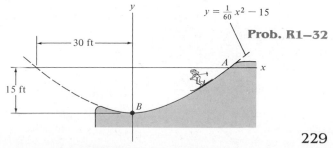

Prob. R1–32

229

Review 1: Kinematics and Kinetics of a Particle

R1–33. The catapulting mechanism is used to propel the 15-kg slider A to the right along the smooth track. The propelling action is obtained by drawing the pulley attached to rod BC rapidly to the left by means of a piston P. If the piston applies a constant force of F = 8 kN to rod BC such that it moves it 0.2 m, determine the speed attained by the slider if it was originally at rest. Neglect the mass of the pulleys, cable, piston, and rod BC.

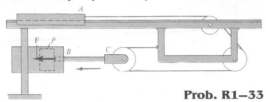

Prob. R1–33

R1–34. A railroad car having a mass of 15 Mg is coasting at 2 m/s on a horizontal track. At the same time another car having a mass of 10 Mg is coasting at 0.75 m/s in the opposite direction. If the cars meet and couple together, determine the speed of both cars just after the coupling. Compute the difference between the total kinetic energy before and after coupling has occurred, and explain qualitatively what happened to this energy.

R1–35. A boy at A throws a ball 45° from the horizontal such that it strikes the slope at B. Determine the speed at which the ball is thrown and the time of flight.

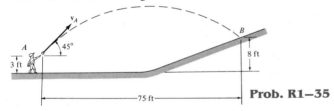

Prob. R1–35

*∎R1–36. Packages having a mass of 2.5 kg ride on the surface of the conveyor belt. If the belt starts from rest and with constant acceleration increases to a speed of 0.75 m/s in 2 s, determine the maximum angle of tilt, θ, so that none of the packages slip on the inclined surface AB of the belt. The coefficient of static friction between the belt and each package is $\mu_s = 0.3$. At what angle φ do the packages first begin to slip off the surface of the belt if the belt is moving at a constant speed of 0.75 m/s?

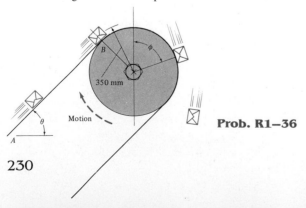

Prob. R1–36

R1–37. To test the manufactured properties of 2-lb steel balls, each ball is released from rest as shown and strikes a 45° inclined surface. If the coefficient of restitution is e = 0.8, determine the distance s to where the ball strikes the horizontal plane at A. At what speed does the ball strike point A?

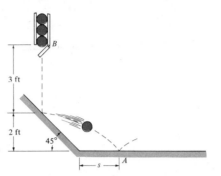

Prob. R1–37

R1–38. A car has a weight of 2500 lb and is traveling forward horizontally at 60 ft/s. If the coefficient of kinetic friction between the tires and the pavement is $\mu_k = 0.6$, determine the time needed to stop the car. The brakes are applied to all four wheels.

R1–39. Packages having a mass of 6 kg slide down a smooth chute and land horizontally with a speed of 3 m/s on the surface of a conveyor belt. If the coefficient of kinetic friction between the belt and a package is $\mu_k = 0.2$, determine the time needed to bring the package to rest on the belt if the belt is moving in the same direction as the package with a speed of v = 1 m/s.

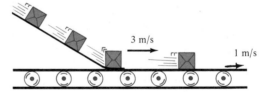

Prob. R1–39

*R1–40. Block B rests on a smooth surface. If the coefficient of static friction between A and B is $\mu_s = 0.4$, determine the acceleration of each block if (a) F = 6 lb, and (b) F = 50 lb.

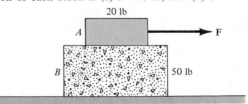

Prob. R1–40

R1–41. The roller-coaster car has a mass of 800 kg, including its passenger, and moves from the top of the hill A with a speed of $v_A = 3$ m/s. Determine the minimum height h of the hill crest so that the car travels around both inside loops without leaving the track. Neglect friction and the size of the car. What is the normal reaction on the car when the car is at B and when it is at C?

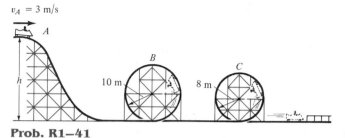

Prob. R1–41

R1–42. A spiral transition curve is used on railroads to connect a straight portion of the track with a curved portion. If the spiral is defined by the equation $y = (10^{-6})x^3$, where x and y are in feet, determine the magnitude of the acceleration of a train engine moving with a constant speed of 40 ft/s when it is at point $x = 600$ ft.

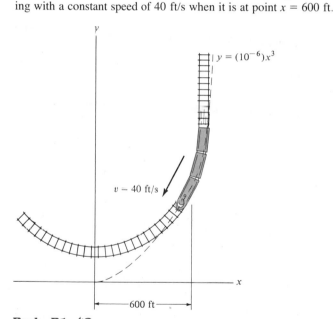

Prob. R1–42

R1–43. Determine the velocity of each block 2 s after the blocks are released from rest. Neglect the mass of the pulleys and cord.

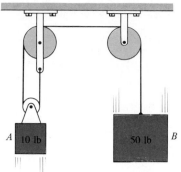

Prob. R1–43

*R1–44. The four 5-lb spheres are rigidly attached to the cross-bar frame. The frame has a negligible weight. If a couple moment $M = (5t)$ lb · ft, where t is in seconds, is applied as shown, determine the speed of each of the spheres in 4 s starting from rest.

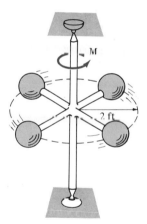

Prob. R1–44

Review 1: Kinematics and Kinetics of a Particle

R1–45. The small 2-lb collar starting from rest at A slides down along the smooth wire. During the motion, the collar is acted upon by a force $\mathbf{F} = \{10\mathbf{i} + 6y\mathbf{j} + 2z\mathbf{k}\}$ lb, where x, y, z are in feet. Determine the collar's speed when it strikes the wall at B.

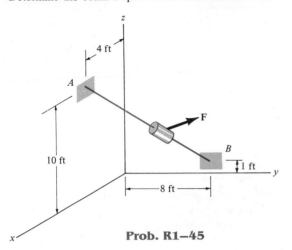

Prob. R1–45

R1–46. In each case, if the end of the cable at A is pulled down with a speed of 2 m/s, determine the speed at which block B rises.

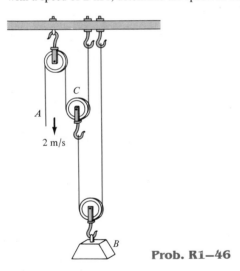

Prob. R1–46

R1–47. A particle is moving along a circular path of 2-m radius such that its position as a function of time is given by $\theta = (5t^2)$ rad, where t is in seconds. Determine the magnitude of the particle's acceleration when $\theta = 30°$. The particle starts from rest when $\theta = 0°$.

***R1–48.** Packages having a weight of 5 lb ride on the surface of the conveyor belt. If the belt travels at a constant speed of 2 ft/s and the coefficient of static friction between the belt and each package is $\mu_s = 0.4$, determine the angle θ at which the packages first begin' to slip off the surface of the belt.

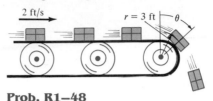

Prob. R1–48

R1–49. Four inelastic cables C are attached to a plate P and hold the spring 6 in. in compression when *no force* acts on the plate. If a block B, having a weight of 5 lb, is placed on the plate and the plate is pushed down 8 in. and released from rest, determine how high the block rises from the point where it was released. Neglect the mass of the plate.

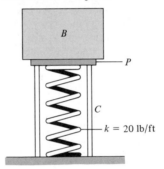

Prob. R1–49

R1–50. The gantry structure is used to test the response of an airplane during a crash. The plane having a mass of 8 Mg is hoisted back until $\theta = 60°$, and then the pull-back cable AC is released when the plane is at rest. Determine the speed of the plane just before crashing into the ground, $\theta = 15°$. What is the maximum tension developed in the supporting cable during the motion? Neglect the effects of lift caused by the wings.

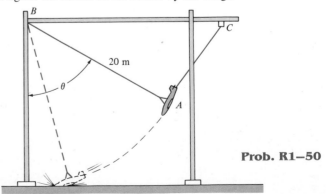

Prob. R1–50

16

Planar Kinematics of a Rigid Body

In this chapter kinematics or the study of the geometry of the planar motion for a rigid body will be discussed. This study is important for the design of gears, cams, and mechanisms used for machine operations. Furthermore, once the kinematics of a rigid body is thoroughly understood, it will be possible to apply the equations of motion, which relate the forces on the body to the body's motion.

A rigid body can be subjected to three types of planar motion, namely, translation, rotation about a fixed axis, and general plane motion. We will first define each of these motions, then discuss the analysis of each separately. In all cases it will be shown that rigid-body planar motion is completely specified provided the motions of any two points on the body are known. The more complex study of three-dimensional kinematics of a rigid body is presented in Chapter 20.

16.1 Rigid-Body Motion

When each of the particles of a rigid body moves along a path which is equidistant from a fixed plane, the body is said to undergo *planar motion*. As previously stated, there are three types of planar motion; in order of increasing complexity, they are:

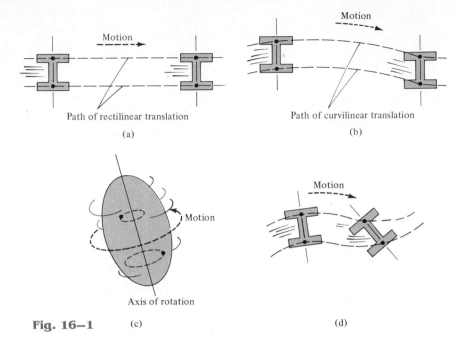

Fig. 16—1   (c)   (d)

1. *Translation*. This type of motion occurs if any line segment on the body remains parallel to its original direction during the motion. When the paths of motion for all the particles of the body are along equidistant straight lines, Fig. 16–1a, the motion is called *rectilinear translation*. However, if the paths of motion are along curved lines which are equidistant, Fig. 16–1b, the motion is called *curvilinear translation*.

2. *Rotation about a fixed axis*. When a rigid body rotates about a fixed axis, all the particles of the body, except those which lie on the axis of rotation, move along circular paths, Fig. 16–1c.

3. *General plane motion*. When a body is subjected to general plane motion, it undergoes a combination of translation *and* rotation, Fig. 16–1d. The translation occurs within a reference plane, and the rotation occurs about an axis perpendicular to the reference plane.

The above planar motions are exemplified by the moving parts of the crank mechanism shown in Fig. 16–2.

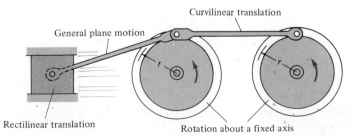

Fig. 16—2

16.2 Translation

Consider a rigid body which is subjected to either rectilinear or curvilinear translation in the x-y plane, Fig. 16–3.

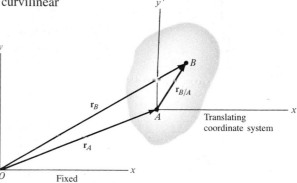

Fixed
coordinate system

Translating
coordinate system

Fig. 16–3

Position

The locations of points A and B in the body are defined from the fixed x, y reference frame by using *position vectors* $\mathbf{r}_A$ and $\mathbf{r}_B$. The translating x', y' coordinate system is *fixed in the body* and has its origin located at A, hereafter referred to as the *base point*. The position of B with respect to A is denoted by the *relative-position vector* $\mathbf{r}_{B/A}$ ("B of B with respect to A"). By vector addition,

$$\mathbf{r}_B = \mathbf{r}_A + \mathbf{r}_{B/A}$$

Velocity

A relationship between the instantaneous velocities of A and B is obtained by taking the time derivative of the position equation, which yields $\mathbf{v}_B = \mathbf{v}_A + d\mathbf{r}_{B/A}/dt$. Here $\mathbf{v}_A$ and $\mathbf{v}_B$ denote *absolute velocities* since these vectors are measured from the x, y axes. The term $d\mathbf{r}_{B/A}/dt = \mathbf{0}$, since the *magnitude* of $\mathbf{r}_{B/A}$ is constant by definition of a rigid body, and because the body is translating the *direction* of $\mathbf{r}_{B/A}$ is constant. Therefore,

$$\mathbf{v}_B = \mathbf{v}_A$$

Acceleration

Taking the time derivative of the velocity equation yields a similar relationship between the instantaneous accelerations of A and B,

$$\mathbf{a}_B = \mathbf{a}_A$$

The above two equations indicate that *all points in a rigid body subjected to either curvilinear or rectilinear translation move with the same velocity and acceleration*. As a result, the kinematics of particle motion, discussed in Chapter 12, may also be applied to specify the kinematics of the particles located in a translating rigid body.

16.3 Rotation About a Fixed Axis

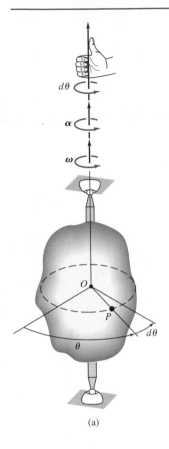

When a body is rotating about a fixed axis, any point P located in the body travels along a *circular path*. This motion depends upon the angular motion of the body about the axis. For this reason, we will first study the angular-motion properties of the body before analyzing the circular motion of P.

Angular Motion

Since a point is without dimension, it has no angular motion. Only lines or bodies undergo angular motion. To study its effects, we will consider the body shown in Fig. 16–4a and the angular motion of a radial line r located within the shaded plane.

Angular Position. At the instant shown, the *angular position* of r is defined by the angle θ, measured between a *fixed* reference line and r. Here r extends perpendicular to the axis of rotation from point O to a point P in the body.

Angular Displacement. The change in the angular position, often measured as a differential $d\theta$, is called the *angular displacement*.* This vector has a *magnitude* of $d\theta$, which is measured in degrees, radians, or revolutions, where 1 rev $= 2\pi$ rad. Since motion is about a *fixed axis,* the direction of $d\theta$ is *always* along the axis. Specifically, the *direction* of $d\theta$ is determined by the right-hand rule; that is, the fingers of the right hand are curled with the sense of rotation, so that in this case the thumb, or $d\theta$, points upward, Fig. 16–4a. In two dimensions, as shown on the top view of the shaded plane, Fig. 16–4b, both θ and $d\theta$ are directed counterclockwise and so the thumb points outward.

Angular Velocity. The time rate of change in the angular position is called the *angular velocity* ω (omega). Since $d\theta$ occurs during an instant of time dt, then,

$(\downarrow+)$

$$\omega = \frac{d\theta}{dt}$$

(16–1)

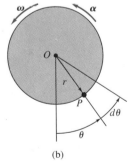

(a)

(b)

Fig. 16–4(a, b)

This vector has a *magnitude* which is often measured in rad/s. It is expressed here in scalar form, since its *direction* is always along the axis of rotation, i.e., in the same direction as $d\theta$, Fig. 16-4a. When indicating the angular motion in the shaded plane, Fig. 16–4b, we can refer to the sense of rotation as clockwise or counterclockwise. Here we have *arbitrarily* chosen counterclockwise rotations as *positive,* and indicated this by the curl shown in parentheses next to Eq. 16–1. Realize, however, that the directional sense of ω is actually outward.

---

*It is shown in Sec. 20.1 that finite rotations or finite angular displacements are *not* vector quantities, although differential rotations $d\theta$ are vectors.

Angular Acceleration. The *angular acceleration* $\boldsymbol{\alpha}$ (alpha) measures the time rate of change of the angular velocity. Hence, the *magnitude* of this vector may be written as

$(\curvearrowright +)$
$$\boxed{\alpha = \frac{d\omega}{dt}}$$
(16–2)

Or taking the second time derivative of Eq. 16–1, it is possible to express α as

$(\curvearrowright +)$
$$\alpha = \frac{d^2\theta}{dt^2}$$
(16–3)

The line of action of $\boldsymbol{\alpha}$ is the same as that for $\boldsymbol{\omega}$, Fig. 16–4a; however, its sense of *direction* depends upon whether $\boldsymbol{\omega}$ is increasing or decreasing with time. In particular, if $\boldsymbol{\omega}$ is decreasing $\boldsymbol{\alpha}$ is called an *angular deceleration* and therefore has a sense of direction which is opposite to $\boldsymbol{\omega}$.

By eliminating dt from Eqs. 16–1 and 16–2, it is possible to obtain a differential relation between the angular acceleration, angular velocity, and angular displacement, namely,

$(\curvearrowright +)$
$$\boxed{\alpha \, d\theta = \omega \, d\omega}$$
(16–4)

The similarity between the differential relations for angular motion and those developed for rectilinear motion of a particle ($v = ds/dt$, $a = dv/dt$, and $a \, ds = v \, dv$) should be apparent.

Constant Angular Acceleration. If the angular acceleration of the body is constant, $\boldsymbol{\alpha} = \boldsymbol{\alpha}_c$, Eqs. 16–1, 16–2, and 16–4, when integrated, yield a set of formulas which relate the body's angular velocity, its angular position, and time. These equations are similar to Eqs. 12–4 to 12–6 used for rectilinear motion. The results are

$(\curvearrowright +)$
$(\curvearrowright +)$
$(\curvearrowright +)$
$$\boxed{\begin{aligned} \omega &= \omega_0 + \alpha_c t \\ \theta &= \theta_0 + \omega_0 t + \tfrac{1}{2}\alpha_c t^2 \\ \omega^2 &= \omega_0^2 + 2\alpha_c(\theta - \theta_0) \\ \text{Constant Angular Acceleration} \end{aligned}}$$

(16–5)
(16–6)
(16–7)

Here θ_0 and ω_0 are the initial values of the body's angular position and angular velocity, respectively.

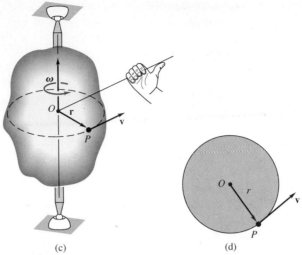

(c) (d)

Fig. 16–4(c, d)

Motion of Point P

As the rigid body in Fig. 16–4c rotates, point P travels along a *circular path* of radius r and with center at point O. This path is contained in the shaded plane shown in top view, Fig. 16–4d.

Position. The position of P is defined by the position vector **r**, which extends from O to P.

Velocity. The velocity of P has a magnitude which can be determined from the circular motion of P using its polar coordinate components $v_r = \dot{r}$ and $v_\theta = r\dot{\theta}$, Eqs. 12–25. Since r is constant, the radial component $v_r = \dot{r} = 0$, so that $v = v_\theta = r\dot{\theta}$. Because $\omega = \dot{\theta}$, Eq. 16–1, the result is

$$v = \omega r \tag{16–8}$$

As shown in Figs. 16–4c and 16–4d, the *direction* of **v** is *tangent* to the circular path.

Both the magnitude and direction of **v** can be accounted for by using the cross product of **ω** and **r** (see Appendix C). We have

$$\mathbf{v} = \boldsymbol{\omega} \times \mathbf{r} \tag{16–9}$$

The order of the vectors in this formulation is important, since the cross product is not commutative, i.e., $\boldsymbol{\omega} \times \mathbf{r} \neq \mathbf{r} \times \boldsymbol{\omega}$. In this regard, notice in Fig. 16–4c how the correct direction of **v** is established by the right-hand rule. The fingers of the right hand are curled from **ω** to **r** (**ω** "cross" **r**). The thumb indicates the correct direction of **v**, that is, tangent to the path in the direction of motion.

Acceleration. For convenience, the acceleration of P will be expressed in terms of its normal and tangential components.* Using $a_t = dv/dt$ and $a_n = v^2/\rho$, noting that $\rho = r$, $v = \omega r$, and $\alpha = d\omega/dt$, we have

$$a_t = \alpha r \tag{16-10}$$

$$a_n = \omega^2 r \tag{16-11}$$

The *tangential component of acceleration*, Fig. 16–4e and f, represents the time rate of change in the velocity's magnitude. If the speed of P is increasing, then $\mathbf{a}_t$ acts in the same direction as $\mathbf{v}$; if the speed is decreasing, $\mathbf{a}_t$ acts in the opposite direction of $\mathbf{v}$, and finally, if the speed is constant, $\mathbf{a}_t$ is zero.

The *normal component of acceleration* represents the time rate of change in the velocity's direction. The *direction* of $\mathbf{a}_n$ is always towards O, the center of the circular path, Fig. 16–4e and f.

Like the velocity, the acceleration of point P may be expressed in terms of the vector cross product. Taking the time derivative of Eq. 16–9 yields

$$\mathbf{a} = \frac{d\mathbf{v}}{dt} = \frac{d\boldsymbol{\omega}}{dt} \times \mathbf{r} + \boldsymbol{\omega} \times \frac{d\mathbf{r}}{dt}$$

Recalling that $\boldsymbol{\alpha} = d\boldsymbol{\omega}/dt$, and using Eq. 16–9 ($d\mathbf{r}/dt = \mathbf{v} = \boldsymbol{\omega} \times \mathbf{r}$), we have

$$\mathbf{a} = \boldsymbol{\alpha} \times \mathbf{r} + \boldsymbol{\omega} \times (\boldsymbol{\omega} \times \mathbf{r}) \tag{16-12}$$

By definition of the cross product, the first term on the right has a magnitude $a_t = \alpha r$, and by the right hand rule, $\boldsymbol{\alpha} \times \mathbf{r}$ is in the direction of $\mathbf{a}_t$, Fig. 16–4e. Likewise, the second term has a magnitude $a_n = \omega^2 r$, and applying the right-hand rule twice, first $\boldsymbol{\omega} \times \mathbf{r}$ then $\boldsymbol{\omega} \times (\boldsymbol{\omega} \times \mathbf{r})$, it can be seen that this result is in the same direction as $\mathbf{a}_n$, Fig. 16–4e. Noting that this is also the *same* direction as $-\mathbf{r}$, we can express $\mathbf{a}_n$ in a much simpler form as $\mathbf{a}_n = -\omega^2\mathbf{r}$. Hence, Equation 16–12 becomes

$$\begin{aligned} \mathbf{a} &= \mathbf{a}_t + \mathbf{a}_n \\ &= \boldsymbol{\alpha} \times \mathbf{r} - \omega^2 \mathbf{r} \end{aligned} \tag{16-13}$$

Since $\mathbf{a}_t$ and $\mathbf{a}_n$ are perpendicular, Fig. 16–4e and f, if needed the magnitude of acceleration can be determined from the Pythagorean theorem; namely, $a = \sqrt{a_n^2 + a_t^2}$.

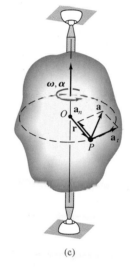

(e)

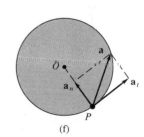

(f)

Fig. 16–4(e, f)

---

*Polar coordinates can also be used. Since $a_r = \ddot{r} - r\dot{\theta}^2$ and $a_\theta = r\ddot{\theta} + 2\dot{r}\dot{\theta}$, substituting $\dot{r} = \ddot{r} = 0$, $\dot{\theta} = \omega$, $\ddot{\theta} = \alpha$, we obtain Eqs. 16–10 and 16–11.

PROCEDURE FOR ANALYSIS

In order to determine the velocity and acceleration of a point located in a rigid body that is rotating about a fixed axis, it is first necessary to know the body's angular velocity and angular acceleration. If Eqs. 16–1 to 16–7 are used to obtain ω and α, it is important that a *positive direction* along the axis of rotation be established. By doing so, the sense of θ, ω, and α can be determined from the algebraic signs of their numerical quantities. In the examples that follow, the positive sense will be indicated by a curl alongside each kinematic equation as it is applied.

Angular Motion. If $\boldsymbol{\alpha}$ or $\boldsymbol{\omega}$ are unknown, then the relationships between the angular motions are defined by the differential equations:

$$\omega = \frac{d\theta}{dt} \qquad \alpha = \frac{d\omega}{dt} \qquad \alpha\, d\theta = \omega\, d\omega$$

If one is *absolutely certain* that the body's angular acceleration is *constant,* the following equations can be used:

$$\omega = \omega_0 + \alpha_c t$$
$$\theta = \theta_0 + \omega_0 t + \tfrac{1}{2}\alpha_c t^2$$
$$\omega^2 = \omega_0^2 + 2\alpha_c(\theta - \theta_0)$$

Motion of P. When the motion of a point P in the body is to be determined, it is suggested that a *kinematic diagram* accompany the problem solution. This diagram is simply a graphical representation showing the motion of the point.

In most cases the velocity and the two components of acceleration can be determined from the scalar equations

$$v = \omega r$$
$$a_t = \alpha r$$
$$a_n = \omega^2 r$$

However, if the geometry of the problem is difficult to visualize, the following vector equations should be used:

$$\mathbf{v} = \boldsymbol{\omega} \times \mathbf{r}$$
$$\mathbf{a}_t = \boldsymbol{\alpha} \times \mathbf{r}$$
$$\mathbf{a}_n = -\omega^2\mathbf{r}$$

For application, $\mathbf{r}$, $\boldsymbol{\omega}$, and $\boldsymbol{\alpha}$ are expressed in terms of their $\mathbf{i}$, $\mathbf{j}$, $\mathbf{k}$ components and, if necessary, the cross products are computed by using a determinant expansion (see Eq. C–12).

The following examples numerically illustrate application of these equations.

Example 16–1

A cord is wrapped around a wheel which is initially at rest as shown in Fig. 16–5. If a force is applied to the cord and gives it an acceleration $a = (4t)\text{m/s}^2$, where t is in seconds, determine as a function of time (a) the angular velocity of the wheel, and (b) the angular position of line OP in radians.

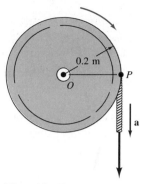

SOLUTION

Part (a). The wheel is subjected to rotation about a fixed axis passing through point O. Thus, point P on the wheel has motion about a circular path, and therefore the acceleration of this point has both tangential and normal components. In particular, $(a_P)_t = (4t) \text{ m/s}^2$, since the cord is connected to the wheel and *tangent* to it at P. Hence, the angular acceleration of the wheel is

$(\curvearrowright +)$

$$(a_P)_t = \alpha r$$
$$4t = \alpha(0.2)$$
$$\alpha = 20t \text{ rad/s}^2$$

Fig. 16–5

The wheel's angular velocity ω can be determined by using $\alpha = d\omega/dt$, since this equation relates α, t, and ω. Integrating, with the initial condition that $\omega = 0$ at $t = 0$, yields

$(\curvearrowright +)$

$$\alpha = \frac{d\omega}{dt} = 20t$$
$$\int_0^\omega d\omega = \int_0^t 20t \, dt$$
$$\omega = 10t^2 \text{ rad/s} \downarrow \qquad\qquad \textit{Ans.}$$

Why is it not possible to use Eq. 16–5 ($\omega = \omega_0 + \alpha_c t$) to obtain this result?

Part (b). The angular position θ of the radial line OP can be computed using $\omega = d\theta/dt$, since this equation relates θ, ω, and t. Using the initial condition $\theta = 0$ at $t = 0$, we have

$(\curvearrowright +)$

$$\frac{d\theta}{dt} = \omega = 10t^2$$
$$\int_0^\theta d\theta = \int_0^t 10t^2 \, dt$$
$$\theta = 3.33 \, t^3 \text{ rad} \qquad\qquad \textit{Ans.}$$

Example 16–2

Disk A, shown in Fig. 16–6a, starts from rest and rotates with a constant angular acceleration of $\alpha_A = 2$ rad/s^2. If disk A is in contact with disk B and no slipping occurs between the disks, determine the angular velocity and angular acceleration of B just after A turns 10 revolutions.

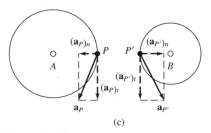

(a)

SOLUTION

The angular velocity of A can be determined using the equations of *constant* angular acceleration since $\alpha_A = 2$ rad/s^2. There are 2π radians to one revolution, so that

$$\theta_A = 10 \text{ rev} \left(\frac{2\pi \text{ rad}}{1 \text{ rev}} \right) = 62.83 \text{ rad}$$

Thus,

$$(\curvearrowright +) \qquad \omega^2 = \omega_0^2 + 2\alpha_c(\theta - \theta_0)$$
$$\omega_A^2 = 0 + 2(2)(62.83 - 0)$$
$$\omega_A = 15.9 \text{ rad/s} \downarrow$$

As shown in Fig. 16–6b, the speed of the contacting point P on the rim of A is

$$(+\downarrow) \qquad v_P = \omega_A r_A = (15.9)(2) = 31.8 \text{ ft/s} \downarrow$$

The velocity is always tangent to the path of motion, so that the speed of point P' on B is the *same* as the speed of P on A. The angular velocity of B is therefore

$$(\curvearrowleft +) \qquad \omega_B = \frac{v_{P'}}{r_B} = \frac{31.8}{1.5} = 21.2 \text{ rad/s} \curvearrowleft \qquad \textit{Ans.}$$

(b)

(c)

Fig. 16–6

As shown in Fig. 16–6c, the *tangential components* of acceleration of both disks are equal, since the disks are in contact with one another. Hence,

$$(a_P)_t = (a_{P'})_t$$
$$\alpha_A r_A = \alpha_B r_B$$

$$\alpha_B = \alpha_A \left(\frac{r_A}{r_B} \right) = 2 \left(\frac{2}{1.5} \right) = 2.67 \text{ rad/s}^2 \curvearrowleft \qquad \textit{Ans.}$$

Notice that the normal components of acceleration $(a_P)_n$ and $(a_{P'})_n$ act in *opposite directions*, since the paths of motion for both points are *different*. Furthermore, $(a_P)_n \neq (a_{P'})_n$, since the *magnitudes* of these components depend upon the radius and angular velocity of each disk, i.e., $(a_P)_n = \omega_A^2 r_A$ and $(a_{P'})_n = \omega_B^2 r_B$.

PROBLEMS

16–1. A disk having a radius of 0.5 ft rotates with an initial angular velocity of 2 rad/s and has a constant angular acceleration of 1 rad/s^2. Determine the magnitude of the velocity and acceleration of a point on the rim of the disk when $t = 2$ s.

16–2. A wheel has its angular speed increased uniformly from 20 rad/s when $t = 0$ to 40 rad/s when $t = 35$ s. If the diameter of the wheel is 1 ft, determine the magnitudes of the normal and tangential components of acceleration of a point on the rim of the wheel when $t = 10$ s, and the total distance the point travels during the 35-s time period.

16–3. If the angular velocity of the drum is increased uniformly from 3 rev/min when $t = 0$ to 10 rev/min when $t = 4$ s, determine the magnitudes of the velocity and acceleration of point A on the drum when $t = 4$ s.

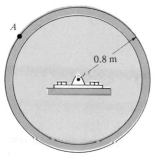

Prob. 16–3

*16–4. The angular position of a disk is defined by $\theta = (t + 4t^2)$ rad, where t is in minutes. Determine the number of revolutions, the angular velocity, and angular acceleration of the disk when $t = 90$ s.

16–5. The disk is originally rotating at $\omega_1 = 8$ rad/s. If it is subjected to a constant angular acceleration of $\alpha = 6$ rad/s^2, determine the magnitudes of the velocity and the n and t components of acceleration of point A at the instant $t = 3$ s.

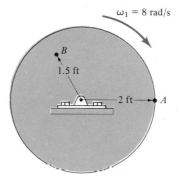

Probs. 16–5/16–6

16–6. The disk is originally rotating at $\omega_1 = 8$ rad/s. If it is subjected to an angular acceleration of $\alpha = 6$ rad/s^2, determine the magnitudes of the velocity and the n and t components of acceleration of point B just after the wheel undergoes 2 revolutions.

16–7. The gear A on the drive shaft of the outboard motor has a radius $r_A = 0.5$ in. and the meshed pinion gear B on the propeller shaft has a radius $r_B = 1.2$ in. Determine the angular velocity of the propeller in $t = 1.5$ s, if the drive shaft rotates with an angular acceleration $\alpha = (400t^3)$ rad/s^2, where t is in seconds. The propeller is originally at rest and the motor frame does not move.

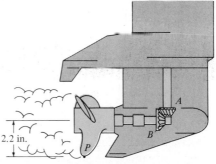

Probs. 16–7/16–8

*16–8. For the outboard motor in Prob. 16–7, determine the magnitudes of the velocity and acceleration of a point P located on the tip of the propeller at the instant $t = 0.75$ s.

16–9. A textile mill uses the belt-and-pulley arrangement shown to transmit power. If an electric motor turns pulley A at 5 rad/s, compute the angular velocities of pulleys B and C. The hub at D is rigidly *connected* to C and turns with it.

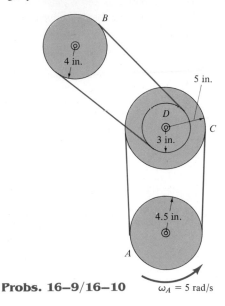

Probs. 16–9/16–10 $\omega_A = 5$ rad/s

16–10. A textile mill uses the belt-and-pulley arrangement shown to transmit power. When $t = 0$ an electric motor is turning pulley A with an angular velocity of $\omega_A = 5$ rad/s. If this pulley is subjected to a constant angular acceleration of 2 rad/s², determine the angular velocity of pulley B after B turns 6 revolutions. The hub at D is rigidly *connected* to pulley C and turns with it.

16–11. If disk A has an initial angular velocity of $\omega_1 = 6$ rad/s and a constant angular acceleration of $\alpha_A = 3$ rad/s², determine the magnitudes of the velocity and acceleration of block B in 2 s.

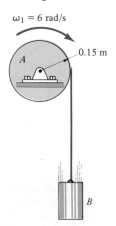

$\omega_1 = 6$ rad/s

0.15 m

A

B

Probs. 16–11/16–12

*16–12. Solve Prob. 16–11 assuming that a motor gives the disk A an angular acceleration of $\alpha_A = (0.6t^2 + 0.75)$ rad/s² and initial angular velocity of $\omega_1 = 6$ rad/s, where t is in seconds.

16–13. A belt is wrapped around the inner hubs of pulleys A and B as shown. Each hub is fixed to its respective pulley and turns with it. If A has a constant angular acceleration $\alpha_A = 6$ rad/s², determine the velocity of block C in 3 s if it starts from rest. How high does it rise in 3 s?

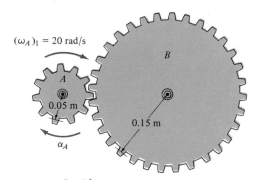

Prob. 16–13

16–14. A motor gives gear A an angular acceleration of $\alpha_A = (0.25\theta^3 + 0.5)$ rad/s², where θ is in radians. If this gear is initially turning at $(\omega_A)_1 = 20$ rad/s, determine the angular velocity of gear B after gear A undergoes an angular displacement of 10 rev.

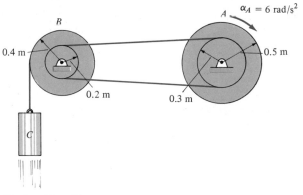

Prob. 16–14

16–15. Determine the distance the load W is lifted in $t = 5$ s using the hoist. The shaft of the motor M turns with an angular velocity $\omega = 100(4 + t)$ rad/s, where t is in seconds.

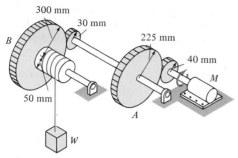

300 mm
30 mm
B
225 mm
40 mm
M
50 mm
A
W

Prob. 16–15

*16–16. Due to an increase in power, the motor M rotates the shaft A with an angular acceleration $\alpha = (0.06\theta^2)$ rad/s^2, where θ is in radians. If the shaft is initially turning at $\omega_1 = 50$ rad/s, determine the angular velocity of gear B after the shaft undergoes an angular displacement $\Delta\theta = 10$ rev.

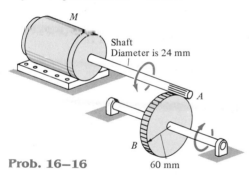

M
Shaft
Diameter is 24 mm
A
B

Prob. 16–16   60 mm

16–17. The angular acceleration of a wheel having a diameter of 2 ft is defined by $\alpha = (3 + 0.1\theta)$ rad/s^2, where θ is in radians. Determine the magnitudes of the velocity and acceleration of a point P on its rim when $\theta = 2$ rad. How much time is needed to reach this angular position? Originally, $\theta = 0$, and $\omega = 0$ when $t = 0$.

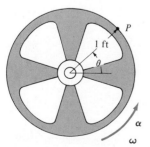

P
1 ft
θ
α
ω **Prob. 16–17**

16–18. Just after the fan is turned on, the motor gives the blade an angular acceleration $\alpha = (20e^{-0.6t})$ rad/s^2, where t is in seconds. Determine the speed of the tip P of one of the blades when $t = 3$ s. How many revolutions has the blade turned in 3 s? When $t = 0$ the blade is at rest.

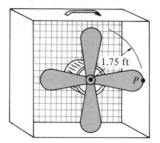

1.75 ft
P

Prob. 16–18

16–19. The rope of diameter d is wrapped around the tapered drum which has the dimensions shown. If the drum is rotating at a constant rate of ω, determine the upward acceleration of the block. Neglect the small horizontal displacement of the block.

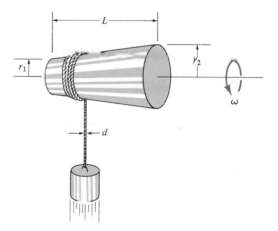

L
r_1
r_2
ω
d

Prob. 16–19

*16.4 Absolute General Plane Motion Analysis

A body subjected to *general plane motion* undergoes a *simultaneous* translation and rotation. If the body is represented by a thin slab, the slab translates in the plane and rotates about an axis perpendicular to the plane. The analysis of this motion can be specified by knowing *both* the angular rotation of a line fixed in the body and the rectilinear motion of a point in the body. One way to define these motions is to use a position coordinate s to specify the location of the point along its path and an angular position coordinate θ to specify the orientation of the line. The two coordinates are then related using the geometry of the problem. By *direct application* of the time-differential equations $v = ds/dt$, $a = dv/dt$, $\omega = d\theta/dt$, and $\alpha = d\omega/dt$, the *motion* of the point and the *angular motion* of the line can then be related. In some cases, this procedure may also be used to relate the motions of one body to those of a connected body.

PROCEDURE FOR ANALYSIS

The following procedure provides a method for relating the velocity and acceleration of a point P undergoing rectilinear motion to the angular velocity and angular acceleration of a line contained in a body.

Position Coordinate Equation. Locate point P using a position coordinate s, which is measured from a *fixed origin* and is *directed along the straight-line path of motion* of point P. Also measure from a fixed reference line the angular position θ of a line lying in the body. From the dimensions of the body, relate s to θ, $s = f(\theta)$, using geometry and/or trigonometry.

Time Derivatives. Take the first derivative of $s = f(\theta)$ with respect to time to get a relationship between v and ω. Take the second time derivative to get a relationship between a and α.

This procedure is illustrated in the following example problems.

Example 16–3

At a given instant, the cylinder of radius r, shown in Fig. 16–7, has an angular velocity $\boldsymbol{\omega}$ and angular acceleration $\boldsymbol{\alpha}$. Determine the velocity and acceleration of its center G if it rolls without slipping.

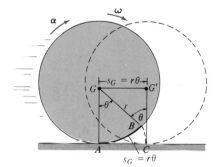

Fig. 16–7

SOLUTION

Position-Coordinate Equation. By inspection, point G moves *horizontally* to the right from G to G' as the cylinder rolls, Fig. 16–7. Consequently its new location G' will be specified by the *horizontal* position coordinate s_G, which is measured from the original position (G) of the cylinder's center. Notice also, that as the cylinder rolls (without slipping), points on its surface contact the ground such that the arc length AB of contact must be equal to the distance s_G. Consequently, motion from the colored to the dashed position requires the radial line GB to rotate θ to the position $G'C$. Since the arc $AB = r\theta$, then G travels a distance

$$s_G = r\theta$$

Time Derivatives. Taking successive time derivatives of this equation, realizing that r is constant, $\omega = d\theta/dt$, and $\alpha = d\omega/dt$, gives the necessary relationships

$$s_G = r\theta$$
$$v_G = r\omega \qquad\qquad \textit{Ans.}$$
$$a_G = r\alpha \qquad\qquad \textit{Ans.}$$

Example 16–4

The end of rod R shown in Fig. 16–8 maintains contact with the cam by means of a spring. If the cam rotates about an axis through point O with an angular acceleration $\boldsymbol{\alpha}$ and angular velocity $\boldsymbol{\omega}$, compute the velocity and acceleration of R when the cam is in the arbitrary position θ.

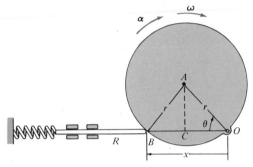

Fig. 16–8

SOLUTION

Position Coordinate Equation. Coordinates x and θ are chosen for the analysis in order to express the motion of the cam, which is defined by the angular motion of line OA, i.e., $\omega = d\theta/dt$ and the rectilinear motion of the rod (or horizontal component of the motion of point B), i.e., $v = dx/dt$. These coordinates are measured from the fixed point O and may be related to each other using trigonometry. Since $OC = CB = r \cos\theta$, Fig. 16–8, then

$$x = 2r \cos\theta$$

Time Derivatives. Using the chain rule of calculus, we have

$$\frac{dx}{dt} = -2r \sin\theta \frac{d\theta}{dt}$$

$$v = -2r\,\omega \sin\theta \qquad\qquad \textit{Ans.}$$

$$\frac{dv}{dt} = -2r\left(\frac{d\omega}{dt}\right)\sin\theta - 2r\omega\left(\cos\theta \frac{d\theta}{dt}\right)$$

$$a = -2r\,\alpha \sin\theta - 2r\omega^2 \cos\theta \qquad\qquad \textit{Ans.}$$

PROBLEMS

*16–20. The *Scotch yoke* is used to convert the constant circular motion of crank *OA* into translating motion of rod *BC*. If *OA* is rotating with a constant angular velocity $\omega = 5$ rad/s, determine the velocity and acceleration of *BC* for any angle θ of the crank.

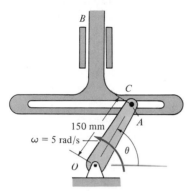

Probs. 16–20/16–21

16–21. The *Scotch yoke* is used to convert the constant circular motion of crank *OA* into translating motion of rod *BC*. When $\theta = 30°$, *OA* is rotating with an angular velocity $\omega = 5$ rad/s and angular acceleration $\alpha = 2$ rad/s². Determine the velocity and acceleration of *BC* at this instant.

16–22. At the instant shown, $\theta = 60°$, and rod *AB* is subjected to a deceleration of 16 m/s² when the velocity is 10 m/s. Determine the angular velocity and angular acceleration of link *CD* at this instant.

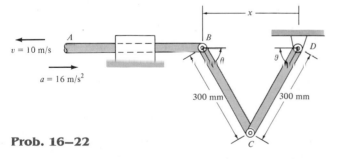

Prob. 16–22

16–23. Compute the angular velocity of rod *CD* at the instant $\theta = 30°$. Rod *AB* moves to the left at a constant rate $v_{AB} = 5$ m/s.

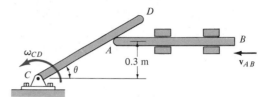

Probs. 16–23/16–24

*16–24. Compute the angular acceleration of rod *CD* at the instant $\theta = 30°$. Rod *AB* has zero velocity, i.e., $v_{AB} = 0$, and an acceleration of $a_{AB} = 2$ rad/s² to the right when $\theta = 30°$.

16–25. Determine the velocity and acceleration of the plate *C* as a function of θ, if the circular cam is rotating about the fixed point *O* with a constant angular velocity $\omega = 4$ rad/s.

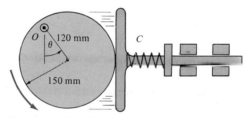

Probs. 16–25/16–26

16–26. Determine the velocity and acceleration of the plate *C* at the instant $\theta = 30°$, if at this instant the circular cam is rotating about the fixed point *O* with an angular velocity $\omega = 4$ rad/s and an angular acceleration $\alpha = 2$ rad/s².

16–27. The bar remains in contact with the floor and with point *A*. If point *B* moves to the right with a constant velocity of v_B, determine the angular velocity and angular acceleration of the bar as a function of *x*.

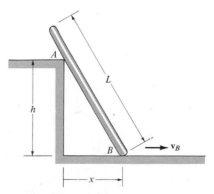

Prob. 16–27

*16–28. The pins at A and B are confined to move in the vertical and horizontal tracks. If the slotted arm is causing A to move downward at $\mathbf{v}_A$, determine the velocity $\mathbf{v}_B$ at the instant shown.

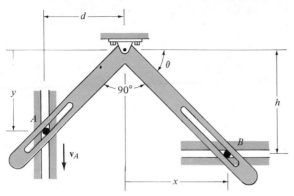

Prob. 16–28

16–29. The end A of the bar is moving downward along the slotted guide with a constant velocity $\mathbf{v}_A$. Determine the angular velocity $\boldsymbol{\omega}$ and angular acceleration $\boldsymbol{\alpha}$ of the bar as a function of its position y.

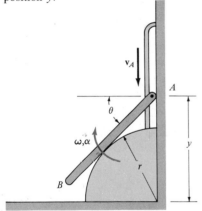

Prob. 16–29

16–30. The block moves to the left with a constant velocity $\mathbf{v}_0$. Determine the angular velocity and angular acceleration of the bar as a function of x.

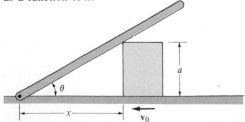

Prob. 16–30

16–31. The bar is confined to move along the vertical and inclined planes. If the velocity of the roller at A is $v_A = 6$ ft/s when $\theta = 45°$, determine the bar's angular velocity and the velocity of roller B at this instant.

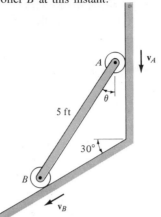

Probs. 16–31/16–32

*16–32. The bar is confined to move along the vertical and inclined planes. If the velocity of the roller at A is $v_A = 6$ ft/s, when $\theta = 45°$, determine the bar's angular acceleration and the acceleration of roller B at this instant.

16–33. The end A of the bar is moving to the left with a constant velocity $\mathbf{v}_A$. Determine the angular velocity $\boldsymbol{\omega}$ and angular acceleration $\boldsymbol{\alpha}$ of the bar as a function of its position x.

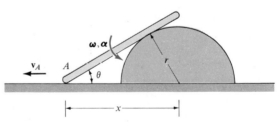

Probs. 16–33/16–34

16–34. The end A of the bar is moving to the left with an acceleration $\mathbf{a}_A$ and at the instant shown it has a velocity $\mathbf{v}_A$. Determine the angular velocity $\boldsymbol{\omega}$ and angular acceleration $\boldsymbol{\alpha}$ of the bar as a function of its position x.

16–35. Wheel *A* rolls without slipping over the surface of the *fixed* cylinder *B*. Determine the angular velocity of *A* if its center *C* has a speed $v_C = 5$ m/s. How many revolutions will *A* have made about its center just after link *DC* completes one revolution?

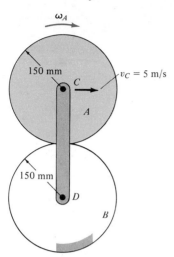

Prob. 16–35

*16–36. Bar *AB* rotates uniformly about the fixed pin *A* with a constant angular velocity ω. Determine the velocity and acceleration of block *C*, at the instant $\theta = 60°$.

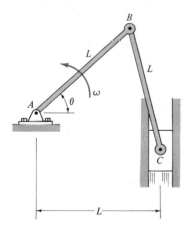

Prob. 16–36

16–37. The slotted yoke is pinned at *A* while end *B* is used to move the ram *R* horizontally. If the disk rotates with a constant angular velocity ω, determine the velocity and acceleration of the ram. The crank pin *C* is fixed to the disk and turns with it.

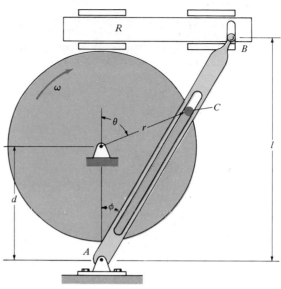

Prob. 16–37

16.5 Relative-Motion Analysis: Velocity

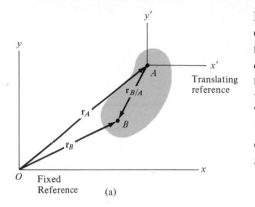

(a)

It was pointed out in Sec. 16.1 that general plane motion of a rigid body consists of a *combination* of translation and rotation. Oftentimes it is convenient to view these "component" motions *separately* using a *relative-motion analysis*. Two sets of coordinate axes will be used to do this. The x, y coordinate system is fixed and will be used to measure the *absolute* positions, velocities, and accelerations of two points A and B in the body, Fig. 16–9a. The origin of the x', y' coordinate system will be attached to the selected "base point" A, which generally has a *known* motion. The axes of the x', y' coordinate system are *not fixed to the body;* rather they will be allowed to *translate* with respect to the fixed frame.

Position

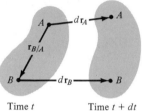

Time t Time $t + dt$

General plane
motion

(b)

The position vector $\mathbf{r}_A$ in Fig. 16–9a specifies the location of the "base point" A, whereas the relative-position vector $\mathbf{r}_{B/A}$ locates point B in the body with respect to A. By vector addition, the *position* of B can be determined from the equation

$$\mathbf{r}_B = \mathbf{r}_A + \mathbf{r}_{B/A}$$

Displacement

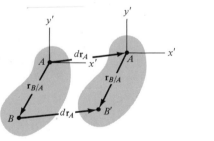

Translation

(c)

During an instant of time dt, points A and B undergo displacements $d\mathbf{r}_A$ and $d\mathbf{r}_B$ as shown in Fig. 16–9b. As stated above, here we will consider the motion by its component parts. For example, the *entire body* first *translates* by an amount $d\mathbf{r}_A$ so that A, the base point, moves to its *final position* and B moves to B', Fig. 16–9c. The body is then rotated by an amount $d\theta$ about A, so that B' undergoes a *relative displacement* $d\mathbf{r}_{B/A}$ and thus moves to its final position B, Fig. 16–9d. Note that since the body is *rigid*, $\mathbf{r}_{B/A}$ has a fixed magnitude and thus $d\mathbf{r}_{B/A}$ accounts only for a change in the *direction* of $\mathbf{r}_{B/A}$. Due to the rotation about A the magnitude of the relative displacement is $dr_{B/A} = r_{B/A}d\theta$ and the displacement of B in Fig. 16–9b is thus

$$d\mathbf{r}_B = d\mathbf{r}_A + d\mathbf{r}_{B/A}$$

Velocity

To determine the relationship between the velocities of points A and B, it is necessary to take the time derivative of the position equation, or simply divide the displacement equation by dt. This yields

$$\frac{d\mathbf{r}_B}{dt} = \frac{d\mathbf{r}_A}{dt} + \frac{d\mathbf{r}_{B/A}}{dt}$$

Rotation about the
base point A

(d)

Fig. 16–9(a–d)

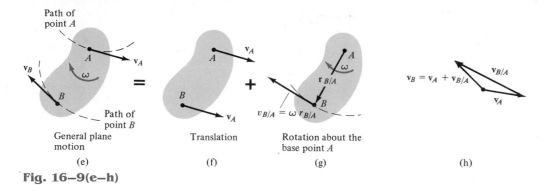

General plane motion (e) = Translation (f) + Rotation about the base point A (g)

$v_B = v_A + v_{B/A}$ (h)

Fig. 16–9(e–h)

The terms $d\mathbf{r}_B/dt = \mathbf{v}_B$ and $d\mathbf{r}_A/dt = \mathbf{v}_A$ are measured from the fixed x, y axes and represent the *absolute velocities* of points A and B, respectively. The magnitude of the third term is $d(r_{B/A}\theta)/dt = r_{B/A}\dot{\theta} = r_{B/A}\omega$, where ω is the angular velocity of the body at the instant considered. Note that this rotation is *absolute* since it is the *same* when measured from either the fixed x, y or translating x', y' axes. We will denote this third term as the *relative velocity* $\mathbf{v}_{B/A}$, since it represents the velocity of B with respect to A as measured by an observer fixed to the translating x', y' axes. Since the body is rigid, it is important to realize that this observer sees point B move along a *circular arc* that has a radius of curvature $r_{B/A}$. In other words, *the body appears to move as if it were rotating with an angular velocity* $\boldsymbol{\omega}$ *about the z' axis passing through A.* Consequently, $\mathbf{v}_{B/A}$ has a magnitude of $v_{B/A} = \omega r_{B/A}$ and a *direction* which is perpendicular to $r_{B/A}$. We therefore have

$$\boxed{\mathbf{v}_B = \mathbf{v}_A + \mathbf{v}_{B/A}} \qquad (16\text{–}14)$$

where

$\mathbf{v}_B$ = velocity of point B

$\mathbf{v}_A$ = velocity of the base point A

$\mathbf{v}_{B/A}$ = relative velocity of "B with respect to A."

    Since the relative motion is *circular*, the *magnitude* of $\mathbf{v}_{B/A}$ is $v_{B/A} = \omega r_{B/A}$ and the *direction* is perpendicular to $\mathbf{r}_{B/A}$.

Each of the three terms in Eq. 16–14 is represented graphically on the *kinematic diagrams* in Fig. 16–9e, f, and g. Here it is seen that at a given instant the velocity of B, Fig. 16–9e, is determined by considering the entire body to translate with a velocity of $\mathbf{v}_A$, Fig. 16–9f, and then rotate about the base point A with an angular velocity $\boldsymbol{\omega}$, Fig. 16–9g. Vector addition of these two effects, applied to B, yields $\mathbf{v}_B$, as shown in Fig. 16–9h.

Since the relative velocity $\mathbf{v}_{B/A}$ represents the effect of *circular motion*, observed from translating axes having their origin at the base point A, this term can be expressed by the cross product $\mathbf{v}_{B/A} = \boldsymbol{\omega} \times \mathbf{r}_{B/A}$, Eq. 16–9.

Hence, for application, we can also write Eq. 16–14 as

$$\boxed{\mathbf{v}_B = \mathbf{v}_A + \boldsymbol{\omega} \times \mathbf{r}_{B/A}} \qquad (16\text{–}15)$$

where

$\mathbf{v}_B$ = velocity of B

$\mathbf{v}_A$ = velocity of the base point A

$\boldsymbol{\omega}$ = angular velocity of the body

$\mathbf{r}_{B/A}$ = relative-position vector drawn from A to B

To apply this equation, all of these terms must be represented in Cartesian vector form.

The velocity equation 16–14 or 16–15 may be used in a practical manner to study the motion of a rigid body which is either pin-connected to or in contact with other moving bodies. To obtain the necessary data when applying this equation, points A and B should generally be selected at joints which are pin-connected, or at points in contact with adjacent bodies which have a *known motion*. For example, both points A and B on link AB, Fig. 16–10a, have circular paths of motion. Hence, at the instant shown, the magnitudes of their velocities are determined by the angular motion of the wheel and link BC so that $v_A = \omega_A r$ and $v_B = \omega_{BC} l$. The *directions* of $\mathbf{v}_A$ and $\mathbf{v}_B$ are always *tangent* to their paths of motion, Fig. 16–10b. In the case of the wheel in Fig. 16–11, which rolls without slipping, point A can be selected at the ground. Here A momentarily has zero velocity since the ground does not move. Furthermore, the center of the wheel, B, moves along a horizontal path so that $\mathbf{v}_B$ is horizontal.

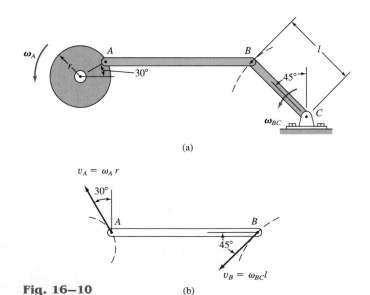

(a)

(b)

Fig. 16–10

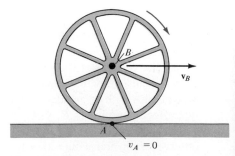

Fig. 16–11

PROCEDURE FOR ANALYSIS

Equation 16–14 relates the velocities of any two points A and B located on the *same* rigid body. It can be applied either by using Cartesian vector analysis, in the form of Eq. 16–15, or by writing the x and y scalar component equations directly. For application, it is suggested that the following procedure be used.

Kinematic Diagram. Establish the directions of the fixed x, y coordinates, and draw a kinematic diagram of the body. Indicate on it the velocities $\mathbf{v}_A$, $\mathbf{v}_B$ of points A and B, the angular velocity $\boldsymbol{\omega}$, and the relative position vector $\mathbf{r}_{B/A}$. Normally the base point A is selected as a point having a known velocity. Identify the *two* unknowns on the diagram. If the magnitudes of $\mathbf{v}_A$, $\mathbf{v}_B$, or $\boldsymbol{\omega}$ are unknown, the sense of direction of these vectors may be assumed.

If the velocity equation is to be applied *without* using a Cartesian vector analysis, then the magnitude and direction of the relative velocity $\mathbf{v}_{B/A}$ must be established. This can be done by drawing a kinematic diagram such as shown in Fig. 16–9g. Since the body is considered to be "pinned" momentarily at the base point A, the *magnitude* of $\mathbf{v}_{B/A}$ is $v_{B/A} = \omega r_{B/A}$. The *direction* is established from the diagram, such that $\mathbf{v}_{B/A}$ acts perpendicular to $\mathbf{r}_{B/A}$ in accordance with the rotational motion $\boldsymbol{\omega}$ of the body.*

Velocity Equation. To apply Eq. 16–15, express the vectors in Cartesian vector form and substitute them into the equation. Evaluate the cross product and then equate the respective $\mathbf{i}$ and $\mathbf{j}$ components to obtain two scalar equations. To obtain these scalar equations *directly*, write Eq. 16–14 in symbolic form, $\mathbf{v}_B = \mathbf{v}_A + \mathbf{v}_{B/A}$, and underneath each of the terms represent the vectors by their magnitudes and directions. To do this, use the data tabulated on the kinematic diagrams. The scalar equations are determined from the x and y components of these vectors. Solve these equations for the two unknowns. If the solution yields a *negative* answer for an *unknown*, it indicates the sense of direction of the vector is opposite to that shown on the kinematic diagram.

The following example problems numerically illustrate both methods of application.

*The notation $\mathbf{v}_B = \mathbf{v}_A + \mathbf{v}_{B/A\text{(pin)}}$ may be helpful in recalling that A is "pinned."

Example 16–5

The link shown in Fig. 16–12a is guided by two blocks at A and B, which move in the fixed slots. If the velocity of A is 2 m/s downward, determine the velocity of B at the instant $\theta = 45°$.

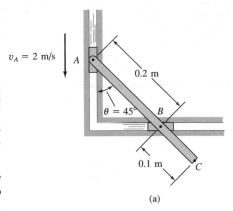

(a)

SOLUTION (VECTOR ANALYSIS)

Kinematic Diagram. Since points A and B are restricted to move along the fixed slots and $\mathbf{v}_A$ is directed downward, the velocity $\mathbf{v}_B$ must be directed horizontally to the right, Fig. 16 12b. This motion causes the link to rotate counterclockwise; that is, by the right-hand rule the angular velocity $\boldsymbol{\omega}$ is directed outward, perpendicular to the page (the plane of motion). Knowing the magnitude and direction of $\mathbf{v}_A$ and the lines of action of $\mathbf{v}_B$ and $\boldsymbol{\omega}$, it is possible to apply the velocity equation $\mathbf{v}_B = \mathbf{v}_A + \boldsymbol{\omega} \times \mathbf{r}_{B/A}$ to points A and B in order to solve for the two unknown magnitudes v_B and ω. Since $\mathbf{r}_{B/A}$ is needed, it is also shown in Fig. 16–12b.

Velocity Equation. Expressing each of the vectors in Fig. 16–12b in terms of their **i**, **j**, and **k** components, we have

$$\mathbf{v}_A = \{-2\mathbf{j}\} \text{ m/s}, \qquad \mathbf{v}_B = v_B\mathbf{i}, \qquad \boldsymbol{\omega} = \omega\mathbf{k}$$
$$\mathbf{r}_{B/A} = \{0.2 \sin 45°\mathbf{i} - 0.2 \cos 45°\mathbf{j}\} \text{ m}$$

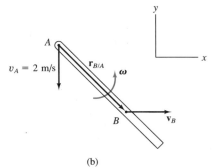

(b)

Fig. 16–12

Applying Eq. 16–15 to A, the base point, and B, we have

$$\mathbf{v}_B = \mathbf{v}_A + \boldsymbol{\omega} \times \mathbf{r}_{B/A}$$
$$v_B\mathbf{i} = -2\mathbf{j} + [\omega\mathbf{k} \times (0.2 \sin 45°\mathbf{i} - 0.2 \cos 45°\mathbf{j})]$$

or

$$v_B\mathbf{i} = -2\mathbf{j} + 0.2\omega \sin 45°\mathbf{j} + 0.2\omega \cos 45°\mathbf{i}$$

Equating the **i** and **j** components gives

$$v_B = 0.2\omega \cos 45°, \qquad 0 = -2 + 0.2\omega \sin 45°$$

Thus,

$$\omega = 14.1 \text{ rad/s} \curvearrowleft$$
$$v_B = 2 \text{ m/s} \rightarrow \qquad\qquad\qquad Ans.$$

Since both results are *positive*, the *directions* of $\mathbf{v}_B$ and $\boldsymbol{\omega}$ are indeed *correct* as shown in Fig. 16–12b. It should be emphasized that these results are *valid only* at the instant $\theta = 45°$. A recalculation for $\theta = 44°$ yields $v_B = 2.07$ m/s and $\omega = 14.4$ rad/s; whereas when $\theta = 46°$, $v_B = 1.93$ m/s and $\omega = 13.9$ rad/s, etc.

Having determined $\boldsymbol{\omega}$, as an exercise apply Eq. 16–15 to points A and C and show that $v_C = 3.16$ m/s, directed at $\theta = 18.4°$ up from the horizontal.

Example 16–6

The cylinder shown in Fig. 16–13a rolls freely on the surface of a conveyor belt which is moving at 2 ft/s. Assuming that no slipping occurs between the cylinder and the belt, determine the velocity of point A. The cylinder has a clockwise angular velocity $\omega = 15$ rad/s at the instant shown.

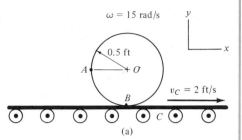

(a)

SOLUTION I (VECTOR ANALYSIS)

Kinematic Diagram. Since no slipping occurs, point B on the cylinder has the same velocity as the conveyor, Fig. 16–13b. Knowing the angular velocity of the cylinder, we will apply the velocity equation to B, the base point, and A to determine the x and y components of $\mathbf{v}_A$, i.e., $\mathbf{v}_A = \mathbf{v}_B + \boldsymbol{\omega} \times \mathbf{r}_{A/B}$.

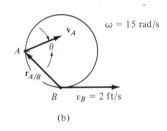

(b)

Velocity Equation

$$\mathbf{v}_A = \mathbf{v}_B + \boldsymbol{\omega} \times \mathbf{r}_{A/B}$$
$$(v_A)_x\mathbf{i} + (v_A)_y\mathbf{j} = 2\mathbf{i} + (-15\mathbf{k}) \times (-0.5\mathbf{i} + 0.5\mathbf{j})$$
$$(v_A)_x\mathbf{i} + (v_A)_y\mathbf{j} = 2\mathbf{i} + 7.50\mathbf{j} + 7.50\mathbf{i}$$

so that

$$(v_A)_x = 2 + 7.50 = 9.50 \text{ ft/s} \qquad (1)$$
$$(v_A)_y = 7.50 \text{ ft/s} \qquad (2)$$

Thus,

$$v_A = \sqrt{(9.50)^2 + (7.50)^2} = 12.1 \text{ ft/s} \qquad \textit{Ans.}$$
$$\theta = \tan^{-1}\frac{7.50}{9.50} = 38.3° \qquad \textit{Ans.}$$

SOLUTION II (SCALAR COMPONENTS)

As an alternative procedure, the scalar components of $\mathbf{v}_A = \mathbf{v}_B + \mathbf{v}_{A/B}$ can be obtained directly. From the kinematic diagram showing the relative "circular" motion $\mathbf{v}_{A/B}$, Fig. 16–13c, we have

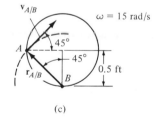

(c)

Fig. 16–13

$$v_{A/B} = \omega r_{A/B} = (15 \text{ rad/s})\left(\frac{0.5 \text{ ft}}{\cos 45°}\right) = 10.6 \text{ ft/s} \quad \measuredangle^{45°}$$

Thus,

$$v_A = v_B + v_{A/B}$$
$$\begin{bmatrix} (v_A)_x \\ \rightarrow \end{bmatrix} + \begin{bmatrix} (v_A)_y \\ \uparrow \end{bmatrix} = \begin{bmatrix} 2 \text{ ft/s} \\ \rightarrow \end{bmatrix} + \begin{bmatrix} 10.6 \text{ ft/s} \\ \measuredangle^{45°} \end{bmatrix}$$

Equating the x and y components gives the same results as before, namely,

$$(\xrightarrow{+}) \qquad (v_A)_x = 2 + 10.6 \cos 45° = 9.50 \text{ ft/s}$$
$$(+\uparrow) \qquad (v_A)_y = 0 + 10.6 \sin 45° = 7.50 \text{ ft/s}$$

Example 16–7

The collar C in Fig. 16–14a is moving downward with a velocity of 2 m/s. Determine the angular velocity of CB at this instant.

SOLUTION I (VECTOR ANALYSIS)
Kinematic Diagram. The downward motion of C causes B to move to the right. Also, CB rotates counterclockwise. It is subjected to general plane motion, Fig. 16–14b.

Velocity Equation

$$\mathbf{v}_B = \mathbf{v}_C + \boldsymbol{\omega}_{CB} \times \mathbf{r}_{B/C}$$
$$v_B\mathbf{i} = -2\mathbf{j} + \omega_{CB}\mathbf{k} \times (0.2\mathbf{i} - 0.2\mathbf{j})$$
$$v_B\mathbf{i} = -2\mathbf{j} + 0.2\omega_{CB}\mathbf{j} + 0.2\omega_{CB}\mathbf{i}$$

Equating the $\mathbf{i}$ and $\mathbf{j}$ components yields

$$v_B = 0.2\omega_{CB} \qquad\qquad (1)$$
$$0 = -2 + 0.2\omega_{CB} \qquad\qquad (2)$$

So that

$$\omega_{CB} = 10 \text{ rad/s} \nwarrow \qquad\qquad \textit{Ans.}$$
$$v_B = 2 \text{ m/s} \rightarrow$$

SOLUTION II (SCALAR COMPONENTS)
The scalar component equations of $\mathbf{v}_B = \mathbf{v}_C + \mathbf{v}_{B/C}$ can be obtained directly. The kinematic diagram in Fig. 16–14c shows the relative "circular" motion $\mathbf{v}_{B/C}$. We have

$$\mathbf{v}_B = \mathbf{v}_C + \mathbf{v}_{B/C}$$
$$\begin{bmatrix} v_B \\ \rightarrow \end{bmatrix} = \begin{bmatrix} 2 \text{ m/s} \\ \downarrow \end{bmatrix} + \begin{bmatrix} \omega_{CB}(0.2\sqrt{2} \text{ m}) \\ \angle^{45°} \end{bmatrix}$$

Resolving these vectors in the x and y directions yields

$(\overset{+}{\rightarrow})$ $\qquad\qquad v_B = 0 + \omega_{CB}(0.2\sqrt{2}\cos 45°)$

$(+\uparrow)$ $\qquad\qquad 0 = -2 + \omega_{CB}(0.2\sqrt{2}\sin 45°)$

which is the same as Eqs. (1) and (2).

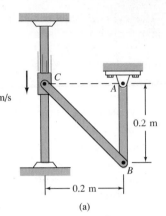

$v_C = 2$ m/s

0.2 m

0.2 m

(a)

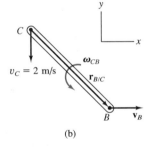

$v_C = 2$ m/s

(b)

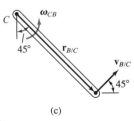

(c)

Fig. 16–14

Example 16–8

The bar AB of the linkage shown in Fig. 16–15a has a clockwise angular velocity of 30 rad/s when $\theta = 60°$. Compute the angular velocities of members BC and DC at this instant.

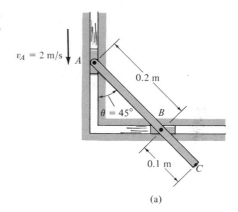

(a)

SOLUTION (VECTOR ANALYSIS)
Kinematic Diagram. The velocities of points B and C are defined by the rotation of links AB and DC about their fixed axes. The relative-position vectors and the angular velocity of each link are shown on the kinematic diagram in Fig. 16–15b. Here we will apply a Cartesian vector analysis to each link.

Velocity Equation. Expressing each vector in Cartesian vector form, we have

$$\boldsymbol{\omega}_{AB} = \{-30\mathbf{k}\}\text{ rad/s}, \qquad \boldsymbol{\omega}_{BC} = \omega_{BC}\mathbf{k}, \qquad \boldsymbol{\omega}_{DC} = \omega_{DC}\mathbf{k}$$

$$\mathbf{r}_{B/A} = \{0.2\cos 60°\mathbf{i} + 0.2\sin 60°\mathbf{j}\}\text{ m}, \qquad \mathbf{r}_{C/B} = \{0.2\mathbf{i}\}\text{ m},$$

$$\mathbf{r}_{C/D} = \{-0.1\mathbf{j}\}\text{ m}$$

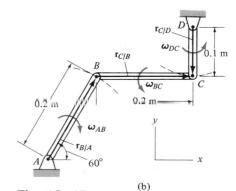

Fig. 16–15   (b)

For link AB (rotation about a fixed axis):

$$\mathbf{v}_B = \boldsymbol{\omega}_{AB} \times \mathbf{r}_{B/A}$$
$$= (-30\mathbf{k}) \times (0.2\cos 60°\mathbf{i} + 0.2\sin 60°\mathbf{j})$$
$$= \{5.20\mathbf{i} - 3.0\mathbf{j}\}\text{ m/s}$$

For link BC (general plane motion):

$$\mathbf{v}_C = \mathbf{v}_B + \boldsymbol{\omega}_{BC} \times \mathbf{r}_{C/B}$$
$$= 5.20\mathbf{i} - 3.0\mathbf{j} + (\omega_{BC}\mathbf{k}) \times (0.2\mathbf{i})$$
$$= 5.20\mathbf{i} + (0.2\omega_{BC} - 3.0)\mathbf{j}$$

For link DC (rotation about a fixed axis):

$$\mathbf{v}_C = \boldsymbol{\omega}_{DC} \times \mathbf{r}_{C/D}$$
$$5.20\mathbf{i} + (0.2\omega_{BC} - 3.0)\mathbf{j} = (\omega_{DC}\mathbf{k}) \times (-0.1\mathbf{j})$$
$$5.20\mathbf{i} + (0.2\omega_{BC} - 3.0)\mathbf{j} = 0.1\omega_{DC}\mathbf{i}$$

Equating the respective $\mathbf{i}$ and $\mathbf{j}$ components and solving, we have

$$\omega_{DC} = 52\text{ rad/s} \; \curvearrowleft \qquad\qquad \textit{Ans.}$$
$$\omega_{BC} = 15\text{ rad/s} \; \curvearrowleft \qquad\qquad \textit{Ans.}$$

Note that by inspection, Fig. 16–15a, $v_B = (0.2)(30) = 6$ m/s, $\searrow^{30°}$ and $\mathbf{v}_C$ is directed to the right. As an exercise, obtain ω_{BC} by applying $\mathbf{v}_C = \mathbf{v}_B + \mathbf{v}_{C/B}$ using scalar components.

PROBLEMS

16–38. The crankshaft AB is rotating at 500 rad/s about a fixed axis passing through A. Determine the speed of the piston P at the instant it is in the position shown.

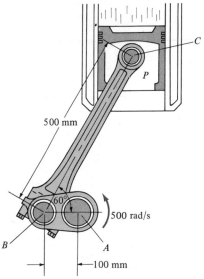

Prob. 16–38

16–39. A bowling ball is cast on the "alley" with a backspin of $\omega = 10$ rad/s while its center O has a forward velocity of $v_O = 8$ m/s. Determine the velocity of the contact point A touching the alley.

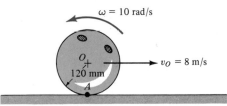

Prob. 16–39

*16–40. The cylinder B rolls on the *fixed cylinder A* without slipping. If the connected bar CD is rotating with an angular velocity of $\omega_{CD} = 5$ rad/s, determine the angular velocity of cylinder B. Point C is a fixed point.

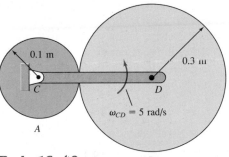

Prob. 16–40

16–41. At the instant shown the boomerang has an angular velocity of $\omega = 4$ rad/s, and its mass center G has a velocity of $v_G = 6$ in./s. Determine the velocity of point A at this instant.

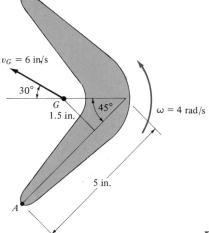

Prob. 16–41

16–42. The wheel is rotating with an angular velocity $\omega = 8$ rad/s. Determine the velocity of the collar A at this instant.

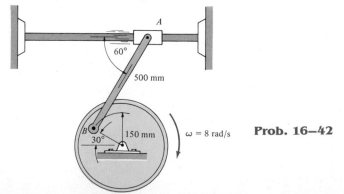

Prob. 16–42

16–43. If the angular velocity of link AB is $\omega_{AB} = 3$ rad/s, determine the velocity of the block at C and the angular velocity of the connecting link CB at the instant shown.

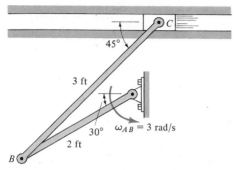

Prob. 16–43

*16–44. Part of an automatic transmission consists of a ring gear R, three equal planet gears P, the sun gear S, and the planet carrier C, which is shaded. If the sun gear is *fixed* and the ring gear is rotating at $\omega_R = 5$ rad/s, determine the angular velocity of the *planet carrier*. Note that C is pin-connected to the center of each of the planet gears.

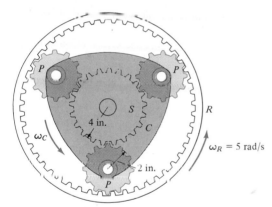

Prob. 16–44

16–45. When the crank on the Chinese windlass is turning, the rope on shaft A unwinds while that on shaft B winds up. Determine the speed at which the block lowers if the crank is turning with an angular velocity $\omega = 4$ rad/s. What is the angular velocity of the pulley at C? The rope segments on each side of the pulley are both parallel and vertical, and the rope does not slip on the pulley.

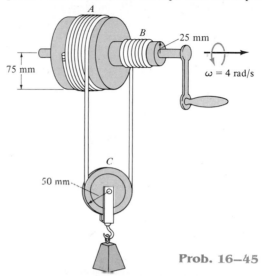

Prob. 16–45

16–46. If the end of the cord is pulled downward with a speed $v_C = 120$ mm/s, determine the angular velocities of pulleys A and B and the speed of block D. Assume that the cord does not slip on the pulleys.

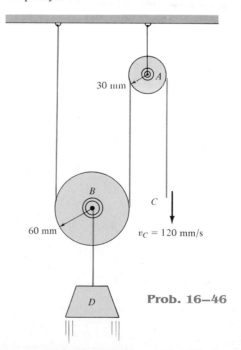

Prob. 16–46

16–47. If rod *AB* is rotating with an angular velocity $\omega_{AB} = 12$ rad/s, determine the angular velocity of rods *BC* and *CD* at the instant shown.

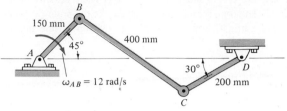

Prob. 16–47

16–50. If disk *A* has an angular velocity $\omega_A = 4$ rad/s, determine the angular velocity of disk *D* at the instant shown.

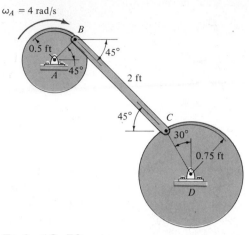

Prob. 16–50

*16–48. If the slider block *A* is moving downward at $v_A = 4$ m/s, determine the velocity of blocks *B* and *C* at the instant shown.

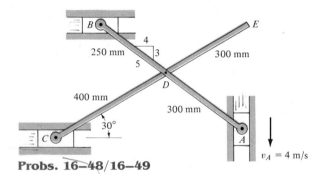

Probs. 16–48/16–49

16–49. If the slider block *A* is moving downward at $v_A = 4$ m/s, determine the velocity of point *E* at the instant shown.

16–51. If the link *AB* is rotating about the pin at *A* with an angular velocity of $\omega_{AB} = 5$ rad/s, determine the velocity of blocks *C* and *E* at the instant shown.

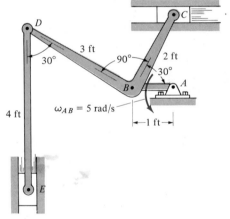

Prob. 16–51

262

16.6 Instantaneous Center of Zero Velocity

The velocity of any point B located in a rigid body can be obtained in a simple way if one chooses the base point A to be a point that has *zero velocity* at the instant considered. In this case, $\mathbf{v}_A = \mathbf{0}$, and therefore the velocity equation, $\mathbf{v}_B = \mathbf{v}_A + \boldsymbol{\omega} \times \mathbf{r}_{B/A}$, becomes $\mathbf{v}_B = \boldsymbol{\omega} \times \mathbf{r}_{B/A}$. For a body having general plane motion, point A so chosen is called the *instantaneous center of zero velocity* (*IC*) and it lies on the *instantaneous axis of zero velocity* (*IA*), Fig. 16–17a. The *IA* is always perpendicular to the plane used to represent the motion, and the intersection of the *IA* with this plane defines the location of the *IC*. Since point A is coincident with the *IC*, $\mathbf{v}_B = \boldsymbol{\omega} \times \mathbf{r}_{B/IC}$ and so point B moves momentarily about the *IC* in a *circular path*; in other words, the body appears to rotate about the *IA*. If the relative-position vector $\mathbf{r}_{B/IC}$ is established from the *IC* to point B as shown, then the *magnitude* of $\mathbf{v}_B$ is simply $v_B = \omega r_{B/IC}$, where ω is the angular velocity of the body. Due to the circular motion, the *direction* of $\mathbf{v}_B$ must always be *perpendicular* to $\mathbf{r}_{B/IC}$. For example, consider a wheel which rolls *without slipping*, Fig. 16–16a. In this case the point of *contact* with the ground has *zero velocity*. Hence this point represents the *IC* for the wheel, Fig. 16–16b. If it is imagined that the wheel is momentarily pinned at this point, the velocities of points A, B, O, and so on, can be found using $v = \omega r$. Here the radial distances $r_{A/IC}$, $r_{B/IC}$, and $r_{O/IC}$, shown in Fig. 16–16b, must be determined from the geometry of the wheel.

When a body is subjected to general plane motion, the point determined as the instantaneous center of zero velocity for the body can only be used for an *instant of time*. Since the body changes its position from one instant to the next, then for each position of the body a unique instantaneous center must be determined. The locus of points which define the *IC* during the body's motion is called a *centrode*, Fig. 16–17a. Thus, each point on the centrode acts as the *IC* for the body only for an instant of time.

Although the *IC* may be conveniently used to determine the velocity of any point in a body, it generally *does not have zero acceleration* and therefore it *should not* be used for finding the accelerations of points in a body.

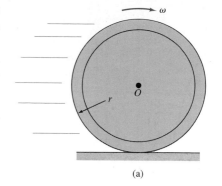

(a)

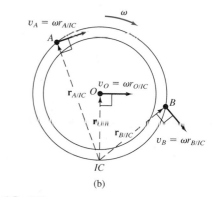

(b)

Fig. 16–16

Location of the IC

If the location of the *IC* is unknown, it may be determined by using the fact that the *relative-position vector* extending from the *IC* to a point is *always perpendicular* to the *velocity* of the point. Several possibilities exist:

1. *Given the velocity of a point in the body and the angular velocity of the body.* In this case if $\mathbf{v}_B$ and $\boldsymbol{\omega}$ are known, the *IC* is located along the line drawn perpendicular to $\mathbf{v}_B$ at B, such that the distance from B to the *IC* is $r_{B/IC} = v_B/\omega$, Fig. 16–17a. Note that the *IC* lies on that side of B which causes rotation about the *IC* which is consistent with the direction of motion caused by $\boldsymbol{\omega}$ and $\mathbf{v}_B$.

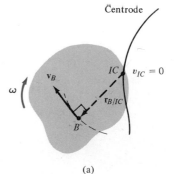

(a)

Fig. 16–17a

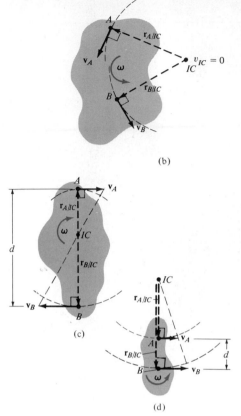

(b)

(c)

(d)

Fig. 16–17(b–d)

2. *Given the lines of action of two nonparallel velocities.* Consider the body in Fig. 16–17b, where the lines of action of the velocities $\mathbf{v}_A$ and $\mathbf{v}_B$ are known. From each of these lines construct at points A and B perpendicular line segments which then define the lines of action of $\mathbf{r}_{A/IC}$ and $\mathbf{r}_{B/IC}$, respectively. Extending these perpendiculars to their *point of intersection* as shown locates the IC at the instant considered. The magnitudes of $\mathbf{r}_{A/IC}$ and $\mathbf{r}_{B/IC}$ are generally determined from the geometry of the body and trigonometry. Furthermore, if the magnitude and sense of $\mathbf{v}_A$ are known, then the angular velocity of the body is determined from $\omega = v_A/r_{A/IC}$. Once computed, ω can then be used to determine $v_B = \omega r_{B/IC}$.

3. *Given the magnitude and direction of two parallel velocities.* When the velocities of points A and B are parallel and have known magnitudes v_A and v_B, then the location of the IC is determined by proportional triangles. Examples are shown in Fig. 16–17c and d. In both cases $v_A = \omega r_{A/IC}$ and $v_B = \omega r_{B/IC}$ so that, $r_{A/IC} = r_{B/IC}(v_A/v_B)$. If d is the known distance between points A and B, then in Fig. 16–17c, $r_{A/IC} = d - r_{B/IC}$ and in Fig. 16–17d, $r_{A/IC} = r_{B/IC} - d$. Substituting each of these equations into the previous equation yields a solution for $r_{B/IC}$. As a special case, note that if the body is *translating*, $\mathbf{v}_A = \mathbf{v}_B$ and the IC would be located at infinity, in which case $r_{A/IC} = r_{B/IC} \rightarrow \infty$. This being the case, $\omega = v_A/\infty = v_B/\infty = 0$, as expected.

PROCEDURE FOR ANALYSIS

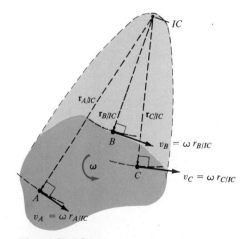

Fig. 16–18

The velocity of a point in a body which is subjected to general plane motion can be determined with reference to its instantaneous center of zero velocity, provided the location of the IC is first established. This is done by using one of the methods described above. As shown on the kinematic diagram in Fig. 16–18, the body is imagined as "extended and pinned" at the IC such that it rotates about this pin with its instantaneous angular velocity $\boldsymbol{\omega}$. The *magnitude* of velocity for the arbitrary points A, B, and C in the body can then be determined by using the equation $v = \omega r$, where r is the radial line drawn from the IC to the point. The line of action of each velocity vector is *perpendicular* to its associated radial line, and the velocity has a *direction* which tends to move the point in a manner consistent with the angular rotation of the radial line, Fig. 16–18.

The following examples illustrate the method of determining the IC and computing the velocities of points in a rigid body subjected to general plane motion.

Example 16–9

Determine the location of the instantaneous center of zero velocity for (a) the crankshaft BC shown in Fig. 16–19a; and (b) the link CB shown in Fig. 16–19b.

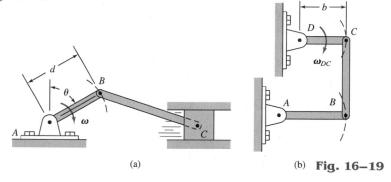

(a)

(b) **Fig. 16–19**

SOLUTION

Part (a). As shown in Fig. 16–19a, link AB rotates about the fixed pin A. Thus, point B has a speed of $v_B = \omega d$, caused by the clockwise rotation of AB. Also, $\mathbf{v}_B$ is perpendicular to AB, so that it acts at an angle θ from the horizontal as shown in Fig. 16–19c. The motion of point B causes the piston to move forward *horizontally* with a velocity $\mathbf{v}_C$. Consequently, point C on the crankshaft moves horizontally with this same velocity. When lines are drawn perpendicular to $\mathbf{v}_B$ and $\mathbf{v}_C$, Fig. 16–19c, they intersect at the IC. The magnitudes of $\mathbf{r}_{B/IC}$ and $\mathbf{r}_{C/IC}$ are determined strictly from the geometry of construction.

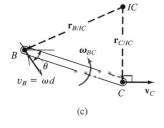

(c)

Part (b). Points B and C follow circular paths of motion since rods AB and DC are each subjected to rotation about a fixed axis, Fig. 16–19b. In particular, $v_C = \omega_{DC}b$. Since the velocity is always tangent to the path, at the instant considered, $\mathbf{v}_C$ on rod DC and $\mathbf{v}_B$ on rod AB are both directed vertically downward, along the axis of link CB, Fig. 16–19d. Furthermore, since CB is *rigid*, no relative displacement occurs between points B and C, so that $\mathbf{v}_B = \mathbf{v}_C$. Radial lines drawn perpendicular to these two velocities form parallel lines which intersect at "infinity;" i.e., $r_{C/IC} \to \infty$ and $r_{B/IC} \to \infty$. Thus, $\omega_{CB} = v_C/r_{C/IC} = (\omega_{DC}b)/\infty = 0$. As a result, rod CB momentarily *translates* with a speed of $v_C = \omega_{DC}b$. An instant later, however, CB will move to a new position, causing the instantaneous center to move to some finite location.

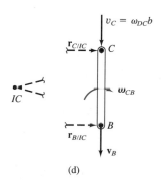

(d)

Example 16–10

Block D shown in Fig. 16–20a moves with a speed of 3 m/s. Determine the angular velocities of links BD and AB, and the velocity of point B at the instant shown.

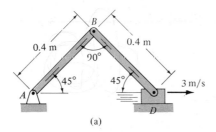

(a)

Fig. 16–20a

SOLUTION

Since D moves to the right at 3 m/s, it causes arm AB to rotate about point A in a clockwise direction. Hence, $\mathbf{v}_B$ is directed perpendicular to AB as shown in Fig. 16–20b. The instantaneous center of zero velocity for BD is located at the intersection of the line segments drawn perpendicular to $\mathbf{v}_B$ and $\mathbf{v}_D$, Fig. 16–20b. From the geometry,

$$r_{B/IC} = 0.4 \tan 45° \text{ m} = 0.4 \text{ m}$$

$$r_{D/IC} = \frac{0.4 \text{ m}}{\cos 45°} = 0.566 \text{ m}$$

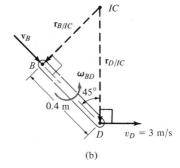

(b)

Since the magnitude of $\mathbf{v}_D$ is known, the angular velocity of the link is

$$\omega_{BD} = \frac{v_D}{r_{D/IC}} = \frac{3 \text{ m/s}}{0.566 \text{ m}} = 5.30 \text{ rad/s} \uparrow \qquad \textit{Ans.}$$

The velocity of B is therefore

$$v_B = \omega_{BD}(r_{B/IC}) = 5.30 \text{ rad/s}(0.4 \text{ m}) = 2.12 \text{ m/s} \quad \searrow 45° \qquad \textit{Ans.}$$

Link AB is subjected to rotation about a fixed axis passing through A, Fig. 16–20c, and since v_B is known, the angular velocity of AB is

$$\omega_{AB} = \frac{v_B}{r_{B/A}} = \frac{2.12 \text{ m/s}}{0.4 \text{ m}} = 5.30 \text{ rad/s} \downarrow \qquad \textit{Ans.}$$

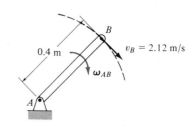

(c)

Fig. 16–20(b, c)

Example 16–11

The cylinder shown in Fig. 16–21a rolls without slipping between the two moving plates E and D. Determine the angular velocity of the cylinder and the velocity of its center C at the instant shown.

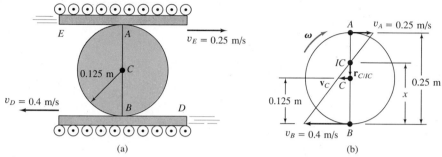

Fig. 16–21

SOLUTION

Since no slipping occurs, the contact points A and B on the cylinder have the same velocity as the plates E and D. Furthermore, the velocities $\mathbf{v}_A$ and $\mathbf{v}_B$ are *parallel*, so that by the proportionality of right triangles the IC is located at a point on line AB, Fig. 16–21b. Assuming this point to be a distance x from B, we have

$$v_B = \omega x; \qquad\qquad 0.4 = \omega x$$
$$v_A = \omega(0.25 - x); \qquad 0.25 = \omega(0.25 - x)$$

Dividing one of these equations into the other eliminates ω and yields

$$0.4(0.25 - x) = 0.25x$$
$$x = \frac{0.1}{0.65} - 0.154 \text{ m}$$

Hence, the angular velocity is

$$\omega = \frac{v_B}{x} = \frac{0.4}{0.154} = 2.60 \text{ rad/s} \downarrow \qquad\qquad\qquad Ans.$$

The velocity of point C is therefore

$$v_C = \omega r_{C/IC} = 2.60(0.154 - 0.125)$$
$$= 0.0754 \text{ m/s} \leftarrow \qquad\qquad\qquad Ans.$$

PROBLEMS

*16–52. Solve Prob. 16–40 using the method of instantaneous center of zero velocity.

16–53. Solve Prob. 16–38 using the method of instantaneous center of zero velocity.

16–54. Solve Prob. 16–42 using the method of instantaneous center of zero velocity.

16–55. Solve Prob. 16–43 using the method of instantaneous center of zero velocity.

*16–56. Solve Prob. 16–44 using the method of instantaneous center of zero velocity.

16–57. Solve Prob. 16–47 using the method of instantaneous center of zero velocity.

16–58. In each case show graphically how to locate the instantaneous center of zero velocity of link AB. Assume the geometry is known.

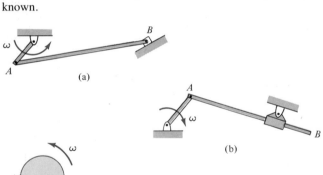

(a)

(b)

(c)

Prob. 16–58

16–59. The conveyor belt is moving to the right at $v = 8$ ft/s, and at the same instant the cylinder is rolling counterclockwise at $\omega = 2$ rad/s without slipping. Determine the velocities of the cylinder's center C and point B at this instant.

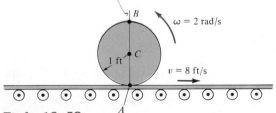

Prob. 16–59

*16–60. At the instant shown, the disk is rotating at $\omega = 4$ rad/s. If the end of the cord wrapped around the disk is fixed at D, determine the velocities of points A, B, and C.

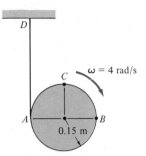

Prob. 16–60

16–61. As the car travels forward at 80 ft/s on a wet road, due to slipping the rear wheels have an angular velocity $\omega = 100$ rad/s. Determine the speeds of points A, B, and C caused by the motion.

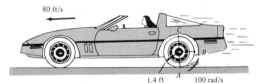

Prob. 16–61

16–62. Determine the velocity of block C at the instant shown, if link AB is rotating at 8 rad/s.

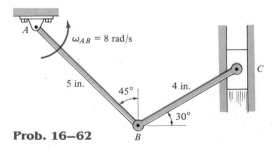

Prob. 16–62

16–63. The disk of radius r is confined to roll without slipping at A and B. If the plates have the velocities shown, determine the angular velocity of the disk.

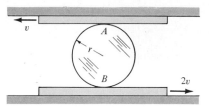

Prob. 16–63

*16–64. If link AB is rotating at $\omega_{AB} = 6$ rad/s, determine the angular velocities of links BC and CD at the instant shown.

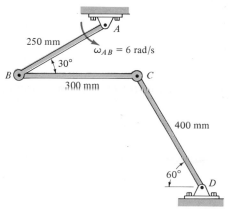

Prob. 16–64

16–65. Due to slipping points A and B on the rim of the disk have the velocities shown. Determine the velocity of the center point C and point D at this instant.

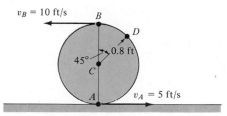

Prob. 16–65

16–66. The planet gear A is pin-connected to the end of the link BC. If the link rotates about the fixed point B at 4 rad/s, determine the angular velocity of the ring gear R. The sun gear D is fixed from rotating.

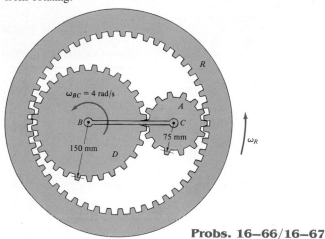

Probs. 16–66/16–67

16–67. Solve Prob. 16–66 if the sun gear D is rotating clockwise at $\omega_D = 5$ rad/s while link BC rotates counterclockwise at $\omega_{BC} = 4$ rad/s.

*16–68. If the hub gear H and ring gear R have angular velocities $\omega_H = 5$ rad/s and $\omega_R = 20$ rad/s, respectively, determine the angular velocity ω_S of the spur gear S and the angular velocity of its attached arm OA.

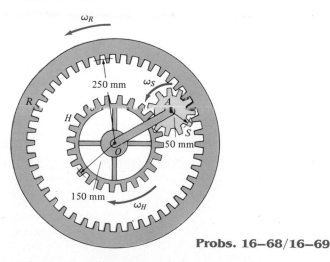

Probs. 16–68/16–69

16–69. If the hub gear H has an angular velocity $\omega_H = 5$ rad/s, determine the angular velocity of the ring gear R so that the arm OA attached to the spur gear S remains stationary ($\omega_{OA} = 0$). What is the angular velocity of the spur gear?

16–70. If bar *AB* has an angular velocity ω_{AB} = 4 rad/s, determine the velocity of the slider block *C* at the instant shown.

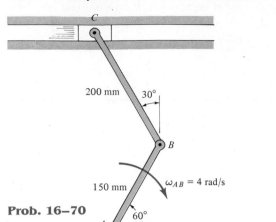

200 mm

30°

B

150 mm

ω_{AB} = 4 rad/s

60°

A

Prob. 16–70

*16–72. If the slider block *A* is moving to the right at v_A = 8 ft/s, determine the velocity of blocks *B* and *C* at the instant shown.

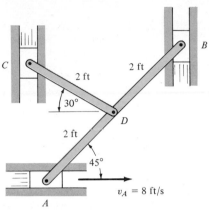

C

2 ft

2 ft

B

30°

D

2 ft

45°

v_A = 8 ft/s

A

Prob. 16–72

16–71. Mechanical toy animals often use a walking mechanism as shown idealized in the figure. If the driving crank *AB* is propelled by a spring motor such that ω_{AB} = 5 rad/s, determine the velocity of the rear foot *E* at the instant shown.

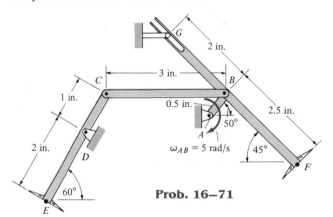

G

2 in.

3 in.

C

B

1 in.

0.5 in.

2.5 in.

A

50°

ω_{AB} = 5 rad/s

45°

F

2 in.

D

60°

E

Prob. 16–71

16.7 Relative-Motion Analysis: Acceleration

An equation that relates the accelerations of two points in a rigid body subjected to general plane motion may be determined by differentiating the velocity equation $\mathbf{v}_B = \mathbf{v}_A + \mathbf{v}_{B/A}$ with respect to time. Thus,

$$\frac{d\mathbf{v}_B}{dt} = \frac{d\mathbf{v}_A}{dt} + \frac{d\mathbf{v}_{B/A}}{dt}$$

The terms $d\mathbf{v}_B/dt = \mathbf{a}_B$ and $d\mathbf{v}_A/dt = \mathbf{a}_A$ are measured from a set of *fixed x, y axes* and represent the *absolute accelerations* of points *B* and *A*. The last term represents the acceleration of B with respect to A as measured by an observer fixed to the translating x', y' axes having their origin at the base point A. In Sec. 16.5, it was shown that to this observer point B appears to move along a *circular arc* that has a radius of curvature $r_{B/A}$. In other words, *the body appears to move as if it were rotating about the z' axis passing through point A*. Consequently, $\mathbf{a}_{B/A}$ can be expressed in terms of its tangential and normal components of motion; i.e., $\mathbf{a}_{B/A} = (\mathbf{a}_{B/A})_t + (\mathbf{a}_{B/A})_n$, where $(a_{B/A})_t = \alpha r_{B/A}$ and $(a_{B/A})_n = \omega^2 r_{B/A}$. Hence, the above equation can be written in the form

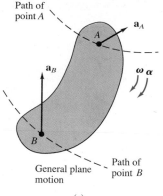

$$\boxed{\mathbf{a}_B = \mathbf{a}_A + (\mathbf{a}_{B/A})_t + (\mathbf{a}_{B/A})_n} \qquad (16\text{-}16)$$

where

$\mathbf{a}_B$ = acceleration of point B

$\mathbf{a}_A$ = acceleration of point A

$(\mathbf{a}_{B/A})_t$ = relative tangential acceleration component of "B with respect to A." Since the relative motion is circular, the *magnitude* of $(\mathbf{a}_{B/A})_t$ is $(a_{B/A})_t = \alpha r_{B/A}$ and the *direction* is perpendicular to $\mathbf{r}_{B/A}$

$(\mathbf{a}_{B/A})_n$ = relative normal acceleration component of "B with respect to A." Since the relative motion is circular, the *magnitude* of $(\mathbf{a}_{B/A})_n$ is $(a_{B/A})_n = \omega^2 r_{B/A}$ and the *direction* is always from B towards A

Each of the four terms in Eq. 16–16 is represented graphically on the *kinematic diagrams* shown in Fig. 16–22. Here it is seen that at a given instant the acceleration of B, Fig. 16–22a, is determined by considering the body to translate with an acceleration $\mathbf{a}_A$, Fig. 16–22b, and simultaneously rotate about the base point A with an instantaneous angular velocity $\boldsymbol{\omega}$ and angular acceleration $\boldsymbol{\alpha}$, Fig. 16–22c. Vector addition of these two effects, applied to B, yields $\mathbf{a}_B$, as shown in Fig. 16–22d. It should be noted from Fig. 16–22a that points A and B move along *curved paths*, and as a result the accelerations of these points have *both tangential and normal components*. (Recall that the acceleration of a point is *tangent to the path only* when the path is *rectilinear* or when it is an inflection point.)

Since the relative-acceleration components represent the effect of *circular motion* observed from translating axes having their origin at the base point A, these terms can be expressed as $(\mathbf{a}_{B/A})_t = \boldsymbol{\alpha} \times \mathbf{r}_{B/A}$ and $(\mathbf{a}_{B/A})_n = -\omega^2 \mathbf{r}_{B/A}$, Eq. 16–13. Hence, Eq. 16–16 becomes

$$\boxed{\mathbf{a}_B = \mathbf{a}_A + \boldsymbol{\alpha} \times \mathbf{r}_{B/A} - \omega^2 \mathbf{r}_{B/A}} \qquad (16\text{-}17)$$

where

$\mathbf{a}_B$ = acceleration of point B

$\mathbf{a}_A$ = acceleration of the base point A

$\boldsymbol{\alpha}$ = angular acceleration of the body

$\boldsymbol{\omega}$ = angular velocity of the body

$\mathbf{r}_{B/A}$ = relative-position vector drawn from A to B

To apply this equation, all of these terms must be represented in Cartesian vector form.

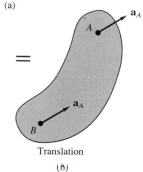

Translation

(b)

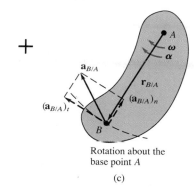

Rotation about the base point A

(c)

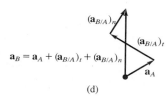

$\mathbf{a}_B = \mathbf{a}_A + (\mathbf{a}_{B/A})_t + (\mathbf{a}_{B/A})_n$

(d)

Fig. 16–22

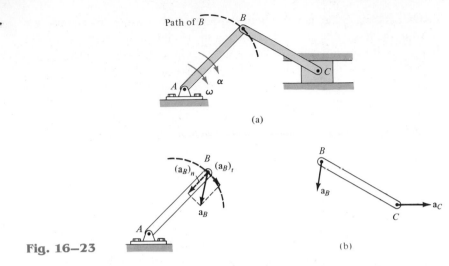

Path of B

(a)

Fig. 16-23

(b)

If Eq. 16–16 or 16–17 is applied in a practical manner to study the accelerated motion of a rigid body which is pin-connected to two other bodies, it should be realized that points which are *coincident at the pin* move with the *same acceleration,* since the path of motion over which they travel is the *same.* For example, point B lying on either rod AB or BC of the crank mechanism shown in Fig. 16–23a has the same acceleration, since the rods are pin-connected at B. Here the motion of B is along a *curved path,* so that $\mathbf{a}_B$ is calculated on the basis of its tangential and normal components $(a_B)_t = \alpha r_{B/A}$ and $(a_B)_n = \omega^2 r_{B/A}$, which are defined by the angular motion of AB. At the other end of rod BC, however, point C moves along a *rectilinear path,* which is defined by the piston. Hence, in this case, the acceleration $\mathbf{a}_C$ is directed along the path, Fig. 16–23b.

If two bodies contact one another *without slipping,* and the *points in contact* move along *different paths,* the *tangential components* of acceleration of the points will be the *same;* however, the *normal components* will *not* be the same. For example, consider the two meshed gears in Fig. 16–24a. Point A is located on gear B and a coincident point A' is located on gear C. Due to the rotational motion, $(\mathbf{a}_A)_t = (\mathbf{a}_{A'})_t$; however, since both points follow different curved paths, $(\mathbf{a}_A)_n \neq (\mathbf{a}_{A'})_n$ and therefore $\mathbf{a}_A \neq \mathbf{a}_{A'}$, Fig. 16–24b.

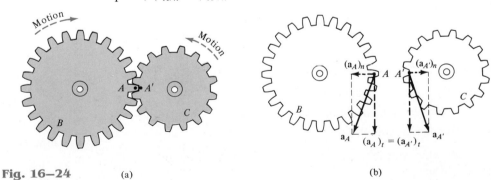

Fig. 16-24   (a)   (b)

PROCEDURE FOR ANALYSIS

Equation 16–16 relates the accelerations of any two points A and B located on the *same* rigid body. Like the velocity equation, Eq. 16–14, it can be applied either by using a Cartesian vector analysis, in the form of Eq. 16–17, or by writing the x and y scalar component equations directly. For application, it is suggested that the following procedure be used.

Velocity Analysis. If the angular velocity $\boldsymbol{\omega}$ of the body is unknown, determine it by using a velocity analysis as discussed in Sec. 16.5 or 16.6. Determine also the velocities $\mathbf{v}_A$ and $\mathbf{v}_B$ of points A and B *if these points move along curved paths.*

Kinematic Diagrams. Establish the directions of the fixed x, y coordinates, and draw the kinematic diagram of the body. Indicate on it the accelerations of points A and B, $\mathbf{a}_A$ and $\mathbf{a}_B$, the angular velocity $\boldsymbol{\omega}$, the angular acceleration $\boldsymbol{\alpha}$, and the relative position vector $\mathbf{r}_{B/A}$. Normally the base point A is selected as a point having a known acceleration. In particular, if points A and B move along *curved paths*, their accelerations should be expressed in terms of their tangential and normal components, i.e., $\mathbf{a}_A = (\mathbf{a}_A)_t + (\mathbf{a}_A)_n$ and $\mathbf{a}_B = (\mathbf{a}_B)_t + (\mathbf{a}_B)_n$. Here $(a_A)_n = (v_A)^2/\rho_A$ and $(a_B)_n = (v_B)^2/\rho_B$, where ρ_A and ρ_B define the radii of curvature of the paths of points A and B, respectively. Identify the two unknowns on the diagram. If the magnitudes of $(\mathbf{a}_A)_t$, $(\mathbf{a}_B)_t$, or $\boldsymbol{\alpha}$ are unknown, the sense of direction of these vectors may be assumed.

If the acceleration equation is applied *without* using a Cartesian vector analysis, then the magnitude and direction of the relative acceleration components $(\mathbf{a}_{B/A})_t$ and $(\mathbf{a}_{B/A})_n$ must be established. This can be done by drawing a kinematic diagram such as shown in Fig. 16–22c. Since the body is considered to be momentarily "pinned" at the base point A, these components have *magnitudes* of $(a_{B/A})_t = \alpha r_{B/A}$ and $(a_{B/A})_n = \omega^2 r_{B/A}$. Their *directions* are established from the diagram such that $(\mathbf{a}_{B/A})_t$ acts perpendicular to $\mathbf{r}_{B/A}$, in accordance with the rotational motion $\boldsymbol{\alpha}$ of the body, and $(\mathbf{a}_{B/A})_n$ is directed from B toward A.*

Acceleration Equation. To apply Eq. 16–17, express the vectors in Cartesian vector form and substitute them into the equation. Evaluate the cross product and then equate the respective **i** and **j** components to obtain two scalar equations. To obtain these scalar equations *directly* write Eq. 16–16 in symbolic form, $\mathbf{a}_B = \mathbf{a}_A + (\mathbf{a}_{B/A})_t + (\mathbf{a}_{B/A})_n$, and underneath each of the terms represent the vectors by their magnitudes and directions. To do this, use the data tabulated on the kinematic diagrams. The scalar equations are determined from the x and y components of these vectors. Solve these equations for the two unknowns. If the solution yields a *negative* answer for an *unknown*, it indicates the sense of direction of the vector is opposite to that shown on the kinematic diagram.

The following example problems numerically illustrate both methods of application.

---

*Perhaps the notation $\mathbf{a}_B = \mathbf{a}_A + (\mathbf{a}_{B/A(\text{pin})})_t + (\mathbf{a}_{B/A(\text{pin})})_n$ may be helpful in recalling that A is pinned.

273

Example 16–12

The rod AB shown in Fig. 16–25a is confined to move along the inclined planes at A and B. If point A has an acceleration of 3 m/s² and a velocity of 2 m/s, both directed down the plane at the instant the rod becomes horizontal, determine the angular acceleration of the rod at this instant.

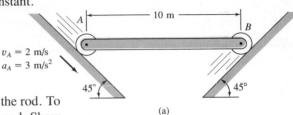

(a)

SOLUTION I (VECTOR ANALYSIS)

We will apply the acceleration equation to points A and B on the rod. To do so it is first necessary to determine the angular velocity of the rod. Show that it is $\omega = 0.283$ rad/s $\curvearrowleft$ using either the velocity equation or the method of instantaneous centers.

Kinematic Diagram. The x, y axes are established in Fig. 16–25b. Since points A and B both move along straight-line paths, they have *no* components of acceleration in the normal direction. Why? There are two unknowns in Fig. 16–25b, namely, a_B and α.

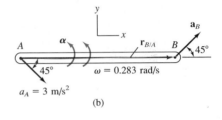

(b)

Acceleration Equation. Expressing each of the vectors in Cartesian vector form, we have

$$\mathbf{r}_{B/A} = \{10\mathbf{i}\} \text{ m} \qquad \boldsymbol{\omega} = \{0.283\mathbf{k}\} \text{ rad/s} \qquad \boldsymbol{\alpha} = \alpha\mathbf{k}$$
$$\mathbf{a}_A = \{3 \cos 45°\mathbf{i} - 3 \sin 45°\mathbf{j}\} \text{ m/s}^2 \qquad \mathbf{a}_B = a_B \cos 45°\mathbf{i} + a_B \sin 45°\mathbf{j}$$

Applying Eq. 16–17 to points A and B on the rod, we have

$$\mathbf{a}_B = \mathbf{a}_A + \boldsymbol{\alpha} \times \mathbf{r}_{B/A} - \omega^2\mathbf{r}_{B/A}$$

$a_B \cos 45°\mathbf{i} + a_B \sin 45°\mathbf{j}$
$$= 3 \cos 45°\mathbf{i} - 3 \sin 45°\mathbf{j} + (\alpha\mathbf{k}) \times (10\mathbf{i}) - (0.283)^2(10\mathbf{i})$$

Carrying out the cross product and equating the $\mathbf{i}$ and $\mathbf{j}$ components yields

$$a_B \cos 45° = 3 \cos 45° - (0.283)^2(10) \qquad (1)$$
$$a_B \sin 45° = -3 \sin 45° + \alpha(10) \qquad (2)$$

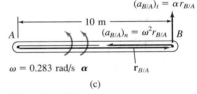

(c)

Fig. 16–25

Solving, we have

$$a_B = 1.87 \text{ m/s}^2 \quad \angle^{45°}$$
$$\alpha = 0.344 \text{ rad/s}^2 \curvearrowleft \qquad\qquad Ans.$$

SOLUTION II (SCALAR COMPONENTS)

As an alternative procedure, the scalar component equations (1) and (2) can be obtained directly. From the kinematic diagram, showing the relative acceleration components $(\mathbf{a}_{B/A})_t$ and $(\mathbf{a}_{B/A})_n$, Fig. 16–25c, we have

$$\mathbf{a}_B = \mathbf{a}_A + (\mathbf{a}_{B/A})_t + (\mathbf{a}_{B/A})_n$$

$$\begin{bmatrix} a_B \\ \angle^{45°} \end{bmatrix} = \begin{bmatrix} 3 \text{ m/s}^2 \\ \searrow^{45°} \end{bmatrix} + \begin{bmatrix} \alpha(10 \text{ m}) \\ \uparrow \end{bmatrix} + \begin{bmatrix} (0.283 \text{ rad/s})^2(10 \text{ m}) \\ \leftarrow \end{bmatrix}$$

Equating the x and y components yields Eqs. (1) and (2) and the solution proceeds as before.

Example 16–13

At a given instant, the cylinder of radius r, shown in Fig. 16–26a, has an angular velocity $\boldsymbol{\omega}$ and angular acceleration $\boldsymbol{\alpha}$. Determine the velocity and acceleration of its center G if it rolls without slipping.

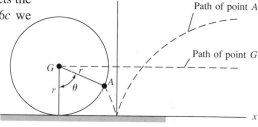

(a)

SOLUTION

As the cylinder rolls, point G moves along a straight line, and point A, located on the rim of the cylinder, moves along a curved path called a *cycloid*,* Fig. 16–26b. We will apply the velocity and acceleration equations to these points.

Velocity Analysis. Since no slipping occurs, at the instant A contacts the ground, $\mathbf{v}_A = \mathbf{0}$. Thus, from the kinematic diagram in Fig. 16–26c we have

$$\mathbf{v}_G = \mathbf{v}_A + \boldsymbol{\omega} \times \mathbf{r}_{G/A}$$
$$v_G\mathbf{i} = \mathbf{0} + (-\omega\mathbf{k}) \times (r\mathbf{j})$$
$$v_G = \omega r \qquad\qquad (1) \quad Ans.$$

This same result can also be obtained directly by noting that point A represents the instantaneous center of zero velocity.

Kinematic Diagram. The acceleration of point G is horizontal since it moves along a straight-line path. *Just before* point A touches the ground, its velocity is directed downward along the y axis, Fig. 16–26b, and just after contact, its velocity is directed upward.† For this reason, point A begins to accelerate upward when it leaves the ground at A, Fig. 16–26d. No motion occurs in the x direction during the instant of ground contact. The magnitudes of $\mathbf{a}_A$ and $\mathbf{a}_G$ are unknown.

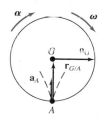

(c)

Acceleration Equation

$$\mathbf{a}_G = \mathbf{a}_A + \boldsymbol{\alpha} \times \mathbf{r}_{G/A} - \omega^2\mathbf{r}_{G/A}$$
$$a_G\mathbf{i} = a_A\mathbf{j} + (-\alpha\mathbf{k}) \times (r\mathbf{j}) - \omega^2(r\mathbf{j})$$

Evaluating the cross product and equating the $\mathbf{i}$ and $\mathbf{j}$ components yields

$$a_G = \alpha r \qquad\qquad (2) \quad Ans.$$
$$a_A = \omega^2 r \qquad\qquad (3)$$

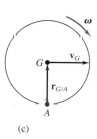

Fig. 16–26 (d)

These important results, that $v_G = \omega r$ and $a_G = \alpha r$, were also obtained in Example 16–3. They apply to any circular object, such as a ball, pulley, disk, etc., that rolls *without* slipping. Also, the fact that $a_A = \omega^2 r$ indicates that the instantaneous center of zero velocity, point A, is not a point of zero acceleration.

---

*Although it is not necessary here, one can show that the equation of the path (cycloid) can be written in terms of θ as $x = r(\theta - \sin\theta)$ and $y = r(1 - \cos\theta)$.

†The slope of the path, dy/dx, is along the y axis at $\theta = 90°$.

Example 16–14

The spool shown in Fig. 16–27a unravels from the cord, such that at the instant shown it has an angular velocity of 3 rad/s and an angular acceleration of 4 rad/s². Determine the acceleration of point B.

SOLUTION I (VECTOR ANALYSIS)

The spool "appears" to be rolling downward without slipping at point A. Therefore, we can use the results of Example 16–13 to determine the acceleration of point G, i.e.,

$$a_G = \alpha r = 4(0.5) = 2 \text{ ft/s}^2$$

We will apply the acceleration equation between points G and B.

Kinematic Diagram. Point B moves along a *curved path* having an unknown radius of curvature.* Its acceleration is represented by its unknown x and y components as shown in Fig. 16–27b.

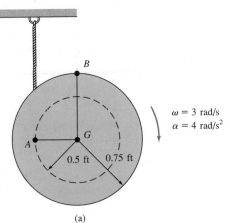

(a)

Acceleration Equation

$$\mathbf{a}_B = \mathbf{a}_G + \boldsymbol{\alpha} \times \mathbf{r}_{B/G} - \omega^2 \mathbf{r}_{B/G}$$

$$(a_B)_x \mathbf{i} + (a_B)_y \mathbf{j} = -2\mathbf{j} + (-4\mathbf{k}) \times (0.75\mathbf{j}) - (3)^2(0.75\mathbf{j})$$

Therefore, the component equations are

$$(a_B)_x = 4(0.75) = 3 \text{ ft/s}^2 \qquad (1)$$
$$(a_B)_y = -2 - 6.75 = -8.75 \text{ ft/s}^2 = 8.75 \text{ ft/s}^2 \qquad (2)$$

The magnitude and direction of $\mathbf{a}_B$ are therefore

$$a_B = \sqrt{(3)^2 + (8.75)^2} = 9.25 \text{ ft/s}^2 \qquad \textit{Ans.}$$

$$\theta = \tan^{-1} \frac{8.75}{3} = 71.1° \qquad \textit{Ans.}$$

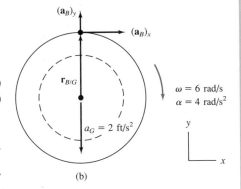

(b)

SOLUTION II (SCALAR COMPONENTS)

This problem may be solved by writing the scalar component equations directly. The kinematic diagram in Fig. 16–27c shows the relative acceleration components $(\mathbf{a}_{B/G})_t$ and $(\mathbf{a}_{B/G})_n$. Thus,

$$\mathbf{a}_B = \mathbf{a}_G + (\mathbf{a}_{B/G})_t + (\mathbf{a}_{B/G})_n$$

$$\begin{bmatrix} (a_B)_x \\ \rightarrow \end{bmatrix} + \begin{bmatrix} (a_B)_y \\ \uparrow \end{bmatrix}$$
$$= \begin{bmatrix} 2 \text{ ft/s}^2 \\ \downarrow \end{bmatrix} + \begin{bmatrix} 4 \text{ rad/s}^2(0.75 \text{ ft}) \\ \rightarrow \end{bmatrix} + \begin{bmatrix} (3 \text{ rad/s})^2(0.75 \text{ ft}) \\ \downarrow \end{bmatrix}$$

The x and y components yield Eqs. (1) and (2) shown above.

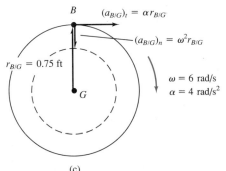

(c)

Fig. 16–27

*Realize that the path's radius of curvature ρ is *not* equal to the radius of the cylinder since the cylinder is *not* rotating about point G. Furthermore, ρ is *not* defined as the distance from A (IC) to B, since the location of the IC depends only on the velocity of a point and *not* the geometry of its path.

Example 16–15

The collar C in Fig. 16–28a is moving downward with an acceleration of 1 m/s². At the instant shown, it has a speed of 2 m/s which gives links CB and AB an angular velocity $\omega_{AB} = \omega_{CB} = 10$ rad/s. (See Example 16–7.) Determine the angular accelerations of CB and AB at this instant.

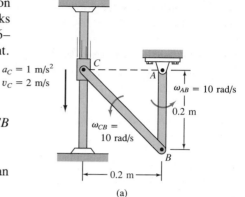

$a_C = 1$ m/s²
$v_C = 2$ m/s

$\omega_{AB} = 10$ rad/s

0.2 m

$\omega_{CB} = 10$ rad/s

0.2 m

(a)

SOLUTION (VECTOR ANALYSIS)

Kinematic Diagrams. The kinematic diagrams of *both* links AB and CB are shown in Fig. 16–28b.

Acceleration Equations. Expressing each of the vectors in Cartesian vector form, we have

$$\mathbf{r}_{B/A} = \{-0.2\mathbf{j}\} \text{ m} \qquad \mathbf{r}_{B/C} = \{0.2\mathbf{i} - 0.2\mathbf{j}\} \text{ m}$$

$$\boldsymbol{\omega}_{AB} = \{10\mathbf{k}\} \text{ rad/s} \qquad \boldsymbol{\omega}_{CB} = \{10\mathbf{k}\} \text{ rad/s}$$

$$\boldsymbol{\alpha}_{AB} = \alpha_{AB}\mathbf{k} \qquad \boldsymbol{\alpha}_{CB} = \alpha_{CB}\mathbf{k}$$

$$\mathbf{a}_C = \{-1\mathbf{j}\} \text{ m/s}^2$$

Link AB is rotating about a fixed axis. Applying Eq. 16–13 to point B, we have

$$\mathbf{a}_B = \boldsymbol{\alpha}_{AB} \times \mathbf{r}_{B/A} - \omega_{AB}^2 \mathbf{r}_{B/A}$$

$$\mathbf{a}_B = (\alpha_{AB}\mathbf{k}) \times (-0.2\mathbf{j}) - (10)^2(-0.2\mathbf{j})$$

$$\mathbf{a}_B = 0.2\alpha_{AB}\mathbf{i} + 20\mathbf{j}$$

Link BC is subjected to general plane motion. Applying Eq. 16–17 to points B and C, using the above value for $\mathbf{a}_B$, yields

$$\mathbf{a}_B = \mathbf{a}_C + \boldsymbol{\alpha}_{CB} \times \mathbf{r}_{B/C} - \omega_{CB}^2 \mathbf{r}_{B/C}$$

$$0.2\alpha_{AB}\mathbf{i} + 20\mathbf{j} = -1\mathbf{j} + (\alpha_{CB}\mathbf{k}) \times (0.2\mathbf{i} - 0.2\mathbf{j}) - (10)^2(0.2\mathbf{i} - 0.2\mathbf{j})$$

$$0.2\alpha_{AB}\mathbf{i} + 20\mathbf{j} = -1\mathbf{j} + 0.2\alpha_{CB}\mathbf{j} + 0.2\alpha_{CB}\mathbf{i} - 20\mathbf{i} + 20\mathbf{j}$$

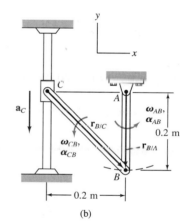

a_C

$\mathbf{r}_{B/C}$

ω_{CB}, α_{CB}

ω_{AB}, α_{AB}

0.2 m

$\mathbf{r}_{B/A}$

0.2 m

(b)

Fig. 16–28

Equating the respective $\mathbf{i}$ and $\mathbf{j}$ components, we obtain

$$0.2\alpha_{AB} = 0.2\alpha_{CB} - 20 \qquad\qquad (1)$$

$$20 = -1 + 0.2\alpha_{CB} + 20 \qquad\qquad (2)$$

Hence,

$$\alpha_{CB} = 5 \text{ rad/s}^2$$

$$\alpha_{AB} = -95 \text{ rad/s}^2$$

Thus,

$$\boldsymbol{\alpha}_{CB} = \{5\mathbf{k}\} \text{ rad/s}^2 \qquad\qquad Ans.$$

$$\boldsymbol{\alpha}_{AB} = \{-95\mathbf{k}\} \text{ rad/s}^2 \qquad\qquad Ans.$$

Example 16–16

The crankshaft AB of an engine turns with a clockwise angular acceleration of 20 rad/s², Fig. 16–29a. Determine the acceleration of the piston at the instant AB is in the position shown. At this instant $\omega_{AB} = 10$ rad/s and $\omega_{BC} = 2.43$ rad/s.

SOLUTION (VECTOR ANALYSIS)

Kinematic Diagrams. The kinematic diagrams for both AB and BC are shown in Fig. 16–29b.

Acceleration Equations. Expressing each of the vectors in Cartesian vector form yields

$$\mathbf{r}_{B/A} = \{-0.25 \sin 45°\mathbf{i} + 0.25 \cos 45°\mathbf{j}\} \text{ ft} = \{-0.177\mathbf{i} + 0.177\mathbf{j}\} \text{ ft}$$
$$\mathbf{r}_{C/B} = \{0.75 \sin 13.6°\mathbf{i} + 0.75 \cos 13.6°\mathbf{j}\} \text{ ft} = \{0.176\mathbf{i} + 0.729\mathbf{j}\} \text{ ft}$$

$$\boldsymbol{\omega}_{AB} = \{-10\mathbf{k}\} \text{ rad/s} \qquad \boldsymbol{\alpha}_{AB} = \{-20\mathbf{k}\} \text{ rad/s}^2$$
$$\boldsymbol{\omega}_{BC} = \{2.43\mathbf{k}\} \text{ rad/s} \qquad \boldsymbol{\alpha}_{BC} = \alpha_{BC}\mathbf{k}$$
$$\mathbf{a}_C = a_C\mathbf{j}$$

The connecting rod AB is subjected to rotation about a fixed axis. Applying Eq. 16–13 to point B, we have

$$\mathbf{a}_B = \boldsymbol{\alpha}_{AB} \times \mathbf{r}_{B/A} - \omega_{AB}^2\mathbf{r}_{B/A}$$
$$= (-20\mathbf{k}) \times (-0.177\mathbf{i} + 0.177\mathbf{j}) - (10)^2(-0.177\mathbf{i} + 0.177\mathbf{j})$$
$$= \{21.24\mathbf{i} - 14.16\mathbf{j}\} \text{ ft/s}^2$$

The crankshaft is subjected to general plane motion. Applying Eq. 16–17 to points B and C yields

$$\mathbf{a}_C = \mathbf{a}_B + \boldsymbol{\alpha}_{BC} \times \mathbf{r}_{C/B} - \omega_{BC}^2\mathbf{r}_{C/B}$$
$$a_C\mathbf{j} = 21.24\mathbf{i} - 14.16\mathbf{j} + (\alpha_{BC}\mathbf{k}) \times (0.176\mathbf{i} + 0.729\mathbf{j})$$
$$- (2.43)^2(0.176\mathbf{i} + 0.729\mathbf{j})$$
$$a_C\mathbf{j} = 21.24\mathbf{i} - 14.16\mathbf{j} + 0.176\alpha_{BC}\mathbf{j} - 0.729\alpha_{BC}\mathbf{i} - 1.04\mathbf{i} - 4.30\mathbf{j}$$

Equating the respective $\mathbf{i}$ and $\mathbf{j}$ components, we have

$$0 = 20.20 - 0.729\alpha_{BC}$$
$$a_C = 0.176\alpha_{BC} - 18.46$$

Solving yields

$$\alpha_{BC} = 27.7 \text{ rad/s}^2 \text{ }\nwarrow$$
$$a_C = -13.6 \text{ ft/s}^2 \qquad\qquad Ans.$$

The negative sign indicates that the piston is decelerating, i.e., $\mathbf{a}_C = \{-13.6\mathbf{j}\}$ ft/s². This causes the speed of the piston to decrease until the crankshaft AB becomes vertical, at which time the piston is momentarily at rest.

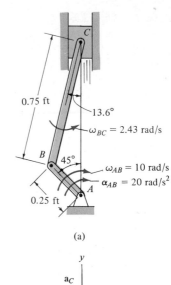

(a)

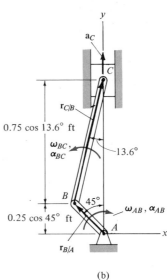

(b)

Fig. 16–29

PROBLEMS

16–73. At a given instant the top A of the bar has the velocity and acceleration shown. Determine the acceleration of the bottom B of the bar, and the bar's angular acceleration at this instant.

$v_A = 5$ ft/s
$a_A = 7$ ft/s^2

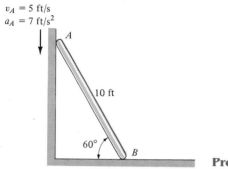

Prob. 16–73

16–74. The disk is moving to the left such that is has an angular acceleration $\alpha = 8$ rad/s^2 and angular velocity $\omega = 3$ rad/s at the instant shown. If it does not slip at A, determine the acceleration of point B.

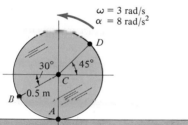

Probs. 16–74/16–75

16–75. The disk is moving to the left such that is has an angular acceleration $\alpha = 8$ rad/s^2 and angular velocity $\omega = 3$ rad/s at the instant shown. If it does not slip at A, determine the acceleration of point D.

*16–76. A cord is wrapped around the inner spool of the gear. If it is pulled with a constant velocity $\mathbf{v}$, determine the velocity and acceleration of points A and B. The gear rolls on the fixed gear rack.

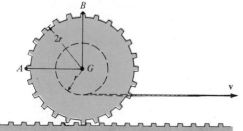

Prob. 16–76

16–77. The hoop is cast on the rough surface such that it has an angular velocity of $\omega = 4$ rad/s and an angular deceleration of $\alpha = 5$ rad/s^2. Also, its center has a velocity of $v_O = 5$ m/s and a deceleration of $a_O = 2$ m/s^2. Determine the acceleration of point A at this instant.

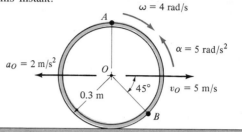

Probs. 16–77/16–78

16–78. The hoop is cast on the rough surface such that it has an angular velocity of $\omega = 4$ rad/s and an angular deceleration of $\alpha = 5$ rad/s^2. Also, its center has a velocity of $v_O = 5$ m/s and a deceleration of $a_O = 2$ m/s^2. Determine the acceleration of point B at this instant.

16–79. The disk rotates with an angular velocity $\omega = 5$ rad/s and an angular acceleration $\alpha = 6$ rad/s^2. Determine the angular accelerations of links AB and BC at this instant.

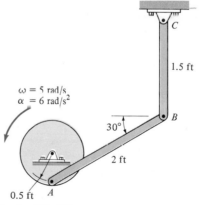

Prob. 16–79

*16–80. At a given instant the wheel is rotating with the angular velocity and angular acceleration shown. Determine the acceleration of block *B* at this instant.

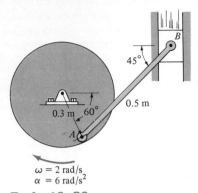

Prob. 16–80

16–81. If rod *AB* has the angular motion at the instant shown, determine the acceleration of block *C* at this instant.

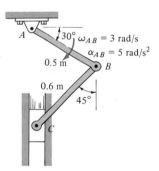

Prob. 16–81

16–82. At a given instant, link *AB* has an angular velocity $\omega_{AB} = 4$ rad/s and an angular acceleration $\alpha_{AB} = 12$ rad/s². Determine the angular velocity and angular acceleration of links *BC* and *CD* at this instant.

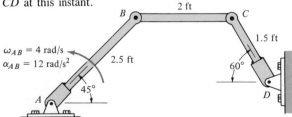

Prob. 16–82

16–83. At a given instant gears *A* and *B* have the angular motions shown. Determine the angular acceleration of gear *C* and the acceleration of its center point *D* at this instant. Note that the inner hub of gear *C* is in mesh with gear *A* and its outer rim is in mesh with gear *B*.

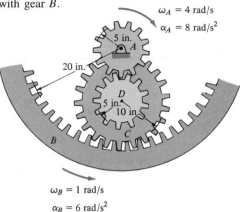

Prob. 16–83

*16–84. The wheel rolls without slipping such that at the instant shown it has an angular velocity ω and angular acceleration α. Determine the angular velocity and angular acceleration of the rod at this instant.

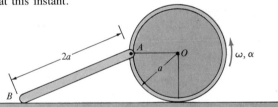

Prob. 16–84

16–85. The ends of the bar *AB* are confined to move along the paths shown. At a given instant, *A* has a velocity of $v_A = 4$ ft/s and an acceleration of $a_A = 7$ ft/s². Determine the angular velocity and angular acceleration of *AB* at this instant.

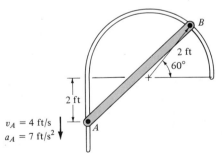

Prob. 16–85

16.8 Relative-Motion Analysis Using Rotating Axes

In the previous sections, the relative-motion analysis for velocity and acceleration was described using a translating coordinate system. This type of analysis is useful for determining the motion of points in the *same* rigid body, or the motion of points located in several pin-connected rigid bodies. In some problems, however, rigid bodies (mechanisms) are constructed such that *sliding* will occur at their connections. The kinematic analysis for such cases is best performed if the motion is analyzed using a coordinate system which both *translates* and *rotates*. Furthermore, this frame of reference is useful for analyzing the motions of two points on a mechanism which are *not* located in the *same* rigid body and for specifying the kinematics of particle motion when one of the particles is moving along a rotating path.

In the following analysis two equations are developed which relate the velocity and acceleration of a point to the origin of a moving frame of reference subjected to both a translation and a rotation in the plane.* Due to the generality in the derivation which follows, these two points may represent either two particles moving independently of one another or two points located in the same (or different) rigid bodies.

Position

Consider the two points A and B shown in Fig. 16–30a. Their location is specified by the position vectors $\mathbf{r}_A$ and $\mathbf{r}_B$, which are measured from the fixed X, Y, Z coordinate system. As shown in the figure, the "base point" A represents the origin of the x, y, z coordinate system, which is assumed to be both translating and rotating with respect to the X, Y, Z system. The position of B with respect to A is specified by the relative-position vector $\mathbf{r}_{B/A}$. The components of this vector may be expressed either in terms of unit vectors along the X, Y axes, i.e., $\mathbf{I}$ and $\mathbf{J}$, or by unit vectors along the x, y axes, i.e., $\mathbf{i}$ and $\mathbf{j}$. Although the magnitude of $\mathbf{r}_{B/A}$ is the same when measured in both coordinate systems, the direction of this vector will be measured differently if the x, y axes are not parallel to the X, Y axes. For the proof which follows, $\mathbf{r}_{B/A}$ will be measured relative to the moving x, y frame of reference. Thus, if B has coordinates (x_B, y_B), Fig. 16–30a, then

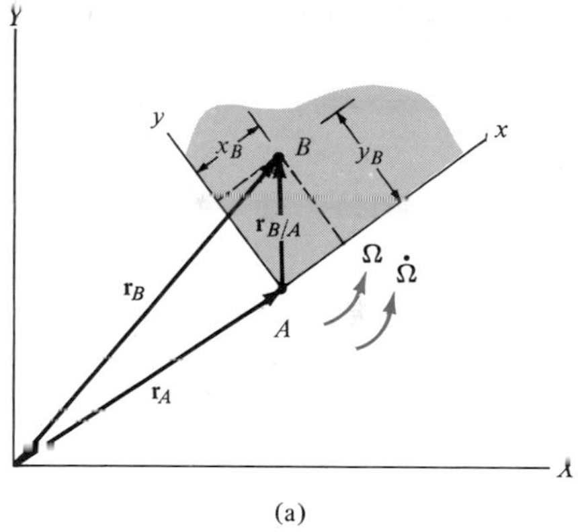

$$\mathbf{r}_{B/A} = x_B\mathbf{i} + y_B\mathbf{j}$$

Using vector addition, the three position vectors in Fig. 16–30a may be related by the equation

$$\boxed{\mathbf{r}_B = \mathbf{r}_A + \mathbf{r}_{B/A}} \qquad (16\text{–}18)$$

Fig. 16–30a

---

*The more general, three-dimensional motion of the points is developed in Sec. 20.4.

At the instant considered, point A has a velocity $\mathbf{v}_A$ and an acceleration $\mathbf{a}_A$, while the angular velocity and angular acceleration of the x, y axes are $\boldsymbol{\Omega}$ (omega) and $\dot{\boldsymbol{\Omega}} = d\boldsymbol{\Omega}/dt$, respectively. All these vectors are measured from the X, Y, Z frame of reference, although they may be expressed either in terms of $\mathbf{I}$, $\mathbf{J}$, $\mathbf{K}$ or $\mathbf{i}$, $\mathbf{j}$, $\mathbf{k}$ components. Since planar motion is specified, then by the right-hand rule $\boldsymbol{\Omega}$ and $\dot{\boldsymbol{\Omega}}$ are always directed *perpendicular* to the reference plane of motion, whereas $\mathbf{v}_A$ and $\mathbf{a}_A$ lie in this plane.

Velocity

The velocity of point B is determined by taking the time derivative of Eq. 16–18, which yields

$$\mathbf{v}_B = \mathbf{v}_A + \frac{d\mathbf{r}_{B/A}}{dt} \tag{16–19}$$

The last term in this equation is evaluated as follows:

$$
\begin{aligned}
\frac{d\mathbf{r}_{B/A}}{dt} &= \frac{d}{dt}(x_B\mathbf{i} + y_B\mathbf{j}) \\
&= \frac{dx_B}{dt}\mathbf{i} + x_B\frac{d\mathbf{i}}{dt} + \frac{dy_B}{dt}\mathbf{j} + y_B\frac{d\mathbf{j}}{dt} \\
&= \left(\frac{dx_B}{dt}\mathbf{i} + \frac{dy_B}{dt}\mathbf{j}\right) + \left(x_B\frac{d\mathbf{i}}{dt} + y_B\frac{d\mathbf{j}}{dt}\right)
\end{aligned}
\tag{16–20}
$$

The two terms in the first set of parentheses represent the components of velocity of point B as measured by an observer attached to the moving coordinate system. These terms will be denoted by vector $(\mathbf{v}_{B/A})_{\text{rel}}$. In the second set of parentheses the instantaneous time rate of change of the unit vectors $\mathbf{i}$ and $\mathbf{j}$ is measured by an observer located in the fixed X, Y, Z coordinate system. These changes, $d\mathbf{i}$ and $d\mathbf{j}$, are due *only* to the instantaneous *rotation* $d\theta$ of the x, y, z axes, Fig. 16–30b. As shown, the *magnitudes* of both $d\mathbf{i}$ and $d\mathbf{j}$ equal 1 ($d\theta$), since $i = j = 1$. The *direction* of $d\mathbf{i}$ is defined by $+\mathbf{j}$, since $d\mathbf{i}$ is tangent to the path described by the tip of $\mathbf{i}$ in the limit as $\Delta t \rightarrow dt$. Likewise, $d\mathbf{j}$ acts in the $-\mathbf{i}$ direction Fig. 16–30b. Hence,

$$\frac{d\mathbf{i}}{dt} = \frac{d\theta}{dt}(\mathbf{j}) = \Omega\mathbf{j} \qquad \frac{d\mathbf{j}}{dt} = \frac{d\theta}{dt}(-\mathbf{i}) = -\Omega\mathbf{i}$$

Viewing the axes in three dimensions, Fig. 16–30c, and noting that $\boldsymbol{\Omega} = \Omega\mathbf{k}$, we can express the above derivatives in terms of the cross product as

$$\frac{d\mathbf{i}}{dt} = \boldsymbol{\Omega} \times \mathbf{i} \qquad \frac{d\mathbf{j}}{dt} = \boldsymbol{\Omega} \times \mathbf{j} \tag{16–21}$$

Substituting these results into Eq. 16–20 and using the distributive property of the vector cross product, we obtain

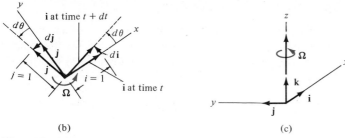

Fig. 16—30(b, c)

$$\frac{d\mathbf{r}_{B/A}}{dt} = \mathbf{v}_{B/A} + \mathbf{\Omega} \times (x_B\mathbf{i} + y_B\mathbf{j}) = (\mathbf{v}_{B/A})_{\text{rel}} + \mathbf{\Omega} \times \mathbf{r}_{B/A} \quad (16\text{–}22)$$

Hence, Eq. 16–19 becomes

$$\boxed{\mathbf{v}_B = \mathbf{v}_A + \mathbf{\Omega} \times \mathbf{r}_{B/A} + (\mathbf{v}_{B/A})_{\text{rel}}} \qquad (16\text{–}23)$$

where

$\mathbf{v}_B$ = velocity of B, measured from the X, Y, Z reference

$\mathbf{v}_A$ = velocity of the origin A of the x, y, z reference, measured from the X, Y, Z reference

$(\mathbf{v}_{B/A})_{\text{rel}}$ = relative velocity of "B with respect to A," as measured by an observer attached to the *rotating* x, y, z reference

$\mathbf{\Omega}$ = angular velocity of the x, y, z reference, measured from the X, Y, Z reference

$\mathbf{r}_{B/A}$ = relative position of "B with respect to A"

Comparing Eq. 16–23 with Eq. 16–15 ($\mathbf{v}_B = \mathbf{v}_A + \mathbf{\Omega} \times \mathbf{r}_{B/A}$), which is valid for a translating frame of reference, it can be seen that the only difference between the equations is represented by the term $(\mathbf{v}_{B/A})_{\text{rel}}$.

When applying Eq. 16–23 it is often useful to understand what each of the terms represents. In order of appearance, they are as follows:

$\mathbf{v}_B$ ⎰absolute velocity of B

(equals)

$\mathbf{v}_A$ ⎰absolute velocity of origin ⎱of x, y, z frame

(plus)

$\mathbf{\Omega} \times \mathbf{r}_{B/A}$ ⎰angular velocity effect caused ⎱by rotation of x, y, z frame ⎱motion of x, y, z frame

(plus)

$(\mathbf{v}_{B/A})_{\text{rel}}$ ⎰relative velocity of B ⎱with respect to A ⎰motion of B within ⎱x, y, z frame

Acceleration

The acceleration of point B, observed from the X, Y, Z coordinate system, may be expressed in terms of its motion measured with respect to the rotating or moving system of coordinates by taking the time derivative of Eq. 16–23, i.e.,

$$\frac{d\mathbf{v}_B}{dt} = \frac{d\mathbf{v}_A}{dt} + \frac{d\mathbf{\Omega}}{dt} \times \mathbf{r}_{B/A} + \mathbf{\Omega} \times \frac{d\mathbf{r}_{B/A}}{dt} + \frac{d(\mathbf{v}_{B/A})_{\text{rel}}}{dt}$$

$$\mathbf{a}_B = \mathbf{a}_A + \dot{\mathbf{\Omega}} \times \mathbf{r}_{B/A} + \mathbf{\Omega} \times \frac{d\mathbf{r}_{B/A}}{dt} + \frac{d(\mathbf{v}_{B/A})_{\text{rel}}}{dt} \qquad (16\text{–}24)$$

Here $\dot{\mathbf{\Omega}} = d\mathbf{\Omega}/dt$ is the angular acceleration of the x, y, z coordinate system. For planar motion $\mathbf{\Omega}$ is always perpendicular to the plane of motion, and therefore $\dot{\mathbf{\Omega}}$ measures *only the change in magnitude* of $\mathbf{\Omega}$. The derivative $d\mathbf{r}_{B/A}/dt$ in Eq. 16–24 is defined by Eq. 16–22, so that

$$\mathbf{\Omega} \times \frac{d\mathbf{r}_{B/A}}{dt} = \mathbf{\Omega} \times (\mathbf{v}_{B/A})_{\text{rel}} + \mathbf{\Omega} \times (\mathbf{\Omega} \times \mathbf{r}_{B/A}) \qquad (16\text{–}25)$$

Computing the time derivative of $(\mathbf{v}_{B/A})_{\text{rel}} = (v_{B/A})_x\mathbf{i} + (v_{B/A})_y\mathbf{j}$, we have

$$\frac{d(\mathbf{v}_{B/A})_{\text{rel}}}{dt} = \left[\frac{d(v_{B/A})_x}{dt}\mathbf{i} + \frac{d(v_{B/A})_y}{dt}\mathbf{j} \right] + \left[(v_{B/A})_x\frac{d\mathbf{i}}{dt} + (v_{B/A})_y\frac{d\mathbf{j}}{dt} \right]$$

The two terms in the first set of brackets represent the components of acceleration of point B as measured by an observer attached to the moving coordinate system. These terms will be denoted by vector $(\mathbf{a}_{B/A})_{\text{rel}}$. The terms in the second set of brackets can be simplified using Eqs. 16–21. Hence,

$$\frac{d(\mathbf{v}_{B/A})_{\text{rel}}}{dt} = (\mathbf{a}_{B/A})_{\text{rel}} + \mathbf{\Omega} \times (\mathbf{v}_{B/A})_{\text{rel}}$$

Substituting this equation and Eq. 16–25 into Eq. 16–24 and rearranging terms yields

$$\boxed{\mathbf{a}_B = \mathbf{a}_A + \dot{\mathbf{\Omega}} \times \mathbf{r}_{B/A} + \mathbf{\Omega} \times (\mathbf{\Omega} \times \mathbf{r}_{B/A}) + 2\mathbf{\Omega} \times (\mathbf{v}_{B/A})_{\text{rel}} + (\mathbf{a}_{B/A})_{\text{rel}}}$$

$$(16\text{–}26)$$

where

$\mathbf{a}_B$ = acceleration of B, measured from the X, Y, Z reference

$\mathbf{a}_A$ = acceleration of the origin A of the x, y, z reference, measured from the X, Y, Z reference

$(\mathbf{a}_{B/A})_{\text{rel}}, (\mathbf{v}_{B/A})_{\text{rel}}$ = relative acceleration and relative velocity of "B with respect to A," as measured by an observer attached to the *rotating* x, y, z reference

$\dot{\mathbf{\Omega}}, \mathbf{\Omega}$ = angular acceleration and angular velocity of the x, y, z reference, measured from the X, Y, Z reference

$\mathbf{r}_{B/A}$ = relative position of "B with respect to A"

If the motions of points A and B are along *curved paths,* it is often convenient to express the accelerations $\mathbf{a}_B$, $\mathbf{a}_A$, and $(\mathbf{a}_{B/A})_{rel}$ in Eq. 16–26 in terms of their normal and tangential components. If Eq. 16–26 is compared with Eq. 16–17, written in the form $\mathbf{a}_B = \mathbf{a}_A + \dot{\mathbf{\Omega}} \times \mathbf{r}_{B/A} + \mathbf{\Omega} \times (\mathbf{\Omega} \times \mathbf{r}_{B/A})$, which is valid for a translating frame of reference, it can be seen that the difference between the equations is represented by the terms $2\mathbf{\Omega} \times (\mathbf{v}_{B/A})_{rel}$ and $(\mathbf{a}_{B/A})_{rel}$. In partic- ular, $2\mathbf{\Omega} \times (\mathbf{v}_{B/A})_{rel}$ is called the *Coriolis acceleration,* named after the French engineer G. C. Coriolis (1792–1843), who was the first to determine it. This term represents the difference in the acceleration of B as measured from non- rotating and rotating x, y, z axes. As indicated by the vector cross product, the Coriolis acceleration will *always* be perpendicular to both $\mathbf{\Omega}$ and $(\mathbf{v}_{B/A})_{rel}$. It is an important component of the acceleration which must be considered whenever rotating reference frames are used. This often occurs, for example, when studying the accelerations and forces which act on rockets, long-range projectiles, or other bodies having motions which are largely affected by the rotation of the earth.

The following interpretation of the terms in Eq. 16–26 may be useful when applying this equation to the solution of problems.

$\mathbf{a}_B$ {absolute acceleration of B

(equals)

$\mathbf{a}_A$ {absolute acceleration of origin of x, y, z frame

(plus)

$\dot{\mathbf{\Omega}} \times \mathbf{r}_{B/A}$ {angular acceleration effect caused by rotation of x, y, z frame | motion of x, y, z frame

(plus)

$\mathbf{\Omega} \times (\mathbf{\Omega} \times \mathbf{r}_{B/A})$ {angular velocity effect caused by rotation of x, y, z frame

(plus)

$2\mathbf{\Omega} \times (\mathbf{v}_{B/A})_{rel}$ {combined effect of B moving relative to x, y, z coordinates and rotation of x, y, z frame} interacting motion

(plus)

$(\mathbf{a}_{B/A})_{rel}$ {relative acceleration of B with respect to A} motion of B within x, y, z frame

PROCEDURE FOR ANALYSIS

The following procedure provides a method for applying Eqs. 16–23 and 16–26 to the solution of problems involving the planar motion of particles or rigid bodies.

Coordinate Axes. Choose an appropriate location for the origin and proper orientation of the axes for both the X, Y, Z and moving x, y, z reference frames. Most often solutions are easily obtained if at the instant considered: (1) the origins are coincident, (2) the axes are collinear, and/or (3) the axes are parallel. The moving frame should be selected fixed to the body or device where the relative motion occurs.

Kinematic Equations. After defining the origin A of the moving reference and specifying the moving point B, Eqs. 16–23 and 16–26 should be written in symbolic form

$$\mathbf{v}_B = \mathbf{v}_A + \boldsymbol{\Omega} \times \mathbf{r}_{B/A} + (\mathbf{v}_{B/A})_{\text{rel}}$$
$$\mathbf{a}_B = \mathbf{a}_A + \dot{\boldsymbol{\Omega}} \times \mathbf{r}_{B/A} + \boldsymbol{\Omega} \times (\boldsymbol{\Omega} \times \mathbf{r}_{B/A}) + 2\boldsymbol{\Omega} \times (\mathbf{v}_{B/A})_{\text{rel}} + (\mathbf{a}_{B/A})_{\text{rel}}$$

Each of the vectors in these equations should be defined from the problem data and expressed in Cartesian vector form. This essentially requires a determination of (1) the motion of the moving reference, i.e., $\mathbf{v}_A$, $\mathbf{a}_A$, $\boldsymbol{\Omega}$, and $\dot{\boldsymbol{\Omega}}$; and (2) the motion of B measured with respect to the moving reference, i.e., $\mathbf{r}_{B/A}$, $(\mathbf{v}_{B/A})_{\text{rel}}$, and $(\mathbf{a}_{B/A})_{\text{rel}}$. The components of all these vectors may be selected along either the X, Y, Z axes or the x, y, z axes. The choice is arbitrary provided a consistent set of unit vectors is used. Finally, substitute the data into the kinematic equations and perform the vector operations.

The following examples numerically illustrate this procedure.

Example 16–17

At the instant $\theta = 60°$, the rod in Fig. 16–31 has an angular velocity of 3 rad/s and an angular acceleration of 2 rad/s². At this same instant, the collar C is traveling outward along the rod such that when $x = 0.2$ m the velocity is 2 m/s and the acceleration is 3 m/s², both measured relative to the rod. Determine the Coriolis acceleration and the velocity and acceleration of the collar at this instant.

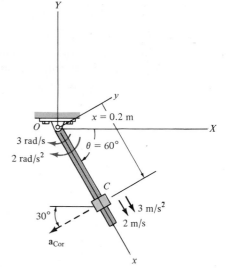

Fig. 16–31

SOLUTION

Coordinate Axes. The origin of both coordinate systems is located at point O, Fig. 16–31. Since motion of the collar is reported relative to the rod, the moving x, y, z frame of reference is *attached* to the rod.

Kinematic Equations

$$\mathbf{v}_C = \mathbf{v}_O + \mathbf{\Omega} \times \mathbf{r}_{C/O} + (\mathbf{v}_{C/O})_{rel} \tag{1}$$

$$\mathbf{a}_C = \mathbf{a}_O + \dot{\mathbf{\Omega}} \times \mathbf{r}_{C/O} + \mathbf{\Omega} \times (\mathbf{\Omega} \times \mathbf{r}_{C/O}) + 2\mathbf{\Omega} \times (\mathbf{v}_{C/O})_{rel} + (\mathbf{a}_{C/O})_{rel} \tag{2}$$

It will be simpler to express the data in terms of $\mathbf{i}$, $\mathbf{j}$, $\mathbf{k}$ component vectors rather than $\mathbf{I}$, $\mathbf{J}$, $\mathbf{K}$ components. Hence,

Motion *of moving reference*	*Motion of C with respect* *to moving reference*
$\mathbf{v}_O = \mathbf{0}$	$\mathbf{r}_{C/O} = \{0.2\mathbf{i}\}$ m
$\mathbf{a}_O = \mathbf{0}$	$(\mathbf{v}_{C/O})_{rel} = \{2\mathbf{i}\}$ m/s
$\mathbf{\Omega} = \{-3\mathbf{k}\}$ rad/s	$(\mathbf{a}_{C/O})_{rel} = \{3\mathbf{i}\}$ m/s²
$\dot{\mathbf{\Omega}} = \{-2\mathbf{k}\}$ rad/s²	

From Eq. (2) the Coriolis acceleration is defined as

$$\mathbf{a}_{Cor} = 2\mathbf{\Omega} \times (\mathbf{v}_{C/O})_{rel} = 2(-3\mathbf{k}) \times (2\mathbf{i}) = \{-12\mathbf{j}\} \text{ m/s}^2 \qquad \textit{Ans.}$$

This vector is shown dashed in Fig. 16–31. If desired, it may be resolved into $\mathbf{I}$, $\mathbf{J}$ components acting along the X and Y axes, respectively.

The velocity and acceleration of the collar are determined by substituting the data into Eqs. (1) and (2) and evaluating the cross products, which yields

$$\mathbf{v}_C = \mathbf{v}_O + \mathbf{\Omega} \times \mathbf{r}_{C/O} + (\mathbf{v}_{C/O})_{rel}$$
$$= \mathbf{0} + (-3\mathbf{k}) \times (0.2\mathbf{i}) + 2\mathbf{i}$$
$$= \{2\mathbf{i} - 0.6\mathbf{j}\} \text{ m/s} \qquad \textit{Ans.}$$

$$\mathbf{a}_C = \mathbf{a}_O + \dot{\mathbf{\Omega}} \times \mathbf{r}_{C/O} + \mathbf{\Omega} \times (\mathbf{\Omega} \times \mathbf{r}_{C/O}) + 2\mathbf{\Omega} \times (\mathbf{v}_{C/O})_{rel} + (\mathbf{a}_{C/O})_{rel}$$
$$= \mathbf{0} + (-2\mathbf{k}) \times (0.2\mathbf{i}) + (-3\mathbf{k}) \times [(-3\mathbf{k}) \times (0.2\mathbf{i})] + 2(-3\mathbf{k}) \times (2\mathbf{i}) + 3\mathbf{i}$$
$$= \mathbf{0} - 0.4\mathbf{j} - 1.80\mathbf{i} - 12\mathbf{j} + 3\mathbf{i}$$
$$= \{1.20\mathbf{i} - 12.4\mathbf{j}\} \text{ m/s}^2 \qquad \textit{Ans.}$$

Example 16–18

The rod AB, shown in Fig. 16–32, rotates clockwise such that it has an angular velocity $\omega_{AB} = 2$ rad/s and angular acceleration $\alpha_{AB} = 4$ rad/s^2 when $\theta = 45°$. Determine the angular motion of rod DE at this instant. The collar at C is pin-connected to AB and slides over rod DE.

SOLUTION

Coordinate Axes. The origins of both the fixed and moving frames of reference are located at D, Fig. 16–32. Furthermore, the x, y, z reference is attached to and rotates with rod DE.

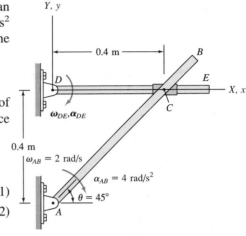

Kinematic Equations

$$\mathbf{v}_C = \mathbf{v}_D + \mathbf{\Omega} \times \mathbf{r}_{C/D} + (\mathbf{v}_{C/D})_{\text{rel}} \qquad (1)$$
$$\mathbf{a}_C = \mathbf{a}_D + \dot{\mathbf{\Omega}} \times \mathbf{r}_{C/D} + \mathbf{\Omega} \times (\mathbf{\Omega} \times \mathbf{r}_{C/D}) + 2\mathbf{\Omega} \times (\mathbf{v}_{C/D})_{\text{rel}} + (\mathbf{a}_{A/B})_{\text{rel}} \qquad (2)$$

All vectors will be expressed in terms of $\mathbf{i}$, $\mathbf{j}$, $\mathbf{k}$ components.

Motion of moving reference	*Motion of C with respect to moving reference*
$\mathbf{v}_D = \mathbf{0}$	$\mathbf{r}_{C/D} = \{0.4\mathbf{i}\}$ m
$\mathbf{a}_D = \mathbf{0}$	$(\mathbf{v}_{C/D})_{\text{rel}} = (v_{C/D})_{\text{rel}}\mathbf{i}$
$\mathbf{\Omega} = -\omega_{DE}\mathbf{k}$	$(\mathbf{a}_{C/D})_{\text{rel}} = (a_{C/D})_{\text{rel}}\mathbf{i}$
$\dot{\mathbf{\Omega}} = -\alpha_{DE}\mathbf{k}$	

Fig. 16–32

Motion of C: Since the collar moves along a *circular path*, its velocity and acceleration can be determined using Eqs. 16–9 and 16–13.

$$\mathbf{v}_C = \boldsymbol{\omega}_{AB} \times \mathbf{r}_{C/A} = (-2\mathbf{k}) \times (0.4\mathbf{i} + 0.4\mathbf{j}) = \{0.8\mathbf{i} - 0.8\mathbf{j}\} \text{ m/s}$$
$$\mathbf{a}_C = \boldsymbol{\alpha}_{AB} \times \mathbf{r}_{C/A} - \omega_{AB}^2 \mathbf{r}_{C/A} = (-4\mathbf{k}) \times (0.4\mathbf{i} + 0.4\mathbf{j}) - (2)^2(0.4\mathbf{i} + 0.4\mathbf{j}) = \{-3.2\mathbf{j}\} \text{ m/s}^2$$

Substituting the data into Eqs. (1) and (2), we have

$$\mathbf{v}_C = \mathbf{v}_D + \mathbf{\Omega} \times \mathbf{r}_{C/D} + (\mathbf{v}_{C/D})_{\text{rel}}$$
$$0.8\mathbf{i} - 0.8\mathbf{j} = \mathbf{0} + (-\omega_{DE}\mathbf{k}) \times (0.4\mathbf{i}) + (v_{C/D})_{\text{rel}}\mathbf{i}$$
$$0.8\mathbf{i} - 0.8\mathbf{j} = \mathbf{0} - 0.4\omega_{DE}\mathbf{j} + (v_{C/D})_{\text{rel}}\mathbf{i}$$
$$(v_{C/D})_{\text{rel}} = 0.8 \text{ m/s}, \quad \omega_{DE} = 2 \text{ rad/s} \qquad \textit{Ans.}$$

$$\mathbf{a}_C = \mathbf{a}_D + \dot{\mathbf{\Omega}} \times \mathbf{r}_{C/D} + \mathbf{\Omega} \times (\mathbf{\Omega} \times \mathbf{r}_{C/D}) + 2\mathbf{\Omega} \times (\mathbf{v}_{C/D})_{\text{rel}} + (\mathbf{a}_{C/D})_{\text{rel}}$$
$$-3.2\mathbf{j} = \mathbf{0} + (-\alpha_{DE}\mathbf{k}) \times (0.4\mathbf{i}) + (-2\mathbf{k}) \times [(-2\mathbf{k}) \times (0.4\mathbf{i})] + 2(-2\mathbf{k}) \times (0.8\mathbf{i}) + (a_{C/D})_{\text{rel}}\mathbf{i}$$
$$-3.2\mathbf{j} = -0.4\alpha_{DE}\mathbf{j} + 0.4\alpha_{DE}\mathbf{i} - 1.6\mathbf{i} - 1.6\mathbf{j} - 3.2\mathbf{j} + (a_{C/D})_{\text{rel}}\mathbf{i}$$
$$(a_{C/D})_{\text{rel}} = 1.6 \text{ m/s}^2 \qquad \alpha_{DE} = 0 \qquad \textit{Ans.}$$

Example 16–19

Two planes are flying at the same elevation and have the motions shown in Fig. 16–33. Determine the velocity and acceleration of A as measured by the pilot of B.

SOLUTION

Coordinate Axes. Since the relative motion of A with respect to B is being sought, the x, y, z axes are attached to plane B, Fig. 16–33. At the *instant* considered, the origin B coincides with the origin of the fixed X, Y, Z frame.

Kinematic Equations

$$\mathbf{v}_A = \mathbf{v}_B + \mathbf{\Omega} \times \mathbf{r}_{A/B} + (\mathbf{v}_{A/B})_{\text{rel}} \tag{1}$$

$$\mathbf{a}_A = \mathbf{a}_B + \dot{\mathbf{\Omega}} \times \mathbf{r}_{A/B} + \mathbf{\Omega} \times (\mathbf{\Omega} \times \mathbf{r}_{A/B}) + 2\mathbf{\Omega} \times (\mathbf{v}_{A/B})_{\text{rel}} + (\mathbf{a}_{A/B})_{\text{rel}} \tag{2}$$

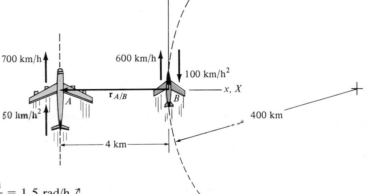

Motion of moving reference:

$$\mathbf{v}_B = \{600\mathbf{j}\} \text{ km/h}$$

$$\mathbf{a}_B = (\mathbf{a}_B)_n + (\mathbf{a}_B)_t = \{900\mathbf{i} - 100\mathbf{j}\} \text{ km/h}^2$$

$$(a_B)_n = \frac{v_B^2}{\rho} = \frac{(600)^2}{400} = 900 \text{ km/h}^2$$

$$\mathbf{\Omega} = \{-1.5\mathbf{k}\} \text{ rad/h} \qquad \Omega = \frac{v_B}{\rho} = \frac{600 \text{ km/h}}{400 \text{ km}} = 1.5 \text{ rad/h} \; \nearrow$$

$$\dot{\mathbf{\Omega}} = \{0.25\mathbf{k}\} \text{ rad/h}^2 \qquad \dot{\Omega} = \frac{(a_B)_t}{\rho} = \frac{100 \text{ km/h}^2}{400 \text{ km}} = 0.25 \text{ rad/h}^2 \; \nwarrow$$

Fig. 16–33

Motion of A with respect to moving reference:

$$\mathbf{r}_{A/B} = \{-4\mathbf{i}\} \text{ km} \qquad (\mathbf{v}_{A/B})_{\text{rel}} = ? \qquad (\mathbf{a}_{A/B})_{\text{rel}} = ?$$

Substituting this data into Eqs. (1) and (2), realizing that $\mathbf{v}_A = \{700\mathbf{j}\}$ km/h and $\mathbf{a}_A = \{50\mathbf{j}\}$ km/h^2, we have

$$\mathbf{v}_A = \mathbf{v}_B + \mathbf{\Omega} \times \mathbf{r}_{A/B} + (\mathbf{v}_{A/B})_{\text{rel}}$$

$$700\mathbf{j} = 600\mathbf{j} + (-1.5\mathbf{k}) \times (-4\mathbf{i}) + (\mathbf{v}_{A/B})_{\text{rel}}$$

$$(\mathbf{v}_{A/B})_{\text{rel}} = \{94\mathbf{j}\} \text{ km/h} \qquad\qquad\qquad Ans.$$

$$\mathbf{a}_A = \mathbf{a}_B + \dot{\mathbf{\Omega}} \times \mathbf{r}_{A/B} + \mathbf{\Omega} \times (\mathbf{\Omega} \times \mathbf{r}_{A/B}) + 2\mathbf{\Omega} \times (\mathbf{v}_{A/B})_{\text{rel}} + (\mathbf{a}_{A/B})_{\text{rel}}$$

$$50\mathbf{j} = (900\mathbf{i} - 100\mathbf{j}) + (0.25\mathbf{k}) \times (-4\mathbf{i})$$

$$+ (-1.5\mathbf{k}) \times [(-1.5\mathbf{k}) \times (-4\mathbf{i})] + 2(-1.5\mathbf{k}) \times (94\mathbf{j}) + (\mathbf{a}_{A/B})_{\text{rel}}$$

$$(\mathbf{a}_{A/B})_{\text{rel}} = \{-1190\mathbf{i} + 151\mathbf{j}\} \text{ km/h}^2 \qquad\qquad\qquad Ans.$$

PROBLEMS

16–86. A ball B of negligible size rolls through the tube such that at the instant shown it has a velocity of 5 ft/s and an acceleration of 3 ft/s^2, measured relative to the tube. If the tube has an angular velocity of $\omega = 3$ rad/s and an angular acceleration of $\alpha = 5$ rad/s^2 at this same instant, determine the velocity and acceleration of the ball.

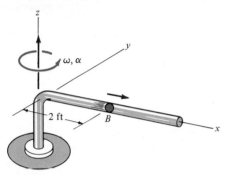

Prob. 16–86

16–87. The man stands on the platform at O and runs out toward the edge such that when he is at A, $y = 5$ ft, his mass center has a velocity of 2 ft/s and an acceleration of 3 ft/s^2, both measured with respect to the platform and directed along the y axis. If the platform has the angular motions shown, determine the velocity and acceleration of his mass center at this instant.

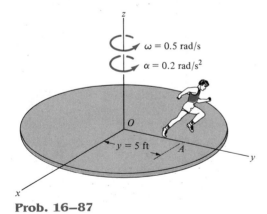

Prob. 16–87

*16–88. Block B moves along the slot in the platform with a constant speed of 2 ft/s, measured relative to the platform in the direction shown. If the platform is rotating at a constant rate of $\omega = 5$ rad/s, determine the velocity and acceleration of the block at the instant $\theta = 90°$.

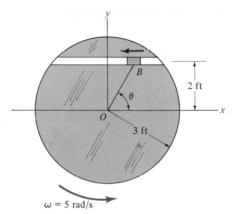

Probs. 16–88/16–89/16–90

16–89. Solve Prob. 16–88 if when $\theta = 90°$ the block has a velocity of 2 ft/s and an acceleration of 4 ft/s^2, measured relative to the platform in the direction shown, and at this same instant the platform has an angular velocity of $\omega = 5$ rad/s and an angular acceleration of $\alpha = 8$ rad/s^2.

16–90. Block B moves along the slot in the platform with a constant speed of 2 ft/s, measured relative to the platform in the direction shown. If the platform is rotating at a constant rate of $\omega = 5$ rad/s, determine the velocity and acceleration of the block at the instant $\theta = 60°$.

16–91. Rod *AB* rotates counterclockwise with a constant angular velocity of $\omega = 3$ rad/s. Determine the velocity of point *C* located on the double collar when $\theta = 30°$. The collar consists of two pin-connected slider blocks which are constrained to move along the circular path and the rod *AB*.

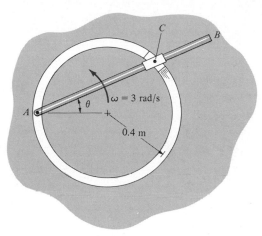

Probs. 16–91/16–92

*16–92. Rod *AB* rotates counterclockwise with a constant angular velocity of $\omega = 3$ rad/s. Determine the velocity and acceleration of point *C* located on the double collar when $\theta = 45°$. The collar consists of two pin-connected slider blocks which are constrained to move along the circular path and the rod *AB*.

16–93. A ride in an amusement park consists of a rotating arm *AB* having a constant angular velocity $\omega_{AB} = 2$ rad/s about point *A* and a car mounted at the end of the arm which has a constant angular velocity of $\omega' = \{-0.5\mathbf{k}\}$ rad/s, measured relative to the arm. At the instant shown, determine the velocity and acceleration of the passenger at *C*.

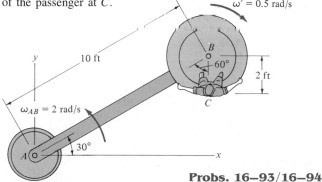

Probs. 16–93/16–94

16–94. Solve Prob. 16–93 if at the instant shown the arm has an angular acceleration of $\alpha_{AB} = 1$ rad/s² when $\omega_{AB} = 2$ rad/s and the car has a relative angular acceleration of $\alpha' = \{-0.6\mathbf{k}\}$ rad/s² when $\omega' = \{-0.5\mathbf{k}\}$ rad/s.

16–95. At the instant shown, the robotic arm *AB* is rotating counterclockwise at $\omega = 5$ rad/s and has an angular acceleration of $\alpha = 2$ rad/s². Simultaneously, the grip *BC* is rotating counterclockwise at $\omega' = 6$ rad/s and $\alpha' = 2$ rad/s², both measured with respect to a *fixed* reference. Determine the velocity and acceleration of the object held at the grip, *C*.

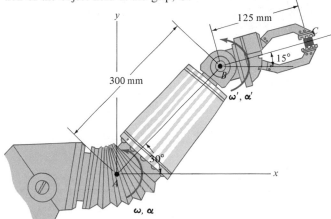

Prob. 16–95

*16–96. The "quick return" mechanism consists of a crank *AB*, slider block *B*, and slotted link *CD*. If the crank has the angular motions shown, determine the angular velocity and angular acceleration of the slotted link at this instant.

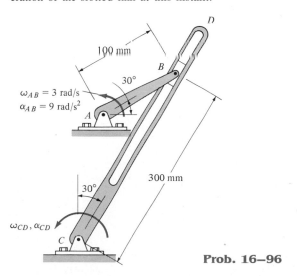

Prob. 16–96

291

16–97. At a given instant rod AB has the angular motions shown. Determine the angular velocity and angular acceleration of rod CD at this instant. There is a collar at D.

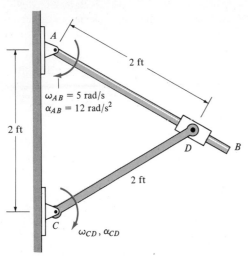

Prob. 16–97

16–98. At the instant $\theta = 60°$ rod AB has an angular velocity $\omega_{AB} = 3$ rad/s and an angular acceleration $\alpha_{AB} = 5$ rad/s². Determine the angular velocity and angular acceleration of rod CD at this instant. The collar at C is pin-connected to CD and slides over AB.

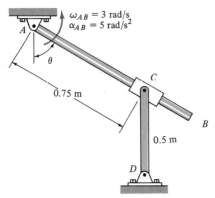

Prob. 16–98

16–99. The disk rotates with the angular motion shown. Determine the angular velocity and angular acceleration of the slotted link AC at this instant. The peg at B is fixed to the disk.

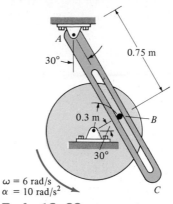

Prob. 16–99

*16–100. The two-link mechanism serves to amplify angular motion. Link AB has a pin at B which is confined to move within the slot of link CD. If at the instant shown AB (input) has an angular velocity $\omega_{AB} = 3$ rad/s and an angular acceleration $\alpha_{AB} = 2$ rad/s², determine the angular velocity and angular acceleration of CD (output) at this instant.

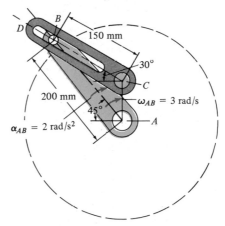

Prob. 16–100

17

Planar Kinetics of a Rigid Body: Force and Acceleration

In Chapter 16 the planar kinematics of rigid-body motion was presented in order of increasing complexity, that is, translation, rotation about a fixed axis, and general plane motion. The study of rigid-body kinetics in this chapter will be presented in somewhat the same order. The chapter begins by introducing a property of a body called the moment of inertia. Afterwards, a derivation of the equations of general plane motion for a symmetric rigid body is given. These equations are then applied to specific problems of rigid-body translation, rotation about a fixed axis, and finally general plane motion. A kinetic study of these motions is referred to as the kinetics of planar motions or simply *planar kinetics*. The more complex study of three-dimensional rigid-body kinetics, which includes planar motion of unsymmetrical rigid bodies, is presented in Chapter 21.

17.1 Moment of Inertia

Since a body has a definite size and shape, an applied nonconcurrent force system may cause the body to both translate and rotate. The translational aspects of the motion were studied in Chapter 13 and are governed by the equation $\mathbf{F} = m\mathbf{a}$. It will be shown in Sec. 17.2 that the rotational aspects, caused by the moment $\mathbf{M}$, are governed by an equation of the form $\mathbf{M} = I\boldsymbol{\alpha}$. The symbol I in this equation is termed the moment of inertia. By comparison, the *moment of inertia* is a measure of the resistance of a body to *angular acceleration* ($\mathbf{M} = I\boldsymbol{\alpha}$) in the same way that *mass* is a measure of the body's resistance to *acceleration* ($\mathbf{F} = m\mathbf{a}$).

We define the *moment of inertia* as the integral of the "second moment" about an axis of all the differential-size elements of mass dm which compose the body.* For example, consider the rigid body shown in Fig. 17–1. The body's moment of inertia about the z axis is

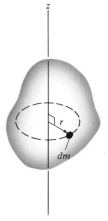

Fig. 17–1

$$I = \int_m r^2 \, dm \qquad (17\text{–}1)$$

Here the "moment arm" r is the perpendicular distance from the axis to the arbitrary element dm. Since the formulation involves r, the value of I is different for each axis about which it is computed. For example, if the axis coincides with the longitudinal axis for a slender rod, I will be small, since r is small for each element of the rod. If the axis is perpendicular to the rod, I will be large, because the rod will have a larger mass distributed farther from the axis. In the study of planar kinetics, the axis which is generally chosen for analysis passes through the body's mass center G and is always perpendicular to the plane of motion. The moment of inertia computed about this axis will be denoted as I_G. Note that because r is squared in Eq. 17–1 the mass moment of inertia is always a positive quantity. Common units used for its measurement are $\text{kg} \cdot \text{m}^2$ or $\text{slug} \cdot \text{ft}^2$.

---

*Another property of the body, which measures the symmetry of the body's mass with respect to a coordinate system, is the product of inertia. This property applies to the three-dimensional motion of a body and will be discussed in Chapter 21.

PROCEDURE FOR ANALYSIS

In this treatment, we will only consider symmetric bodies having surfaces which are generated by revolving a curve about an axis. An example of such a body of which the volume of revolution is generated about the z axis is shown in Fig. 17–2. If the body consists of material having a variable density, $\rho = \rho(x, y, z)$, the elemental mass dm of the body may be expressed in terms of its density and volume as $dm = \rho\, dV$. Substituting dm in Eq. 17–1, the body's moment of inertia is then computed using *volume elements* for integration, i.e.,

$$I = \int_V r^2 \rho\, dV \qquad (17\text{–}2)$$

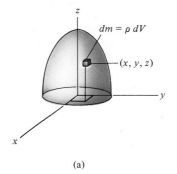

(a)

In the special case of ρ being a *constant*, this term may be factored out of the integral and the integration is then purely a function of geometry,

$$I = \rho \int_V r^2\, dV \qquad (17\text{–}3)$$

When the elemental volume chosen for integration has infinitesimal dimensions in all three directions, e.g., $dV = dx\, dy\, dz$, Fig. 17–2a, the moment of inertia of the body must be computed using "triple integration." The integration process can, however, be simplified to a *single integration* provided the chosen elemental volume has a differential size or thickness in only *one direction*. Shell or disk elements are often used for this purpose.

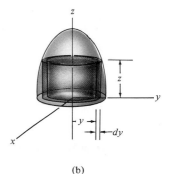

(b)

Shell Element. If a *shell element* having a height z, radius $r - y$, and thickness dy is chosen for integration, Fig. 17–2b, then the volume is $dV = (2\pi y)(z)\, dy$. This element may be used in Eq. 17–2 or 17–3 for computing the moment of inertia I_z of the body about the z axis, since the *entire element*, due to its "thinness," lies at the *same* perpendicular distance $r = y$ from the z axis (see Example 17–1).

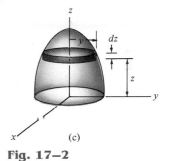

(c)

Fig. 17–2

Disk Element. If a disk element having a radius $r = y$ and a thickness dz is chosen for integration, Fig. 17–2c, then the volume is $dV = (\pi y^2)\, dz$. In this case, however, the element is *finite* in the radial direction, and consequently parts of it *do not* all lie at the *same radial distance* r from the z axis. As a result, Eq. 17–2 or 17–3 *cannot* be used to determine I_z directly. Instead, to perform the integration using this element, it is first necessary to determine the moment of inertia *of the element* about the z axis and then integrate this result (see Example 17–2).

Example 17–1

Determine the moment of inertia of the right circular cylinder shown in Fig. 17–3a about the z axis. The density ρ of the material is constant.

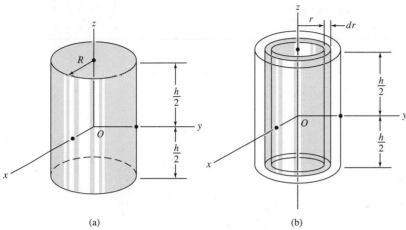

Fig. 17–3

SOLUTION

Shell Element. This problem may be solved using the *shell element* in Fig. 17–3b and single integration. The volume of the element is $dV = (2\pi r)(h)\, dr$, so that the mass is $d\dot{m} = \rho\, dV = \rho(2\pi hr\, dr)$. Since the *entire element* lies at the same distance r from the z axis, the moment of inertia *of the element* is

$$dI_z = r^2\, dm = \rho 2\pi hr^3\, dr$$

Integrating over the entire region of the cylinder yields

$$I_z = \int_m r^2\, dm = \rho 2\pi h \int_0^R r^3\, dr = \frac{\rho\pi}{2}R^4 h$$

The mass of the cylinder is

$$m = \int_m dm = \rho 2\pi h \int_0^R r\, dr = \rho\pi hR^2$$

so that

$$I_z = \frac{1}{2}mR^2 \qquad\qquad\qquad Ans.$$

Example 17–2

A solid is formed by revolving the shaded area shown in Fig. 17–4a about the y axis. If the density of the material is 5 slug/ft³, determine the moment of inertia about the y axis.

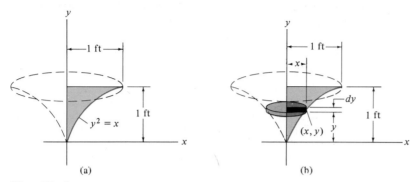

Fig. 17–4

SOLUTION

Disk Element. The moment of inertia will be computed using a *disk element*, as shown in Fig. 17–4b. Here the element intersects the curve at the arbitrary point (x, y) and has a mass

$$dm = \rho \, dV = \rho(\pi x^2) \, dy$$

Although all portions of the element are *not* located at the same distance from the y axis, it is still possible to determine the moment of inertia dI_y *of the element* about the y axis. In the preceding example it was shown that the moment of inertia of a cylinder about its longitudinal axis is $I = \frac{1}{2}mR^2$, where m and R are the mass and radius of the cylinder. Since the height of the cylinder is not involved in this formula, the moment of inertia of the disk element in Fig. 17–4b is

$$dI_y = \tfrac{1}{2}(dm)x^2 = \tfrac{1}{2}[\rho(\pi x^2) \, dy]x^2$$

Substituting $x = y^2$, $\rho = 5$ slug/ft³, and integrating with respect to y, from $y = 0$ to $y = 1$ ft, yields the moment of inertia for the entire solid.

$$I_y = \frac{\pi(5)}{2} \int_0^1 x^4 \, dy = \frac{\pi(5)}{2} \int_0^1 y^8 \, dy = 0.873 \text{ slug} \cdot \text{ft}^2 \quad \textit{Ans.}$$

Parallel-Axis Theorem

If the moment of inertia of the body about an axis passing through the body's mass center is known, then the body's moment of inertia may be determined about any other *parallel axis* by using the *parallel-axis theorem*. This theorem can be derived by considering the body shown in Fig. 17–5. The z' axis passes through the body's mass center G, whereas the corresponding *parallel axis z* lies at a constant distance d away. Selecting the differential element of mass dm, which is located at point (x', y'), and using the Pythagorean theorem, $r^2 = (d + x')^2 + y'^2$, we can express the moment of inertia of the body about the z axis as

$$I = \int_m r^2\, dm = \int_m [(d + x')^2 + y'^2]\, dm$$

$$= \int_m (x'^2 + y'^2)\, dm + 2d \int_m x'\, dm + d^2 \int_m dm$$

Since $r'^2 = x'^2 + y'^2$, the first integral represents I_G. The second integral equals *zero*, since the z' axis passes through the body's mass center, i.e., $\int x'\, dm = \bar{x} \int dm = 0$ since $\bar{x} = 0$. Finally, the third integral represents the total mass m of the body. Hence, the moment of inertia about the z axis can be written as

$$\boxed{I = I_G + md^2} \tag{17-4}$$

where

I_G = moment of inertia about the z' axis passing through the mass center G
m = mass of the body
d = perpendicular distance between the parallel axes

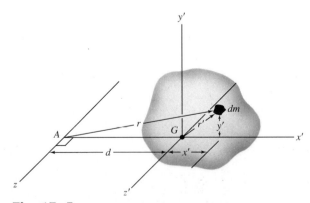

Fig. 17–5

Radius of Gyration

Occasionally, the moment of inertia of a body about a specified axis is reported in handbooks using the *radius of gyration, k.* When this length k and the body's mass m are known, the body's moment of inertia is determined from the equation

$$I = mk^2 \quad \text{or} \quad k = \sqrt{\frac{I}{m}} \tag{17–5}$$

Note the *similarity* between the definition of k in this formula and r in the equation $dI = r^2\, dm$, which defines the moment of inertia of an elemental mass dm of the body about an axis.

Composite Bodies

If a body is constructed of a number of simple shapes such as disks, spheres, and rods, the moment of inertia of the body about any axis z can be determined by adding algebraically the moments of inertia of all the composite shapes computed about the z axis. Algebraic addition is necessary since a composite part must be considered as a negative quantity if it has already been counted as part of another part—for example a "hole" subtracted from a solid plate. The parallel-axis theorem is needed for the calculations if the center of mass of each composite part does not lie on the z axis. For the calculation then $I = \Sigma(I_G + md^2)$. Here I_G for each of the composite parts is either known or can be computed by integration.*

---

*See the table given on the inside back cover of this book.

Example 17–3

If the plate shown in Fig. 17–6a has a density of 8000 kg/m^3 and a thickness of 10 mm, compute its moment of inertia about an axis directed perpendicular to the page and passing through point O.

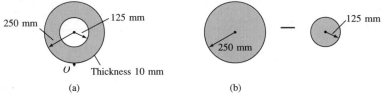

250 mm · 125 mm · O · Thickness 10 mm · 250 mm · 125 mm

(a)   (b)

Fig. 17–6

SOLUTION

The plate consists of two composite parts, the 250-mm-radius disk *minus* a 125-mm-radius disk, Fig. 17–6b. The moment of inertia about O can be determined by computing the moment of inertia of each of these parts about O and then *algebraically* adding the results. The computations are performed by using the parallel-axis theorem in conjunction with the data listed in the table on the inside back cover.

Disk. The moment of inertia of a thin disk about an axis perpendicular to the plane of the disk is $I_G = \frac{1}{2}mr^2$. The mass center of the disk is located at a distance of 0.25 m from point O. Thus,

$$m_d = \rho_d V_d = 8000 \text{ kg/m}^3[\pi(0.25 \text{ m})^2(0.01 \text{ m})] = 15.71 \text{ kg}$$
$$(I_O)_d = \tfrac{1}{2}m_d r_d^2 + m_d d^2$$
$$= \frac{1}{2}(15.71 \text{ kg})(0.25 \text{ m})^2 + (15.71 \text{ kg})(0.25 \text{ m})^2$$
$$= 1.473 \text{ kg} \cdot \text{m}^2$$

Hole. For the 125-mm-radius disk (hole), we have

$$m_h = \rho_h V_h = 8000 \text{ kg/m}^3[\pi(0.125 \text{ m})^2(0.01 \text{ m})] = 3.93 \text{ kg}$$
$$(I_O)_h = \tfrac{1}{2}m_h r_h^2 + m_h d^2$$
$$= \frac{1}{2}(3.93 \text{ kg})(0.125 \text{ m})^2 + (3.93 \text{ kg})(0.25 \text{ m})^2$$
$$= 0.276 \text{ kg} \cdot \text{m}^2$$

The moment of inertia of the plate about point O is therefore

$$I_O = (I_O)_d - (I_O)_h$$
$$= 1.473 - 0.276$$
$$= 1.20 \text{ kg} \cdot \text{m}^2 \qquad \qquad \textit{Ans.}$$

Example 17–4

The pendulum consists of two thin rods, each having a weight of 10 lb and suspended from point O as shown in Fig. 17–7. Compute the pendulum's moment of inertia about an axis passing through (a) the pin at O, and (b) the mass center G of the pendulum.

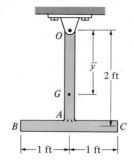

Fig. 17–7

SOLUTION

Part (a). Using the table on the inside back cover, the moment of inertia of rod OA about an axis perpendicular to the page and passing through the end point O of the rod is $I_O = \frac{1}{3}ml^2$. Hence,

$$(I_O)_{OA} = \frac{1}{3}ml^2 = \frac{1}{3}\left(\frac{10}{32.2}\right)(2)^2 = 0.414 \text{ slug} \cdot \text{ft}^2$$

The same value is obtained using $I_G = \frac{1}{12}ml^2$ and the parallel-axis theorem

$$(I_O)_{OA} = \frac{1}{12}ml^2 + md^2 = \frac{1}{12}\left(\frac{10}{32.2}\right)(2)^2 + \left(\frac{10}{32.2}\right)(1)^2$$

$$= 0.414 \text{ slug} \cdot \text{ft}^2$$

For rod BC we have

$$(I_O)_{BC} = \frac{1}{12}ml^2 + md^2 = \frac{1}{12}\left(\frac{10}{32.2}\right)(2)^2 + \left(\frac{10}{32.2}\right)(2)^2$$

$$= 1.346 \text{ slug} \cdot \text{ft}^2$$

The moment of inertia of the pendulum is therefore

$$I_O = 0.414 + 1.346 = 1.76 \text{ slug} \cdot \text{ft}^2 \qquad \textit{Ans.}$$

Part (b). The mass center G will be located relative to the pin at O. Assuming this distance to be $\bar{y}$, Fig. 17–7, and using the formula for determining the mass center, we have

$$\bar{y} = \frac{\Sigma \tilde{y}m}{\Sigma m} = \frac{1\left(\dfrac{10}{32.2}\right) + 2\left(\dfrac{10}{32.2}\right)}{\left(\dfrac{10}{32.2}\right) + \left(\dfrac{10}{32.2}\right)} = 1.50 \text{ ft}$$

The moment of inertia I_G may be computed in the same manner as I_O, which requires successive applications of the parallel-axis theorem to transfer the moments of inertia of rods OA and BC to G. A more direct solution, however, involves using the result for I_O, i.e.,

$$I_O = I_G + md^2; \qquad 1.76 = I_G + \left(\frac{20}{32.2}\right)(1.50)^2$$

$$I_G = 0.362 \text{ slug} \cdot \text{ft}^2 \qquad \textit{Ans.}$$

PROBLEMS

17-1. Determine the moment of inertia I_y for the slender rod. The rod's density ρ and cross-sectional area A are constant. Express the result in terms of the rod's total mass m.

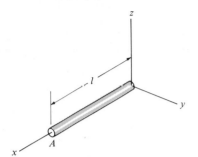

Prob. 17-1

17-2. Determine the moment of inertia of the thin ring about the z axis. The ring has a mass m.

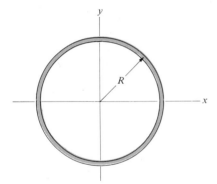

Prob. 17-2

17-3. The right circular cone is formed by revolving the shaded area around the x axis. Determine the moment of inertia I_x and express the result in terms of the total mass m of the cone. The cone has a constant density ρ.

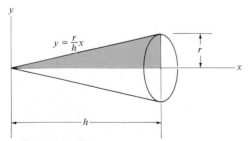

$$y = \frac{r}{h}x$$

Prob. 17-3

*17-4. The paraboloid is formed by revolving the shaded area around the x axis. Determine the radius of gyration k_x. The material has a constant density ρ.

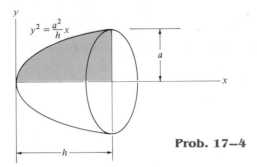

$$y^2 = \frac{a^2}{h}x$$

Prob. 17-4

17-5. A semiellipsoid is formed by rotating the shaded area about the x axis. Determine the moment of inertia of this solid with respect to the x axis and express the result in terms of the mass m of the solid. The material has a constant density ρ.

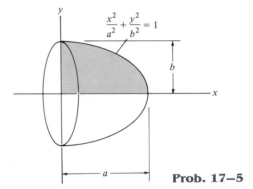

$$\frac{x^2}{a^2} + \frac{y^2}{b^2} = 1$$

Prob. 17-5

17-6. The sphere is formed by revolving the shaded area around the x axis. Determine the moment of inertia I_x and express the result in terms of the total mass m of the sphere. The sphere has a constant density ρ.

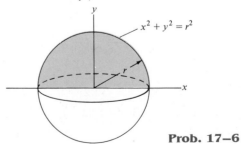

$$x^2 + y^2 = r^2$$

Prob. 17-6

17–7. The solid is formed by revolving the shaded area around the x axis. Determine the radius of gyration k_x. The specific weight of the material is $\gamma = 380$ lb/ft^3.

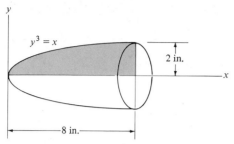

Prob. 17–7

*17–8. Determine the moment of inertia of the homogeneous triangular prism with respect to the y axis. Express the result in terms of the mass m of the prism. *Hint:* For integration, use thin plate elements parallel to the x-y plane and having a thickness of dz.

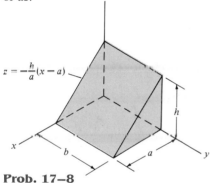

Prob. 17–8

17–9. The concrete solid is formed by rotating the shaded area about the y axis. Determine the moment of inertia I_y. The specific weight of concrete is $\gamma = 150$ lb/ft^3.

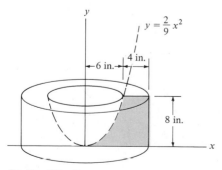

Prob. 17–9

17–10. The slender rods have a weight of 3 lb/ft. Determine the moment of inertia of the assembly about an axis perpendicular to the page and passing through point A.

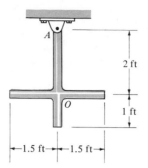

Prob. 17–10

17–11. The wheel consists of a thin ring having a mass of 10 kg and *four* spokes made from slender rods and each having a mass of 2 kg. Determine the wheel's moment of inertia about an axis perpendicular to the page and passing through point A.

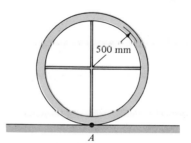

Prob. 17–11

*17–12. Determine the moment of inertia of the solid steel assembly about the xx axis. Steel has a specific weight of $\gamma = 490$ lb/ft^3.

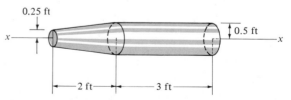

Prob. 17–12

17–13. Determine the moment of inertia I_z of the frustum of the cone which has a conical depression. The material has a density of 200 kg/m³.

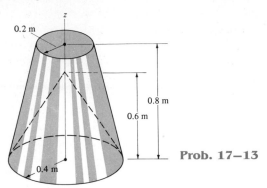

Prob. 17–13

17–14. Determine the moment of inertia of the wheel about an axis which is perpendicular to the page and passes through the center of mass G. The material has a specific weight of $\gamma =$ 90 lb/ft³.

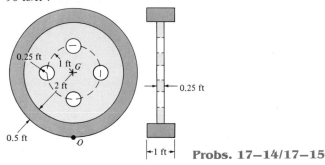

Probs. 17–14/17–15

17–15. Determine the moment of inertia of the wheel about an axis which is perpendicular to the page and passes through point O. The material has a specific weight of $\gamma = 90$ lb/ft³.

***17–16.** The pendulum consists of a disk having a mass of 6 kg and slender rods AB and DC which have a mass of 2 kg/m. Determine the length L of DC so that the center of mass is at the bearing O. What is the moment of inertia of the assembly about an axis perpendicular to the page and passing through point O?

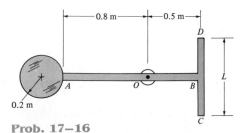

Prob. 17–16

17–17. Determine the moment of inertia of the wire triangle about an axis perpendicular to the page and passing through point O. Also, locate the mass center G and determine the moment of inertia about an axis perpendicular to the page and passing through point G. The wire has a mass of 0.3 kg/m.

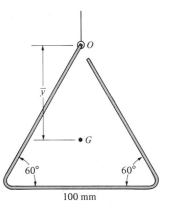

Prob. 17–17

17–18. The pendulum consists of two slender rods AB and OC which have a mass of 3 kg/m. The thin plate has a mass of 12 kg/m². Determine the location $\bar{y}$ of the center of mass G of the pendulum, then calculate the moment of inertia of the pendulum about an axis perpendicular to the page and passing through G.

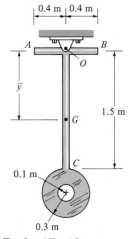

Prob. 17–18

17.2 Planar Kinetic Equations of Motion

In the following analysis we will limit our study of planar kinetics to rigid bodies which, along with their loadings, are considered to be *symmetrical* with respect to a fixed reference plane.* The plane motion of such a rigid body may be analyzed in a fixed reference plane because the path of motion of each particle of the body is a plane curve parallel to the reference plane. Thus, for the analysis, the motion of the body may be represented by a *slab* having the same mass and exterior outline as the body. All the forces (and couples) acting on the body may then be projected onto the plane of this slab. An example of an arbitrary slab (body) of this type is shown in Fig. 17–8a. Here the *inertial frame of reference x, y, z* has its origin *coincident* with the arbitrary point *P* in the body. By definition, *these axes do not rotate and are either fixed or translate with constant velocity.*

Equation of Translational Motion

The external forces shown on the body in Fig. 17–8a symbolically represent the effect of gravitational, electrical, magnetic, or contact forces between adjacent bodies. Since this force system has been considered previously in Sec. 13.3, for the analysis of a system of particles, the results may be used here, in which case the particles are contained within the boundary of the body (or slab). Hence, if the equation of motion is applied to each of the particles of the body, and the results added vectorially, it may be concluded that

$$\Sigma \mathbf{F} = m\mathbf{a}_G$$

This equation is referred to as the *equation of motion* for the mass center of a rigid body. It states that *the sum of all the external forces acting on the body is equal to the body's mass times the acceleration of the mass center G.*

For motion of the body (or slab) in the *x-y* plane, the equation of motion for *G* may be written in the form of two independent scalar equations, namely,

$$\Sigma F_x = m(a_G)_x$$
$$\Sigma F_y = m(a_G)_y$$

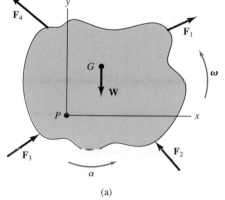

Fig. 17–8a

(a)

---

*By doing this, the rotational equation of motion reduces to a rather simplified form. The more general case of body shape and loading is considered in Chapter 21.

305

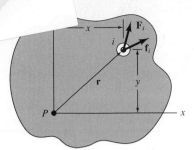

Particle free-body diagram

(b)

=

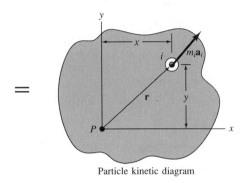

Particle kinetic diagram

(c)

Fig. 17–8(b, c)

Equation of Rotational Motion

We will now determine the effects caused by the moments of the external force system computed about an axis perpendicular to the plane of motion (the z axis) and passing through point P. As shown on the free-body diagram of the ith particle, Fig. 17–8b, $\mathbf{F}_i$ represents the *resultant external force* acting on the particle, and $\mathbf{f}_i$ is the *resultant of the internal forces* caused by interactions with adjacent particles. If the particle has a mass m_i and at the instant considered its acceleration is $\mathbf{a}_i$, then the kinetic diagram is constructed as shown in Fig. 17–8c. If moments of the forces acting on the particle are summed about point P, then we require

$$\mathbf{r} \times \mathbf{F}_i + \mathbf{r} \times \mathbf{f}_i = \mathbf{r} \times m_i\mathbf{a}_i$$

or

$$(\mathbf{M}_P)_i = \mathbf{r} \times m_i\mathbf{a}_i$$

We will now write this equation in terms of the acceleration $\mathbf{a}_P$ of point P. If the body has an angular acceleration $\boldsymbol{\alpha}$ and angular velocity $\boldsymbol{\omega}$, Fig. 17–8d, then using Eq. 16–17 we have

$$(\mathbf{M}_P)_i = m_i\mathbf{r} \times (\mathbf{a}_P - \omega^2\mathbf{r} + \boldsymbol{\alpha} \times \mathbf{r})$$
$$= m_i[\mathbf{r} \times \mathbf{a}_P - \omega^2(\mathbf{r} \times \mathbf{r}) + \mathbf{r} \times (\boldsymbol{\alpha} \times \mathbf{r})]$$

The second term on the right side is zero since $\mathbf{r} \times \mathbf{r} = \mathbf{0}.$ Expressing the vectors with Cartesian components and carrying out the cross-product operations yields

$$(M_P)_i\,\mathbf{k} = m_i\{(x\mathbf{i} + y\mathbf{j}) \times [(a_P)_x\mathbf{i} + (a_P)_y\mathbf{j}]$$
$$+ (x\mathbf{i} + y\mathbf{j}) \times [\alpha\mathbf{k} \times (x\mathbf{i} + y\mathbf{j})]\}$$
$$(M_P)_i\mathbf{k} = m_i[-y(a_P)_x + x(a_P)_y + \alpha x^2 + \alpha y^2]\mathbf{k}$$
$$(M_P)_i = m_i[-y(a_P)_x + x(a_P)_y + \alpha r^2]$$

Letting $m_i \rightarrow dm$, and integrating with respect to the entire mass m of the body, we obtain the resultant moment equation

$$\Sigma M_P = -\left(\int_m y\,dm\right)(a_P)_x + \left(\int_m x\,dm\right)(a_P)_y + \left(\int_m r^2\,dm\right)\alpha$$

Here ΣM_P represents only the moment of the *external forces* acting on the body about point P. The resultant moment of the internal forces is zero, since for the entire body they occur in equal and opposite collinear pairs and thus the moment of each pair of forces about point P cancels. The integrals in the first and second terms on the right are used to locate the body's center of mass G with respect to P, since $\bar{y}m = \int y\,dm$ and $\bar{x}m = \int x\,dm$, Fig. 17–8d. Also, the last integral represents the body's moment of inertia computed about the z axis, i.e., $I_P = \int r^2\,dm$. Thus,

$$\Sigma M_P = -\bar{y}m(a_P)_x + \bar{x}m(a_P)_y + I_P\alpha \qquad (17\text{–}6)$$

It is possible to reduce this equation to a simpler form if point P coincides with the mass center G for the body.* If this is the case, then $\bar{x} = \bar{y} = 0$, and therefore

$$\Sigma M_G = I_G\alpha \qquad (17\text{–}7)$$

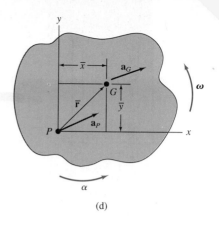

This rotational equation of motion states that the sum of the moments of all the external forces computed about the body's mass center G is equal to the product of the moment of inertia of the body about an axis passing through G and the body's angular acceleration.

Equation 17–6 can also be rewritten in terms of the x and y components of $\mathbf{a}_G$ and the body's moment of inertia I_G. If point G is located at point $(\bar{x},\bar{y})$, Fig. 17–8d, then by the parallel-axis theorem, $I_P = I_G + m(\bar{x}^2 + \bar{y}^2)$. Substituting into Eq. 17–6 and rearranging terms, we get

$$\Sigma M_P = \bar{y}m[-(a_P)_x + \bar{y}\alpha] + \bar{x}m[(a_P)_y + \bar{x}\alpha] + I_G\alpha \qquad (17\text{–}8)$$

(d)

From the kinematic diagram of Fig. 17–8d, $\mathbf{a}_P$ can be expressed in terms of $\mathbf{a}_G$ as

$$\mathbf{a}_G = \mathbf{a}_P - \omega^2\bar{\mathbf{r}} + \alpha \times \bar{\mathbf{r}}$$
$$(a_G)_x\mathbf{i} + (a_G)_y\mathbf{j} = (a_P)_x\mathbf{i} + (a_P)_y\mathbf{j} - \omega^2(\bar{x}\mathbf{i} + \bar{y}\mathbf{j}) + \alpha\mathbf{k} \times (\bar{x}\mathbf{i} + \bar{y}\mathbf{j})$$

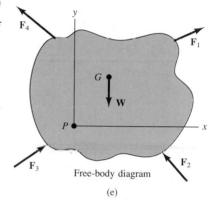

Carrying out the cross product and equating the respective $\mathbf{i}$ and $\mathbf{j}$ components yields the two scalar equations.

$$(a_G)_x = (a_P)_x - \bar{x}\omega^2 - \bar{y}\alpha$$
$$(a_G)_y = (a_P)_y - \bar{y}\omega^2 + \bar{x}\alpha$$

From these equations, $[-(a_P)_x + \bar{y}\alpha] = [-(a_G)_x - \bar{x}\omega^2]$ and $[(a_P)_y + \bar{x}\alpha] = [(a_G)_y + \bar{y}\omega^2]$. Substituting these results into Eq. 17–8 and simplifying gives

Free-body diagram

(e)

$$\Sigma M_P = -\bar{y}m(a_G)_x + \bar{x}m(a_G)_y + I_G\alpha \qquad (17\text{–}9)$$

This important result indicates that when moments of the external forces acting on the free-body diagram are summed about point P, Fig. 17–8e, they are equivalent to the sum of the "kinetic moments" of the components of $m\mathbf{a}_G$ about P plus the "kinetic moment" of $I_G\boldsymbol{\alpha}$, Fig. 17–8f. In other words, when the "kinetic moments," $\Sigma(\mathcal{M}_k)_P$, are computed, Fig. 17–8f, the vectors $m(\mathbf{a}_G)_x$ and $m(\mathbf{a}_G)_y$ are treated as sliding vectors, that is, they can act at any point along their line of action. In a similar manner, $I_G\boldsymbol{\alpha}$ can be treated as a free vector and can therefore act at any point. It is important to keep in mind that $m\mathbf{a}_G$ and $I_G\boldsymbol{\alpha}$ are not the same as a force or a couple. Instead, they are caused by the effects of forces and moments acting on the body. With this in mind we can therefore write Eq. 17–9 in a more general form as

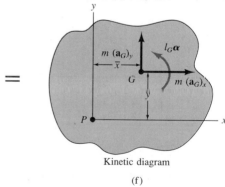

Kinetic diagram

(f)

Fig. 17–8(d–f)

$$\Sigma M_P = \Sigma(\mathcal{M}_k)_P \qquad (17\text{–}10)$$

---

*It also reduces to this same simple form $\Sigma M_P = I_P\alpha$ if point P is a *fixed point* (see Eq. 17–16) or the acceleration of point P is directed along the line PG.

General Application of the Equations of Motion

To summarize the above analysis, *three* scalar equations may be written to describe the general plane motion of a symmetrical rigid body.

$$\Sigma F_x = m(a_G)_x$$
$$\Sigma F_y = m(a_G)_y \qquad\qquad (17\text{--}11)$$
$$\Sigma M_G = I_G\alpha \quad \text{or} \quad \Sigma M_P = \Sigma(\mathcal{M}_k)_P$$

When applying these equations, one should *always* draw a free-body diagram, Fig. 17–8e, in order to account for the terms involved in ΣF_x, ΣF_y, ΣM_G or ΣM_P. In some problems it may also be helpful to draw the *kinetic diagram* for the body. This diagram graphically accounts for the terms $m(\mathbf{a}_G)_x$, $m(\mathbf{a}_G)_y$, and $I_G\boldsymbol{\alpha}$, and it is especially convenient when used to determine the components of $m\mathbf{a}_G$ and the moment terms in $\Sigma(\mathcal{M}_k)_P$.*

17.3 Equations of Motion: Translation

When a rigid body undergoes a *translation*, Fig. 17–9a, all the particles of the body have the *same acceleration*, so that $\mathbf{a}_G = \mathbf{a}$. Furthermore, $\boldsymbol{\alpha} = \mathbf{0}$, in which case the rotational equation of motion applied at point G reduces to a simplified form, $\Sigma M_G = 0$. Application of this and the translational equations of motion will now be discussed for each of the two types of translation presented in Chapter 16.

Rectilinear Translation

When a body is subjected to *rectilinear translation,* all the particles of the body (slab) travel along parallel straight-line paths. The free-body and kinetic diagrams are shown in Fig. 17–9b. Since $I_G\boldsymbol{\alpha} = \mathbf{0}$, only $m\mathbf{a}_G$ is shown on the kinetic diagram. Hence, the equations of motion which apply in this case become

$$\boxed{\begin{aligned} \Sigma F_x &= m(a_G)_x \\ \Sigma F_y &= m(a_G)_y \qquad\qquad (17\text{--}12) \\ \Sigma M_G &= 0 \end{aligned}}$$

The last equation requires that the sum of the moments of all the external forces computed about the body's center of mass be equal to zero. It is possible, of course, to sum moments about other points on or off the body, in which case the moment of $m\mathbf{a}_G$ must be taken into account. For example, if

---

*For this reason, the kinetic diagram will be used in the solution of an example problem whenever $\Sigma M_P = \Sigma(\mathcal{M}_k)_P$ is applied.

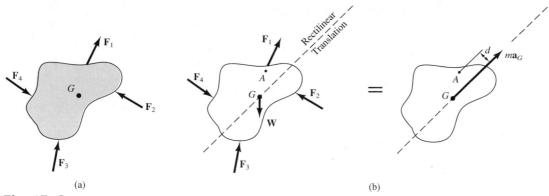

(a)   (b)

Fig. 17–9

point A is chosen, which lies at a perpendicular distance d from the line of action of $m\mathbf{a}_G$, the following moment equation applies:

$$\zeta + \Sigma M_A = \Sigma(\mathcal{M}_k)_A; \qquad \Sigma M_A = (ma_G)d$$

Here the moment of the external forces about A (ΣM_A, free-body diagram) equals the moment of $m\mathbf{a}_G$ about A ($\Sigma(\mathcal{M}_k)_A$, kinetic diagram).

Curvilinear Translation

When a rigid body is subjected to *curvilinear translation,* all the particles of the body travel along *parallel curved paths.* For analysis, it is often convenient to use an inertial coordinate system having axes which are oriented in the normal and tangential directions of the path of motion, Fig. 17–9c. The three scalar equations of motion are then

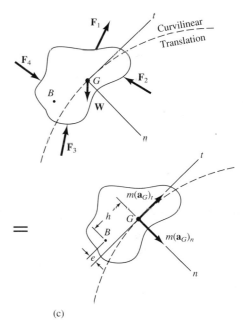

(c)

$$\boxed{\begin{aligned} \Sigma F_n &= m(a_G)_n \\ \Sigma F_t &= m(a_G)_t \\ \Sigma M_G &= 0 \end{aligned}}$$

$$(17\text{–}13)$$

where $(a_G)_t$ and $(a_G)_n$ represent, respectively, the magnitudes of the tangential and normal components of acceleration of point G.

If the moment equation $\Sigma M_G = 0$ is replaced by a moment summation about the arbitrary point B, Fig. 17–9c, it is necessary to account for the moments, $\Sigma(\mathcal{M}_k)_B$, of the two components $m(\mathbf{a}_G)_n$ and $m(\mathbf{a}_G)_t$ about this point. From the kinetic diagram, h and e represent the perpendicular distances (or "moment arms") from B to the lines of action of the components. If positive moments are assumed to be clockwise, the required moment equation becomes

$$\zeta + \Sigma M_B = \Sigma(\mathcal{M}_k)_B; \qquad \Sigma M_B = h[m(a_G)_n] - e[m(a_G)_t]$$

PROCEDURE FOR ANALYSIS

The following procedure provides a method for solving kinetic problems involving rigid-body *translation*.

Free-Body Diagram. Establish the *x*, *y* or *n*, *t* inertial coordinate system and draw the free-body diagram in order to account for all the external forces and moments that act on the body. The direction and sense of the acceleration of the body's mass center $\mathbf{a}_G$ should also be established. If the sense of its components *cannot* be determined, assume they are in the direction of the positive inertial coordinate axes. Identify the unknowns in the problem. If it is decided that the rotational equation of motion $\Sigma M_P = \Sigma(\mathcal{M}_k)_P$ is to be used in the solution, then it may be beneficial to draw the kinetic diagram, since it graphically accounts for the components $m(\mathbf{a}_G)_x$, $m(\mathbf{a}_G)_y$ or $m(\mathbf{a}_G)_t$, $m(\mathbf{a}_G)_n$ and is therefore convenient for "visualizing" the terms needed in the moment sum $\Sigma(\mathcal{M}_k)_P$.

Equations of Motion. Apply the three equations of motion, Eqs. 17–12 or Eqs. 17–13. To simplify the analysis, the moment equation $\Sigma M_G = 0$ can be replaced by the more general equation $\Sigma M_P = \Sigma(\mathcal{M}_k)_P$, where point *P* is usually located at the intersection of the lines of action of as many unknown forces as possible.

If the body is in contact with a *rough surface* and slipping occurs, use the frictional equation $F = \mu_k N$ to relate the normal force **N** to its associated frictional force **F**.

Kinematics. Use kinematics if the velocity and position of the body are to be determined. For *rectilinear translation* with *variable acceleration* use

$$a_G = \frac{dv_G}{dt} \qquad a_G\, ds_G = v_G\, dv_G \qquad v_G = \frac{ds_G}{dt}$$

For *rectilinear translation* with *constant acceleration*, use

$$v_G = (v_G)_0 + a_G t$$
$$v_G^2 = (v_G)_0^2 + 2a_G[s_G - (s_G)_0]$$
$$s_G = (s_G)_0 + (v_G)_0 t + \tfrac{1}{2}a_G t^2$$

For *curvilinear translation*, use

$$(a_G)_n = \frac{v_G^2}{\rho} = \omega^2\rho$$

$$(a_G)_t = \frac{dv_G}{dt} \qquad (a_G)_t\, ds_G = v_G\, dv_G \qquad (a_G)_t = \alpha\rho$$

The following examples numerically illustrate application of this procedure.

Example 17–5

The car shown in Fig. 17–10a has a mass of 2 Mg and a center of mass at G. Determine the car's acceleration if the "driving" wheels in the back are always slipping, whereas the front wheels freely rotate. Neglect the mass of the wheels. The coefficient of kinetic friction between the wheels and the road is $\mu_k = 0.25$.

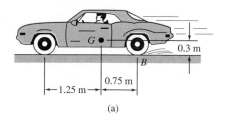

(a)

SOLUTION I

Free-Body Diagram. As shown in Fig. 17–10b, the rear-wheel frictional force $\mathbf{F}_B$ pushes the car forward, and since *slipping occurs* the magnitude of this force is related to the magnitude of its associated normal force N_B by $F_B = 0.25N_B$. The frictional forces acting on the *front wheels* are *zero*, since these wheels have negligible mass.* There are three unknowns in the problem, namely N_A, N_B, and a_G. Here we will sum moments about the mass center. The car (point G) is assumed to accelerate to the left, i.e., in the negative x direction, Fig. 17–10b.

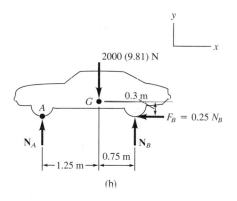

(b)

Equations of Motion

$$\xrightarrow{+} \Sigma F_x = m(a_G)_x; \qquad -0.25N_B = -2000a_G \qquad (1)$$

$$+\uparrow \Sigma F_y = m(a_G)_y; \quad N_A + N_B - 2000(9.81) = 0 \qquad (2)$$

$$\zeta+\Sigma M_G = 0; \quad -N_A(1.25) - 0.25N_B(0.3) + N_B(0.75) = 0 \qquad (3)$$

Solving,

$$a_G = 1.59 \text{ m/s}^2 \quad \text{Ans.}$$
$$N_A = 6.88 \text{ kN}$$
$$N_B = 12.7 \text{ kN}$$

SOLUTION II

Free-Body and Kinetic Diagrams. If the "moment" equation is applied about point A, then the unknown N_A will be eliminated from the equation. To "visualize" the moment of $m\mathbf{a}_G$ about A we will include the kinetic diagram as part of the analysis, Fig. 17–10c.

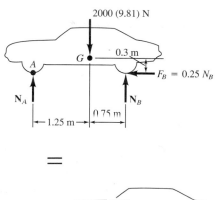

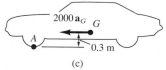

(c)

Equation of Motion. We require

$$\zeta+\Sigma M_A = \Sigma(\mathcal{M}_k)_A; \quad N_B(2) - 2000(9.81)(1.25) = 2000a_G(0.3)$$

Solving this and Eq. (1) for a_G leads to a simpler solution than that obtained from Eqs. (1) to (3).

*If the mass of the front wheels were to be included in the analysis, the frictional force acting at A would be *directed to the right* to create the necessary counterclockwise rotation of the wheels. The problem solution for this case would be more involved since a general-plane-motion analysis of the wheels would have to be considered (see Sec. 17.5).

Fig. 17–10

Example 17–6

The motorcycle shown in Fig. 17–11a has a mass of 125 kg and a center of mass at G_1, while the rider has a mass of 75 kg and a center of mass at G_2. If the coefficient of static friction between the wheels and the pavement is $\mu_s = 0.8$, determine if it is possible for the rider to do a "wheely," i.e., lift the front wheel off the ground. What acceleration is necessary to do this? Neglect the mass of the wheels and assume that the front wheel is free to roll.

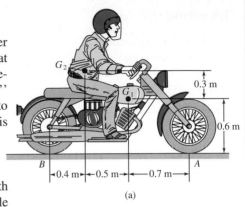

(a)

SOLUTION

Free-Body and Kinetic Diagrams. In this problem we will consider both the motorcycle and the rider as the "system" to be analyzed. It is possible first to determine the location of the center of mass for this "system" by using the equations $\bar{x} = \Sigma \tilde{x}m/\Sigma m$ and $\bar{y} = \Sigma \tilde{y}m/\Sigma m$. Here, however, we will consider the separate weight and mass of each of its *component parts* as shown on the free-body and kinetic diagrams, Fig. 17–11b. Both parts move with the *same* acceleration and we have assumed that the front wheel is about to leave the ground, so that the normal reaction $N_A \approx 0$. In order for this to happen it is required that the friction force $F_B \leq 0.8N_B$, otherwise slipping will occur. The three unknowns in the problem are N_B, F_B, and a_G.

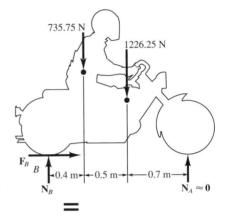

Equations of Motion

$$\xrightarrow{+} \Sigma F_x = m(a_G)_x; \quad F_B = (75 + 125)a_G \quad\quad (1)$$

$$+\uparrow \Sigma F_y = m(a_G)_y; \quad N_B - 735.75 - 1226.25 = 0 \quad\quad (2)$$

$$\zeta + \Sigma M_B = \Sigma(\mathcal{M}_k)_B; \quad (735.75)(0.4) + (1226.25)(0.9) = (75a_G)(0.9) + (125a_G)(0.6)$$

Solving,

$$a_G = 9.81 \text{ m/s}^2 \rightarrow \quad\quad \textit{Ans.}$$

$$N_B = 1.96 \text{ kN}$$

$$F_B = 1.96 \text{ kN}$$

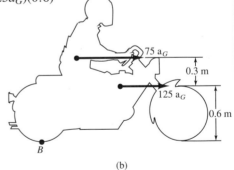

(b)

Fig. 17–11

The maximum frictional force that can be developed at B is

$$(F_B)_{max} = \mu_s N_B = 0.8(1.96) = 1.57 \text{ kN}$$

Since this force is less than that required ($F_B = 1.96$ kN), it is *not possible* to lift the front wheel off the ground.

Note: If the maximum acceleration of the motorcycle is to be obtained, then $F_B = 0.8N_B$ and $N_A \neq 0$. The problem would have to be solved in a manner similar to that of Example 17–5. Try it, and show that $(a_G)_{max} = 6.76$ m/s^2.

Example 17–7

A uniform 50-kg crate rests on a horizontal surface for which the coefficient of kinetic friction is $\mu_k = 0.2$. Determine the crate's acceleration if a force of $P = 600$ N is applied to the crate as shown in Fig. 17–12a.

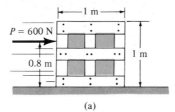

(a)

Fig. 17–12

SOLUTION

Free-Body Diagram.   The force **P** can either cause the crate to slide or to tip over. As shown in Fig. 17–12b, it is assumed that the crate slides, so that $F = \mu_k N_C = 0.2N_C$. Also, the resultant normal force N_C acts at O, a distance x (where $0 < x \leq 0.5$ m) from the crate's center line.* The three unknowns are N_C, x, and a_G.

Equations of Motion

$$\xrightarrow{+} \Sigma F_x = m(a_G)_x; \qquad 600 - 0.2N_C = 50a_G \qquad (1)$$

$$+ \uparrow \Sigma F_y = m(a_G)_y; \qquad N_C - 490.5 = 0 \qquad (2)$$

$$\zeta + \Sigma M_G = 0; \qquad -600(0.3) + N_C(x) - 0.2N_C(0.5) = 0 \qquad (3)$$

Solving, we obtain

$$N_C = 490 \text{ N}$$

$$x = 0.467 \text{ m}$$

$$a_G = 10.0 \text{ m/s}^2 \rightarrow \qquad \qquad \textit{Ans.}$$

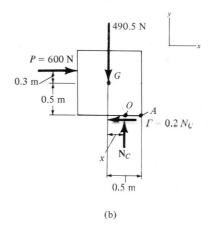

(b)

Since $x = 0.467$ m < 0.5 m, indeed the crate slides as originally assumed. If the solution had given a value of $x > 0.5$ m, the problem would have to be reworked with the assumption that tipping occurred. If this were the case, N_C would act at the *corner point A* and $F \leq 0.2N_C$.

---

*The line of action of N_C does not necessarily pass through the mass center G ($x = 0$), since N_C must counteract the tendency for tipping caused by **P**. See Sec. 8.1 of *Engineering Mechanics: Statics*.

Example 17–8

The 100-kg beam BD shown in Fig. 17–13a is supported by two rods having negligible mass. Determine the force created in each rod if at the instant $\theta = 30°$ the rods are both rotating with an angular velocity of $\omega = 6$ rad/s.

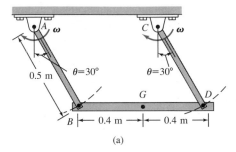

(a)　　　　　　　　　　　　　　**Fig. 17–13**

SOLUTION

Free-Body Diagram. The beam moves with *curvilinear translation* since points B and D and the center of mass G move along circular paths, each path having the same radius of 0.5 m. Using normal and tangential coordinates, the free-body diagram for the beam is shown in Fig. 17–13b. Because of the *translation*, G has the *same* motion as the pin at B, which is connected to both the rod and the beam. By studying the angular motion of rod AB, Fig. 17–13c, note that the tangential component of acceleration acts downward to the left due to the clockwise direction of $\boldsymbol{\alpha}$. Furthermore, the normal component of acceleration is *always* directed toward the center of curvature (toward point A for rod AB). Since the angular velocity of AB is 6 rad/s, then

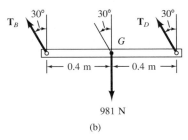

(b)

$$(a_G)_n = \omega^2 r = (6)^2(0.5) = 18 \text{ m/s}^2$$

The three unknowns are T_B, T_D, and $(a_G)_t$. The kinetic diagram has not been shown, since the directions of $(\mathbf{a}_G)_n$ and $(\mathbf{a}_G)_t$ have been established, and we will sum moments about G.

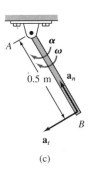

(c)

Equations of Motion

$$+\nwarrow \Sigma F_n = m(a_G)_n; \quad T_B + T_D - 981 \cos 30° = 100(18) \tag{1}$$
$$+\swarrow \Sigma F_t = m(a_G)_t; \quad 981 \sin 30° = 100(a_G)_t \tag{2}$$
$$\zeta + \Sigma M_G = 0; \quad -(T_B \cos 30°)0.4 + (T_D \cos 30°)0.4 = 0 \tag{3}$$

Simultaneous solution of these three equations gives

$$T_B = T_D = 1.33 \text{ kN} \quad \nwarrow_{30°} \qquad \qquad Ans.$$
$$(a_G)_t = 4.90 \text{ m/s}^2$$

PROBLEMS

17-19. The desk has a weight of 75 lb and a center of gravity at G. Determine its initial acceleration if a man pushes on it at C with a force of 60 lb. Also, find the vertical reactions on each of the two legs at A and at B. The coefficient of kinetic friction at A and B is $\mu_k = 0.2$.

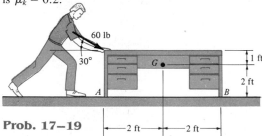

Prob. 17-19

*17-20. The handcart has a mass of 200 kg and center of mass at G. Determine the normal reactions at each of the wheels A and B if a force of P = 50 N is applied to the handle. Neglect the mass and rolling resistance of the wheels.

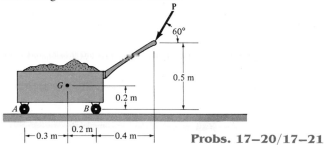

Probs. 17-20/17-21

17-21. The handcart has a mass of 200 kg and center of mass at G. Determine the magnitude of the largest force P that can be applied to the handle so that the wheels at A or B continue to maintain contact with the ground. Neglect the mass and rolling resistance of the wheels.

17-22. The smooth 180-lb pipe has a length of 20 ft and a negligible diameter. It is carried on a truck as shown. Determine the maximum acceleration which the truck can have without causing the normal reaction at A to be zero. Also determine the horizontal and vertical components of force which the truck exerts on the pipe at B.

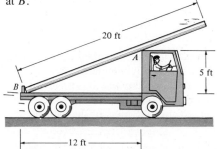

Prob. 17-22

17-23. The smooth 180-lb pipe has a length of 20 ft and a negligible diameter. It is carried on a truck as shown. If the truck accelerates at $a = 5$ ft/s², determine the normal reaction at A and the horizontal and vertical components of force which the truck exerts on the pipe at B.

*17-24. The jet aircraft has a total mass of 22 Mg and a center of mass at G. Initially at take-off the engines provide a thrust of $2T = 4$ kN and $T' = 1.5$ kN. Determine the acceleration of the plane and the normal reactions on the nose wheel and each of the *two* wing wheels located at B. Neglect the mass of the wheels and, due to low velocity, neglect any lift caused by the wings.

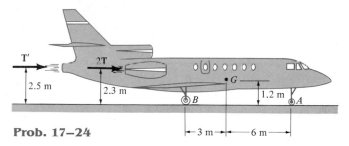

Prob. 17-24

17-25. The dresser has a weight of 80 lb and is pushed along the floor. If the coefficient of static friction at A and B is $\mu_s = 0.3$ and the coefficient of kinetic friction is $\mu_k = 0.2$, determine the smallest horizontal force **P** needed to cause motion. If this force is increased slightly, determine the acceleration of the dresser. Also, what are the normal reactions at A and B when it begins to move?

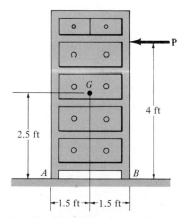

Prob. 17-25

315

17–26. The dresser has a weight of 80 lb and is pushed along the floor. If the coefficient of static friction at A and B is $\mu_s = 0.3$ and the coefficient of kinetic friction is $\mu_k = 0.2$, determine the maximum force horizontal **P** that can be applied without causing the dresser to tip over.

17–27. The truck carries the 800-lb crate which has a center of gravity at G_c. Determine the largest acceleration **a** of the truck so that the crate will not slip or tip on the truck bed. The coefficient of static friction between the crate and the truck is $\mu_s = 0.6$.

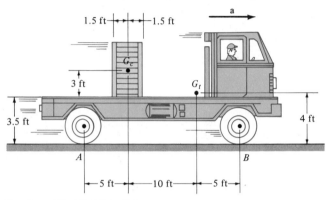

Probs. 17–27/17–28

*17–28. The truck has a weight of 8000 lb and center of gravity at G_t. It carries the 800-lb crate which has a center of gravity at G_c. Determine the normal reaction at *each* of its four tires if it accelerates at $a = 0.5$ ft/s². Also, what is the frictional force acting between the crate and the truck, and between each of the rear tires and the road? Assume that power is delivered only to the rear tires. The front tires are free to roll. Neglect the mass of the tires. The crate does not slip on the truck.

17–29. The sports car has been designed so that the 260-kg engine E and 165-kg transmission T have been placed over the front and rear wheels, respectively. Their mass centers are located at G_E and G_T. The remaining 512-kg mass of the body and frame is located at G_b. If power is supplied to the rear wheels only, determine the shortest time it takes for the car to reach a speed of 80 km/h, starting from rest. The front wheels are free rolling, and the coefficient of kinetic friction between the wheels and the road is $\mu_k = 0.4$. Neglect the mass of the wheels and driver.

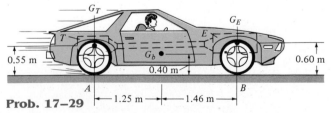

Prob. 17–29

17–30. The pipe has a mass of 800 kg and is being towed behind the truck. If the acceleration of the truck is $a_t = 0.5$ m/s², determine the angle θ and the tension in the cable. The coefficient of kinetic friction between the pipe and the ground is $\mu_k = 0.1$.

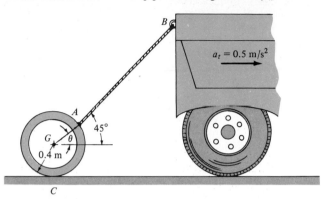

Prob. 17–30

17–31. The bicycle and rider have a mass of 80 kg with center of mass located at G. If the coefficient of kinetic friction at the rear tire is $\mu_B = 0.8$, determine the normal reactions at the tires A and B, and the deceleration of the rider, when the rear wheel locks for braking. What is the normal reaction at the rear wheel when the bicycle is traveling at constant velocity and the brakes are not applied? Neglect the mass of the wheels.

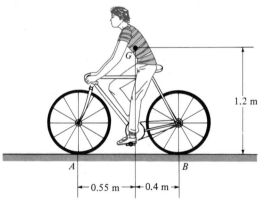

Probs. 17–31/17–32

*17–32. The bicycle and rider have a mass of 80 kg with center of mass located at G. Determine the minimum coefficient of kinetic friction between the road and the wheels so that the rear wheel B starts to lift off the ground when the rider applies the brakes to the front wheel. Neglect the mass of the wheels.

17–33. The van has a weight of 4500 lb and center of gravity at G_v. It carries a fixed 800-lb load which has a center of gravity at G_l. If the van is traveling at 40 ft/s, determine the shortest distance it skids before stopping at the moment the driver applies the brakes to *all* the wheels and causes each wheel to slip. The coefficient of kinetic friction between the wheels and the pavement is $\mu_k = 0.3$. Compare this distance with that of the van being empty. Neglect the mass of the wheels.

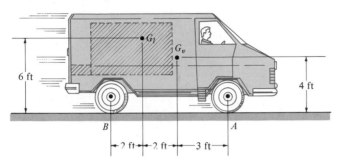

Probs. 17–33/17–34

17–34. The van has a weight of 4500 lb and center of gravity at G_v. It carries a fixed 800-lb load which has a center of gravity at G_l. If power is applied only to the rear wheels B, causing them to slip on the pavement, determine the acceleration of the van. The coefficient of kinetic friction between the wheels and the pavement is $\mu_k = 0.3$. Neglect the mass of the wheels. The front wheels are free to roll.

17–35. The portable crane has a mass of 500 kg and mass center at G. It is used to lift the 50-kg crate when the boom is in the horizontal position shown. Determine the largest initial angular acceleration of the boom, so that the front wheels E do not lose contact with the ground. Neglect the mass of the boom and hydraulic cylinder AC. The wheels at H are fixed from rolling.

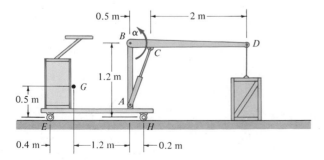

Prob. 17–35

*17–36. The two 3-lb rods EF and HI are fixed (welded) to the link AC at E. Determine the internal axial force E_x, shear force E_y, and moment M_E, which the bar AC exerts on FE at E if at the instant $\theta = 30°$ link AB has an angular velocity $\omega = 5$ rad/s and an angular acceleration $\alpha = 8$ rad/s^2 as shown.

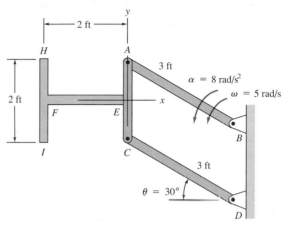

Prob. 17–36

17–37. The 1.6-Mg car shown has been "raked" by increasing the height $h = 0.2$ m of its center of mass. This was done by raising the springs on the rear axle. If the coefficient of kinetic friction between the rear wheels and the ground is $\mu_k = 0.3$, show that the car can accelerate slightly faster than its counterpart for which $h = 0$. Neglect the mass of the wheels and driver and assume the front wheels at B are free to roll while the rear wheels slip.

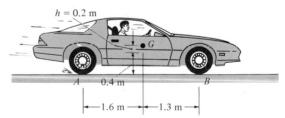

Prob. 17–37

17–38. The dragster has a mass of 1500 kg and a center of mass at G. If the coefficient of kinetic friction between the rear wheels and the pavement is $\mu_k = 0.6$, determine if it is possible for the driver to lift the front wheels, A, off the ground while the rear wheels are slipping. If so, what acceleration is necessary to do this? Neglect the mass of the wheels and assume that the front wheels are free to roll.

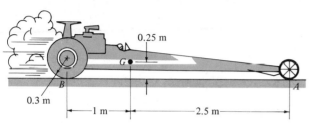

Probs. 17–38/17–39

17–39. The dragster has a mass of 1500 kg and a center of mass at G. If no slipping occurs, determine the friction force $\mathbf{F}_B$ which must be applied to each of the rear wheels B in order to develop an acceleration of $a = 6$ m/s^2. What are the normal reactions of each wheel on the ground? Neglect the mass of the wheels and assume that the front wheels are free to roll.

*17–40. The uniform bar BC has a weight of 40 lb and is pin-connected to the two links which have negligible mass. Determine the moment $\mathbf{M}$ that must be applied to link DC so that DC has the angular motion shown at the instant $\theta = 30°$. Also, what is the force developed in member AB?

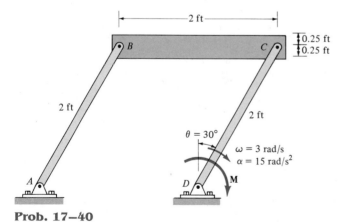

Prob. 17–40

17–41. The crate C has a weight of 150 lb and rests on the floor of a truck elevator for which the coefficient of static friction is $\mu_s = 0.4$. Determine the largest initial angular acceleration α, starting from rest, which the parallel links AB and DE can have without causing the crate to slip. No tipping occurs.

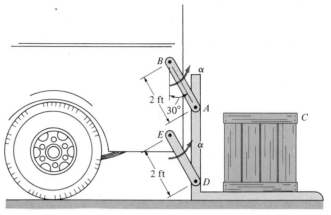

Prob. 17–41

17–42. The car having a mass of 1.40 Mg and mass center at G_c pulls a loaded trailer having a mass of 0.800 Mg and mass center at G_t. Determine the normal reactions at each of the car and trailer wheels if the driver applies the car's brakes and causes the rear wheels C car to skid. The coefficient of kinetic friction is $\mu_C = 0.4$. Assume the hitch at A is a pin or ball-and-socket joint. The wheels at B and D are free to roll. Neglect their mass and the mass of the driver.

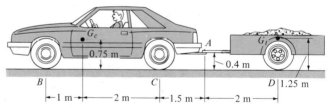

Prob. 17–42

$\sqrt{}$ 17.4 Equations of Motion: Rotation About a Fixed Axis

Consider the rigid body (or slab) shown in Fig. 17–14a, which is constrained to rotate in the vertical plane about a fixed axis perpendicular to the page and passing through the pin at O. The angular velocity and angular acceleration are caused by the external force system acting on the body. Because the body's center of mass G moves in a *circular path,* the acceleration of this point is represented by its tangential and normal components. The *tangential component of acceleration* has a *magnitude* of $(a_G)_t = \alpha r_G$ and must act in a *direction* which is *consistent* with the angular acceleration $\boldsymbol{\alpha}$. The *magnitude* of the *normal component of acceleration* is $(a_G)_n = \omega^2 r_G$. This component is *always directed* from point G to O regardless of the direction of $\boldsymbol{\omega}$.

The free-body and kinetic diagrams for the body are shown in Fig. 17–14b. The weight of the body, $W = mg$, and the pin reaction $\mathbf{F}_O$ are included on the free-body diagram since they represent external forces acting on the body. The two components $m(\mathbf{a}_G)_t$ and $m(\mathbf{a}_G)_n$, shown on the kinetic diagram, are associated with the tangential and normal acceleration components of the body's mass center. These vectors act in the same *direction* as the acceleration components and have *magnitudes* of $m(a_G)_t$ and $m(a_G)_n$. The $I_G\boldsymbol{\alpha}$ vector acts in the same *direction* as $\boldsymbol{\alpha}$ and has a *magnitude* of $I_G\alpha$, where I_G is the body's moment of inertia calculated about an axis which is perpendicular to the page and passing through G. From the derivation given in Sec. 17.2, the equations of motion which apply to the body may be written in the form

$$\Sigma F_n = m(a_G)_n = m\omega^2 r_G$$
$$\Sigma F_t = m(a_G)_t = m\alpha r_G \qquad (17\text{–}14)$$
$$\Sigma M_G = I_G\alpha$$

The moment equation may be replaced by a moment summation about any arbitrary point P on or off the body provided one accounts for the moments $\Sigma(\mathcal{M}_k)_P$ produced by $I_G\boldsymbol{\alpha}$, $m(\mathbf{a}_G)_t$, and $m(\mathbf{a}_G)_n$ about the point. In many problems it is convenient to sum moments about the pin at O in order to eliminate the *unknown* force $\mathbf{F}_O$. From the kinetic diagram, Fig. 17–14b, this requires

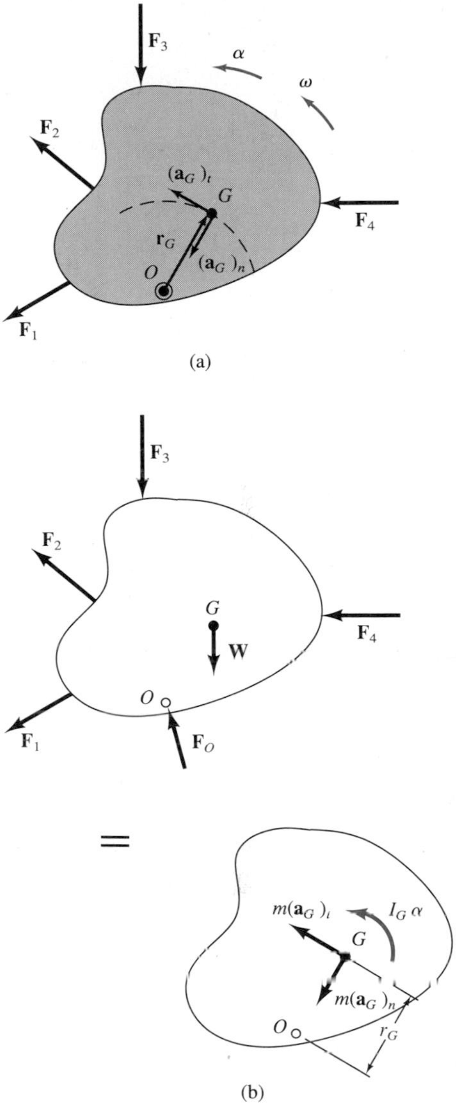

$$\zeta + \Sigma M_O = \Sigma(\mathcal{M}_k)_O; \qquad \Sigma M_O = r_G m(a_G)_t + I_G\alpha \qquad (17\text{–}15)$$

Note that the moment of $m(\mathbf{a}_G)_n$ is not included in the summation since the line of action of this vector passes through O. Substituting $(a_G)_t = r_G\alpha$, we may rewrite the above equation as $\zeta + \Sigma M_O = (I_G + mr_G^2)\alpha$. From the parallel-axis theorem, $I_O = I_G + md^2$, the term in parentheses represents the *moment of inertia of the body about the fixed axis of rotation passing through O.* Therefore, we can write the three equations of motion for the body as*

---

*The result $\Sigma M_O = I_O\alpha$ can also be obtained *directly* from Eq. 17–6 by selecting point P to coincide with O, realizing that $(a_P)_x = (a_P)_y = 0$.

(a)

(b)

Fig. 17–14

319

$$\Sigma F_n = m(a_G)_n = m\omega^2 r_G$$
$$\Sigma F_t = m(a_G)_t = m\alpha r_G \qquad\qquad (17\text{--}16)$$
$$\Sigma M_O = I_O\alpha$$

For applications, one should remember that "$I_O\alpha$" accounts for the "moment" of *both* $m(\mathbf{a}_G)_t$ *and* $I_G\alpha$ about point O, Fig. 17–14b. In other words, $\Sigma M_O = \Sigma(\mathcal{M}_k)_O = I_O\alpha$, as indicated by Eqs. 17–15 and 17–16.

PROCEDURE FOR ANALYSIS

The following procedure provides a method for solving kinetic problems which involve the rotation of a body about a fixed axis.

Free-Body Diagram. Establish the inertial x, y or n, t coordinate system, and specify the direction and sense of the accelerations $(\mathbf{a}_G)_n$ and $(\mathbf{a}_G)_t$, and the angular acceleration $\boldsymbol{\alpha}$ of the body. If the sense of these vectors *cannot* be established, assume they are in the direction of the positive coordinate axes. Draw the free-body diagram to account for all the external forces that act on the body and compute the moment of inertia I_G or I_O. Identify the unknowns in the problem. If it is decided that the rotational equation of motion $\Sigma M_P = \Sigma(\mathcal{M}_k)_P$ is to be used, i.e., moments will *not* be summed about point G or O, then consider drawing the kinetic diagram in order to help "visualize" the "moment" developed by the components $m(\mathbf{a}_G)_n$, $m(\mathbf{a}_G)_t$, and $I_G\boldsymbol{\alpha}$ when writing the terms for the moment sum $\Sigma(\mathcal{M}_k)_P$.

Equations of Motion. Apply the three equations of motion, Eqs. 17–14 or 17–16.

Kinematics. Use kinematics if a complete solution cannot be obtained strictly from the equations of motion. In this regard, if the *angular acceleration is variable*, use

$$\alpha = \frac{d\omega}{dt}, \qquad \alpha\, d\theta = \omega\, d\omega \qquad \omega = \frac{d\theta}{dt}$$

If the *angular acceleration is constant*, use

$$\omega = \omega_0 + \alpha_c t$$
$$\theta = \theta_0 + \omega_0 t + \tfrac{1}{2}\alpha_c t^2$$
$$\omega^2 = \omega_0^2 + 2\alpha_c(\theta - \theta_0)$$

The following examples numerically illustrate application of this procedure.

Example 17–9

The 30-kg disk shown in Fig. 17–15a is pin-supported at its center. If it starts from rest, determine the number of revolutions it must make to attain an angular velocity of 20 rad/s. Also, what are the reactions at the pin? The disk is acted upon by a constant force $F = 10$ N, which is applied to a cord wrapped around its periphery, and a constant couple moment $M = 5$ N · m. Neglect the mass of the cord in the calculation.

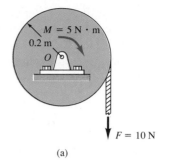

(a)

SOLUTION

Free-Body Diagram. Fig. 17–15b. Note that the mass center is not subjected to an acceleration; however, the disk has a clockwise angular acceleration.

The moment of inertia of the disk is

$$I_O = \tfrac{1}{2}mr^2 = \frac{1}{2}(30)(0.2)^2 = 0.6 \text{ kg} \cdot \text{m}^2$$

The three unknowns are O_x, O_y, and α.

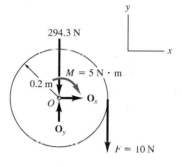

Equations of Motion

$$\xrightarrow{+} \Sigma F_x = m(a_G)_x; \qquad\qquad O_x = 0 \qquad\qquad\qquad Ans.$$

$$+\uparrow \Sigma F_y = m(a_G)_y; \qquad O_y - 294.3 - 10 = 0$$

$$O_y = 304 \text{ N} \qquad\qquad Ans.$$

$$\zeta + \Sigma M_O = I_O\alpha; \qquad -10(0.2) - 5 = 0.6\alpha \qquad \alpha = 11.7 \text{ rad/s}^2 \; \downarrow$$

(b)

Fig. 17–15

Kinematics. Since α is constant and rotates clockwise as assumed, the number of radians the disk must turn to obtain a clockwise angular velocity of 20 rad/s is

$$\zeta + \qquad\qquad \omega^2 = \omega_0^2 + 2\alpha_c(\theta - \theta_0)$$

$$(-20)^2 = 0 + 2(-11.7)(\theta - 0)$$

$$\theta = -17.1 \text{ rad} = 17.1 \text{ rad} \; \downarrow$$

Hence,

$$\theta = 17.1 \text{ rad}\left(\frac{1 \text{ rev}}{2\pi \text{ rad}}\right) = 2.73 \text{ rev} \; \downarrow \qquad\qquad Ans.$$

Example 17–10

The drum shown in Fig. 17–16a has a mass of 60 kg and a radius of gyration $k_O = 0.25$ m. A cord of negligible mass is wrapped around the periphery of the drum and attached to a crate having a mass of 20 kg. If the crate is released, determine the drum's angular acceleration.

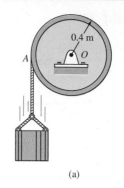

(a)

SOLUTION I

Free-Body Diagram. Here we will consider the drum and crate separately, Fig. 17–16b. Assuming the crate accelerates *downward* at **a**, it creates a *counterclockwise* angular acceleration α of the drum.

The moment of inertia of the drum is

$$I_O = mk_O^2 = (60)(0.25)^2 = 3.75 \text{ kg} \cdot \text{m}^2$$

There are five unknowns, namely O_x, O_y, T, a and α.

Equations of Motion. Applying the translational equations of motion $\Sigma F_x = m(a_G)_x$ and $\Sigma F_y = m(a_G)_y$ to the drum is of no consequence to the solution, since these equations involve the unknowns O_x and O_y. Thus, for the drum and crate, respectively

$$\zeta + \Sigma M_O = I_O \alpha; \qquad T(0.4) = 3.75\alpha \qquad (1)$$
$$+ \uparrow \Sigma F_y = m(a_G)_y; \qquad -20(9.81) + T = -20a \qquad (2)$$

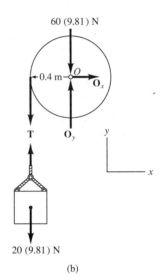

Kinematics. Since the point of contact A between the cord and drum has a tangential component of acceleration **a**, Fig. 17–16a, then

$$\zeta + a = \alpha r; \qquad a = \alpha(0.4) \qquad (3)$$

Solving the above equations

$$T = 106 \text{ N}$$
$$a = 4.52 \text{ m/s}^2$$
$$\alpha = 11.3 \text{ rad/s}^2 \text{ }\rceil \qquad \qquad Ans.$$

(b)

SOLUTION II

Free-Body and Kinetic Diagrams. The cable tension T can be eliminated from the analysis by considering the drum and crate as a single system, Fig. 17–16c. The kinetic diagram is shown since moments will be summed about point O.

Equations of Motion. Using Eq. (3) and applying the moment equation about point O to eliminate the unknowns O_x and O_y, we have

$$\zeta + \Sigma M_O = \Sigma(\mathcal{M}_k)_O; \qquad 20(9.81)(0.4) = 3.75\alpha + [20(0.4\alpha)]0.4$$
$$\alpha = 11.3 \text{ rad/s}^2 \text{ }\rceil \qquad \qquad Ans.$$

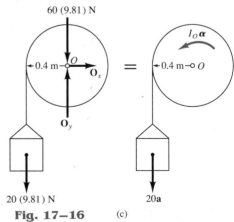

Fig. 17–16 (c)

Example 17–11

The unbalanced 50-lb flywheel shown in Fig. 17–17a has a radius of gyration of $k_G = 0.6$ ft about an axis passing through its mass center G. If it has a clockwise angular velocity of 8 rad/s at the instant shown, determine the horizontal and vertical components of reaction at the pin O.

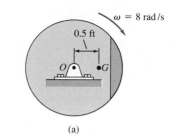

(a)

SOLUTION

Free-Body and Kinetic Diagrams. Since G moves in a *circular path*, $(a_G)_y = \alpha r_G$ and $(a_G)_x = \omega^2 r_G$. By inspection, the weight creates a clockwise angular acceleration $\boldsymbol{\alpha}$, so $I_G\alpha$ acts in a clockwise sense and $m(\mathbf{a}_G)_y$ acts downward, in accordance with the direction of $\boldsymbol{\alpha}$, Fig. 17–17b. Furthermore, $m(\mathbf{a}_G)_x$ acts to the left, toward the center of curvature O.

The moment of inertia of the flywheel about its mass center is determined from the radius of gyration and the mass; i.e., $I_G = mk_G^2 = (50/32.2)(0.6)^2 = 0.559$ slug · ft².

The three unknowns are O_x, O_y, and α.

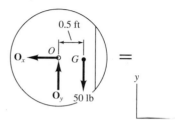

Equations of Motion

$$\xleftarrow{+}\Sigma F_x = m\omega^2 r_G; \qquad O_x = \left(\frac{50}{32.2}\right)(8)^2(0.5) \qquad (1)$$

$$+\uparrow \Sigma F_y = m\alpha r_G; \qquad O_y - 50 - \left(\frac{50}{32.2}\right)(\alpha)(0.5) \qquad (2)$$

$$\zeta+\Sigma M_G = I_G\alpha; \qquad O_y(0.5) = 0.559\alpha \qquad (3)$$

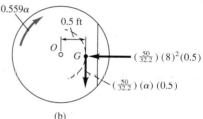

(b)

Fig. 17–17

Solving,

$$\alpha = 26.4 \text{ rad/s}^2, \quad O_x = 49.7 \text{ lb}, \quad O_y - 29.5 \text{ lb} \qquad \textit{Ans.}$$

Moments can also be summed about point O in order to eliminate $\mathbf{O}_x$ and $\mathbf{O}_y$ and thereby obtain a *direct solution* for $\boldsymbol{\alpha}$, Fig. 17–17b. This can be done in one of *two* ways, i.e., by using either $\Sigma M_O = \Sigma(\mathcal{M}_k)_O$ or $\Sigma M_O = I_O\alpha$. If the first of these equations is applied, then,

$$\zeta+\Sigma M_O = \Sigma(\mathcal{M}_k)_O; \quad 50(0.5) = 0.559\alpha + \left[\left(\frac{50}{32.2}\right)\alpha(0.5)\right](0.5)$$

$$50(0.5) = 0.947\alpha \qquad (4)$$

If $\Sigma M_O = I_O\alpha$ is applied, then by the parallel-axis theorem the moment of inertia of the flywheel about O is

$$I_O = I_G + mr_G^2 = 0.559 + \left(\frac{50}{32.2}\right)(0.5)^2 = 0.947 \text{ slug} \cdot \text{ft}^2$$

Hence, from the free-body diagram, Fig. 17–17b, we require

$$\zeta+\Sigma M_O = I_O\alpha; \qquad 50(0.5) = 0.947\alpha$$

which is the same as Eq. (4). Solving for α and substituting into Eq. (2) yields the answer for O_y obtained previously.

Example 17–12

The slender rod shown in Fig. 17–18a has a mass m and length l and is released from rest when $\theta = 0°$. Determine the horizontal and vertical components of force which the pin at O exerts on the rod at the instant $\theta = 90°$.

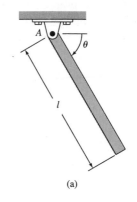

(a)

SOLUTION

Free-Body Diagram. The free-body diagram for the rod is shown when the rod is in the general position θ, Fig. 17–18b. For convenience, the force components at A are shown acting in the n and t directions. Note that α acts clockwise.

The moment of inertia of the rod about point A is

$$I_A = \tfrac{1}{3}ml^2$$

Equations of Motion. Moments will be summed about A in order to eliminate the reactive forces there.*

$+\nwarrow\Sigma F_n = m\omega^2 r_G;$ $\qquad A_n - mg \sin \theta = m\omega^2(l/2)$ $\qquad$ (1)

$+\swarrow\Sigma F_t = m\alpha r_G;$ $\qquad A_t + mg \cos \theta = m\alpha(l/2)$ $\qquad$ (2)

$\curvearrowright+\Sigma M_A = I_A\alpha;$ $\qquad mg \cos \theta(l/2) = (\tfrac{1}{3}ml^2)\alpha$ $\qquad$ (3)

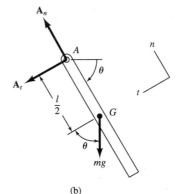

(b)

Fig. 17–18

Kinematics. For a given angle θ there are four unknowns in the above three equations: A_n, A_t, ω, and α. As shown by Eq. (3) α is *not constant*; rather, it depends upon the position θ of the rod. Using kinematics, α and ω can be related to θ by the equation

$\curvearrowright+$ $\qquad\qquad\qquad\qquad \omega \, d\omega = \alpha \, d\theta$ $\qquad$ (4)

In order to solve for ω at $\theta = 90°$, eliminate α from Eqs. (3) and (4), which yields

$$\omega \, d\omega = (1.5 \; g/l) \cos \theta \, d\theta$$

Since $\omega = 0$ at $\theta = 0°$, we have

$$\int_0^\omega \omega \, d\omega = (1.5 \; g/l) \int_{0°}^{90°} \cos \theta \, d\theta$$

$$\omega^2 = 3 \; g/l$$

Substituting this value into Eq. (1) with $\theta = 90°$ and solving Eqs. (1) to (3) yields

$$\alpha = 0, \quad A_t = 0, \quad A_n = 2.5 \; mg \qquad\qquad\qquad \textit{Ans.}$$

---

*If $\Sigma M_A = \Sigma(\mathcal{M}_k)_A$ is used, one must account for the moments of $I_G\alpha$ and $m(\mathbf{a}_G)_t$ about A. Refer to Eq. 17–15.

PROBLEMS

17–43. The 10-lb rod is pin-connected to its support and has an angular velocity $\omega = 4$ rad/s when it is in the horizontal position shown. Determine its angular acceleration and the horizontal and vertical components of reaction which the pin exerts on the rod at this instant.

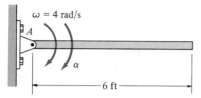

Prob. 17–43

*17–44. The 10-kg wheel has a radius of gyration $k_A = 0.2$ m. If the wheel is subjected to a moment of $M = 50$ N · m, determine the magnitudes of the velocity and the normal and tangential components of acceleration of point P on its rim after it has rotated 4 revolutions, starting from rest. Also, compute the reactions which the fixed pin A exerts on the wheel during the motion.

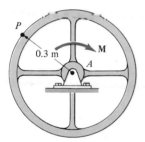

Prob. 17–44

17–45. The fan blade has a mass of 2 kg and a moment of inertia $I_O = 0.18$ kg · m² about an axis passing through its center O. If it is subjected to a moment of $M = 3(1 - e^{-0.2t})$ N · m, where t is in seconds, determine its angular velocity in $t = 4$ s starting from rest.

Prob. 17–45

17–46. The cord is wrapped around the inner core of the spool. If a 5-lb weight B is suspended from the cord and released from rest, determine the spool's angular velocity in $t = 3$ s. Neglect the mass of the cord. The spool has a weight of 180 lb and the radius of gyration about the axle A is $k_A = 1.25$ ft. Solve the problem in two ways. First by considering the "system" consisting of the block and spool, and then by considering the block and spool separately.

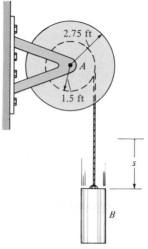

Prob. 17–46

17–47. The 80-kg disk is supported by a pin at A. If it is released from rest from the position shown, determine the initial horizontal and vertical components of reaction at the pin.

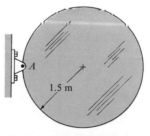

Probs. 17–47/17–48

*17–48. The 80-kg disk is supported by a pin at A. If it is rotating clockwise at $\omega = 0.5$ rad/s when it is in the position shown, determine the horizontal and vertical components of reaction at the pin at this instant.

17–49. The two blocks A and B have a mass of m_A and m_B, respectively, where $m_B > m_A$. If the pulley can be treated as a disk of mass M, determine the acceleration of block A. Neglect the mass of the cord and any slipping on the pulley.

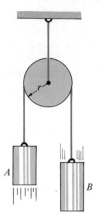

Prob. 17–49

17–50. The spool is supported on small rollers at A and B. Determine the angular acceleration of the spool and the normal forces at A and B if a vertical force of $P = 80$ N is applied to the cable. The spool has a mass of 60 kg and a centroidal radius of gyration $k_O = 0.65$ m. For the calculation neglect the mass of the cable and the mass of the rollers at A and B.

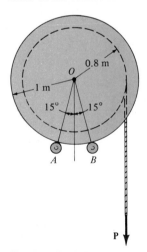

Prob. 17–50

17–51. The drum has a weight of 80 lb and a radius of gyration of $k_O = 0.4$ ft. If the cable, which is wrapped around the drum, is subjected to a vertical force of $P = 15$ lb, determine the time needed to increase the drum's angular velocity from $\omega_1 = 5$ rad/s to $\omega_2 = 25$ rad/s. Neglect the mass of the cable.

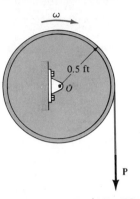

Probs. 17–51/17–52

*17–52. The drum has a weight of 80 lb and a radius of gyration of $k_O = 0.4$ ft. If the cable, which is wrapped around the drum, is subjected to a vertical force of $P = (3t^2 + 5)$ lb, where t is in seconds, determine the angular velocity of the drum in $t = 2$ s if at $t = 0$, $\omega_1 = 5$ rad/s. Neglect the mass of the cable.

17–53. The 20-kg roll of paper has a radius of gyration $k_A = 90$ mm about an axis passing through point A. It is pin-supported at both ends by two brackets AB. If the roll rests against a wall for which the coefficient of kinetic friction is $\mu_C = 0.2$ and a vertical force of 30 N is applied to the end of the paper, determine the angular acceleration of the roll as the paper unrolls.

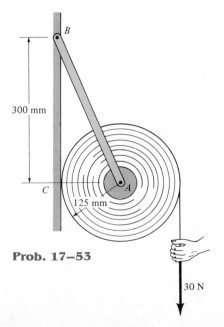

Prob. 17–53

17–54. The pendulum consists of a uniform 5-kg plate and a 2-kg slender rod. Compute the horizontal and vertical components of reaction that the pin O exerts on the rod at the instant $\theta = 30°$, at which time its angular velocity is $\omega = 3$ rad/s.

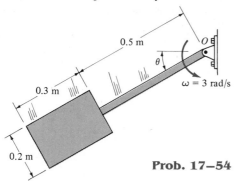

0.5 m

θ

$\omega = 3$ rad/s

0.3 m

0.2 m

Prob. 17–54

17–55. The kinetic diagram representing the general rotational motion of a rigid body about a fixed axis at O is shown in the figure. Show that $I_G\alpha$ may be eliminated by moving the vectors $m(\mathbf{a}_G)_t$ and $m(\mathbf{a}_G)_n$ to point P, located a distance $r_{GP} = k_G^2/r_{OG}$ from the center of mass G of the body. Here k_G represents the radius of gyration of the body about G. The point P is called the *center of percussion* of the body.

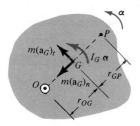

Prob. 17–55

*17–56. Determine the position r_P of the center of percussion P of the 10-lb slender bar. (See Prob. 17–55.) What is the horizontal force $\mathbf{A}_x$ at the pin when the bar is struck at P with a force of $F = 20$ lb?

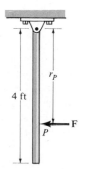

r_P

4 ft

$\mathbf{F}$

P

Prob. 17–56

17–57. The two blocks A and B having a weight of 10 lb and 5 lb, respectively, are attached to the ends of a cord which passes over a 3-lb pulley which has a radius of gyration of $k_O = 0.5$ ft. If the blocks are released from rest, determine their speed just after A descends 3 ft. The cord does not slip on the pulley. Neglect the mass of the cord. Solve the problem in two ways. First by considering the "system" consisting of both the blocks and the pulley, and then by considering the blocks and pulley separately.

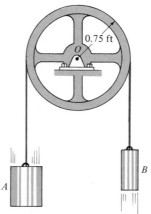

0.75 ft

O

B

A

Prob. 17–57

17–58. The 10-lb bar is pinned at its center O and connected to a torsional spring. The spring has a stiffness $k = 5$ lb · ft/rad, so that the torque $\mathbf{M}$ developed is $M = (5\theta)$ lb · ft, where θ is in radians. If the bar is released from rest when it is vertical at $\theta = 90°$, determine its angular velocity at the instant $\theta = 0°$.

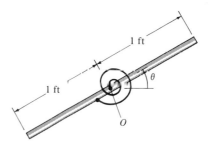

1 ft

1 ft

θ

O

Prob. 17–58

17–59. The armature (slender rod) AB has a mass of 0.2 kg and can pivot about the pin at A. Movement is controlled by the electromagnet E, which exerts a horizontal attractive force on the armature at B of $F_B = (0.2(10^{-3})l^{-2})$ N, where l in meters is the gap between the armature and the magnet at any instant. If the armature lies in the horizontal plane, and is originally at rest, determine the speed of the contact at B the instant $l = 0.01$ m. Originally $l = 0.02$ m.

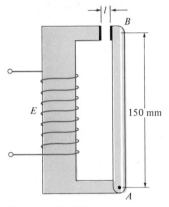

Prob. 17–59

*17–60. Disk A has a weight of 5 lb and disk B has a weight of 10 lb. If no slipping occurs between them, determine the couple **M** which must be applied to disk A to give it an angular acceleration of 4 rad/s².

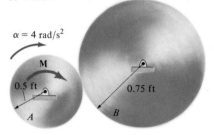

Prob. 17–60

17–61. Determine the angular acceleration of the 25-kg diving board and the horizontal and vertical components of reaction at the pin A the instant the man jumps off. Assume that the board is thin, uniform, and rigid, and that at the instant the man jumps off the spring is compressed a maximum amount of 200 mm, $\omega = 0$, and the board is horizontal.

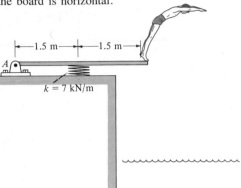

Prob. 17–61

17–62. The assembly consists of the 30-lb blocks A and B and a pulley which has a weight of 20 lb and a radius of gyration of $k_C = 0.5$ ft. Assuming the cord does not slip over the surface of the pulley, determine the force in the horizontal and vertical segments of the cord. Neglect both the mass of the cord and friction between the block at A and the horizontal plane.

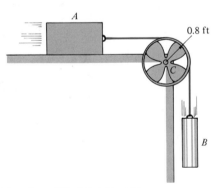

Probs. 17–62/17–63

17–63. The assembly consists of the 30-lb blocks A and B and a pulley which has a weight of 20 lb and a radius of gyration of $k_C = 0.5$ ft. Assuming the cord does not slip over the surface of the pulley, determine the force in the horizontal and vertical segments of the cord. Neglect the mass of the cord and take the coefficient of kinetic friction between the block at A and the horizontal plane to be $\mu_k = 0.3$.

*17–64. The cylinder has a radius r and mass m and rests in the trough for which the coefficient of kinetic friction at A and B is μ. If a horizontal force $\mathbf{P}$ is applied to the cylinder, determine the cylinder's angular acceleration if it begins to spin.

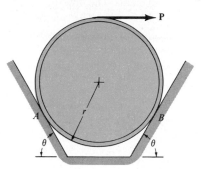

Prob. 17–64

17–65. The rod has a length L and mass m. If it is released from rest when $\theta \approx 0°$, determine its angular velocity as a function of θ. Also, express the horizontal and vertical components of reaction at the pin O as a function of θ.

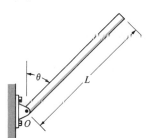

Prob. 17–65

17–66. The slender rod of length L and mass m is released from rest when $\theta = 0°$. Determine as a function of θ the normal and frictional forces which are exerted on the ledge at A as it falls downward. At what angle θ does it begin to slip if the coefficient of static friction at A is μ?

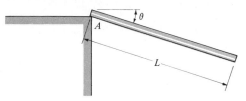

Prob. 17–66

■17–67. A 40-kg boy sits on top of the large wheel which has a mass of 400 kg and a radius of gyration of $k_G = 5.5$ m. If the boy essentially starts from rest at $\theta = 0°$, and the wheel begins to freely rotate, determine the angle at which the boy begins to slip. The coefficient of static friction between the wheel and the boy is $\mu_s = 0.5$. Neglect the size of the boy in the calculation.

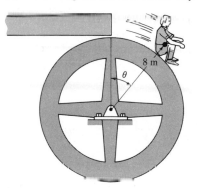

Prob. 17–67

Equations of Motion: General Plane Motion

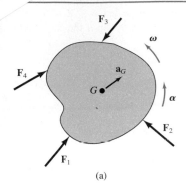

(a)

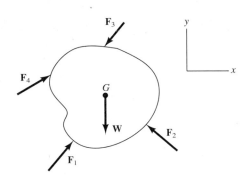

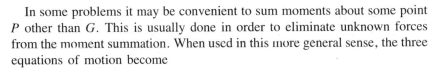

The rigid body (or slab) shown in Fig. 17–19a is subjected to general plane motion caused by the externally applied force system. The free-body and kinetic diagrams for the body are shown in Fig. 17–19b. The vector $m\mathbf{a}_G$ (shown dashed) has the same *direction* as the acceleration of the body's mass center, and $I_G\boldsymbol{\alpha}$ acts in the same *direction* as the angular acceleration. If an x and y inertial coordinate system is chosen as shown, the three equations of motion may be written as

$$\begin{aligned}\Sigma F_x &= m(a_G)_x \\ \Sigma F_y &= m(a_G)_y \\ \Sigma M_G &= I_G\alpha \end{aligned} \qquad (17\text{--}17)$$

In some problems it may be convenient to sum moments about some point P other than G. This is usually done in order to eliminate unknown forces from the moment summation. When used in this more general sense, the three equations of motion become

$$\begin{aligned}\Sigma F_x &= m(a_G)_x \\ \Sigma F_y &= m(a_G)_y \\ \Sigma M_P &= \Sigma(\mathcal{M}_k)_P \end{aligned} \qquad (17\text{--}18)$$

where $\Sigma(\mathcal{M}_k)_P$ represents the moment sum of $I_G\boldsymbol{\alpha}$ and $m\mathbf{a}_G$ (or its components) about P as determined by the data on the kinetic diagram.

PROCEDURE FOR ANALYSIS

The following procedure provides a method for solving kinetic problems involving general plane motion of a rigid body.

Free-Body Diagram. Establish the x, y inertial coordinate system and draw the free-body diagram for the body. Specify the direction and sense of the acceleration of the mass center, $\mathbf{a}_G$, and the angular acceleration $\boldsymbol{\alpha}$ of the body. Also, compute the moment of inertia I_G. Identify the unknowns in the problem. If it is decided that the rotational equation of motion $\Sigma M_P = \Sigma(\mathcal{M}_k)_P$ is to be used, then consider drawing the kinetic diagram in order to help "visualize" the "moments" developed by the components $m(\mathbf{a}_G)_x$, $m(\mathbf{a}_G)_y$, and $I_G\boldsymbol{\alpha}$ when writing the terms in the moment sum $\Sigma(\mathcal{M}_k)_P$.

Equations of Motion. Apply the three equations of motion, Eqs. 17–17 or 17–18.

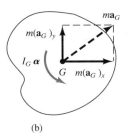

Fig. 17–19

(b)

Kinematics. Use kinematics if a complete solution cannot be obtained strictly from the equations of motion. In particular, if the body's motion is *constrained* due to its supports, additional equations may be obtained by using $\mathbf{a}_B = \mathbf{a}_A + \mathbf{a}_{B/A}$, which relate the accelerations of any two points A and B on the body (see Example 17–16).

Frictional Rolling Problems

There is a class of planar kinetics problems which deserves special mention. These problems involve wheels, cylinders, or bodies of similar shape, which roll on a *rough* plane surface. Because of the applied loadings, it may not be known if the body *rolls without slipping,* or if it *slides as it rolls.* For example, consider the homogeneous disk shown in Fig. 17–20a, which has a mass m and is subjected to a known horizontal force **P.** Following the procedure outlined above, the free-body diagram is shown in Fig. 17–20b. Since $\mathbf{a}_G$ is directed to the right and $\boldsymbol{\alpha}$ is clockwise, we have

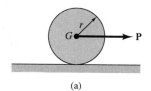

(a)

$$\xrightarrow{+}\Sigma F_x = m(a_G)_x; \qquad P - F = ma_G \qquad (17\text{-}19)$$

$$+\uparrow\Sigma F_y = m(a_G)_y; \qquad N - mg = 0 \qquad (17\text{-}20)$$

$$\zeta+\Sigma M_G = I_G\alpha; \qquad Fr = I_G\alpha \qquad (17\text{-}21)$$

A fourth equation is needed since these *three equations* contain *four unknowns: F, N, α,* and a_G.

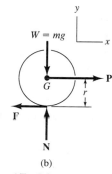

(b)
Fig. 17–20

No Slipping. If the frictional force **F** is great enough to allow the disk to roll *without slipping,* then a_G may be related to α by the *kinematic equation**

$$(\zeta+) \qquad\qquad a_G = \alpha r \qquad\qquad (17\text{-}22)$$

The four unknowns are determined by *solving simultaneously* Eqs. 17–19 to 17–22. When the solution is obtained, the assumption of no slipping must be *checked.* Recall that no slipping occurs provided $F \leqslant \mu_s N$, where μ_s is the coefficient of static friction. If the inequality is satisfied, the problem is solved. However, if $F > \mu_s N$, the problem must be *reworked* since then the disk slips as it rolls.

Slipping. In the case of slipping, α and a_G are *independent of one another* so that Eq. 17–22 does not apply. Instead, the magnitude of the frictional force is related to the magnitude of the normal force using the coefficient of kinetic friction μ_k, i.e.,

$$F = \mu_k N \qquad\qquad (17\text{-}23)$$

Here **F** acts to the left on the disk to prevent the slipping motion to the right, Fig. 17–20b. In this case Eqs. 17–19 to 17–21 and 17–23 are used for the solution. Examples 17–14 and 17–15 illustrate these concepts numerically.

*See Example 16–3 or 16-13.

Example 17–13

The spool in Fig. 17–21a has a mass of 8 kg and a radius of gyration of $k_G = 0.35$ m. If cords of negligible mass are wrapped around its inner hub and outer rim as shown, determine the spool's angular velocity 3 s after it is released from rest.

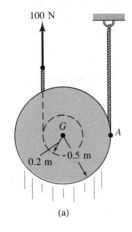

100 N

(a)

SOLUTION

Free-Body Diagram. Fig. 17–21b.   The 100-N force causes $\mathbf{a}_G$ to act upward. Also, $\boldsymbol{\alpha}$ acts clockwise since the spool winds around the cord at A.

There are three unknowns, namely T, a_G, and α. The moment of inertia of the spool about its mass center is

$$I_G = mk_G^2 = 8(0.35)^2 = 0.980 \text{ kg} \cdot \text{m}^2$$

Equations of Motion

$$+\uparrow \Sigma F_y = m(a_G)_y; \qquad T + 100 - 78.48 = 8a_G \qquad (1)$$

$$\zeta+\Sigma M_G = I_G\alpha; \qquad 100(0.2) - T(0.5) = 0.980\alpha \qquad (2)$$

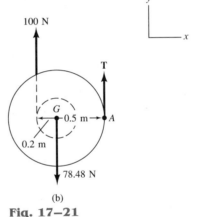

100 N

(b)

Fig. 17–21

Kinematics.   A complete solution is obtained if kinematics is used to relate a_G to α. In this case the spool "rolls without slipping" on the cord at A. Hence, we can use the results of Example 16–3 or 16–13, so that

$$(\zeta+)a_G = r\alpha; \qquad a_G = 0.5\alpha \qquad (3)$$

Solving Eqs. (1) to (3), we have

$$\alpha = 10.3 \text{ rad/s}^2$$
$$a_G = 5.16 \text{ m/s}^2$$
$$T = 19.8 \text{ N}$$

Since α is constant, the angular velocity in 3 s is

$(\zeta+)$
$$\omega = \omega_0 + \alpha_c t$$
$$= 0 + 10.3(3)$$
$$= 30.9 \text{ rad/s} \; \zeta \qquad\qquad \textit{Ans.}$$

Example 17–14

The 50-lb wheel shown in Fig. 17–22a has a radius of gyration $k_G = 0.70$ ft. If a 35-lb · ft couple moment is applied to the wheel, determine the acceleration of its mass center G. The coefficients of static and kinetic friction between the wheel and the plane at A are $\mu_s = 0.3$ and $\mu_k = 0.25$, respectively.

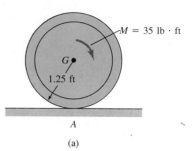

(a)

SOLUTION

Free-Body Diagram. By inspection of Fig. 17–22b, it is seen that the couple moment causes the wheel to have a clockwise angular acceleration of α. As a result, the acceleration of the mass center, $\mathbf{a}_G$, is directed to the right. The moment of inertia is

$$I_G = mk_G^2 = \frac{50}{32.2}(0.70)^2 = 0.761 \text{ slug} \cdot \text{ft}^2$$

The unknowns are N_A, F_A, a_G, and α.

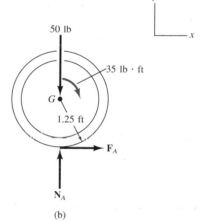

Equations of Motion

$$\xrightarrow{+}\Sigma F_x = m(a_G)_x; \qquad F_A = \frac{50}{32.2}a_G \qquad (1)$$

$$+\uparrow \Sigma F_y = m(a_G)_y; \qquad N_A - 50 = 0 \qquad (2)$$

$$\zeta+\Sigma M_G = I_G\alpha; \qquad 35 - 1.25(F_A) = 0.761\alpha \qquad (3)$$

Kinematics (No Slipping). If this assumption is made, then

$$(\zeta+) \qquad\qquad a_G = 1.25\alpha \qquad (4) \qquad \textbf{Fig. 17–22}$$

Solving Eqs. (1) to (4),

$$N_A = 50.0 \text{ lb} \qquad F_A - 21.3 \text{ lb}$$
$$\alpha = 11.0 \text{ rad/s}^2 \qquad a_G = 13.7 \text{ ft/s}^2$$

The original assumption of no slipping requires $F_A \leq \mu_s N_A$. However, since 21.3 lb $> 0.3(50$ lb$) = 15$ lb, the wheel slips as it rolls.

(Slipping). This requires $F_A = \mu_k N_A$, or

$$F_A = 0.25N_A \qquad (5)$$

Solving Eqs. (1) to (3) and (5) yields

$$N_A = 50.0 \text{ lb} \qquad F_A = 12.5 \text{ lb}$$
$$\alpha = 25.5 \text{ rad/s}^2$$
$$a_G = 8.05 \text{ ft/s}^2 \rightarrow \qquad\qquad \textit{Ans.}$$

Example 17–15

The uniform slender pole shown in Fig. 17–23a has a mass of 100 kg and a moment of inertia $I_G = 75$ kg · m². If the coefficients of static and kinetic friction between the end of the pole and the surface are $\mu_s = 0.3$ and $\mu_k = 0.25$, respectively, determine the pole's angular acceleration at the instant the 400-N horizontal force is applied. The pole is originally at rest.

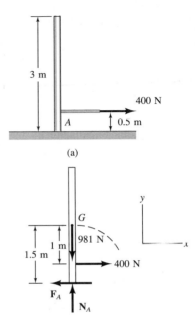

(a)

(b)

Fig. 17–23

SOLUTION

Free-Body Diagram. Figure 17–23b. The path of motion of the mass center G will be along an unknown (dashed) curved path having a radius of curvature ρ which is initially parallel to the y axis. There is no normal or y component of acceleration since the pole is originally at rest, i.e., $\mathbf{v}_G = \mathbf{0}$, so that $(a_G)_y = v_G^2/\rho = 0$. We will assume the mass center accelerates to the right, and the pole has a clockwise angular acceleration of α. The unknowns are N_A, F_A, a_G, and α.

Equations of Motion

$$\xrightarrow{+} \Sigma F_x = m(a_G)_x; \qquad 400 - F_A = 100 a_G \qquad (1)$$

$$+\uparrow \Sigma F_y = m(a_G)_y; \qquad N_A - 981 = 0 \qquad (2)$$

$$\zeta + \Sigma M_G = I_G \alpha; \qquad F_A(1.5) - 400(1) = 75\alpha \qquad (3)$$

A fourth equation is needed for a complete solution.

Kinematics (No Slipping). In this case point A acts as a "pivot" so that

$$\zeta + a_G = \alpha r_{AG}; \qquad a_G = 1.5\alpha \qquad (4)$$

Solving Eqs. (1) to (4) yields

$$N_A = 981 \text{ N} \qquad F_A = 300 \text{ N}$$
$$a_G = 1 \text{ m/s}^2 \qquad \alpha = 0.667 \text{ rad/s}^2$$

Testing the original assumption of no slipping requires $F_A \leq \mu_s N_A$. However, $300 > 0.3(981) = 294$ N. (Slips at A.)

(Slipping). For this case Eq. (4) does *not* apply. Instead, the frictional equation $F_A = \mu_k N_A$ is used, i.e.,

$$F_A = 0.25 N_A \qquad (5)$$

Solving Eqs. (1) to (3) and (5) simultaneously yields

$$N_A = 981 \text{ N} \qquad F_A = 245 \text{ N} \qquad a_G = 1.55 \text{ m/s}^2$$
$$\alpha = -0.428 \text{ rad/s}^2 = 0.428 \text{ rad/s}^2 \; \zeta \qquad \textbf{\textit{Ans.}}$$

Example 17–16

The 30-kg wheel shown in Fig. 17–24a has a mass center at G and a radius of gyration $k_G = 0.15$ m. If the wheel is originally at rest and released from the position shown, determine its angular acceleration. No slipping occurs.

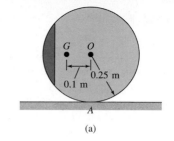

(a)

SOLUTION

Free-Body and Kinetic Diagrams. The two unknowns $\mathbf{F}_A$ and $\mathbf{N}_A$ shown on the free-body diagram, Fig. 17–24b, can be eliminated from the analysis by summing moments about point A. The kinetic diagram accompanies the solution in order to illustrate application of $\Sigma(\mathcal{M}_k)_A$. Since the *path of motion* of G is *unknown*, the two components $m(\mathbf{a}_G)_x$ and $m(\mathbf{a}_G)_y$ must be shown on the kinetic diagram, Fig. 17–24b.

The moment of inertia is

$$I_G = mk_G^2 = 30(0.15)^2 = 0.675 \text{ kg} \cdot \text{m}^2$$

There are five unknowns, namely, N_A, F_A, $(a_G)_x$, $(a_G)_y$, and α.

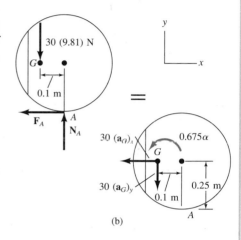

(b)

Equation of Motion. Applying the rotational equation of motion about point A, to eliminate N_A, and F_A, we have

$$\zeta + \Sigma M_A = \Sigma(\mathcal{M}_k)_A;$$
$$30(9.81)(0.1) = 0.675\alpha + 30(a_G)_x(0.25) + 30(a_G)_y(0.1) \quad (1)$$

There are three unknowns in this equation: $(a_G)_x$, $(a_G)_y$, and α.

Kinematics. Using kinematics, $(a_G)_x$, $(a_G)_y$, and α can be related. Since no slipping occurs, $a_O = \alpha r = \alpha(0.25)$, directed to the left, Fig. 17–24c. Also, $\boldsymbol{\omega} = \mathbf{0}$ since the wheel is originally at rest. Using Eq. 16–16, with point O as the base point, Fig. 17–24c and d, we have

$$\mathbf{a}_G = \mathbf{a}_O + (\mathbf{a}_{G/O})_t + (\mathbf{a}_{G/O})_n$$

$$\begin{bmatrix} (a_G)_x \\ \leftarrow \end{bmatrix} + \begin{bmatrix} (a_G)_y \\ \downarrow \end{bmatrix} = \begin{bmatrix} \alpha(0.25) \\ \leftarrow \end{bmatrix} + \begin{bmatrix} \alpha(0.1) \\ \downarrow \end{bmatrix} + \begin{bmatrix} 0 \end{bmatrix}$$

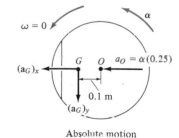

Absolute motion

(c)

Equating the respective horizontal and vertical components,

$$(a_G)_x = \alpha(0.25) \quad (2)$$
$$(a_G)_y = \alpha(0.1) \quad (3)$$

Solving Eqs. (1) to (3) yields

$$\alpha = 10.3 \text{ rad/s}^2 \, \zeta \qquad \textit{Ans.}$$
$$(a_G)_x = 2.58 \text{ m/s}^2$$
$$(a_G)_y = 1.03 \text{ m/s}^2$$

As an exercise, show that $F_A = 77.4$ N and $N_A = 263$ N.

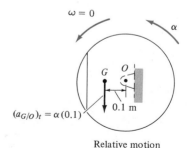

Relative motion

(d)

Fig. 17–24

PROBLEMS

*17–68. If the disk in Fig. 17–20*a rolls without slipping*, show that when moments are summed about the instantaneous center of zero velocity, *IC*, it is possible to use the moment equation $\Sigma M_{IC} = I_{IC}\alpha$, where I_{IC} represents the moment of inertia of the disk calculated about the instantaneous axis of zero velocity.

17–69. The uniform 50-lb board is suspended from cords at *C* and *D*. If these cords are subjected to constant forces of 30 lb and 45 lb, respectively, determine the acceleration of the board's center and its angular acceleration. Assume the board is a thin plate. Neglect the mass of the pulleys at *E* and *F*.

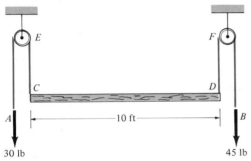

Probs. 17–69/17–70

17–70. Solve Prob. 17–69 if blocks having a weight of $W_A =$ 30 lb and $W_B =$ 45 lb are suspended from the cords at *A* and *B*.

17–71. The wheel has a mass of 80 kg and a radius of gyration $k_G = 0.25$ m. If it is subjected to a couple moment of $M =$ 50 N · m, determine its angular acceleration. The coefficients of static and kinetic friction between the ground and the wheel are $\mu_s = 0.2$ and $\mu_k = 0.15$, respectively.

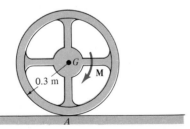

Prob. 17–71

*17–72. The spool and wire wrapped around its core have a mass of 20 kg and a centroidal radius of gyration $k_G = 250$ mm. If the coefficient of kinetic friction at the ground is $\mu_B = 0.1$, determine the angular acceleration of the spool when the 30-N · m couple is applied.

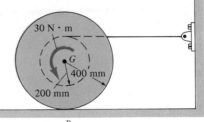

Prob. 17–72

17–73. The spool has a mass of 100 kg and a radius of gyration of $k_G = 0.3$ m. If the coefficients of static and kinetic friction at *A* are $\mu_s = 0.2$ and $\mu_k = 0.15$, respectively, determine the angular acceleration of the spool if $P = 50$ N.

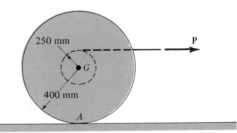

Probs. 17–73/17–74/17–75

17–74. Solve Prob. 17–73 if the cord and force $P = 50$ N are directed vertically upwards.

17–75. The spool has a mass of 50 kg and a radius of gyration of $k_G = 0.3$ m. If the coefficients of static and kinetic friction at *A* are $\mu_s = 0.2$ and $\mu_k = 0.15$, respectively, determine the angular acceleration of the spool if $P = 600$ N.

*17–76. The spool has a mass of 500 kg and a radius of gyration $k_G = 0.380$ m. It rests on the inclined surface for which the coefficient of kinetic friction is $\mu_A = 0.15$. If the spool is released from rest, determine the initial tension in the cord and the angular acceleration of the spool.

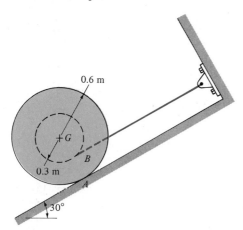

Prob. 17–76

17–77. The 20-kg punching bag has a radius of gyration about its center of mass G of $k_{ty} = 0.4$ m. If it is subjected to a horizontal force of $F = 30$ N, determine the initial angular acceleration of the bag and the tension in the supporting cable AB.

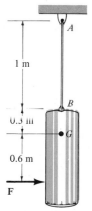

Prob. 17–77

17–78. The 15-lb circular plate is suspended from a pin at A. If the pin is connected to a track which is given an acceleration of $a_A = 3$ ft/s^2, determine the horizontal and vertical components of reaction at A and the acceleration of the mass center G. The plate is originally at rest.

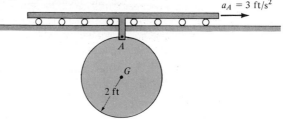

Prob. 17–78

17–79. The assembly consists of an 8-kg disk and a 10-kg bar which is pin-connected to the disk. If the system is released from rest, determine the initial angular acceleration of the disk. The coefficients of static and kinetic friction between the disk and the inclined plane are $\mu_s = 0.6$ and $\mu_k = 0.4$, respectively. Neglect friction at B.

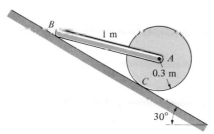

Probs. 17–79/17–80

*17–80. Solve Prob. 17–79 if the bar is removed.

17–81. By pressing down with the finger at B, a thin ring having a mass m is given an initial velocity v_1 and a backspin ω_1 when the finger is released. If the coefficient of kinetic friction between the table and the ring is μ, determine the distance the ring travels forward before backspinning stops.

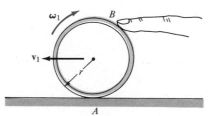

Prob. 17–81

17-82. A biomechanical model of the head and neck of a restrained passenger in an automobile is shown in the figure. Measurements indicate that the head plus half the neck, assumed pinned at A, have a mass of 0.28 slug and a moment of inertia about an axis through the center of gravity of $I_G = 0.024$ slug · ft². If an automobile and the passenger's torso are subjected to a horizontal deceleration $a_A = 15g = 483$ ft/s², when the head is in the position shown, determine the horizontal and vertical components of force which the neck exerts on the head at A at this instant. The restraining moment $\mathbf{M}_A$ is required to hold the head in the *nonrotating* equilibrium position shown, prior to the deceleration. Assume this moment remains constant during the deceleration.

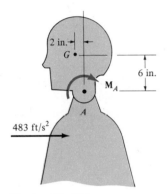

Prob. 17-82

17-83. The uniform beam has a weight W. If it is originally at rest while being supported at A and B by cables, determine the tension in cable A if cable B suddenly fails. Assume the beam is a slender rod.

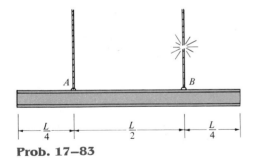

Prob. 17-83

*****17-84.** A cord is wrapped around each of the two disks. If disks A and B each have a mass of 10 kg and they are released from rest, determine the angular acceleration of each disk and the tension in cord C.

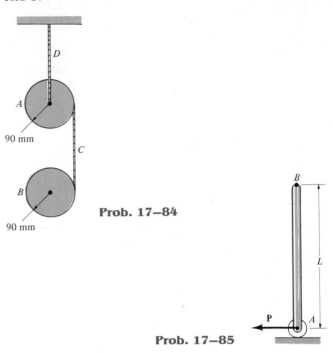

Prob. 17-84

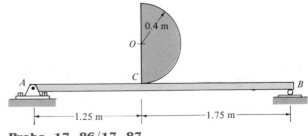

Prob. 17-85

17-85. The uniform bar of mass m and length L is balanced in the vertical position when the horizontal force $\mathbf{P}$ is applied to the roller at A. Determine the bar's initial angular acceleration and the acceleration of its top point B.

17-86. The semicircular disk has a mass of 50 kg and is released from rest from the position shown. The coefficients of static and kinetic friction between the disk and the beam are $\mu_s = 0.5$ and $\mu_k = 0.3$, respectively. Determine the initial reactions at the pin A and roller B used to support the beam. Neglect the mass of the beam for the calculation.

Probs. 17-86/17-87

17–87. The semicircular disk has a mass of 50 kg and is released from rest from the position shown. The coefficients of static and kinetic friction between the disk and the beam are $\mu_s = 0.2$ and $\mu_k = 0.1$, respectively. Determine the initial reactions at the pin A and roller B used to support the beam. Neglect the mass of the beam for the calculation.

*17–88. The 2-kg slender bar is supported by cord BC and then released from rest at A. Determine the initial angular acceleration of the bar and the tension in the cord.

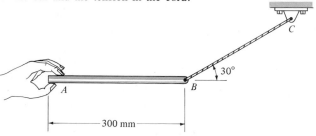

Prob. 17–88

17–89. The two pin-connected bars each have a weight of 10 lb/ft. If a moment of $M = 60$ lb · ft is applied to bar AB, determine the initial vertical reaction at C and the horizontal and vertical components of reaction at B. Neglect the size of the roller at C. The bars are initially at rest.

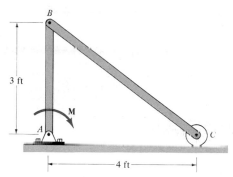

Prob. 17–89

17–90. The 10-lb disk and 4-lb block are released from rest. Determine the velocity of the block when $t = 3$ s. The coefficients of static and kinetic friction at A are $\mu_s = 0.2$ and $\mu_k = 0.15$, respectively. Neglect the mass of the cord and pulleys.

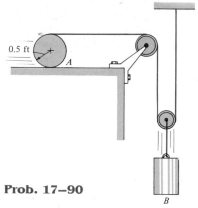

Prob. 17–90

17–91. The disk of mass m and radius r rolls without slipping on the circular path. Determine the normal force which the path exerts on the disk and the disk's angular acceleration if at the instant shown the disk has an angular velocity of $\boldsymbol{\omega}$.

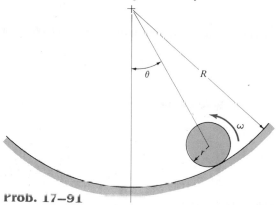

Prob. 17–91

*17–92. The semicircular disk has a mass of 10 kg. If it is rotating at $\omega = 4$ rad/s at the instant $\theta = 60°$, determine the normal and frictional forces it exerts on the ground at this instant. Assume the disk does not slip as it rolls.

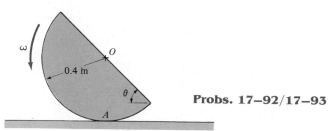

Probs. 17–92/17–93

17–93. The semicircular disk having a mass of 10 kg is rotating at $\omega = 4$ rad/s at the instant $\theta = 60°$. If the coefficient of static friction at A is $\mu_s = 0.5$, determine if the disk slips at this instant.

17–94. The wheel has a mass of 50 kg and a radius of gyration $k_G = 0.4$ m. If it rolls without slipping down the inclined plank, determine the forces which the supports at A and B exert on the plank at the instant the wheel is located at the midpoint of the plank. The plank is uniform and has a mass of 20 kg. Assume that horizontal and vertical reactions occur at A, and a normal reaction occurs at B.

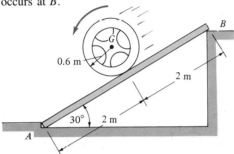

Prob. 17–94

■17–95. The cylinder has a mass m and is released from rest at $\theta = 0°$. If it rolls down the surface of the fixed semicylinder, determine the angle θ at which the cylinder just begins to slip on the surface. The coefficient of static friction between the cylinder and track is $\mu_s = 0.3$.

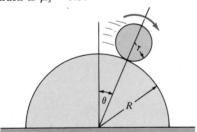

Prob. 17–95

*17–96. The 15-lb disk rests on a 5-lb plate. A cord is wrapped around the periphery of the disk and attached to the wall at B. If a torque $M = 40$ lb · ft is applied to the disk, determine the angular acceleration of the disk and the time needed for the end C of the plate to travel 3 ft and strike the wall. The coefficient of kinetic friction between the plate and disk is $\mu_k = 0.2$, and between the plate and floor $\mu'_k = 0.1$. Neglect the mass of the cord.

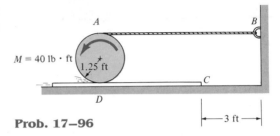

Prob. 17–96

17–97. The slender rod of length L and mass m is released from rest when $\theta = 0°$. If the ground is smooth, determine the angular acceleration of the rod just before $\theta = 90°$. Also, what is the reaction of the ground on the rod at this instant?

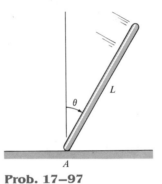

Prob. 17–97

17–98. The slender bar has a mass m and rests on the smooth floor A and against the smooth wall B. Determine its *initial* angular acceleration when it is released from rest from the position θ and allowed to slide downward.

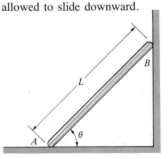

Prob. 17–98

18

Planar Kinetics of a Rigid Body: Work and Energy

It was shown in Chapter 14 that problems which involve force, velocity, and displacement can conveniently be solved by using the principle of work and energy, or if the force system is "conservative," the conservation of energy theorem. In this chapter we will apply work and energy methods to solve problems involving the planar motion of a rigid body. The more general discussion as applied to the three-dimensional motion of a rigid body is presented in Chapter 21.

Before discussing the principle of work and energy for a body, however, the methods for obtaining the body's kinetic energy when it is subjected to translation, rotation about a fixed axis, or general plane motion will be developed.

●18.1 Kinetic Energy

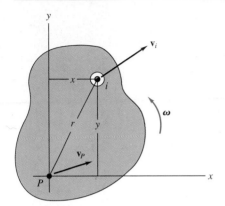

Fig. 18–1

Consider the rigid body shown in Fig. 18–1, which is represented here by a *slab* moving in the inertial *x-y* reference plane. An arbitrary *i*th particle of the body, having a mass dm, is located at r from the arbitrary point P. If at the *instant* shown the particle has a velocity $\mathbf{v}_i$, then the particle's kinetic energy is $T_i = \frac{1}{2}dm\, v_i^2$. The kinetic energy of the entire body is determined by writing similar expressions for each particle of the body and integrating the results, i.e.,

$$T = \frac{1}{2}\int_m dm\, v_i^2$$

This equation may be written in another manner by expressing $\mathbf{v}_i$ in terms of the velocity of point P, i.e., $\mathbf{v}_P$. If the body has an angular velocity $\boldsymbol{\omega}$, then from Fig. 18–1 we have

$$\mathbf{v}_i = \mathbf{v}_P + \mathbf{v}_{i/P}$$
$$= (v_P)_x\mathbf{i} + (v_P)_y\mathbf{j} + \omega\mathbf{k} \times (x\mathbf{i} + y\mathbf{j})$$
$$= ((v_P)_x - \omega y)\mathbf{i} + ((v_P)_y + \omega x)\mathbf{j}$$

The square of the magnitude of $\mathbf{v}_i$ is thus

$$\mathbf{v}_i \cdot \mathbf{v}_i = v_i^2 = [(v_P)_x - \omega y]^2 + [(v_P)_y + \omega x]^2$$
$$= (v_P^2)_x - 2(v_P)_x\omega y + \omega^2 y^2 + (v_P^2)_y + 2(v_P)_y\omega x + \omega^2 x^2$$
$$= v_P^2 - 2(v_P)_x\omega y + 2(v_P)_y\omega x + \omega^2 r^2$$

Substituting into the equation of kinetic energy yields

$$T = \frac{1}{2}\left(\int_m dm\right)v_P^2 - (v_P)_x\omega\left(\int_m y\, dm\right) + (v_P)_y\omega\left(\int_m x\, dm\right) + \frac{1}{2}\omega^2\left(\int_m r^2\, dm\right)$$

The first integral on the right represents the entire mass m of the body. Since $\bar{x}m = \int x\, dm$ and $\bar{y}m = \int y\, dm$, the second and third integrals locate the body's center of mass G with respect to P. The last integral represents the body's moment of inertia I_P, computed about the z axis passing through point P. Thus,

$$T = \frac{1}{2}mv_P^2 - (v_P)_x\,\omega\bar{y}\,m + (v_P)_y\,\omega\bar{x}\,m + \frac{1}{2}I_P\,\omega^2 \qquad (18\text{–}1)$$

This equation reduces to a simpler form if point P coincides with the mass center G for the body, in which case $\bar{x} = \bar{y} = 0$, and therefore

$$T = \frac{1}{2}mv_G^2 + \frac{1}{2}I_G\omega^2 \qquad (18\text{–}2)$$

Here I_G is the mass moment of inertia of the body about an axis which is perpendicular to the plane of motion and passes through the mass center. Both terms on the right side are always *positive*, since the velocities are squared. Furthermore, it may be verified that these terms have units of length times force, common units being m · N or ft · lb. Recall, however, that in the SI system the unit of energy is the joule (J), where 1 J = 1 m · N.

Translation

When a rigid body of mass m is subjected to either rectilinear or curvilinear *translation,* the kinetic energy due to rotation is zero, since $\boldsymbol{\omega} = \mathbf{0}$. From Eq. 18–2, the kinetic energy of the body is therefore

$$T = \tfrac{1}{2}mv_G^2$$

where v_G is the magnitude of the translational velocity at the instant considered, Fig. 18–2.

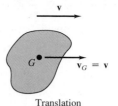

Translation

Fig. 18–2

Rotation About a Fixed Axis

When a rigid body is *rotating about a fixed axis* passing through point O, Fig. 18–3, the body has both *translational* and *rotational* kinetic energy as defined by Eq. 18–2, i.e.,

$$T = \tfrac{1}{2}mv_G^2 + \tfrac{1}{2}I_G\omega^2 \qquad (18\text{–}3)$$

The body's kinetic energy may be formulated in another manner by noting that $v_G = r_{G/O}\omega$, in which case $T = \tfrac{1}{2}(I_G + mr_{G/O}^2)\omega^2$. By the parallel-axis theorem, the terms inside the parentheses represent the moment of inertia I_O of the body about an axis perpendicular to the plane of motion and passing through point O. Hence,*

$$T = \tfrac{1}{2}I_O\omega^2 \qquad (18\text{–}4)$$

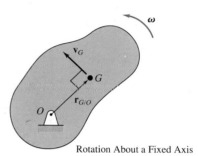

Rotation About a Fixed Axis

Fig. 18–3

From the derivation, this equation may be substituted for Eq. 18–3, since it accounts for *both* the translational kinetic energy of the body's mass center and the rotational kinetic energy of the body computed about the mass center.

General Plane Motion

When a rigid body is subjected to general plane motion, Fig. 18–4, it has an angular velocity $\boldsymbol{\omega}$ and its mass center has a velocity $\mathbf{v}_G$. Hence, the kinetic energy is defined by Eq. 18–2, i.e.,

$$T = \tfrac{1}{2}mv_G^2 + \tfrac{1}{2}I_G\omega^2 \qquad (18\text{–}5)$$

Here it is seen that the total kinetic energy of the body consists of the *scalar* sum of the body's *translational* kinetic energy, $\tfrac{1}{2}mv_G^2$, and *rotational* kinetic energy about its mass center, $\tfrac{1}{2}I_G\omega^2$.

Because energy is a scalar quantity, the total kinetic energy for a system of *connected* rigid bodies is the sum of the kinetic energies of all the moving parts. Depending upon the type of motion, the kinetic energy of *each body* is found by applying Eq. 18–2 or the alternative forms mentioned above.

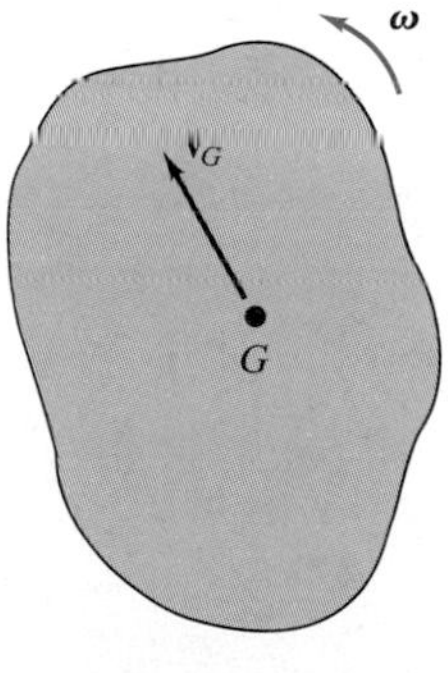

General plane motion

Fig. 18–4

*The similarity between this derivation and that of $\Sigma M_O = I_O\alpha$, Eq. 17–16, should be noted. Also note that the same result can be obtained from Eq. 18–1 by selecting point P at O, realizing that $\mathbf{v}_O = \mathbf{0}$.

Example 18–1

The system of three elements shown in Fig. 18–5a consists of a 6-kg block B, a 10-kg disk D, and a 12-kg cylinder C. A continuous cord of negligible mass is wrapped around the cylinder, passes over the disk, and is then attached to the block. If the block is moving downward with a speed of 0.8 m/s and the cylinder rolls without slipping, determine the total kinetic energy of the system at this instant.

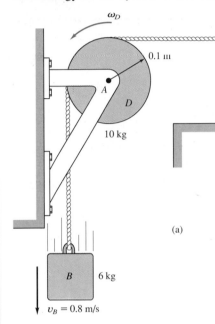

(a)

Fig. 18–5   (b)

SOLUTION

By inspection, the block is translating, the disk rotates about a fixed axis, and the cylinder has general plane motion. Hence, in order to compute the kinetic energy of the disk and cylinder, it is first necessary to determine ω_D, ω_C, and v_G, Fig. 18–5a. From the *kinematics* of the disk,

$$v_B = r_D\omega_D; \qquad 0.8 \text{ m/s} = (0.1 \text{ m})\omega_D; \qquad \omega_D = 8 \text{ rad/s}$$

Since the cylinder rolls without slipping, the instantaneous center of zero velocity is at the point of contact with the ground, Fig. 18–5b, hence,

$$v_E = r_{E/IC}\omega_C; \qquad 0.8 \text{ m/s} = (0.2 \text{ m})\omega_C; \qquad \omega_C = 4 \text{ rad/s}$$

$$v_G = r_{G/IC}\omega_C; \qquad v_G = (0.1 \text{ m})(4 \text{ rad/s}) = 0.4 \text{ m/s}$$

The kinetic energy of the block is

$$T_B = \tfrac{1}{2}m_B v_B^2; \quad T_B = \frac{1}{2}(6 \text{ kg})(0.8 \text{ m/s})^2 = 1.92 \text{ J}$$

The kinetic energy of the disk is

$$T_D = \tfrac{1}{2}I_D\omega_D^2; \quad T_D = \tfrac{1}{2}(\tfrac{1}{2}m_D r_D^2)\omega_D^2$$

$$= \frac{1}{2}\left[\frac{1}{2}(10 \text{ kg})(0.1 \text{ m})^2\right](8 \text{ rad/s})^2 = 1.60 \text{ J}$$

Finally, the kinetic energy of the cylinder is

$$T_C = \tfrac{1}{2}mv_G^2 + \tfrac{1}{2}I_G\omega_C^2; \quad T_C = \tfrac{1}{2}mv_G^2 + \tfrac{1}{2}(\tfrac{1}{2}m_C r_C^2)\omega_C^2$$

$$= \frac{1}{2}(12 \text{ kg})(0.4 \text{ m/s})^2 + \frac{1}{2}\left[\frac{1}{2}(12 \text{ kg})(0.1 \text{ m})^2\right](4 \text{ rad/s})^2 = 1.44 \text{ J}$$

The total kinetic energy of the system is therefore

$$T = T_B + T_D + T_C$$
$$= 1.92 \text{ J} + 1.60 \text{ J} + 1.44 \text{ J} = 4.96 \text{ J} \qquad \textit{Ans.}$$

18.2 The Work of a Force

Several types of forces are often encountered in planar kinetics problems involving a rigid body. The work of each of these forces has been presented in Sec. 14.1 and is listed below as a summary.

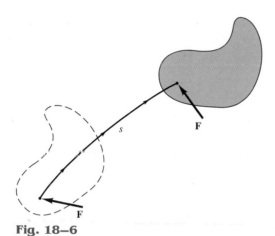

Fig. 18–6

Work of a Variable Force

If an external force **F** acts on a rigid body the work done by the force when it moves along the path s, Fig. 18–6, is defined as

$$U = \int_s F \cos \theta \, ds \qquad (18–6)$$

Here θ is the angle between the "tails" of the force vector and the differential displacement. In general, the integration must account for the variation of the force's direction and magnitude.

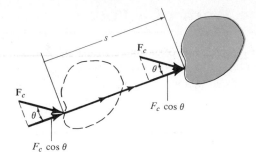

Fig. 18–7 $F_c \cos \theta$

Work of a Constant Force

If an external force $\mathbf{F}_c$ acts on a rigid body, Fig. 18–7, and maintains a constant magnitude F_c and constant direction θ, while the body undergoes a translation s, Eq. 18–6 can be integrated so that the work becomes

$$U_{F_c} = (F_c \cos \theta)s \qquad (18\text{–}7)$$

Here $F_c \cos \theta$ represents the magnitude of the component of force in the direction of displacement.

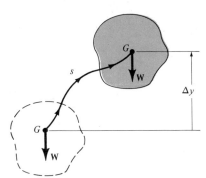

Fig. 18–8

Work of a Weight

The weight of a body does work only when the body's center of mass G undergoes a *vertical displacement* Δy. If this displacement is *upward,* Fig. 18–8, the work is negative, since the weight and displacement are in the opposite directions.

$$U_W = -W\,\Delta y \qquad (18\text{–}8)$$

Likewise, if the displacement is downward $(-\Delta y)$ the work becomes positive. Here the elevation change is considered to be small so that $\mathbf{W}$, which is caused by gravitation, is constant.*

---

*If the difference in elevation is large, use the results of Prob. 14–28.

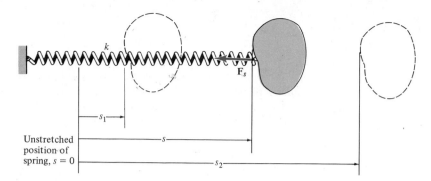

Fig. 18—9

Unstretched position of spring, $s = 0$

Work of a Spring Force

If a linear elastic spring is attached to a body, the spring force $F_s = ks$ *acting on the body* does work when the spring either stretches or compresses from s_1 to a *further* position s_2. In both cases the work will be *negative* since the *displacement of the body* is always in the opposite direction to the force, Fig. 18—9. The work done is

$$U_s = -(\tfrac{1}{2}ks_2^2 - \tfrac{1}{2}ks_1^2)$$

(18—9)

where $|s_2| > |s_1|$.

Forces That Do No Work

There are some external forces that do no work when the body is displaced. These forces can act either at *fixed points* on the body or they can have a direction *perpendicular to their displacement*. Examples include the reactions at a pin support about which a body rotates, the normal reaction acting on a body that moves along a fixed surface, and the weight of a body when the center of gravity of the body moves in a *horizontal plane*. A rolling resistance force $\mathbf{F}_r$ acting on a body as it *rolls without slipping* over a rough surface also does no work, Fig. 18—10.* This is because, during any *instant of time dt*, $\mathbf{F}_r$ acts at a point on the body which has *zero velocity* (instantaneous center, *IC*). In other words, for any differential rolling movement ds_{IC} of the body's *IC*, the work is $dU = F_r \, ds_{IC} = F_r(v_{IC}dt) = 0$, since $v_{IC} = 0$.

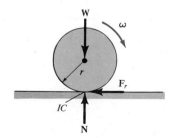

Fig. 18—10

---

*The work done by the frictional force *when the body slips* has been discussed in Sec. 14.3.

18.3 The Work of a Couple

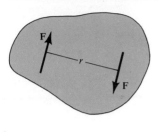

(a)

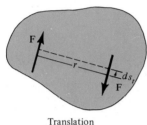

Translation
(b)

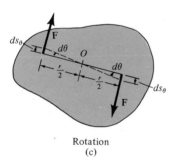

Rotation
(c)

Fig. 18–11

Recall that a *couple* consists of a pair of noncollinear forces which have equal magnitudes and opposite directions. When a body subjected to a couple undergoes general plane motion, the two forces do work *only* when the body undergoes a *rotation*. To show this, consider the body in Fig. 18–11*a*, which is subjected to a couple having a magnitude of $M = Fr$. Any general differential displacement of the body can be considered as a separate translation and rotation. When the body *translates* such that the *component of displacement* along the line of action of the forces is ds_t, Fig. 18–11*b*, clearly the "positive" work of one force *cancels* the "negative" work of the other. Consider now a differential rotation $d\theta$ of the body about an axis which is perpendicular to the plane of the couple and intersects the plane at point O, Fig. 18–11*c*. (For the derivation any other point in the plane may also be considered.) As shown, each force undergoes a displacement $ds_\theta = (r/2)\, d\theta$ in the direction of the force; hence, the total work done is

$$dU_M = F\left(\frac{r}{2}d\theta\right) + F\left(\frac{r}{2}d\theta\right) = (Fr)\, d\theta$$

$$= M\, d\theta$$

Here the line of action of $d\theta$ is parallel to the line of action of M. This is *always the case for general plane motion*, since $\mathbf{M}$ and $d\boldsymbol{\theta}$ are perpendicular to the plane of motion. Furthermore, the resultant work is *positive* when $\mathbf{M}$ and $d\boldsymbol{\theta}$ are in the *same direction* and *negative* if these vectors are in *opposite directions*.

When the body rotates in the plane through a finite angle θ measured in radians, from θ_1 to θ_2, the work of a couple is

$$U_M = \int_{\theta_1}^{\theta_2} M\, d\theta \qquad (18\text{–}10)$$

If the couple moment $\mathbf{M}$ has a *constant magnitude*, then

$$U_M = M(\theta_2 - \theta_1) \qquad (18\text{–}11)$$

Here the work is *positive* provided $\mathbf{M}$ and $(\boldsymbol{\theta}_2 - \boldsymbol{\theta}_1)$ are in the same direction.

18.4 Principle of Work and Energy

In Sec. 14.2 the principle of work and energy was developed for a particle. By applying this principle to each of the particles of a rigid body and adding the results algebraically, since energy is a scalar, the principle of work and energy for a rigid body may be developed. In this regard, the body's initial and final kinetic energies have been defined by the formulations in Sec. 18.1, and the work done by the *external* forces and couples is defined by formulations in Secs. 18.2 and 18.3. Note that the work of the body's *internal forces* does not have to be considered since the body is rigid. These forces occur in equal but opposite collinear pairs, so that when the body moves, the work of one force cancels that of its counterpart. Furthermore, since the body is rigid, *no relative movement* between these forces occurs, so that no internal work is done. Thus the principle of work and energy for a rigid body may be written as

$$T_1 + \Sigma U_{1-2} = T_2 \qquad\qquad (18\text{–}12)$$

This equation states that the body's initial translational *and* rotational kinetic energy plus the work done by all the external forces and couples acting on the body as the body moves from its initial to its final position is equal to the body's final translational *and* rotational kinetic energy.

When several rigid bodies are pin-connected, connected by inextensible cables, or in mesh with one another, this equation may be applied to the entire system of connected bodies. In all these cases the internal forces, which hold the various members together, do no work and hence are eliminated from the analysis.

PROCEDURE FOR ANALYSIS

The principle of work and energy is used to solve kinetics problems that involve *velocity, force,* and *displacement,* since these terms are involved in the formulation. For application, it is suggested that the following procedure be used.

Kinetic Energy (Kinematic Diagrams). Determine the kinetic-energy terms T_1 and T_2 by applying the equation $T = \frac{1}{2}mv_G^2 + \frac{1}{2}I_G\omega^2$ or an appropriate form of this equation developed in Sec. 18.1. In this regard, *kinematic diagrams* for velocity may be useful for determining v_G and ω, or for establishing a *relationship* between v_G and ω.*

Work (Free-Body Diagram). Draw a free-body diagram of the body when it is located at an intermediate point along the path, in order to account for all the forces and couples which do work on the body as it moves along the path. The work of each force and couple can be computed using the appropriate formulations outlined in Secs. 18.2 and 18.3. Since *algebraic addition* of the work terms is required, it is important that the proper sign of each term be specified. Specifically, work is *positive* when the force (couple) is in the *same direction* as its displacement (rotation); otherwise, it is negative.

Principle of Work and Energy. Apply the principle of work and energy, $T_1 + \Sigma U_{1-2} = T_2$. Since this is a scalar equation, it can be used to solve for only one unknown when it is applied to a single rigid body. This is in contrast to the three scalar equations of motion, Eqs. 17–11, which may be written for the same body.

The following example problems numerically illustrate application of this procedure.

* A brief review of Secs. 16.5 to 16.7 may prove helpful in solving problems, since computations for kinetic energy require a kinematic analysis of velocity.

Example 18–2

The 30-kg disk shown in Fig. 18–12a is pin-supported at its center. Determine the number of revolutions it must make to attain an angular velocity of 20 rad/s starting from rest. It is acted upon by a constant force $F = 10$ N, which is applied to a cord wrapped around its periphery, and a constant couple moment $M = 5$ N · m. Neglect the mass of the cord in the calculation.

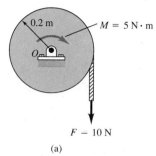

$F = 10$ N

Fig. 18–12

(a)

SOLUTION

Kinetic Energy. Since the disk rotates about a fixed axis, the kinetic energy can be computed using $T = \frac{1}{2}I_O\omega^2$, where the moment of inertia is $I_O = \frac{1}{2}mr^2$. Initially, the disk is at rest, so that

$$T_1 = 0$$

$$T_2 = \tfrac{1}{2}I_O\omega_2^2 = \frac{1}{2}\left[\frac{1}{2}(30)(0.2)^2\right](20)^2 = 120 \text{ J}$$

Work (Free-Body Diagram). As shown in Fig. 18–12b, the pin reactions $\mathbf{O}_x$ and $\mathbf{O}_y$ and the weight (294.3 N) do no work, since they are not displaced. The *couple moment,* having a constant magnitude, does positive work $U_M = M\theta$ as the disk *rotates* through a clockwise angle of θ rad, and the *constant force* $\mathbf{F}$ does positive work $U_{F_c} = Fs$ as the cord *moves* downward $s = \theta r = \theta(0.2)$ m.

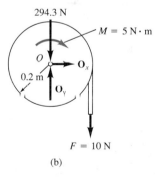

294.3 N

$M = 5$ N · m

O

$\mathbf{O}_x$

0.2 m

$\mathbf{O}_y$

$F = 10$ N

(b)

Principle of Work and Energy

$$\{T_1\} + \{\Sigma U_{1-2}\} = \{T_2\}$$

$$\{T_1\} + \{M\theta + Fs\} = \{T_2\}$$

$$\{0\} + \{5\theta + (10)\theta(0.2)\} = \{120\}$$

$$\theta = 17.1 \text{ rad} = 17.1 \text{ rad}\left(\frac{1 \text{ rev}}{2\pi \text{ rad}}\right) = 2.73 \text{ rev} \qquad Ans.$$

This problem has also been solved in Example 17–9. Compare the two methods of solution and note that since force, velocity, and displacement θ are involved, a work-energy approach yields a more direct solution.

Example 18–3

The uniform 5-kg bar shown in Fig. 18–13a is acted upon by the 30-N force which always acts perpendicular to the bar as shown. If the bar has an initial clockwise angular velocity $\omega_1 = 10$ rad/s when $\theta = 0°$, determine its angular velocity at the instant $\theta = 90°$.

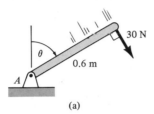

30 N

θ

0.6 m

A

(a)

Fig. 18–13

SOLUTION

Kinetic Energy (Kinematic Diagrams). Two kinematic diagrams of the bar when $\theta = 0°$ (position 1) and $\theta = 90°$ (position 2) are shown in Fig. 18–13b. The initial kinetic energy may be computed with reference to either the fixed point of rotation A or the center of mass G. If A is considered, then

$$T_1 = \tfrac{1}{2}I_A\omega_1^2 = \frac{1}{2}\left[\frac{1}{3}(5)(0.6)^2\right](10)^2 = 30 \text{ J}$$

If point G is considered, then

$$T_1 = \tfrac{1}{2}m(v_G)_1^2 + \tfrac{1}{2}I_G\omega_1^2 = \frac{1}{2}(5)(3)^2 + \frac{1}{2}\left[\frac{1}{12}(5)(0.6)^2\right](10)^2 = 30 \text{ J}$$

In the final position

$$T_2 = \frac{1}{2}\left[\frac{1}{3}(5)(0.6)^2\right]\omega_2^2 = 0.3\omega_2^2$$

Work (Free-Body Diagram). Fig. 18–13c. The reactions $\mathbf{A}_x$ and $\mathbf{A}_y$ do no work, since these forces do not move. The 49.05-N weight, centered at G, moves downward through a vertical distance $\Delta y = 0.3$ m. The 30-N force moves tangent to its path of length $\tfrac{1}{2}\pi(0.6)$ m.

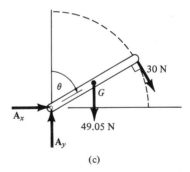

(b)

30 N

θ G

A_x

A_y 49.05 N

(c)

Principle of Work and Energy

$$\{T_1\} + \{\Sigma U_{1-2}\} = \{T_2\}$$
$$\{30\} + \{49.05(0.3) + 30[\tfrac{1}{2}\pi(0.6)]\} = \{0.3\omega_2^2\}$$
$$\omega_2 = 15.6 \text{ rad/s} \;\downarrow \qquad\qquad Ans.$$

Example 18–4

The wheel shown in Fig. 18–14a weighs 40 lb and has a radius of gyration $k_G = 0.6$ ft about its mass center G. If it is subjected to a clockwise couple moment of 15 lb · ft and rolls from rest without slipping, determine its angular velocity after its center G moves 0.5 ft. The spring has a stiffness $k = 10$ lb/ft and is initially unstretched when the couple is applied.

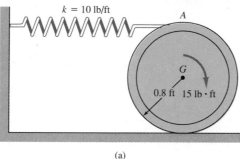

$k = 10$ lb/ft

A

G

0.8 ft 15 lb · ft

Fig. 18–14

(a)

SOLUTION

Kinetic Energy (Kinematic Diagram). Since the wheel is initially at rest,

$$T_1 = 0$$

The kinematic diagram of the wheel when it is in the final position is shown in Fig. 18–14b. The velocity of the mass center $(v_G)_2$ can be related to the angular velocity ω_2 by using the instantaneous center of zero velocity (IC), i.e., $(v_G)_2 = 0.8\omega_2$. Hence, the final kinetic energy is

$$T_2 = \tfrac{1}{2}m(v_G)_2^2 + \tfrac{1}{2}I_G(\omega_2)^2$$

$$= \frac{1}{2}\left(\frac{40}{32.2}\right)(0.8\omega_2)^2 + \frac{1}{2}\left[\frac{40}{32.2}(0.6)^2\right](\omega_2)^2$$

$$= 0.621(\omega_2)^2$$

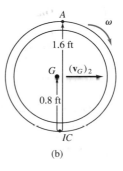

A

ω

1.6 ft

G $(v_G)_2$

0.8 ft

IC

(b)

Work (Free-Body Diagram). As shown in Fig. 18–14c, only the spring force $\mathbf{F}_s$ and the couple moment do work. The normal force does not move along its line of action and the frictional force does *no work*, since the wheel does not slip as it rolls.

The work of $\mathbf{F}_s$ may be computed using $U_s = -\tfrac{1}{2}ks^2$. (Why is the work negative?) Since the wheel does not slip when the center G moves 0.5 ft, the wheel rotates $\theta = s_G/r_{G/IC} = 0.5/0.8 = 0.625$ rad, Fig. 18–14b. Hence, the spring stretches $s_A = \theta\, r_{A/IC} = 0.625(1.6) = 1$ ft.

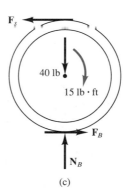

$\mathbf{F}_s$

40 lb

15 lb · ft

$\mathbf{F}_B$

$\mathbf{N}_B$

(c)

Principle of Work and Energy

$$\{T_1\} + \{\Sigma U_{1-2}\} = \{T_2\}$$

$$\{T_1\} + \{M\theta - \tfrac{1}{2}ks^2\} = \{T_2\}$$

$$\{0\} + \left\{15(0.625) - \frac{1}{2}(10)(1)^2\right\} = \{0.621(\omega_2)^2\}$$

$$\omega_2 = 2.65 \text{ rad/s} \downarrow \qquad \textit{Ans.}$$

Example 18–5

The 10-kg rod shown in Fig. 18–15a is constrained so that its ends move along the grooved slots. The rod is initially at rest when $\theta = 0°$. If the slider block at B is acted upon by a horizontal force $P = 50$ N, determine the angular velocity of the rod at the instant $\theta = 45°$. Neglect the mass of blocks A and B. (Why can the principle of work and energy be used to solve this problem?)

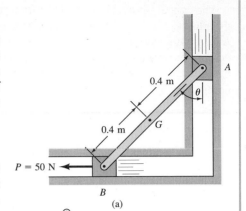

(a)

SOLUTION

Kinetic Energy (Kinematic Diagrams). Two kinematic diagrams of the rod, when it is in the initial position 1 and final position 2, are shown in Fig. 18–15b. When the rod is in position 1, $T_1 = 0$ since $(\mathbf{v}_G)_1 = \boldsymbol{\omega}_1 = \mathbf{0}$. In position 2 the angular velocity is ω_2 and the velocity of the mass center is $(\mathbf{v}_G)_2$. Hence, the kinetic energy is

$$
\begin{aligned}
T_2 &= \tfrac{1}{2}m(v_G)_2^2 + \tfrac{1}{2}I_G(\omega_2)^2 \\
&= \frac{1}{2}(10)(v_G)_2^2 + \frac{1}{2}\left[\frac{1}{12}(10)(0.8)^2\right](\omega_2)^2 \\
&= 5(v_G)_2^2 + 0.267(\omega_2)^2
\end{aligned}
\tag{1}
$$

The two unknowns $(v_G)_2$ and ω_2 may be related via the instantaneous center of zero velocity for the rod, Fig. 18–15b. It is seen that as A moves downward with a velocity $(\mathbf{v}_A)_2$, B moves horizontally to the left with a velocity $(\mathbf{v}_B)_2$. Knowing these directions, the IC may be determined as shown in the figure. Hence,

$$
\begin{aligned}
(v_G)_2 &= r_{G/IC}\omega_2 = (0.4 \tan 45°)\omega_2 \\
&= 0.4\omega_2
\end{aligned}
$$

Substituting into Eq. (1), we have

$$
T_2 = 5(0.4\omega_2)^2 + 0.267(\omega_2)^2 = 1.067(\omega_2)^2
$$

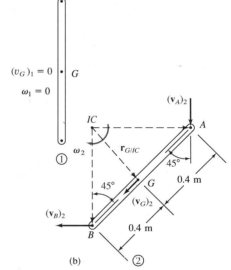

(b)

Work (Free-Body Diagram). Fig. 18–15c. The normal forces N_A and N_B do no work as the rod is displaced. Why? The 98.1-N weight is displaced a vertical distance of $\Delta y = (0.4 - 0.4 \cos 45°)$ m; whereas the 50-N force moves a horizontal distance of $s = (0.8 \sin 45°)$ m.

Principle of Work and Energy

$$
\{T_1\} + \{\Sigma U_{1-2}\} = \{T_2\}
$$
$$
\{T_1\} + \{W\Delta y + Ps\} = \{T_2\}
$$
$$
\{0\} + \{98.1(0.4 - 0.4 \cos 45°) + 50(0.8 \sin 45°)\} = \{1.067(\omega_2)^2\}
$$

Solving for ω_2 gives

$$
\omega_2 = 6.11 \text{ rad/s} \; \downarrow \qquad\qquad\qquad \textit{Ans.}
$$

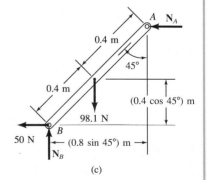

(c)

Fig. 18–15

PROBLEMS

18–1. Solve Prob. 17–44 using the principle of work and energy.

18–2. Solve Prob. 17–57 using the principle of work and energy.

18–3. Solve Prob. 17–58 using the principle of work and energy.

*18–4. At a given instant the body of mass m has an angular velocity ω and its mass center has a velocity v_G. Show that its kinetic energy can be represented as $T = \frac{1}{2} I_{IC} \omega^2$, where I_{IC} is the moment of inertia of the body computed about the instantaneous axis of zero velocity, located at a distance $r_{G/IC}$ from the mass center as shown.

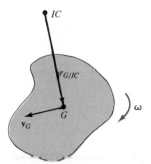

Prob. 18–4

18–5. The wheel is made from a 5-kg thin ring and two 2-kg slender rods. If the torsional spring attached to the wheel's center has a stiffness of $k = 2$ N · m/rad, so that the torque on the center of the wheel is $M = (2\theta)$ N · m, where θ is in radians, determine the maximum angular velocity of the wheel if it is rotated two revolutions and released from rest.

Prob. 18–5

18–6. The 4-kg slender rod is subjected to the force and couple moment. When it is in the position shown it has an angular velocity of $\omega_1 = 6$ rad/s. Determine its angular velocity at the instant it has rotated downward 90°. The force is always applied perpendicular to the axis of the rod. Motion occurs in the vertical plane.

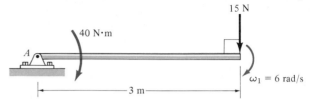

Prob. 18–6

18–7. The soap-box car has a weight of 110 lb, including the passenger but *excluding* its four wheels. Each wheel has a weight of 5 lb and a radius of gyration $k = 0.3$ ft, computed about an axis passing through the wheel's axle. Determine the car's speed after it has traveled 100 ft starting from rest. The wheels roll without slipping. Neglect air resistance.

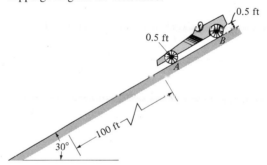

Prob. 18–7

*18–8. The pendulum consists of two slender rods each having a density of 4 kg/m. If it is acted upon by a moment of $M = 50$ N · m and released from the position shown, determine its angular velocity when it has rotated (a) 90° and (b) 180°. Motion occurs in the vertical plane.

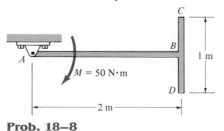

Prob. 18–8

18–9. The revolving door consists of four doors which are attached to an axle AB. Each door can be assumed to be a 50-lb thin plate. Friction at the axle contributes a resultant moment of 2 lb · ft which resists the rotation of the doors. If a man passes through one door by always pushing with a force $P = 15$ lb perpendicular to the plane of the door as shown, determine the door's angular velocity after it has rotated 90°. The doors are originally at rest.

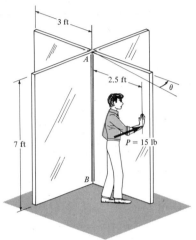

Prob. 18–9

18–10. The linkage consists of two 8-lb rods AB and CD and a 10-lb bar AD. When $\theta = 0°$, rod AB is rotating with an angular velocity $\omega_{AB} = 2$ rad/s. If rod CD is subjected to a couple moment $M = 15$ lb · ft and bar AD is subjected to a horizontal force $P = 20$ lb as shown, determine ω_{AB} at the instant $\theta = 90°$.

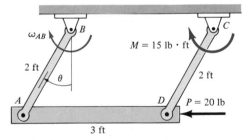

Prob. 18–10

18–11. The 20-kg disk is originally at rest and the spring holds it in equilibrium. A couple moment of $M = 30$ N · m is then applied to the disk as shown. Determine its angular velocity at the instant its mass center G has moved $s = 0.8$ m down along the inclined plane. The disk rolls without slipping.

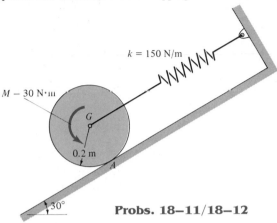

Probs. 18–11/18–12

*18–12. The 20-kg disk is originally at rest and the spring holds it in equilibrium. A couple moment of $M = 30$ N · m is then applied to the disk as shown. Determine how far s the center of mass of the disk travels down along the incline, measured from the equilibrium position, before it stops. The disk rolls without slipping.

18–13. The drum has a mass of 50 kg and a radius of gyration about the pin at O of $k_O = 0.23$ m. Starting from rest, the suspended 15-kg block B is allowed to fall 3 m without applying the brake ACD. Determine the speed of the block at this instant. If the coefficient of kinetic friction at the brake pad C is $\mu_k = 0.5$, determine the force P that must be applied at the brake handle which will then stop the block after it descends *another* 3 m. Neglect the thickness of the handle.

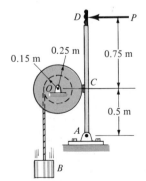

Probs. 18–13/18–14

18–14. The drum has a mass of 50 kg and a radius of gyration about the pin at O of $k_O = 0.23$ m. If the 15-kg block is moving downward at 3 m/s, and a force of $P = 100$ N is applied to the brake arm, determine how far the block descends from the instant the brake is applied until it stops. Neglect the thickness of the handle. The coefficient of kinetic friction at the brake pad is $\mu_k = 0.5$.

18–15. The spool has a weight of 500 lb and a radius of gyration of $k_G = 1.75$ ft. A horizontal force of $P = 15$ lb is applied to a cable wrapped around its inner core. If the spool is originally at rest, determine its angular velocity after the mass center G has moved $s = 6$ ft. The spool rolls without slipping. Neglect the mass of the cable.

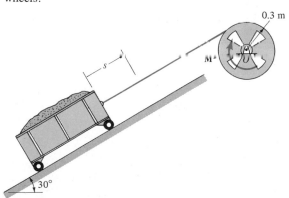

Prob. 18–15

*18–16. The wheel has a mass of 100 kg and a radius of gyration $k_O = 0.40$ m. A motor supplies a torque $M = (40\theta + 900)$ N · m, where θ is in radians, about the drive shaft at O. Determine the speed of the loading car, which has a mass of 300 kg, after it travels $s = 4$ m. Initially the car is at rest when $s = 0$ and $\theta = 0$. Neglect the mass of the attached cable and the mass of the car's wheels.

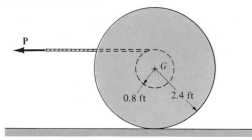

Prob. 18–16

18–17. The spool has a mass of 60 kg and a radius of gyration $k_G = 0.3$ m. If it is released from rest, determine how far it descends down the smooth plane before it attains an angular velocity of $\omega = 6$ rad/s. Neglect friction and the mass of the cord which is wound around the central core.

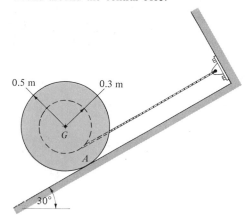

Probs. 18–17/18–18

18–18. Solve Prob. 18–17 if the plane is rough, such that the coefficient of kinetic friction at A is $\mu_A = 0.2$.

18–19. The drum is pinned at its center O and is subjected to a moment of $M = 500$ N · m. If it has a mass of 80 kg and radius of gyration of $k_O = 0.58$ m, determine the speed of the 130-kg car when it has moved up the plane 10 m. The car starts from rest. Neglect the mass of both the pulley at A and the car's wheels.

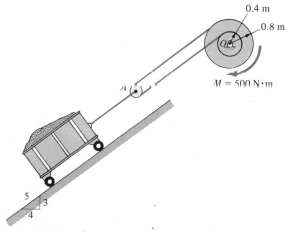

Prob. 18–19

*18–20. The assembly consists of two 15-lb slender rods and a 20-lb disk. If the spring is unstretched when $\theta = 45°$ and the assembly is released from rest at this position, determine the angular velocity of rod AB at the instant $\theta = 0°$. The disk rolls without slipping.

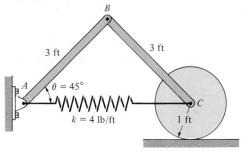

Prob. 18–20

18–21. The 50-lb crate C is lifted using the hoist shown. A motor delivers 50 lb · ft of torque to the pinion gear D, which is in mesh with the larger gear A, and the attached drum over which the cable winds. If A has a weight of 120 lb and a radius of gyration $k_A = 0.6$ ft, determine the velocity of the crate when it has risen 5 ft starting from rest. Neglect the mass of the pinion gear and the cable and pulley at B.

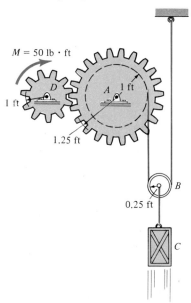

Probs. 18–20/18–21

18–22. Solve Prob. 18–21 assuming that pulley B is a disk having a weight of 20 lb.

18–23. The uniform slender bar has a mass m and a length L. It is subjected to a uniform distributed load w_0 which is always directed perpendicular to the axis of the bar. If it is released from rest from the position shown, determine its angular velocity at the instant it has rotated 90°. Solve the problem for rotation in (a) the horizontal plane, and (b) the vertical plane.

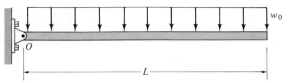

Probs. 18–23/18–24

*18–24. Solve Prob. 18–23 if the distributed load is triangular, which varies from zero at the pin O to w_0 at its other end.

18–25. The two 2-kg gears A and B are attached to the ends of a 3-kg slender bar. The gears roll within the fixed ring gear C, which lies in the horizontal plane. If a 10-N · m torque is applied to the center of the bar as shown, determine the number of revolutions the bar must rotate starting from rest in order for it to have an angular velocity of $\omega_{AB} = 20$ rad/s. For the calculation, assume the gears can be approximated by thin disks. What is the result if the gears lie in the vertical plane?

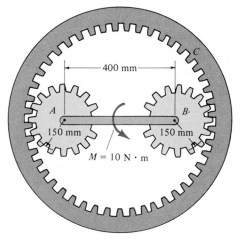

Prob. 18–25

18–26. The two bars are released from rest at the position θ. Determine their angular velocities at the instant they become horizontal. Neglect the mass of the roller at C. Each bar has a mass m and length L.

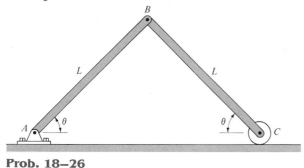

Prob. 18–26

18–27. A ball of mass m and radius r is cast onto the horizontal surface such that it rolls without slipping. Determine the required speed v_G of its mass center G so that it rolls completely around the loop of radius $R + r$ without leaving the track.

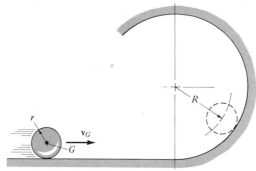

Prob. 18–27

18.5 Conservation of Energy

When a force system acting on a rigid body consists only of *conservative forces*, the conservation of energy theorem may be used to solve a problem which otherwise would be solved using the principle of work and energy. This theorem is often easier to apply since the work of a conservative force is *independent of the path* and only depends upon the initial and final positions of the body. It was shown in Sec. 14.5 that the work of a conservative force may be expressed as the difference in the body's potential energy measured from an arbitrarily selected reference or datum.

Gravitational Potential Energy

Since the total weight of a body can be considered concentrated at its center of gravity, the *gravitational potential energy* of the body is determined by knowing the height of the body's center of gravity above or below a horizontal datum. Measuring y_G as *positive upward*, the gravitational potential energy of the body is thus

$$V_g = W\, y_G \qquad (18\text{–}13)$$

Here the potential energy is *positive* when y_G is positive, since the weight has the ability to do *positive work* when the body is moved back to the datum, Fig. 18–16. Likewise, if the body is located *below* the datum ($-y_G$), the gravitational potential energy is *negative,* since the weight does *negative work* when the body is moved back to the datum.

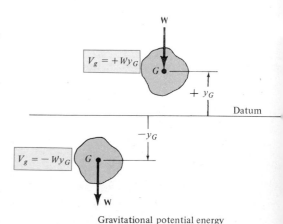

Gravitational potential energy

Fig. 18–16

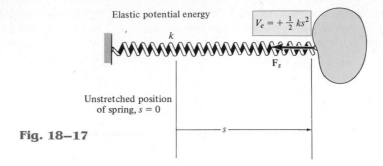

Fig. 18–17

Elastic Potential Energy

The force developed by an elastic spring is also a conservative force. The *elastic potential energy* which a spring imparts to an attached body when the spring is elongated or compressed from an initial undeformed position ($s = 0$) to a final position s, Fig. 18–17, is

$$V_e = +\tfrac{1}{2}ks^2 \qquad (18\text{–}14)$$

In the deformed position, the spring force acting *on the body* always has the capacity for doing positive work when the spring is returned back to its original undeformed position (see Sec. 14.5).

Conservation of Energy

In general, if a body is subjected to both gravitational and elastic forces, the total *potential energy* is expressed as a potential function V represented as the algebraic sum

$$V = V_g + V_e \qquad (18\text{–}15)$$

Here measurement of V depends upon the location of the body with respect to a selected datum in accordance with Eqs. 18–13 and 18–14.

Realizing that the work of conservative forces can be written as a difference in their potential energies, i.e., $(\Sigma U_{1-2})_{\text{cons}} = V_1 - V_2$, Eq. 14–16, we can rewrite the principle of work and energy for a rigid body as

$$T_1 + V_1 + (\Sigma U_{1-2})_{\text{noncons}} = T_2 + V_2 \qquad (18\text{–}16)$$

Here $(\Sigma U_{1-2})_{\text{noncons}}$ represents the work of the nonconservative forces, such as friction, acting on the body. If this term is zero, then

$$T_1 + V_1 = T_2 + V_2 \qquad (18\text{–}17)$$

This is the equation of the conservation of mechanical energy for the body. It states that the *sum* of the potential and kinetic energies of the body remains *constant* when the body moves from one position to another. It also applies to a system of smooth, pin-connected rigid bodies, bodies connected by inextensible cords, and bodies in mesh with other bodies. In all these cases the forces acting at the points of contact are *eliminated* from the analysis, since they occur in equal and opposite collinear pairs and each pair of forces moves through an equal distance when the system undergoes a displacement.

PROCEDURE FOR ANALYSIS

The conservation of energy equation is used to solve problems involving *velocity, displacement,* and *conservative force systems.* For application it is suggested that the following procedure be used.

Potential Energy. Draw two diagrams showing the body located at its initial and final positions along the path. If the center of mass of the body is subjected to a *vertical displacement,* determine where to establish the fixed horizontal datum from which to measure the body's gravitational potential energy V_g. Although the location of the datum is arbitrary, it is advantageous to place it through G when the body is either at its initial or final position, since at the datum $V_g = 0$. Data pertaining to the elevation y of the body's mass center from the datum and the extension or compression of any connecting springs can be determined from the geometry associated with the two diagrams. Recall that the potential energy $V = V_g + V_e$. Here $V_g = Wy_G$, where y_G is positive upward from the datum, and $V_e = \frac{1}{2}ks^2$, which is always positive.

Kinetic Energy. The kinetic-energy terms T_1 and T_2 are determined from $T = \frac{1}{2}mv_G^2 + \frac{1}{2}I_G\omega^2$ or an appropriate form of this equation as developed in Sec. 18.1. In this regard, kinematic diagrams for velocity may be useful for determining v_G and ω, or for establishing a *relationship* between these quantities.

Conservation of Energy. Apply the conservation of energy equation $T_1 + V_1 = T_2 + V_2$.

It is important to remember that *only problems involving conservative force systems may be solved by using this equation.* As stated in Sec. 14.5, friction or other drag-resistant forces, which depend upon velocity or acceleration, are nonconservative. The work of such forces is transformed into thermal energy used to heat up the surfaces of contact, and consequently this energy is dissipated into the surroundings and may not be recovered. Therefore, problems involving frictional forces can either be solved by using the principle of work and energy written in the form of Eq. 18–16, if it applies, or the equations of motion.

The following example problems numerically illustrate application of the above procedure.

Example 18–6

The 10-kg rod AB shown in Fig. 18–18a is confined so that its ends move in the horizontal and vertical slots. The spring has a stiffness of $k = 800$ N/m and is unstretched when $\theta = 0°$. Determine the angular velocity of AB when $\theta = 0°$, if AB is released from rest when $\theta = 30°$. Neglect the mass of the slider blocks.

SOLUTION

Potential Energy. The two diagrams of the rod, when it is located at its initial and final positions, are shown in Fig. 18–18b. The datum, used to measure the gravitational potential energy, is placed in line with the rod when $\theta = 0°$.

When the rod is in position 1, the center of gravity G is located *below the datum* so that the gravitational potential energy is *negative*. Furthermore, (positive) elastic potential energy is stored in the spring, since it is stretched a distance of $s_1 = (0.4 \sin 30°)$ m. Thus,

$$V_1 = -Wy_1 + \tfrac{1}{2}ks_1^2$$

$$= -98.1(0.2 \sin 30°) + \frac{1}{2}(800)(0.4 \sin 30°)^2 = 6.19 \text{ J}$$

When the rod is in position 2, the potential energy of the rod is zero, since the spring is unstretched, $s_2 = 0$, and the center of gravity G is located at the datum. Thus,

$$V_2 = 0$$

Kinetic Energy. The rod is released from rest from position 1, thus $(v_G)_1 = 0$ and $\omega_1 = 0$, and

$$T_1 = 0$$

In position 2 the angular velocity is ω_2 and the rod's mass center has a velocity of $(v_G)_2$. Using *kinematics*, $(v_G)_2$ can be related to ω_2 as shown in Fig. 18–18c. At the instant considered, the instantaneous center of zero velocity (*IC*) for the rod is at point A; hence, $(v_G)_2 = (r_{G/IC})\omega_2 = (0.2)\omega_2$. Thus,

$$T_2 = \tfrac{1}{2}m(v_G)_2^2 + \tfrac{1}{2}I_G(\omega_2)^2$$

$$= \frac{1}{2}(10)(0.2\omega_2)^2 + \frac{1}{2}\left[\frac{1}{12}(10)(0.4)^2\right](\omega_2)^2 = 0.267\omega_2^2$$

Conservation of Energy

$$\{T_1\} + \{V_1\} = \{T_2\} + \{V_2\}$$

$$\{0\} + \{6.19\} = \{0.267\omega_2^2\} + \{0\}$$

$$\omega_2 = 4.82 \text{ rad/s} \;\rotatebox{0}{$\curvearrowleft$}$$

Ans. **Fig. 18–18**

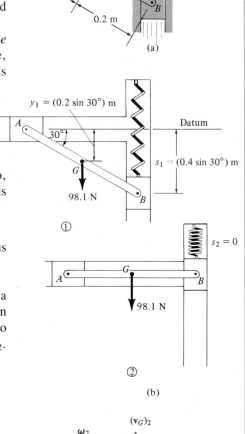

Example 18–7

The disk shown in Fig. 18–19a has a weight of 30 lb and is attached to a spring which has a stiffness $k = 2$ lb/ft and an unstretched length of 1 ft. If the disk is released from rest in the position shown and rolls without slipping, determine its angular velocity at the instant it is displaced 3 ft.

SOLUTION

Potential Energy. Two diagrams of the disk, when it is located in its initial and final positions, are shown in Fig. 18–19b. A gravitational datum is not needed here since the weight is not displaced vertically. From the problem geometry the spring is stretched $s_1 = (\sqrt{3^2 + 4^2} - 1) = 4$ ft and $s_2 = (4 - 1) = 3$ ft in the initial and final positions, respectively. Hence,

$$V_1 = \tfrac{1}{2}ks_1^2 = \frac{1}{2}(2)(4)^2 = 16 \text{ J}$$

$$V_2 = \tfrac{1}{2}ks_2^2 = \frac{1}{2}(2)(3)^2 = 9 \text{ J}$$

Kinetic Energy. The disk is released from rest so that $(\mathbf{v}_G)_1 = \mathbf{0}$, $\boldsymbol{\omega}_1 = \mathbf{0}$, and

$$T_1 = 0$$

Since the disk rolls without slipping, $(\mathbf{v}_G)_2$ can be related to $\boldsymbol{\omega}_2$ from the instantaneous center of zero velocity, Fig. 18–19c. Hence, $(v_G)_2 = 0.75\omega_2$. Thus,

$$
\begin{aligned}
T_2 &= \tfrac{1}{2}m(v_G)_2^2 + \tfrac{1}{2}I_G(\omega_2)^2 \\
&= \frac{1}{2}\left(\frac{30}{32.2}\right)(0.75\omega_2)^2 + \frac{1}{2}\left[\frac{1}{2}\left(\frac{30}{32.2}\right)(0.75)^2\right](\omega_2)^2 \\
&= 0.393(\omega_2)^2
\end{aligned}
$$

Conservation of Energy

$$\{T_1\} + \{V_1\} = \{T_2\} + \{V_2\}$$
$$\{0\} + \{16\} = \{0.393(\omega_2)^2\} + \{9\}$$
$$\omega_2 = 4.22 \text{ rad/s} \,\big\uparrow \qquad\qquad Ans.$$

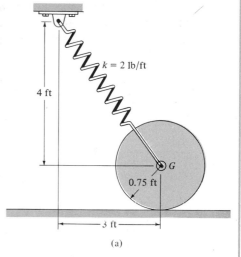

(a)

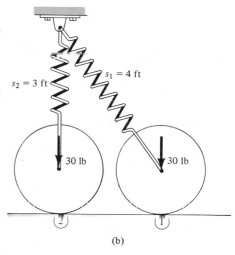

(b)

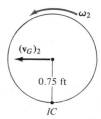

Fig. 18–19

Example 18–8

The 10-kg homogeneous disk shown in Fig. 18–20a is attached to a uniform 5-kg rod AB. If the assembly is released from rest when $\theta = 60°$, determine the angular velocity of the rod when $\theta = 0°$. Assume that the disk rolls without slipping. Neglect friction along the guide and the mass of the collar at B.

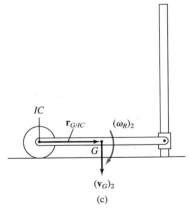

(a)

SOLUTION

Potential Energy. Two diagrams for the rod and disk, when they are located at their initial and final positions, are shown in Fig. 18–20b. For convenience the datum, which is horizontally fixed, passes through point A.

When the system is in position 1, the rod's weight has positive potential energy. Thus,

$$V_1 = W_R y_1 = 49.05(0.3 \sin 60°) = 12.74 \text{ J}$$

When the system is in position 2, both the weight of the rod and the weight of the disk have zero potential energy. Why? Thus,

$$V_2 = 0$$

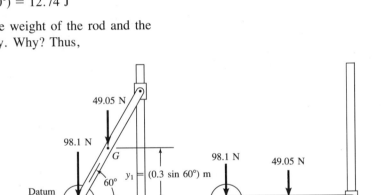

Kinetic Energy. Since the entire system is at rest in the initial position,

$$T_1 = 0$$

(b)

In the final position the rod has an angular velocity $(\omega_R)_2$ and its mass center has a velocity $(v_G)_2$, Fig. 18–20c. Since the rod is *fully extended* in this position, the disk is momentarily at rest, so $(\omega_D)_2 = \mathbf{0}$ and $(v_A)_2 = \mathbf{0}$. For the rod $(v_G)_2$ can be related to $(\omega_R)_2$ from the instantaneous center of zero velocity, which is located at point A, Fig. 18–20c. Hence, $(v_G)_2 = r_{G/IC}(\omega_R)_2$ or $(v_G)_2 = 0.3(\omega_R)_2$. Thus,

$$T_2 = \tfrac{1}{2}m_R(v_G)_2^2 + \tfrac{1}{2}I_G(\omega_R)_2^2 + \tfrac{1}{2}m_D(v_A)_2^2 + \tfrac{1}{2}I_A(\omega_D)_2^2$$

$$= \frac{1}{2}(5)(0.3(\omega_R)_2)^2 + \frac{1}{2}\left[\frac{1}{12}(5)(0.6)^2\right](\omega_R)_2^2 + 0 + 0$$

$$= 0.3(\omega_R)_2^2$$

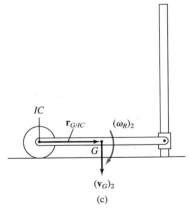

IC

(c)

Conservation of Energy

$$\{T_1\} + \{V_1\} = \{T_2\} + \{V_2\}$$

$$\{0\} + \{12.74\} = \{0.3(\omega_R)_2^2\} + \{0\}$$

$$(\omega_R)_2 = 6.52 \text{ rad/s} \;\downarrow \qquad\qquad \text{Ans.}$$

Fig. 18–20

PROBLEMS

*18–28. Solve Prob. 18–7 using the conservation of energy equation.

18–29. Solve Prob. 18–17 using the conservation of energy equation.

18–30. Solve Prob. 18–26 using the conservation of energy equation.

18–31. Solve Prob. 18–27 using the conservation of energy equation.

*18–32. Solve Prob. 18–20 using the conservation of energy equation.

18–33. The spool has a mass of 50 kg and a radius of gyration $k_O = 0.280$ m. If the 20-kg block A is released from rest, determine its velocity just after it descends 2 m. Neglect the mass of the cable.

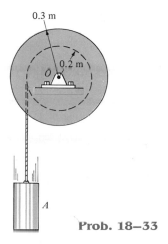

Prob. 18–33

18–34. An automobile tire has a mass of 7 kg and radius of gyration $k_G = 0.3$ m. If it is released from rest at A on the incline, determine its angular velocity when it reaches the horizontal plane. The tire rolls without slipping.

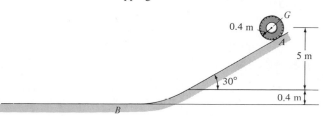

Prob. 18–34

18–35. The jeep and passenger have a total weight of 1600 lb, not including the four wheels. Each wheel has a weight of 20 lb and a radius of gyration $k_G = 0.4$ ft about an axis passing through its center. If the jeep starts from rest and coasts down the hill in neutral gear, determine the distance it must travel to attain a speed of 15 ft/s. The tires do not slip on the ground.

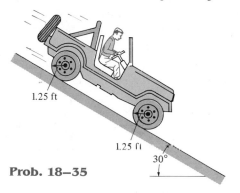

Prob. 18–35

*18–36. When the slender 10-kg bar AB is horizontal it is at rest and the spring is unstretched. Determine the bar's angular velocity when it is released and has rotated downwards 90°.

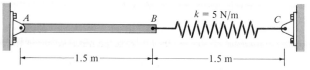

Prob. 18–36

18–37. The 50-lb wheel has a radius of gyration about its mass center G of $k_G = 0.7$ ft. If it rolls without slipping, determine its angular velocity when it has rotated clockwise 90° from the position shown. The spring AB has a stiffness of $k = 1.20$ lb/ft and an unstretched length of 0.5 ft. The wheel is released from rest.

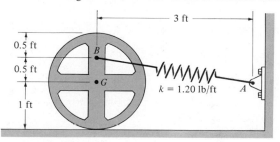

Prob. 18–37

18–38. The 80-lb cylinder is attached to the 10-lb slender rod which is pinned from point A. At the instant $\theta = 30°$ it has an angular velocity of $\omega_1 = 1$ rad/s as shown. Determine the largest angle θ to which the rod swings before it momentarily stops.

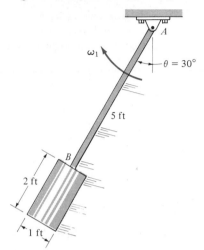

Prob. 18–38

18–39. The 80-lb cylinder is attached to the 10-lb slender rod which is pinned from point A. At the instant $\theta = 30°$ it has an angular velocity of $\omega_1 = 4$ rad/s as shown. Determine the angular velocity of the rod at the instant $\theta = 90°$.

*18–40. The 25-lb slender rod AB is attached to a spring BC which has an unstretched length of 4 ft. If the rod is released from rest when $\theta = 30°$, determine its angular velocity at the instant $\theta = 90°$.

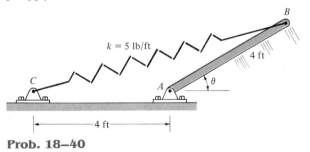

Prob. 18–40

18–41. The assembly consists of two 8-lb bars which are pin-connected to the two 10-lb disks. If the bars are released from rest when $\theta = 60°$, determine their angular velocities at the instant $\theta = 0°$. Assume the disks roll without slipping.

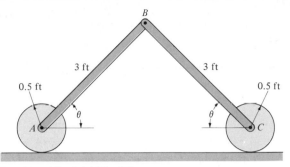

Probs. 18–41/18–42

18–42. The assembly consists of two 8-lb bars which are pin-connected to the two 10-lb disks. If the bars are released from rest when $\theta = 60°$, determine their angular velocities at the instant $\theta = 30°$. Assume the disks roll without slipping.

18–43. The garage door AB has a weight of 80 lb and can be approximated as a thin plate. During operation, a small roller at B is guided along the fixed horizontal track, while the two arms AC located on each side of the door are used to guide the bottom of the door along a circular arc having a center at D and a radius of 4 ft. If the end A of the door is given a horizontal velocity of 3 ft/s to the left and the door has a clockwise angular velocity $\omega_{AB} = 2$ rad/s when $\theta = 0°$ and ADC is vertical, determine the velocity of A at the instant the door becomes fully open, i.e., $\theta = 90°$. Each of the two springs has a stiffness of 50 lb/ft and an unstretched length of 1 ft. Neglect the mass of the arms ADC.

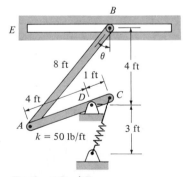

Prob. 18–43

*18–44. The disk is rotating about pin A in the vertical plane with an angular velocity $\omega_1 = 2$ rad/s when $\theta = 0°$. If it has a weight of 15 lb, determine its angular velocity at the instant $\theta = 90°$. Also, compute the horizontal and vertical components of reaction at A at this instant.

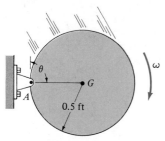

Prob. 18–44

18–45. The uniform window shade AB has a total weight of 0.4 lb. When it is released, it winds around the spring-loaded core O. Motion is caused by a spring within the core, which is coiled so that it exerts a torque $M = (0.3(10^{-3})\theta)$ lb · ft, where θ is in radians, on the core. If the shade is released from rest, determine the angular velocity of the core at the instant the shade is completely rolled up, i.e., after 12 revolutions. When this occurs, the spring becomes uncoiled and the radius of gyration of the shade about the axle at O is $k_O = 0.9$ in. *Note:* The elastic potential energy of a torsional spring is $V_e = \frac{1}{2}k\theta^2$, where $M = k\theta$. Here $k = 0.3(10^{-3})$ lb · ft/rad.

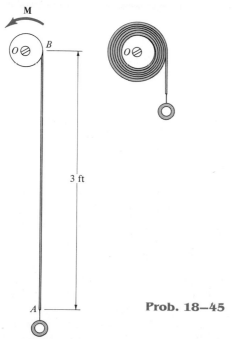

Prob. 18–45

18–46. The spool at A has a mass of 12 kg and a radius of gyration $k_C = 225$ mm. Pulley B has a mass of 7 kg and may be considered as a disk. If the system is released from rest and the cord does not slip, determine the angular velocity of the pulley after the spool descends 0.5 m. Neglect the mass of the cord.

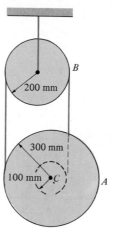

Prob. 18–46

18–47. The slender rod has a mass of 10 kg and is released from rest when it is vertical. Determine its angular velocity at the instant it becomes horizontal. The attached block at D has a mass of 4 kg. The mass of the cable and pulley at C can be neglected.

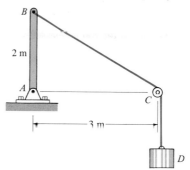

Prob. 18–47

*18–48. The gear has a mass m and may be treated as a uniform disk. If it is released from rest at $\theta = 0°$, and rolls along the circular gear rack, determine the angular velocity of the radial line AB at the instant $\theta = 90°$.

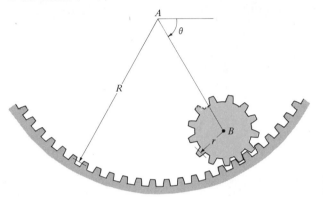

Prob. 18–48

18–49. The assembly consists of a 5-lb slender rod AC to which is pin-connected a 12-lb disk and spring BD. If the rod is brought into the horizontal position $\theta = 0°$, and the disk is given a counterclockwise rotation of 3 rad/s when the rod is released from rest, determine the angular velocity of the rod at the instant $\theta = 90°$. The spring has an unstretched length of 1 ft.

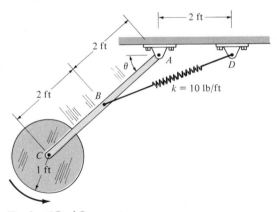

Prob. 18–49

18–50. The garage door CD has a mass of 15 kg and can be treated as a thin plate. If it is released from rest from the horizontal position shown, determine its angular velocity at the instant the two side bars ABC become horizontal. Each of the two side springs has a stiffness of $k = 30$ N/m and is originally unstretched. Neglect the mass of bar ABC.

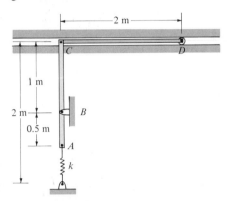

Prob. 18–50

18–51. The slender 15-kg bar is initially at rest and standing in the vertical position when the bottom end A is displaced slightly to the right. If the horizontal surface is smooth, determine the speed at which A strikes the corner D. The bar is constrained to move in the vertical plane. Neglect the mass of the cord BC.

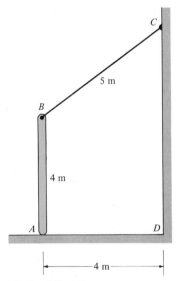

Prob. 18–51

19

Planar Kinetics of a Rigid Body: Impulse and Momentum

In Chapter 15 it was shown that problems which involve force, velocity, and time can conveniently be solved by using the principle of linear or angular impulse and momentum. In this chapter we will extend these concepts somewhat in order to determine the impulse and momentum relationships that apply to a rigid body. The three planar motions which will be considered are translation, rotation about a fixed axis, and general plane motion. The more general discussion of momentum principles applied to the three-dimensional motion of a rigid body is presented in Chapter 21.

*19.1 Linear and Angular Momentum

In this section we will develop formulations for the linear and angular momentum of a rigid body which is symmetric with respect to an inertial *x-y* reference plane. As a result, we will represent the body as a slab having the same mass as the body and moving in the reference plane.

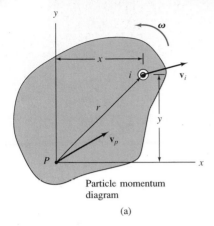

Particle momentum
diagram

(a)

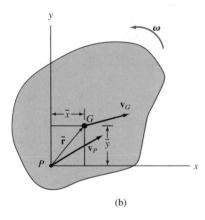

(b)

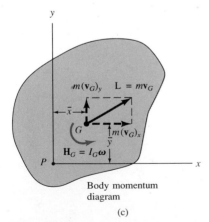

Body momentum
diagram

(c)

Fig. 19–1

Linear Momentum

The linear momentum of a rigid body is determined by summing vectorially the linear momenta of all the particles of the body, i.e., $\mathbf{L} = \Sigma m_i \mathbf{v}_i$. We can simplify this expression by noting that $\Sigma m_i \mathbf{v}_i = m \mathbf{v}_G$ (see Sec. 15.2) so that

$$\mathbf{L} = m \mathbf{v}_G \qquad (19\text{–}1)$$

This equation states that the body's linear momentum is a vector quantity having a *magnitude* $m v_G$, which is commonly measured in units of kg · m/s or slug · ft/s, and a *direction* defined by $\mathbf{v}_G$, the velocity of the mass center.

Angular Momentum

Consider the body in Fig. 19–1a, which is subjected to general plane motion. At the instant shown, the arbitrary point P has a velocity $\mathbf{v}_P$ and the body has an angular velocity $\boldsymbol{\omega}$. If the velocity of the ith particle of the body is to be determined, Fig. 19–1a, then

$$\mathbf{v}_i = \mathbf{v}_P + \mathbf{v}_{i/P}$$

The angular momentum of particle i about point P is equal to the "moment" of the particle's linear momentum about P, Fig. 19–1a. Thus,

$$(\mathbf{H}_P)_i = \mathbf{r} \times m_i \, \mathbf{v}_i$$

Expressing $\mathbf{v}_i$ in terms of $\mathbf{v}_P$ and using Cartesian vectors, we have

$$(H_P)_i \, \mathbf{k} = m_i(x\mathbf{i} + y\mathbf{j}) \times [(v_P)_x\mathbf{i} + (v_P)_y\mathbf{j} + \omega\mathbf{k} \times (x\mathbf{i} + y\mathbf{j})]$$
$$(H_P)_i = -m_i \, y(v_P)_x + m_i \, x(v_P)_y + m_i \, \omega r^2$$

Letting $m_i \rightarrow dm$ and integrating over the entire mass m of the body, we obtain

$$H_P = -\left(\int_m y \, dm\right)(v_P)_x + \left(\int_m x \, dm\right)(v_P)_y + \left(\int_m r^2 \, dm\right)\omega$$

Here H_P represents the angular momentum of the body about an axis (the z axis) perpendicular to the plane of motion and passing through point P. Since $\bar{x}m = \int x \, dm$ and $\bar{y}m = \int y \, dm$, the integrals for the first and second terms on the right are used to locate the body's center of mass G with respect to P. Also, the last integral represents the body's moment of inertia computed about the z axis, i.e., $I_P = \int r^2 \, dm$. Thus,

$$H_P = -\bar{y}m(v_P)_x + \bar{x}m(v_P)_y + I_P\omega \qquad (19\text{–}2)$$

This equation reduces to a simpler form if point P coincides with the mass center G for the body,[*] in which case $\bar{x} = \bar{y} = 0$. Hence, the angular momentum of the body, computed about an axis passing through point G, is therefore

---

[*]It *also* reduces to the same simple form, $H_P = I_P\omega$, if point P is a *fixed point* (see Eq. 19–9) or the velocity of point P is directed along the line PG.

$$\boxed{H_G = I_G\omega} \tag{19-3}$$

This equation states that the angular momentum of the body computed about point G is equal to the product of the moment of inertia of the body about an axis passing through G and the body's angular velocity. It is important to realize that $\mathbf{H}_G$ is a vector quantity having a *magnitude* $I_G\omega$, which is commonly measured in units of $kg \cdot m^2/s$ or $slug \cdot ft^2/s$, and a *direction* defined by $\boldsymbol{\omega}$, which is always perpendicular to the slab.

Equation 19–2 can also be rewritten in terms of the x and y components of the velocity of the body's mass center, $(v_G)_x$ and $(v_G)_y$, and the body's moment of inertia I_G. If point G is located at coordinates $(\bar{x}, \bar{y})$, Fig. 19–1b, then by the parallel-axis theorem, $I_P = I_G + m(\bar{x}^2 + \bar{y}^2)$. Substituting into Eq. 19–2 and rearranging terms, we have

$$H_P = \bar{y}m[-(v_P)_x + \bar{y}\omega] + \bar{x}m[(v_P)_y + \bar{x}\omega] + I_G\omega \tag{19-4}$$

From the kinematic diagram of Fig. 19–1b, $\mathbf{v}_G$ can be expressed in terms of $\mathbf{v}_P$ as

$$\mathbf{v}_G = \mathbf{v}_P + \boldsymbol{\omega} \times \bar{\mathbf{r}}$$
$$(v_G)_x\mathbf{i} + (v_G)_y\mathbf{j} = (v_P)_x\mathbf{i} + (v_P)_y\mathbf{j} + \omega\mathbf{k} \times (\bar{x}\mathbf{i} + \bar{y}\mathbf{j})$$

Carrying out the cross product and equating the respective $\mathbf{i}$ and $\mathbf{j}$ components yields the two scalar equations

$$(v_G)_x = (v_P)_x - \bar{y}\omega$$
$$(v_G)_y = (v_P)_y + \bar{x}\omega$$

Substituting these results into Eq. 19–4 yields

$$H_P = -\bar{y}m(v_G)_x + \bar{x}(v_G)_y + I_G\omega \tag{19-5}$$

With reference to Fig. 19–1c, *this result indicates that when the angular momentum of the body is computed about point P it is equivalent to the moment of the linear momentum* $m\mathbf{v}_G$ *or its components* $m(\mathbf{v}_G)_x$ *and* $m(\mathbf{v}_G)_y$ *about P plus the angular momentum* $I_G\omega$. Note that since $\boldsymbol{\omega}$ is a free vector, $\mathbf{H}_G$ can act at any point on the body (or slab) provided it preserves its same magnitude and direction. Furthermore, since angular momentum is equal to the "moment" of the linear momentum, the *line of action of* $\mathbf{L}$ *must pass through the body's mass center G* in order to preserve the correct magnitude of $\mathbf{H}_P$ when "moments" are computed about P, Fig. 19–1c. As a result of these ideas, we will now consider three types of motion.

Translation

When a rigid body of mass m is subjected to either rectilinear or curvilinear *translation*, Fig. 19–2a, its mass center has a velocity of $\mathbf{v}_G = \mathbf{v}$ and $\boldsymbol{\omega} = \mathbf{0}$. Hence, the linear momentum and the angular momentum, computed about G, become

$$\boxed{\begin{aligned} L &= mv_G \\ H_G &= 0 \end{aligned}}$$

(19–6)

If the angular momentum is computed about any other point A on or off the body, Fig. 19–2a, the "moment" of the linear momentum $\mathbf{L}$ must be computed about the point. Since d is the "moment arm" as shown in the figure, then in accordance with Eq. 19–5, $H_A = (d)(mv_G)$.

Rotation About a Fixed Axis

When a rigid body is *rotating about a fixed axis* passing through point O, Fig. 19–2b, the linear momentum and the angular momentum computed about G are

$$\boxed{\begin{aligned} L &= mv_G \\ H_G &= I_G\omega \end{aligned}}$$

(19–7)

It is sometimes convenient to compute the angular momentum of the body about point O. In this case it is necessary to account for the "moments" of *both* $\mathbf{L}$ and $\mathbf{H}_G$ about O. Noting that $\mathbf{L}$ (or $\mathbf{v}_G$) is always *perpendicular to* $\mathbf{r}_{G/O}$, we have

$$H_O = I_G\omega + r_{G/O}(mv_G)$$

(19–8)

This equation may be *simplified* by first substituting $v_G = r_{G/O}\omega$, in which case $H_O = (I_G + mr_{G/O}^2)\omega$, and, by the parallel-axis theorem, noting that the terms inside the parentheses represent the moment of inertia I_O of the body about an axis perpendicular to the plane of motion and passing through point O. Hence,*

$$\boxed{H_O = I_O\omega}$$

(19–9)

For the computation, then, either Eq. 19–8 or 19–9 can be used.

---

*The similarity between this derivation and that of Eq. 17–16 ($\Sigma M_O = I_O\alpha$) and Eq. 18–4 ($T = \frac{1}{2}I_O\omega^2$) should be noted. Also note that the same result can be obtained from Eq. 19–2 by selecting point P at O, realizing that $(v_O)_x = (v_O)_y = 0$.

General Plane Motion

When a rigid body is subjected to general plane motion, Fig. 19–2c, the linear momentum and the angular momentum computed about G become

$$\boxed{\begin{aligned} L &= m v_G \\ H_G &= I_G \omega \end{aligned}}$$

$(19\text{–}10)$

If the angular momentum is computed about a point A located either on or off the body, Fig. 19–2c, it is necessary to compute the moments of *both* **L** and $\mathbf{H}_G$ about this point. In this case,

$$H_A = I_G \omega + (d)(m v_G)$$

Here d is the moment arm, as shown in the figure.

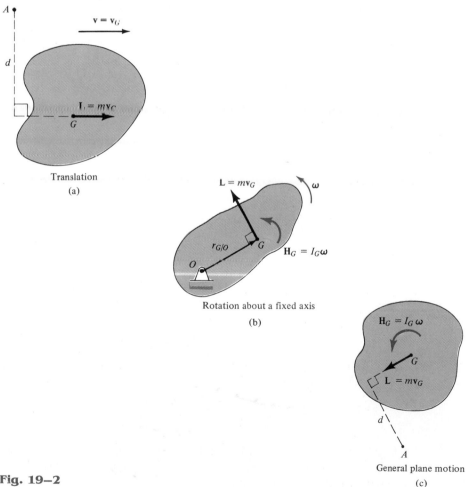

Translation

(a)

Rotation about a fixed axis

(b)

General plane motion

(c)

Fig. 19–2

19.2 Principle of Impulse and Momentum

Like the case for particle motion, the principle of impulse and momentum for a rigid body is developed by *combining* the equation of motion with kinematics. The resulting equation will allow a *direct solution to problems involving force, velocity, and time.*

Principle of Linear Impulse and Momentum

The equation of translational motion for a rigid body can be written as $\Sigma \mathbf{F} = m\mathbf{a}_G = m\,(d\mathbf{v}_G/dt)$. Since the mass of the body is constant,

$$\Sigma \mathbf{F} = \frac{d}{dt}(m\mathbf{v}_G)$$

Multiplying both sides by dt and integrating from $t = t_1$, $\mathbf{v}_G = (\mathbf{v}_G)_1$ to $t = t_2$, $\mathbf{v}_G = (\mathbf{v}_G)_2$ yields

$$\Sigma \int_{t_1}^{t_2} \mathbf{F}\, dt = m(\mathbf{v}_G)_2 - m(\mathbf{v}_G)_1 \qquad (19\text{–}11)$$

This equation is referred to as the *principle of linear impulse and momentum*. It states that the sum of all the impulses created by the *external force system* which acts on the body during the time interval t_1 to t_2 is equal to the change in the linear momentum of the body during the time interval.

Principle of Angular Impulse and Momentum

If the body has *general plane motion* we can write $\Sigma M_G = I_G\alpha = I_G(d\omega/dt)$. Since the moment of inertia is constant,

$$\Sigma M_G = \frac{d}{dt}(I_G\omega)$$

Multiplying both sides by dt and integrating from $t = t_1$, $\omega = \omega_1$ to $t = t_2$, $\omega = \omega_2$ gives

$$\Sigma \int_{t_1}^{t_2} M_G\, dt = I_G\omega_2 - I_G\omega_1 \qquad (19\text{–}12)$$

In a similar manner, for *rotation about a fixed axis* passing through point O, Eq. 17–16 ($\Sigma M_O = I_O\alpha$) when integrated becomes

$$\Sigma \int_{t_1}^{t_2} M_O\, dt = I_O\omega_2 - I_O\omega_1 \qquad (19\text{–}13)$$

Equations 19–12 and 19–13 are referred to as the *principle of angular impulse and momentum*. Both equations state that the sum of the angular impulses acting on the body during the time interval t_1 to t_2 is equal to the change in the body's angular momentum during this time interval. In particular, the angular impulse considered is determined by integrating the moments about point G or O of all the external forces and couples applied to the body.

To summarize the preceding concepts, if motion is occurring in the x-y plane, using impulse and momentum principles the following *three scalar equations* may be written which describe the *planar motion* of the body:

$$m(v_{Gx})_1 + \Sigma \int_{t_1}^{t_2} F_x \, dt = m(v_{Gx})_2$$

$$m(v_{Gy})_1 + \Sigma \int_{t_1}^{t_2} F_y \, dt = m(v_{Gy})_2 \qquad (19\text{--}14)$$

$$I_G\omega_1 + \Sigma \int_{t_1}^{t_2} M_G \, dt = I_G\omega_2$$

The first two of these equations represent the principle of linear impulse and momentum in the x-y plane, Eq. 19–11, and the third equation represents the principle of angular impulse and momentum about the z axis, Eq. 19–12.

The terms in Eqs. 19–14 can be graphically accounted for by drawing a set of impulse and momentum diagrams for the body, Fig. 19–3. Note that the linear momenta $m v_G$ are applied at the body's mass center, Fig. 19–3a and c; whereas the angular momenta $I_G\omega$ are free vectors, and therefore, like a couple moment, they may be applied at any point on the body. When the impulse diagram is constructed, Fig. 19–3b, vectors **F** and **M** which vary with time are indicated by the integrals. However, if **F** and **M** are *constant* from t_1 to t_2, integration of the impulses yields $\mathbf{F}(t_2 - t_1)$ and $\mathbf{M}(t_2 - t_1)$, respectively. Such is the case for the body's weight **W**, Fig. 19–3b.

Equations 19–14 may also be applied to an entire system of connected bodies rather than to each body separately. Doing this eliminates the need to include reactive impulses which occur at the connections since they are *internal to the system*. The resultant equations may be written in symbolic form as

$$\left(\Sigma \frac{\text{syst. linear}}{\text{momentum}} \right)_{x1} + \left(\Sigma \frac{\text{syst. linear}}{\text{impulse}} \right)_{x(1-2)} = \left(\Sigma \frac{\text{syst. linear}}{\text{momentum}} \right)_{x2}$$

$$\left(\Sigma \frac{\text{syst. linear}}{\text{momentum}} \right)_{y1} + \left(\Sigma \frac{\text{syst. linear}}{\text{impulse}} \right)_{y(1-2)} = \left(\Sigma \frac{\text{syst. linear}}{\text{momentum}} \right)_{y2}$$

$$\left(\Sigma \frac{\text{syst. angular}}{\text{momentum}} \right)_{O1} + \left(\Sigma \frac{\text{syst. angular}}{\text{impulse}} \right)_{O(1-2)} = \left(\Sigma \frac{\text{syst. angular}}{\text{momentum}} \right)_{O2}$$

$$(19\text{--}15)$$

As indicated, the system's angular momentum and angular impulse must be computed with respect to the *same fixed reference point O* for all the bodies of the system.

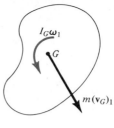

Initial momentum diagram

(a)

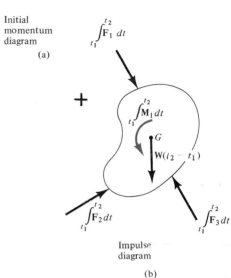

Impulse diagram

(b)

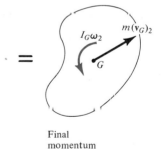

Final momentum diagram

(c)

Fig. 19–3

PROCEDURE FOR ANALYSIS

Impulse and momentum principles are used to solve kinetics problems that involve *velocity, force,* and *time* since these terms are involved in the formulation. For application it is suggested that the following procedure be used.

Free-Body Diagram. Establish the x, y, z inertial frame of reference and draw the free-body diagram in order to account for all the forces that produce impulses on the body. The direction and sense of the initial and final velocity of the body's mass center, $\mathbf{v}_G$, and the body's angular velocity $\boldsymbol{\omega}$ should also be established. Also compute the moment of inertia I_G or I_O.

As an alternative procedure draw the impulse and momentum diagrams for the body or system of bodies. Each of these diagrams represents an outlined shape of the body which graphically accounts for the data required for each of the three terms in Eqs. 19–14 or 19–15, Fig. 19–3. Note that these diagrams are particularly helpful in order to visualize the "moment" terms used in the principle of angular impulse and momentum, when it has been decided that application of this equation is to be about a point other than the body's mass center G or a fixed point O.

Principle of Impulse and Momentum. Apply the three scalar equations 19–14 (or 19–15). In cases where the body is rotating about a fixed axis, Eq. 19–13 may be substituted for the third of Eqs. 19–14.

Kinematics. If more than three equations are needed for a complete solution, it may be possible to relate the velocity of the body's mass center to the body's angular velocity using *kinematics*. If the motion appears to be complicated, kinematic (velocity) diagrams may be helpful in obtaining the necessary relation.

In general, a method to be used for the solution of a particular type of problem should be decided upon *before* attempting to solve the problem. As stated above, the *principle of impulse and momentum* is most suitable for solving problems which involve *velocity, force,* and *time*. For some problems, however, a combination of the equation of motion and its two integrated forms, the principle of work and energy and the principle of impulse and momentum, will yield the most direct solution to the problem.

Example 19–1

The disk shown in Fig. 19–4a weighs 20 lb and is pin-supported at its center. If it is acted upon by a constant couple moment of 4 lb · ft and a force of 10 lb which is applied to a cord wrapped around its periphery, determine the angular velocity of the disk two seconds after starting from rest. What are the force components of reaction at pin A?

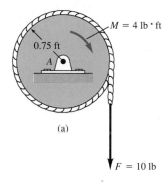

(a)

Fig. 19–4

SOLUTION

Free-Body Diagram Fig. 19–4b. The disk's mass center does not move; however, the loading causes the disk to rotate clockwise.

The moment of inertia of the disk about its fixed axis of rotation is

$$I_A = \tfrac{1}{2}mr^2 = \frac{1}{2}\left(\frac{20}{32.2}\right)(0.75)^2 = 0.175 \text{ slug} \cdot \text{ft}^2$$

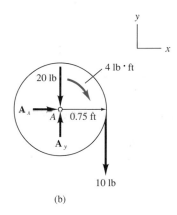

(b)

Principle of Impulse and Momentum

$(\xrightarrow{+})$
$$m(v_{Ax})_1 + \Sigma \int_{t_1}^{t_2} F_x \, dt = m(v_{Ax})_2$$

$$0 + A_x(2) = 0$$

$(+\uparrow)$
$$m(v_{Ay})_1 + \Sigma \int_{t_1}^{t_2} F_y \, dt = m(v_{Ay})_2$$

$$0 + A_y(2) - 20(2) - 10(2) = 0$$

$(\curvearrowright +)$
$$I_A\omega_1 + \Sigma \int_{t_1}^{t_2} M_A \, dt = I_A\omega_2$$

$$0 + 4(2) + [10(2)](0.75) = 0.175\omega_2$$

Solving these equations yields

$$A_x = 0 \qquad\qquad Ans.$$
$$A_y = 30 \text{ lb} \qquad\qquad Ans.$$
$$\omega_2 = 132 \text{ rad/s} \downarrow \qquad\qquad Ans.$$

Example 19–2

The block shown in Fig. 19–5a has a mass of 6 kg. It is attached to a cord which is wrapped around the periphery of a 20-kg disk that has a moment of inertia $I_A = 0.40$ kg · m². If the block is initially moving downward with a speed of 2 m/s, determine its speed in 3 s. Neglect the mass of the cord in the calculation.

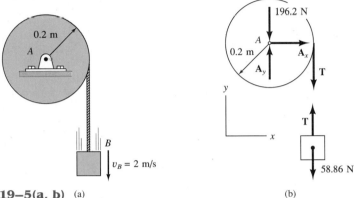

Fig. 19–5(a, b) (a)   (b)

SOLUTION I

Free-Body Diagram. The free-body diagrams of the block and disk are shown in Fig. 19–5b. All the forces are *constant* since the weight of the block causes the motion. The downward motion of the block, $\mathbf{v}_B$, causes $\boldsymbol{\omega}$ of the disk to be clockwise.

Principle of Impulse and Momentum. We can eliminate $\mathbf{A}_x$ and $\mathbf{A}_y$ from the analysis by applying the principle of angular momentum about point A. Hence

Disk
$(\curvearrowright+)$
$$I_A\omega_1 + \Sigma \int M_A\,dt = I_A\omega_2$$
$$0.40(\omega_1) + T(3)(0.2) = 0.4\omega_2$$

Block
$(+\uparrow)$
$$m_B(v_B)_1 + \Sigma \int F_y\,dt = m_B(v_B)_2$$
$$-6(2) + T(3) - 58.86(3) = -6(v_B)_2$$

Kinematics. Since $\omega = v_B/r$, then $\omega_1 = 2/0.2 = 10$ rad/s and $\omega_2 = (v_B)_2/0.2 = 5(v_B)_2$. Substituting and solving the equations simultaneously for $(v_B)_2$ yields

$$(v_B)_2 = 13.0 \text{ m/s} \downarrow \qquad\qquad \textit{Ans.}$$

SOLUTION II

Impulse and Momentum Diagrams. We can obtain $(v_B)_2$ *directly* by considering the *system* consisting of the block, the cord, and the disk. The impulse and momentum diagrams have been drawn to clarify application of the principle of angular impulse and momentum about point A, Fig. 19–5c.

Principle of Angular Impulse and Momentum. Realizing that $\omega_1 = 10$ rad/s and $\omega_2 = 5(v_B)_2$, we have

$$\left(\sum \begin{array}{c} \text{syst. angular} \\ \text{momentum} \end{array} \right)_{A1} + \left(\sum \begin{array}{c} \text{syst. angular} \\ \text{impulse} \end{array} \right)_{A(1-2)} = \left(\sum \begin{array}{c} \text{syst. angular} \\ \text{momentum} \end{array} \right)_{A2}$$

$$(\zeta+) \quad 6(2)(0.2) + 0.4(10) + 58.86(3)(0.2) = 6(v_B)_2(0.2) + 0.4(5(v_B)_2)$$

$$(v_B)_2 = 13.0 \text{ m/s} \downarrow \qquad\qquad Ans.$$

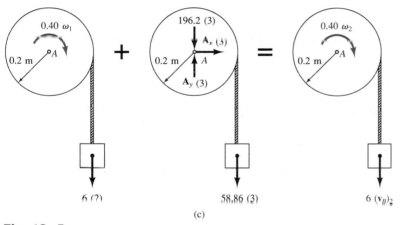

(c)

Fig. 19–5c

Example 19–3

The 100-kg spool shown in Fig. 19–6a has a radius of gyration $k_G = 0.35$ m. A cable is wrapped around the central hub of the spool and a variable horizontal force having a magnitude of $P = (t + 10)$ N is applied, where t is measured in seconds. If the spool is initially at rest, determine its angular velocity in 5 s. Assume that the spool rolls without slipping at A.

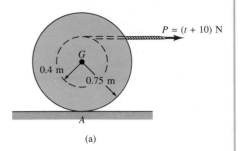

(a)

SOLUTION

Free-Body Diagram. By inspection of the free-body diagram, Fig. 19-6b, the *variable* force **P** will cause the friction force $\mathbf{F}_A$ to be variable, and thus the impulses created by both **P** and $\mathbf{F}_A$ must be determined by integration. The loading causes the mass center to have a velocity $\mathbf{v}_G$ to the right, and the spool has a clockwise angular velocity $\boldsymbol{\omega}$, Fig. 19–6c.

The moment of inertia of the spool about its mass center is

$$I_G = mk_G^2 = 100(0.35)^2 = 12.25 \text{ kg} \cdot \text{m}^2$$

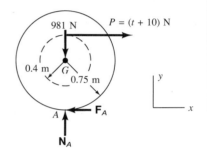

Principle of Impulse and Momentum

$(\xrightarrow{+})$ $\qquad m(v_G)_1 + \Sigma \int F_x \, dt = m(v_G)_2$

$$0 + \int_0^5 (t + 10) \, dt - \int F_A \, dt = 100(v_G)_2 \qquad (1)$$

$(\curvearrowright +)$ $\qquad I_G\omega_1 + \Sigma \int M_G \, dt = I_G\omega_2$

$$0 + \left[\int_0^5 (t + 10) \, dt\right](0.4) + \left(\int F_A \, dt\right)(0.75) = 12.25 \, \omega_2 \qquad (2)$$

Kinematics. Since the spool does not slip, the instantaneous center of zero velocity is at point A, Fig. 19–6c. Hence, the velocity of G can be expressed in terms of the spool's angular velocity as $(v_G)_2 = 0.75\omega_2$. Substituting this, evaluating the integral for the impulse caused by **P** (i.e., 62.5 N · s), and eliminating the unknown impulse $\int F_A \, dt$ between Eqs. (1) and (2), we obtain

$$\omega_2 = 1.05 \text{ rad/s } \curvearrowright \qquad\qquad Ans.$$

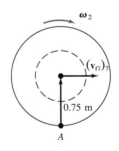

(c)

Fig. 19–6

Example 19–4

The Charpy impact test is used in materials testing to determine the energy absorption characteristics of a material during impact. The test is performed using the pendulum shown in Fig. 19–7a that has a mass m, mass center at G, and a radius of gyration k_G about G. Determine the distance r_P from the pin at A to the point P where the impact with the specimen S should occur so that the horizontal force at the pin is essentially zero during the impact. For the computation, assume the specimen absorbs all the pendulum's kinetic energy during the time it falls and thereby stops the pendulum from swinging at $\theta = 0°$.

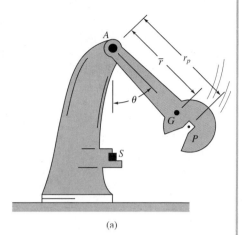

(a)

SOLUTION

Free-Body Diagram. As shown on the free-body diagram, Fig. 19-7b, the conditions of the problem require the horizontal impulse at A to be zero. Just before impact, we assume the pendulum has a clockwise angular velocity $\boldsymbol{\omega}_1$ and the mass center of the pendulum is moving to the left at $(v_G)_1 = \bar{r}\omega_1$.

Principle of Impulse and Momentum. We will apply the principle of angular impulse and momentum about point A. Thus,

$$(\curvearrowright +) \qquad I_A\omega_1 - \left(\int F \, dt\right)r_P = 0$$

$$(\xrightarrow{+}) \qquad -m(\bar{r}\omega_1) + \int F \, dt = 0$$

Eliminating the impulse $\int F \, dt$ and substituting $I_A = mk_G^2 + m\bar{r}^2$ yields

$$[mk_G^2 + m\bar{r}^2]\omega_1 - m(\bar{r}\omega_1)r_P = 0$$

Factoring out $m\omega_1$ and solving for r_P yields

$$r_P = \bar{r} + \frac{k_G^2}{\bar{r}} \qquad\qquad Ans.$$

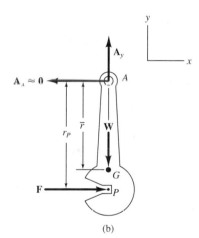

(b)

Fig. 19–7

The point P, so defined, is called the *center of percussion*. By placing the striking point at P, the force developed at the pin will be minimized. Many sports rackets, clubs, etc. are designed so that the object being struck occurs at the center of percussion. As a consequence, no "sting" or little sensation occurs in the hand of the player. (Also see Probs. 17–55 and 19–7.)

PROBLEMS

19–1. Solve Prob. 17–45 using the principle of impulse and momentum.

19–2. Solve Prob. 17–46 using the principle of impulse and momentum.

19–3. Solve Prob. 17–51 using the principle of impulse and momentum.

*19–4. Solve Prob. 17–52 using the principle of impulse and momentum.

19–5. Show that if a slab is rotating about a fixed axis perpendicular to the slab and passing through its mass center G, the angular momentum is the same when computed about any other point P on the slab.

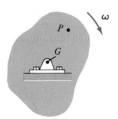

Prob. 19–5

19–6. At a given instant, the body has a linear momentum $\mathbf{L} = m\mathbf{v}_G$ and an angular momentum $\mathbf{H}_G = I_G\boldsymbol{\omega}$ computed about its mass center. Show that the angular momentum of the body computed about the instantaneous center of zero velocity IC can be expressed as $\mathbf{H}_{IC} = I_{IC}\boldsymbol{\omega}$, where I_{IC} represents the body's moment of inertia computed about the instantaneous axis of zero velocity. As shown, the IC is located at a distance of $r_{G/IC}$ away from the mass center G.

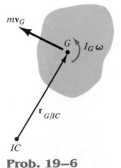

Prob. 19–6

19–7. The rigid body (slab) has a mass m and is rotating with an angular velocity $\boldsymbol{\omega}$ about an axis passing through the fixed point O. Show that the momenta of all the particles composing the body can be represented by a single vector having a magnitude mv_G and acting through point P, called the *center of percussion*, which lies at a distance $r_{P/G} = k_G^2/r_{G/O}$ from the mass center G. Here k_G is the radius of gyration of the body, computed about an axis perpendicular to the plane of motion and passing through G.

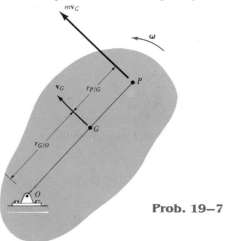

Prob. 19–7

*19–8. The spacecraft has a mass of 1200 kg and a moment of inertia $I_G = 900$ kg · m² about an axis passing through G and directed perpendicular to the page. If it is traveling forward with a speed $v_G = 800$ m/s and executes a turn by means of two jets, which provide a constant thrust of 400 N for 0.3 s, determine the spacecraft's angular velocity just after the jets are turned off.

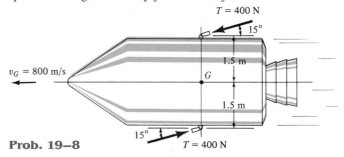

Prob. 19–8

19–9. A flywheel has a mass of 60 kg and a radius of gyration of $k_G = 150$ mm about an axis of rotation passing through its mass center. If a motor supplies a clockwise torque having a magnitude of $M = (5t)$ N · m, where t is in seconds, to the flywheel, determine its angular velocity in $t = 3$ s. Initially the flywheel is rotating clockwise at $\omega_1 = 2$ rad/s.

19–10. The 10-lb rectangular plate is at rest on a smooth *horizontal* floor. If it is given the horizontal impulses shown, determine its angular velocity and the velocity of the mass center.

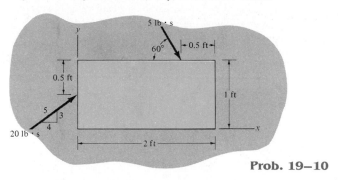

Prob. 19–10

19–11. The spool has a mass of 30 kg and a radius of gyration $k_O = 0.25$ m. Block A has a mass of 25 kg and block B has a mass of 10 kg. If they are released from rest, determine their speed in $t = 2$ s starting from rest. Neglect the mass of the ropes.

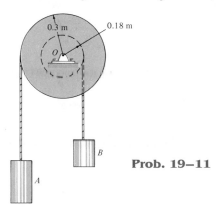

Prob. 19–11

*19–12. The spool has a weight of 30 lb and a radius of gyration $k_O - 0.45$ ft. A cord is wrapped around its inner hub and its end subjected to a horizontal force $P = 5$ lb. Determine the spool's angular velocity in 4 s starting from rest. Assume the spool rolls without slipping.

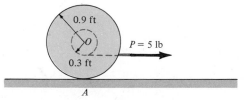

Probs. 19–12/19–13

19–13. Work Prob. 19–12 if the coefficient of static friction is $\mu_A = 0.3$.

19–14. The inner hub of the wheel rests on the inclined track. If it does not slip at A, determine its angular velocity $t = 2$ s after it is released from rest. The wheel has a weight of 30 lb and a radius of gyration of $k_G = 1.30$ ft.

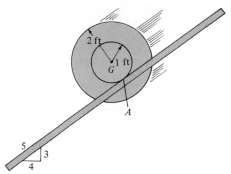

Prob. 19–14

19–15. The spool has a weight of 75 lb and a radius of gyration of $k_O = 1.20$ ft. If the block B weighs 60 lb, and a force of $P = 25$ lb is applied to the cord, determine the speed of the block in $t = 5$ s starting from rest. Neglect the mass of the cord.

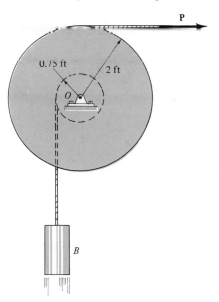

Probs. 19–15/19–16

*19–16. The spool has a weight of 75 lb and a radius of gyration of $k_O = 1.20$ ft. If the block B weighs 60 lb, determine the horizontal force P which must be applied to the cord in order to give the block an upward speed of 10 ft/s in $t = 4$ s, starting from rest. Neglect the mass of the cord.

19–17. Spool B is at rest and spool A is rotating at 6 rad/s when the slack in the cord connecting them is taken up. Determine the angular velocity of each spool immediately after the cord is jerked tight by the spinning of spool A. A and B have weights and radii of gyration $W_A = 30$ lb, $k_A = 0.8$ ft and $W_B = 15$ lb, $k_B = 0.6$ ft, respectively.

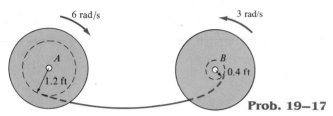

Prob. 19–17

19–18. A torque $M = (0.05e^{0.5t})$ N · m, where t is in seconds, is applied to gear A. If gear A is originally rotating at 5 rad/s, determine its angular velocity when $t = 2$ s. The two smaller *equivalent* gears B and C are pinned at their centers. The mass and centroidal radii of gyration of the gears about axes perpendicular to the page and passing through their centers are given in the figure.

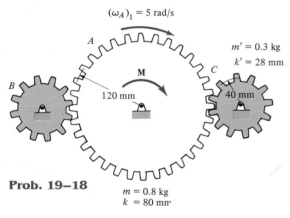

Prob. 19–18

19–19. The two gears A and B have weights and radii of gyration of $W_A = 15$ lb, $k_A = 0.5$ ft, and $W_B = 10$ lb, $k_B = 0.35$ ft, respectively. If a couple moment having a magnitude of $M = 2(1 - e^{-0.5t})$ lb · ft is applied to gear B, determine the angular velocity of each gear in $t = 5$ s, starting from rest.

Prob. 19–19

*19–20. If the hoop has a weight W and radius r and is thrown onto a *rough surface* with a velocity $\mathbf{v}_G$ parallel to the surface, determine the amount of backspin, ω, it must be given so that it stops spinning at the same instant that its forward velocity is zero. It is not necessary to know the coefficient of kinetic friction at A for the calculation.

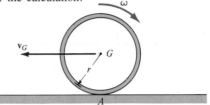

Prob. 19–20

19–21. The ball of mass m and radius r rolls along an inclined plane for which the coefficient of static friction is μ. If the ball is released from rest, determine the maximum angle θ for the incline so that it rolls without slipping at A.

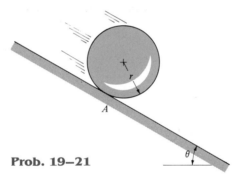

Prob. 19–21

19–22. The slender rod of mass m and length L is released from rest when it is in the vertical position. If it falls and strikes the soft ledge at A without rebounding, determine the impulse which the ledge exerts on the rod.

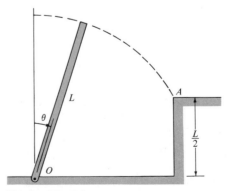

Prob. 19–22

19–23. The square plate has a mass m and is suspended at its corner by a cord. If it receives a horizontal impulse $\mathbf{I}$ at corner B, determine the location y of the point P about which the plate appears to rotate during the impact.

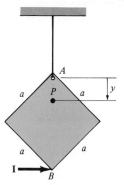

Prob. 19–23

*19–24. The inner hub of the wheel rests on the horizontal track. If it does not slip at A, determine the speed of the 10-lb block when $t = 2$ s after it is released from rest. The wheel has a weight of 30 lb and a radius of gyration $k_G = 1.30$ ft. Neglect the mass of the pulley and cord.

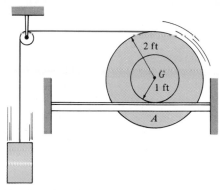

Prob. 19–24

19–25. The rod of length L and mass m lies on a smooth horizontal surface and is subjected to a force $\mathbf{P}$ at its end A as shown. Determine the location d of the point about which the rod begins to turn, i.e., the point that has zero velocity.

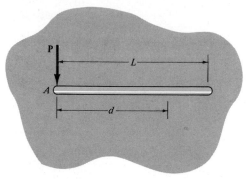

Prob. 19–25

19–26. The 10-lb cylinder rests on the 20-lb dolly. If the system is released from rest, determine the angular velocity of the cylinder in $t = 2$ s. The cylinder does not slip on the dolly. Neglect the mass of the wheels on the dolly.

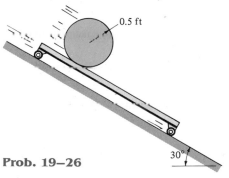

Prob. 19–26

19–27. A cord is wrapped around the rim of each 10-lb disk. If disk B is released from rest, determine the angular velocity of disk A in 2 s. Neglect the mass of the cord.

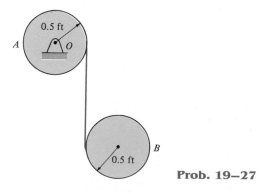

Prob. 19–27

385

19.3 Conservation of Momentum

Conservation of Linear Momentum

If the sum of all the *linear impulses* acting on a system of connected rigid bodies is *zero*, the linear momentum of the system is constant or conserved. Consequently, the first two of Eqs. 19–15 reduce to the form

$$\left(\sum \begin{array}{c} \text{syst. linear} \\ \text{momentum} \end{array} \right)_1 = \left(\sum \begin{array}{c} \text{syst. linear} \\ \text{momentum} \end{array} \right)_2 \qquad (19\text{–}16)$$

This equation is referred to as the *conservation of linear momentum*.

Without inducing appreciable errors in the computations, it may be possible to apply Eq. 19–16 in a specified direction for which the linear impulses are small or *nonimpulsive*. Specifically, nonimpulsive forces occur when small forces act over very short periods of time. For example, the impulse created by the force of a tennis racket hitting a ball during a very short time interval Δt is large, whereas the impulse of the weight of the ball during this time is small by comparison and may therefore be neglected in the motion analysis of the ball during Δt.

Conservation of Angular Momentum

The angular momentum of a system of connected rigid bodies is conserved about the system's center of mass G, or a fixed point O, when the sum of all the angular impulses created by the external forces acting on the system is zero or appreciably small (nonimpulsive) when computed about these points. The third of Eqs. 19–15 then becomes

$$\left(\sum \begin{array}{c} \text{syst. angular} \\ \text{momentum} \end{array} \right)_{O1} = \left(\sum \begin{array}{c} \text{syst. angular} \\ \text{momentum} \end{array} \right)_{O2} \qquad (19\text{–}17)$$

This equation is referred to as the *conservation of angular momentum*. In the case of a single rigid body, Eq. 19–17 applied to point G becomes $(I_G\omega)_1 = (I_G\omega)_2$. To illustrate an application of this equation, consider a swimmer who executes a somersault after jumping off a diving board. By tucking his arms and legs in close to his chest, he *decreases* his body's moment of inertia and thus *increases* his angular velocity ($I_G\omega$ must be constant). If he straightens out just before entering the water, his body's moment of inertia is *increased* and his angular velocity *decreases*. Since the weight of his body creates a linear impulse during the time of motion, this example also illustrates that the angular momentum of a body is conserved and yet the linear momentum is *not*. Such cases occur whenever the external forces creating the linear impulse pass through either the center of mass of the body or a fixed axis of rotation.

PROCEDURE FOR ANALYSIS

Provided the initial velocity of the body is known, the conservation of linear or angular momentum is used to determine the final velocity of a body *just after* the time period considered. Furthermore, by applying these equations to a *system* of bodies, the internal impulses acting within the system, which may be unknown, are eliminated from the analysis, since they occur in equal but opposite collinear pairs. For application it is suggested that the following procedure be used.

Free-Body Diagram. Establish the x, y inertial frame of reference and draw the free-body diagram for the body or system of bodies during the time of impact. From this diagram classify each of the applied forces as being either "impulsive" or "nonimpulsive." In general, for short times, "nonimpulsive forces" consist of the weight of a body, the force of a slightly deformed spring, or any force that is *known to be small* when compared to an impulsive force. It will also be possible to tell if the conservation of linear or angular momentum can be applied. Specifically, the *conservation of linear momentum* applies in a given direction when *no* external impulsive forces act on the body or system in that direction; whereas the *conservation of angular momentum* applies about a fixed point O or at the mass center G of a body or system of bodies when all the external impulsive forces acting on the body or system create zero moment (or zero angular impulse) about O or G.

As an alternative procedure, consider drawing the impulse and momentum diagrams for the body or system of bodies. These diagrams are particularly helpful in order to visualize the "moment" terms used in the conservation of angular momentum equation, when it has been decided that angular momenta are to be computed about a point other than the body's mass center G.

Conservation of Momentum. Apply the conservation of linear or angular momentum in the appropriate directions.

Kinematics. Use *kinematics* if further equations are necessary for the solution of a problem. If the motion appears to be complicated, kinematic (velocity) diagrams may be helpful in obtaining the necessary kinematic relations.

If it is necessary to determine an *internal impulsive force* acting on only one body of a system, the body must be *isolated* (free-body diagram) and the principle of linear or angular impulse and momentum must be applied *to the body*. After the impulse $\int F \, dt$ is calculated, then, provided the time Δt for which the impulse acts is known, the *average impulsive force* F_{avg} can be determined from $F_{\text{avg}} = \int F \, dt/\Delta t$.

The following examples numerically illustrate application of this procedure.

Example 19–5

The 10-kg wheel shown in Fig. 19–8a has a moment of inertia $I_G = 0.156$ kg · m². Assuming that the wheel does not slip or rebound, determine the minimum velocity $\mathbf{v}_G$ it must have to just roll over the obstruction at A.

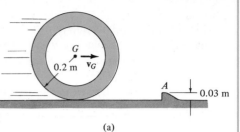

(a)

SOLUTION

Impulse and Momentum Diagrams. Since no slipping or rebounding occurs, the wheel essentially *pivots* about point A during contact. This condition is shown in Fig. 19–8b, which indicates, respectively, the momentum of the wheel *just before impact*, the impulses given to the wheel *during impact*, and the momentum of the wheel *just after impact*. Only two impulses (forces) act on the wheel. By comparison, the impulse at A is much greater than that caused by the weight, and since the time of impact is very short, the weight can be considered nonimpulsive. The impulsive force $\mathbf{F}$ at A has both an unknown magnitude and an unknown direction θ. To eliminate this force from the analysis, note that angular momentum about A is essentially *conserved* since $(98.1\Delta t)d \approx 0$.

Conservation of Angular Momentum. With reference to Fig. 19–8b,

$$(\curvearrowright +) \qquad\qquad (H_A)_1 = (H_A)_2$$

$$r'm(v_G)_1 + I_G\omega_1 = rm(v_G)_2 + I_G\omega_2$$

$$(0.2 - 0.03)(10)(v_G)_1 + (0.156)(\omega_1) = (0.2)(10)(v_G)_2 + (0.156)(\omega_2)$$

Kinematics. Since no slipping occurs, $\omega = v_G/r = v_G/0.2 = 5v_G$. Substituting this into the above equation and simplifying yields

$$(v_G)_2 = 0.892(v_G)_1 \qquad\qquad (1)$$

Conservation of Energy.* In order to roll over the obstruction, the wheel must pass the dashed position 3 shown in Fig. 19–8c. Hence, if $(v_G)_2$ (or $(v_G)_1$) is to be a minimum, it is necessary that the kinetic energy of the wheel at position 2 be equal to the potential energy at position 3. Constructing the datum through the center of gravity, as shown in the figure, and applying the conservation of energy equation, we have

$$\{T_2\} + \{V_2\} = \{T_3\} + \{V_3\}$$

$$\left\{\frac{1}{2}(10)(v_G)_2^2 + \frac{1}{2}(0.156)(\omega_2)^2\right\} + \{0\} = \{0\} + \{(98.1)(0.03)\}$$

Substituting $\omega_2 = 5(v_G)_2$ and Eq. (1) into this equation, and simplifying, yields

$$3.98(v_G)_1^2 + 1.55(v_G)_1^2 = 2.94$$

Thus,

$$(v_G)_1 = 0.729 \text{ m/s} \rightarrow \qquad\qquad Ans.$$

*This principle *does not* apply during *impact*, since energy is *lost* during the collision; however, just after impact it can be used.

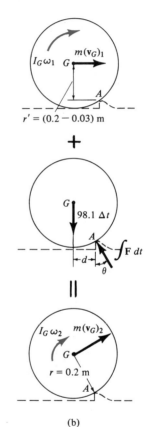

(b)

Fig. 19–8

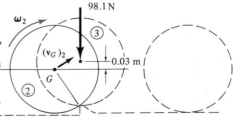

(c)

Example 19–6

The 5-kg slender rod shown in Fig. 19–9a is pinned at O and is initially at rest. If a 4-g bullet is fired into the rod with a velocity of 400 m/s, as shown in the figure, determine the angular velocity of the rod just after the bullet becomes embedded in it.

SOLUTION

Impulse and Momentum Diagrams. The impulse which the bullet exerts on the rod can be eliminated from the analysis and the angular velocity of the rod just after impact can be determined by considering the bullet and rod as a single system. To clarify the principles involved, the impulse and momentum diagrams are shown in Fig. 19–9b. The momentum diagrams are drawn *just before and just after impact*. During impact, the bullet and rod exchange equal but opposite *internal impulses* at A. As shown on the impulse diagram, the impulses that are external to the system are due to the reactions at O and the weights of the bullet and rod. Since the time of impact, Δt, is very short, the rod moves only a slight amount and so the "moments" of these impulses about point O are essentially zero, and therefore angular momentum is conserved about this point.

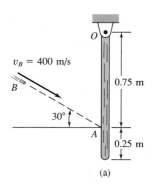

$v_B = 400$ m/s

0.75 m

30°

0.25 m

(a)

Fig. 19–9

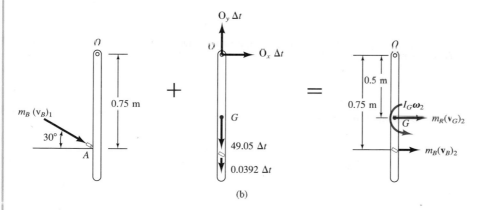

(b)

Conservation of Angular Momentum. From Fig. 19–9b, we have

$(\downarrow+)$ $\qquad \Sigma(H_O)_1 = \Sigma(H_O)_2$

$m_B(v_B)_1 \cos 30°(0.75 \text{ m}) = m_B(v_B)_2(0.75 \text{ m}) + m_R(v_G)_2(0.5 \text{ m}) + I_G\omega_2$

$(0.004)(400 \cos 30°)(0.75) = (0.004)(v_B)_2(0.75) + (5)(v_G)_2(0.5) + \left[\dfrac{1}{12}(5)(1)^2\right]\omega_2$

or

$\qquad 1.039 = 0.003(v_B)_2 + 2.50(v_G)_2 + 0.417\omega_2 \qquad\qquad (1)$

Kinematics. Since the rod is pinned at O, from Fig. 19–9c we have

$\qquad\qquad (v_G)_2 = 0.5\omega_2 \qquad (v_B)_2 = 0.75\omega_2$

Substituting into Eq. (1) and solving yields

$\qquad\qquad \omega_2 = 0.622$ rad/s $\uparrow$ $\qquad\qquad\qquad$ *Ans.*

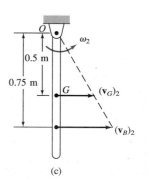

ω_2

0.5 m

0.75 m

G

$(v_G)_2$

$(v_B)_2$

(c)

19.4 Eccentric Impact

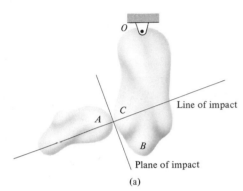

Fig. 19–10a

Impact occurs when two bodies collide with each other during a very *short* interval of time. During the collision very large equal but opposite impulsive forces are created between the two bodies at their point of contact. The concepts involving central and oblique impact of particles have been presented in Sec. 15.4. We will expand this treatment and discuss the eccentric impact of two bodies. *Eccentric impact* occurs when the line connecting the *mass centers of* the two bodies *does not* coincide with the line of impact.* This type of impact often occurs when one or both of the bodies is constrained to rotate about a fixed axis. Consider, for example, the collision between the two bodies A and B, shown in Fig. 19–10a, which collide at point C. Body B rotates about an axis passing through point O, whereas body A both rotates and translates. It is assumed that just before collision B is rotating counterclockwise with an angular velocity $(\omega_B)_1$, and the velocity of the contact point C located on A is $(\mathbf{u}_A)_1$. Kinematic diagrams for both bodies just before collision are shown in Fig. 19–10b. Provided the bodies are smooth at point C, the *impulsive forces* they exert on each other *are directed along the line of impact*. Hence, the component of velocity of point C on body B, which is directed along the line of impact, is $(v_B)_1 = (\omega_B)_1 r$, Fig. 19–10b. Likewise, on body A the component of velocity $(\mathbf{u}_A)_1$ along the line of impact is $(\mathbf{v}_A)_1$. In order for a collision to occur, $(v_A)_1 > (v_B)_1$.

During the impact an equal but opposite impulsive force $\mathbf{P}$ is exerted between the bodies which *deforms* their shapes at the point of contact. The resulting impulse is shown on the impulse diagrams for both bodies, Fig. 19–10c. Note that the impulsive force created at point C, on the rotating body B, creates impulsive pin reactions at the supporting pin O. On these diagrams it is assumed the impact creates forces which are much larger than the weights of the bodies. Hence, the impulses created by the weights are not shown since they are negligible compared to $\int \mathbf{P}\, dt$ and the reactive impulses at O. When the deformation of point C is a maximum, C on both the bodies moves with a common velocity $\mathbf{v}$ along the line of impact, Fig. 19–10d. A period of *restitution* then occurs in which the bodies tend to regain their original shapes. The restitution phase creates an equal but opposite impulsive force $\mathbf{R}$ acting between the bodies as shown on the impulse diagram, Fig. 19–10e. After restitution the bodies move apart such that point C on body B has a velocity $(\mathbf{v}_B)_2$ and point C on body A has a velocity $(\mathbf{u}_A)_2$, Fig. 19–10f, where $(v_B)_2 > (v_A)_2$.

In general, a problem involving the impact of two bodies requires determining the *two unknowns* $(v_A)_2$ and $(v_B)_2$, assuming $(v_A)_1$ and $(v_B)_1$ are known (or can be determined using kinematics, energy methods, the equations of motion, etc.) To solve this problem two equations must be written. The *first equation* generally involves application of *the conservation of angular momentum to the two bodies*. In the case of bodies A and B, we can state that

---

*When these lines coincide, central impact occurs and the problem can be analyzed as discussed in Sec. 15–4.

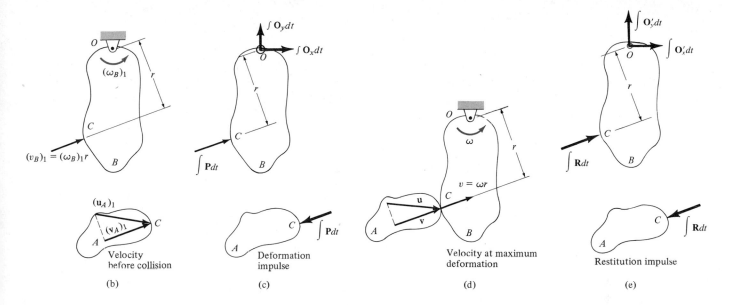

(b) Velocity before collision

(c) Deformation impulse

(d) Velocity at maximum deformation

(e) Restitution impulse

angular momentum is conserved about point O since the impulses at C are internal to the system and the impulses at O create zero moment (or zero angular impulse) about point O. The *second equation* is obtained using the definition of the *coefficient of restitution, e*, which is a ratio of the restitution impulse to the deformation impulse. To establish a useful form of this equation we must first apply the principle of angular impulse and momentum about point O to bodies B and A separately. Combining the results, we then obtain the necessary equation. Proceeding in this manner, the principle of impulse and momentum applied to body B from the time just before the collision to the instant of maximum deformation, Fig. 19–10b, c, and d, becomes

$$(\downarrow+) \qquad I_O(\omega_B)_1 + r \int P \, dt = I_O\omega \qquad (19\text{–}18)$$

Here I_O is the moment of inertia of body B about point O. Similarly, applying the principle of angular impulse and momentum from the instant of maximum deformation to the time just after the impact, Fig. 19–10d, e and f, yields

$$(\downarrow+) \qquad I_O\omega + r \int R \, dt = I_O(\omega_B)_2 \qquad (19\text{–}19)$$

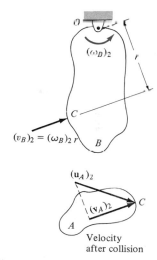

Fig. 19–10(b–f)

(f) Velocity after collision

Solving Eqs. 19–18 and 19–19 for $\int P \, dt$ and $\int R \, dt$, respectively, and formulating e, we have

$$e = \frac{\displaystyle\int R \, dt}{\displaystyle\int P \, dt} = \frac{r(\omega_B)_2 - r\omega}{r\omega - r(\omega_B)_1} = \frac{(v_B)_2 - v}{v - (v_B)_1}$$

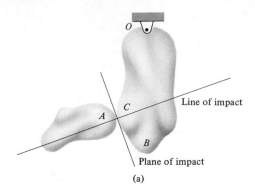

Line of impact

Plane of impact

(a)

Fig. 19–10a

In the same manner, we may write an equation which relates the magnitudes of velocity $(v_A)_1$ and $(v_A)_2$ of body A. The result is

$$e = \frac{(v_A)_2 - v}{v - (v_A)_1}$$

Combining the above equations by eliminating the common velocity v yields the desired result, i.e.,

$(\overset{+}{\rightarrow})$

$$\boxed{e = \frac{(v_B)_2 - (v_A)_2}{(v_A)_1 - (v_B)_1}} \qquad (19\text{--}20)$$

This equation is identical to Eq. 15–11, which was derived for the central impact occurring between two particles. Equation 19–20, however, states that the coefficient of restitution, measured at the points of contact (C) between two colliding bodies, is equal to the ratio of the relative velocity of *separation* of the points *just after impact* to the relative velocity at which the points *approach* one another *just* before impact. In deriving Eq. 19–20, we assumed that the points of contact for both bodies move to the right *both* before and after impact. If motion of any one of the contacting points occurs to the left, the velocity of this point is considered a negative quantity in Eq. 19–20.

As stated previously, when Eq. 19–20 is used in conjunction with the conservation of angular momentum for the bodies, it provides a useful means of obtaining the velocities of two colliding bodies just after collision.

Example 19–7

The 10–lb slender rod is suspended from the pin at A, Fig. 19–11a. If a 2-lb ball B is thrown at the rod and strikes its center with a horizontal velocity of 30 ft/s, determine the angular velocity of the rod just after impact. The coefficient of restitution is $e = 0.4$.

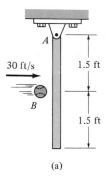

30 ft/s

1.5 ft

B

1.5 ft

Fig. 19–11

(a)

SOLUTION

Conservation of Angular Momentum. Consider the ball and rod as a system, Fig. 19–11b. Angular momentum is conserved about point A since the impulsive force between the rod and ball is *internal*. Also, the *weights* of the ball and rod are *nonimpulsive*. Noting the directions of the velocities of the ball and rod juster after impact as shown on the kinematic diagram, Fig. 19–11c, we require

$$(\downarrow+) \qquad (H_A)_1 = (H_A)_2$$

$$m_B(v_B)_1(1.5) = m_B(v_B)_2(1.5) + m_R(v_G)_2(1.5) + I_G\omega_2$$

$$\left(\frac{2}{32.2}\right)(30)(1.5) = \left(\frac{2}{32.2}\right)(v_B)_2(1.5) + \left(\frac{10}{32.2}\right)(v_G)_2(1.5) + \left(\frac{1}{12}\frac{10}{32.2}(3)^2\right)\omega_2$$

Since $(v_G)_2 = 1.5\omega_2$ then

$$2.795 = 0.09317(v_B)_2 + 0.9317\omega_2 \qquad (1)$$

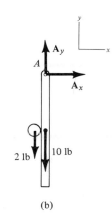

2 lb   10 lb

(b)

Coefficient of Restitution. With reference to Fig. 19–11c, we have

$$e = \frac{(v_G)_2 - (v_B)_2}{(v_B)_1 - (v_G)_1}; \qquad 0.4 = \frac{(1.5\omega_2) - (v_B)_2}{30 - 0}$$

$$12.0 = 1.5\omega_2 - (v_B)_2$$

Solving,

$$(v_B)_2 = -6.52 \text{ ft/s} = 6.52 \text{ ft/s} \leftarrow$$

$$\omega = 3.65 \text{ rad/s} \nwarrow \qquad\qquad\qquad Ans.$$

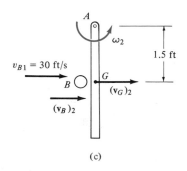

ω_2

1.5 ft

$v_{B1} = 30$ ft/s

B G

$(v_B)_2$ $(v_G)_2$

(c)

PROBLEMS

*19–28. The turntable T of a record player has a mass of 0.75 kg and a radius of gyration $k_z = 125$ mm. It is *turning freely* at $\omega_T = 2$ rad/s when a 50-g record (thin disk) falls on it. Determine the final angular velocity of the turntable just after the record stops slipping on the turntable.

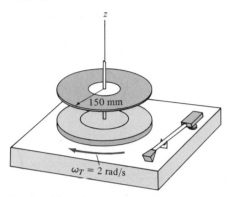

Prob. 19–28

19–29. A thin disk of mass m has an angular velocity ω_1 while rotating on a smooth surface. Determine its new angular velocity just after the hook at its edge strikes the peg P and the disk starts to rotate about P without rebounding.

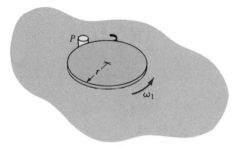

Prob. 19–29

19–30. The 2-kg rod ACB supports the two 4-kg disks at its ends. If both disks are given a clockwise angular velocity of $(\omega_A)_1 = (\omega_B)_1 = 5$ rad/s while the rod is held stationary and then released, determine the angular velocity of the rod after both disks have stopped spinning relative to the rod due to frictional resistance at the pins A and B. Motion is in the *horizontal plane*. Neglect friction at pin C.

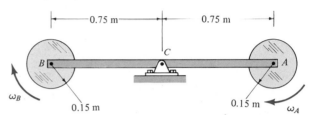

Probs. 19–30/19–31

19–31. Solve Prob. 19–30 if disk A is given an angular velocity of $(\omega_A)_1 = 5$ rad/s clockwise, and disk B has an angular velocity of $(\omega_B)_1 = 3$ rad/s counterclockwise.

*19–32. Determine the height h at which the disk of mass m must be struck with a horizontal force F so that no frictional force develops between it and the ground at A.

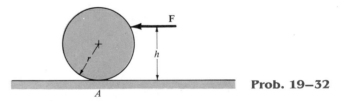

Prob. 19–32

19–33. Each of the two slender rods and the disk have the same mass m. Also, the length of each rod is equal to the diameter d of the disk. If the assembly is rotating with an angular velocity of ω_1 when the rods are directed outwards, determine the angular velocity of the assembly if by internal means the rods are brought to an upright vertical position.

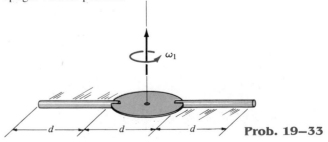

Prob. 19–33

19–34. A horizontal circular platform has a weight of 300 lb and a radius of gyration $k_z = 8$ ft about the z axis passing through its center O. The platform is free to rotate about the z axis and is initially at rest. A man having a weight of 150 lb throws a 15-lb block off the edge of the platform with a horizontal velocity of 5 ft/s, *measured relative to the platform*. Compute the angular velocity of the platform if the block is thrown (a) tangent to the platform, along the $+t$ axis, and (b) outward along a radial line, or $+n$ axis. Neglect the size of the man.

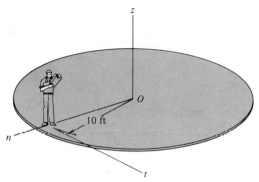

Prob. 19–34

19–35. The disk has a mass m and radius r. If it strikes a rough step having a height $\frac{1}{8}r$ as shown, determine the smallest angular velocity ω_1 the disk can have and not rebound off the step when it strikes it.

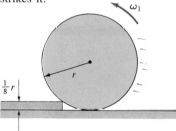

Prob. 19–35

*19–36. The rod of mass m and length L is released from rest without rotating. When it falls a distance L, the end A strikes the hook S, which provides a permanent connection. Determine the angular velocity ω of the rod after it has rotated 90°. Treat the rod's weight during impact as a nonimpulsive force.

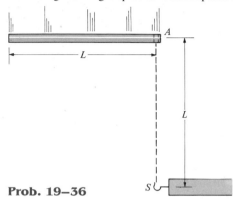

Prob. 19–36

19–37. The solid ball has a mass m and rolls without slipping along a horizontal plane with an angular velocity of ω_1. Provided it does not slip or rebound, determine its angular velocity as it just starts to roll up the inclined plane.

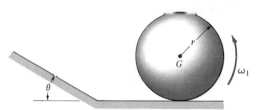

Prob. 19–37

19–38. The plank has a weight of 30 lb, center of gravity at G, and it rests on the two sawhorses at A and B. If the end D is raised 2 ft above the top of the sawhorses and is released from rest, determine how high end C will rise from the top of the sawhorses after the plank falls so that it rotates clockwise about A, strikes and pivots on the sawhorse at B, and rotates clockwise off the sawhorse at A.

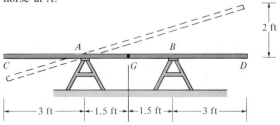

Prob. 19–38

19–39. The solid ball of mass m is dropped with a velocity $\mathbf{v}_1$ onto the edge of the rough step. If it rebounds horizontally off the step with a velocity $\mathbf{v}_2$, determine the angle θ at which contact occurs. Assume no slipping occurs when the ball strikes the step. The coefficient of restitution is e.

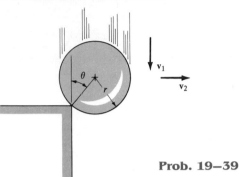

Prob. 19–39

*19–40. The rod has a length L and mass m. A smooth collar having a negligible size and one fourth the mass of the rod is placed on the rod at its midpoint. If the rod is freely rotating at ω about its end and the collar is released, determine the rod's angular velocity at the instant the collar flies off the rod. Also, what is the speed of the collar as it leaves the rod?

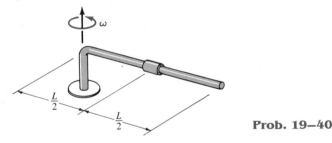

Prob. 19–40

19–41. The pendulum consists of a 10-lb sphere and 4-lb rod. If it is released from rest when $\theta_1 = 90°$, determine the angle θ_2 of rebound after the sphere strikes the floor. Take $e = 0.8$.

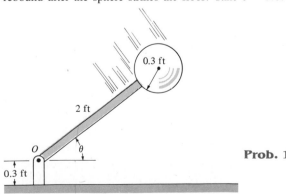

Prob. 19–41

19–42. The pendulum consists of a 10-lb solid ball and 4-lb rod. If it is released from rest when $\theta_1 = 0°$, determine the angle θ_2 of rebound after the ball strikes the wall and the pendulum swings up to the point of momentary rest. Take $e = 0.6$.

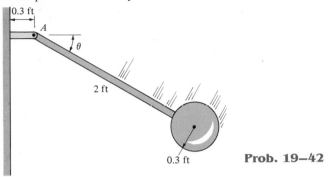

Prob. 19–42

19–43. A 7-g bullet having a velocity of 800 m/s is fired into the edge of the 5-kg disk as shown. Determine the angular velocity of the disk just after the bullet becomes embedded in it. Also, calculate how far θ the disk will swing until it stops. The disk is originally at rest.

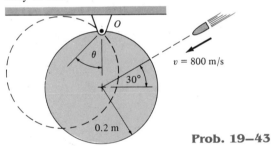

Prob. 19–43

*19–44. The 6-lb slender rod AB is released from rest when it is in the *horizontal position* so that it begins to rotate clockwise. A 1-lb ball is thrown at the rod with a velocity $v = 50$ ft/s. The ball strikes the rod at C at the instant the rod is in the vertical position as shown. Determine the angular velocity of the rod just after the impact. Take $e = 0.7$.

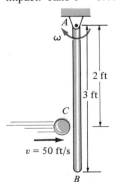

Prob. 19–44

Review 2: Planar Kinematics and Kinetics of a Rigid Body

Having presented the various topics in planar kinematics and kinetics in Chapters 16 through 19, we will now summarize these principles and provide an opportunity for applying them to the solution of various types of problems.

Kinematics

Here we are interested in studying the geometry of motion, without concern for the forces which cause the motion. Before solving a planar kinematics problem, it is *first* necessary to *classify the motion* as being either rectilinear or curvilinear translation, rotation about a fixed axis, or general plane motion. In particular, problems involving general plane motion can be solved either with reference to a fixed axis (absolute motion analysis) or using a translating or rotating frame of reference (relative motion analysis). The choice generally depends upon the type of constraints and the problem geometry. In all cases, application of the necessary equations may be clarified by drawing a kinematic diagram. Remember that the *velocity* of a point is always *tangent* to its path of motion, and the *acceleration* of a point can have *components* in the *n-t* directions when the path is *curved*.

Translation. When the body moves with rectilinear or curvilinear translation, *all* the points on the body have the *same motion*.

$$\mathbf{v}_B = \mathbf{v}_A, \qquad \mathbf{a}_B = \mathbf{a}_A$$

Rotation About a Fixed Axis.

Angular Motion

Variable Angular Acceleration. Provided a mathematical relationship is given between *any two* of the *four* variables θ, ω, α, and t, then a *third* variable can be determined by solving one of the following equations which relate all three variables.

$$\omega = \frac{d\theta}{dt} \qquad \alpha = \frac{d\omega}{dt} \qquad \alpha \, d\theta = \omega \, d\omega$$

Constant Angular Acceleration. The following equations apply when it is *absolutely certain* that the angular acceleration is constant.

$$\theta = \theta_0 + \omega_0 t + \tfrac{1}{2}\alpha_c t^2 \qquad \omega = \omega_0 + \alpha t \qquad \omega^2 = \omega_0^2 + 2\alpha_c(\theta - \theta_0)$$

Motion of Point P

$$v = \omega r \qquad\qquad \mathbf{v} = \boldsymbol{\omega} \times \mathbf{r}$$

$$a_t = \alpha r \quad a_n = \omega^2 r \qquad \mathbf{a} = \boldsymbol{\alpha} \times \mathbf{r} + \boldsymbol{\omega} \times (\boldsymbol{\omega} \times \mathbf{r})$$

General Plane Motion—Relative Motion Analysis.
Recall that when *translating axes* are placed at the "base point" *A*, the *relative motion* of point *B* with respect to *A* is simply *circular motion of B about A*. The following equations apply to two points *A* and *B* located on the *same* rigid body.

$$\mathbf{v}_B = \mathbf{v}_A + \mathbf{v}_{B/A}$$

$$\mathbf{a}_B = \mathbf{a}_A + \mathbf{a}_{B/A}$$

Rotating and translating axes are often used to analyze the motion of rigid bodies which are connected together by collars or slider blocks.

$$\mathbf{v}_B = \mathbf{v}_A + \boldsymbol{\Omega} \times \mathbf{r}_{B/A} + (\mathbf{v}_{B/A})_{\text{rel}}$$

$$\mathbf{a}_B = \mathbf{a}_A + \dot{\boldsymbol{\Omega}} \times \mathbf{r}_{B/A} + \boldsymbol{\Omega} \times (\boldsymbol{\Omega} \times \mathbf{r}_{B/A}) + 2\boldsymbol{\Omega} \times (\mathbf{v}_{B/A})_{\text{rel}} + (\mathbf{a}_{B/A})_{\text{rel}}$$

Kinetics

To analyze the forces which cause the motion we must use the principles of kinetics. When applying the necessary equations, it is important to first establish the inertial coordinate system and define the positive directions of the axes. The *directions* should be the *same* as those selected when writing any equations of kinematics provided *simultaneous solution* of equations becomes necessary.

Equations of Motion. These equations are used to determine accelerated motions or forces causing the motion. If used to determine position, velocity, or time of motion, then kinematics will have to be considered for part of the solution. Before applying the equations of motion, *always draw a free-body diagram* in order to identify all the forces acting on the body. Also, establish the directions of the acceleration of the mass center and the angular acceleration of the body. (A kinetic diagram may also be drawn in order to graphically represent $m\mathbf{a}_G$ and $I_G\boldsymbol{\alpha}$. This diagram is particularly convenient for resolving $m\mathbf{a}_G$ into components and for identifying the terms in the moment sum $\Sigma(\mathcal{M}_k)_P$.)

The three equations of motion are

$$\Sigma F_x = m(a_G)_x$$
$$\Sigma F_y = m(a_G)_y$$
$$\Sigma M_G = I_G\alpha \quad \text{or} \quad \Sigma M_P = \Sigma(\mathcal{M}_k)_P$$

In particular, if the body is rotating about a fixed axis, moments may also be summed about point O on the axis, in which case

$$\Sigma M_O = \Sigma(\mathcal{M}_k)_O = I_O\alpha$$

Work and Energy. *The equation of work and energy is used to solve problems involving force, velocity, and displacement.* Before applying this equation, *always draw a free-body diagram* of the body in order to identify the forces which do work. Recall that the kinetic energy of the body consists of that due to translational motion of the mass center, $\mathbf{v}_G$, *and* rotational motion of the body, $\boldsymbol{\omega}$.

$$T_1 + \Sigma U_{1-2} = T_2$$

where

$$T = \tfrac{1}{2}mv_G^2 + \tfrac{1}{2}I_G\omega^2$$

$$U_F = \int F \cos\theta \, ds \quad \text{(variable force)}$$

$$U_{F_c} = F_c \cos\theta \, (s_2 - s_1) \quad \text{(constant force)}$$
$$U_W = W \, \Delta y \quad \text{(weight)}$$
$$U_s = -(\tfrac{1}{2}ks_2^2 - \tfrac{1}{2}ks_1^2) \quad \text{(spring)}$$
$$U_M = M\theta \quad \text{(constant couple)}$$

If the forces acting on the body are conservative forces, then apply the conservation of energy equation. This equation is easier to use than the equation of work and energy, since it applies only at *two points* on the path and *does not* require calculation of the work done by a force as the body moves along the path.

$$T_1 + V_1 = T_2 + V_2$$

where

$$V_g = Wy \quad \text{(gravitational potential energy)}$$
$$V_e = \tfrac{1}{2}ks^2 \quad \text{(elastic potential energy)}$$

Impulse and Momentum. *The principles of linear and angular impulse and momentum are used to solve problems involving force, velocity, and time.* Before applying the equations, *draw a free-body diagram* in order to identify all the forces which cause linear and angular impulses on the body. Also, establish the directions of the velocity of the mass center and the angular velocity of the body just before and just after the impulses are applied. (As an alternative procedure, the impulse and momentum diagrams may accompany the solution in order to graphically account for the terms in the equations. These diagrams are particularly advantageous when computing the angular impulses and angular momenta about a point other than the body's mass center.)

$$m(\mathbf{v}_G)_1 + \Sigma \int \mathbf{F} \, dt = m(\mathbf{v}_G)_2$$

$$(\mathbf{H}_G)_1 + \Sigma \int \mathbf{M}_G \, dt = (\mathbf{H}_G)_2$$

or

$$(\mathbf{H}_O)_1 + \Sigma \int \mathbf{M}_O \, dt = (\mathbf{H}_O)_2$$

Conservation of Momentum. If nonimpulsive forces or no impulsive forces act on the body in a particular direction, or the motions of several bodies are involved in the problem, then consider applying the conservation of linear or angular momentum for the solution. Investigation of the free-body diagram (or the impulse diagram) will aid in determining the directions for which the impulsive forces are zero, or axes for which the impulsive forces cause zero angular momentum. When this occurs, then

$$m(\mathbf{v}_G)_1 = m(\mathbf{v}_G)_2$$
$$(\mathbf{H}_O)_1 = (\mathbf{H}_O)_2$$

The problems that follow involve application of all the above concepts. They are presented in *random order* so that practice may be gained at identifying the various types of problems and developing the skills necessary for their solution.

PROBLEMS

R2–1. The 40-lb cylinder rests on an inclined surface having a coefficient of kinetic friction $\mu_k = 0.1$. A cord is wrapped around the outer surface of the cylinder. If the cylinder is released from rest, determine the velocity of the center of the cylinder one second after it is released. Neglect the mass of the cord.

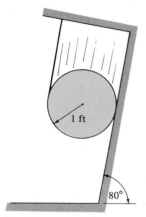

Prob. R2–1

R2–2. The double pendulum consists of two rods. Rod AB has a constant angular velocity of 3 rad/s, and rod BC has a constant angular velocity of 2 rad/s, both measured counterclockwise. Determine the velocity and acceleration of point C at the instant shown.

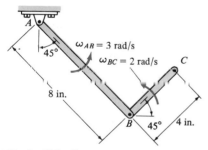

Prob. R2–2

R2–3. A chain that has a negligible mass is draped over a sprocket which has a mass of 2 kg and a radius of gyration of $k_O = 50$ mm. If the 4-kg block A is released from rest in the position shown, $s = 1$ m, determine the angular velocity which the chain imparts to the sprocket when $s = 2$ m.

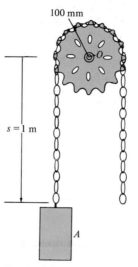

Prob. R2–3

*R2–4. A tape having a thickness s wraps around the wheel which is turning at a constant rate $\boldsymbol{\omega}$. Assuming the unwrapped portion of tape remains horizontal, determine the acceleration of point P on the tape when the radius is r. *Hint:* Since $v_P = \omega r$, take the time derivative and note that $dr/dt = \omega(s/2\pi)$.

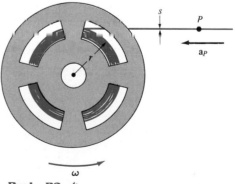

Prob. R2–4

R2–5. The 2-lb bottle rests on the checkout conveyor at a grocery store. If the coefficient of static friction is $\mu_s = 0.2$, determine the largest acceleration which the conveyor can have without causing the bottle to slip or tip. The center of mass is at G.

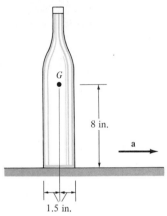

8 in.

a

1.5 in.

Prob. R2–5

R2–6. The 20-lb solid ball is cast on the floor such that it has a backspin of $\omega = 15$ rad/s and its center has an initial horizontal velocity of $v_G = 20$ ft/s. If the coefficient of kinetic friction between the floor and the ball is $\mu_A = 0.3$, determine the distance it travels before it stops spinning.

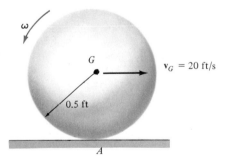

$v_G = 20$ ft/s

0.5 ft

A

Probs. R2–6/R2–7

R2–7. Determine the backspin ω which should be given to the 20-lb ball so that when its center is given an initial horizontal velocity of $v_G = 20$ ft/s it stops spinning and translating at the same instant. The coefficient of kinetic friction is $\mu_A = 0.3$.

*R2–8. Gear A is pinned at B and rotates along the periphery of the circular gear rack R. If A has a weight of 4 lb and a radius of gyration of $k_B = 0.5$ ft, determine the angular momentum of gear A about point C when $\omega_{CB} = 30$ rad/s and (a) $\omega_R = 0$, (b) $\omega_R = 20$ rad/s.

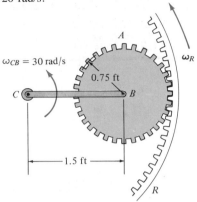

$\omega_{CB} = 30$ rad/s

A

0.75 ft

ω_R

C

B

1.5 ft

R

Prob. R2–8

R2–9. A large roll of paper having a mass of 20 kg and radius $r = 150$ mm is resting over the edge of a table. If the roll is disturbed slightly from its equilibrium position, determine the angle θ at which it begins to leave the table edge A as it falls. The centroidal radius of gyration of the roll is $k_G = 75$ mm.

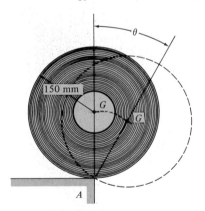

θ

150 mm

G

G

A

Prob. R2–9

R2–10. Knowing that the angular velocity of link CB is $\omega_{CB} = 6$ rad/s, determine the velocity of the collar A at the instant shown. Use Eq. 16–14.

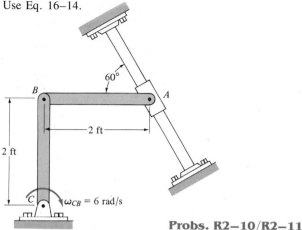

Probs. R2–10/R2–11

R2–11. Solve Prob. R2–10 using the method of instantaneous center of zero velocity.

*R2–12.** A 20-kg roll of paper, originally at rest, is pin-supported at its ends to bracket AB. The roll rests against a wall for which the coefficient of kinetic friction at C is $\mu_C = 0.3$. If a force of 40 N is applied uniformly to the end of the sheet, determine the initial angular acceleration of the roll and the tension in the bracket as the paper unwraps. For the calculation, treat the roll as a cylinder.

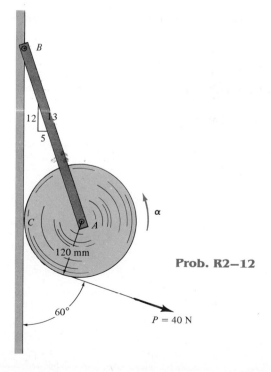

Prob. R2–12

R2–13. The spool and wire wrapped around its core have a mass of 20 kg and a centroidal radius of gyration $k_G = 250$ mm. If the coefficient of kinetic friction at the ground is $\mu_B = 0.1$, determine the angular acceleration of the spool when the 30-N · m couple moment is applied.

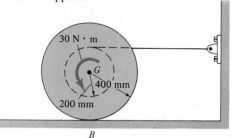

Prob. R2–13

R2–14. The spool has a weight of 30 lb and a radius of gyration of $k_O = 1.40$ ft. If a force of 40 lb is applied to the supporting cord at A as shown, determine the angular velocity of the spool in $t = 3$ s starting from rest. Neglect the mass of the pulley and cord.

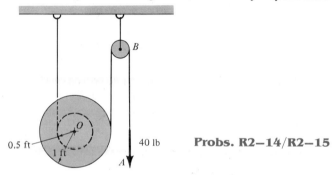

Probs. R2–14/R2–15

R2–15. Solve Prob. R2–14 if a 40-lb block is suspended from the cord at A, rather than applying the 40-lb force.

*R2–16.** If the support at B is suddenly removed, determine the initial horizontal and vertical components of reaction which the pin A exerts on the rod ACB. Segments AC and CB each have a weight of 10 lb.

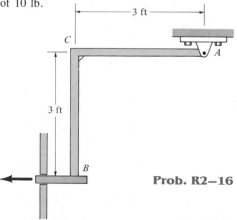

Prob. R2–16

R2–17. Locate the center of mass G of the baseball bat by determining $\bar{y}$. Then, compute the moment of inertia of the bat about an axis which passes through G and is perpendicular to the plane of the page. For the calculation, consider the bat to be composed of a truncated cone and cylinder. Neglect the size of the lip at A. The density of wood is $\rho_w = 750$ kg/m^3.

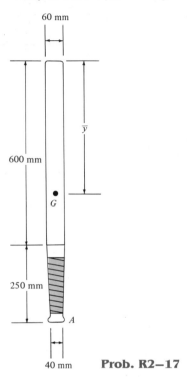

Prob. R2–17

R2–18. Link CD has an angular velocity of 3 rad/s at the instant shown. Determine the angular velocity of link AB at this instant.

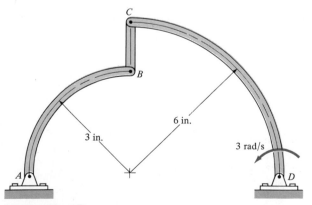

Prob. R2–18

R2–19. The assembly consists of a uniform rod AB having a mass of 0.2 kg and two solid spheres C and D having a mass of 0.4 kg and 0.6 kg, respectively. Determine $\bar{x}$, which locates the center of mass G, and then calculate the moment of inertia about an axis perpendicular to the page and passing through G.

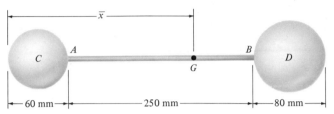

Prob. R2–19

*R2–20. The cart and its contents have a mass of 40 kg and a mass center at G, excluding the wheels. Each of the two wheels has a mass of 2 kg and a radius of gyration of $k_O = 0.120$ m. If the cart is released from rest from the position shown, determine its speed after it travels 4 m down the incline. The coefficient of kinetic friction is $\mu_A = 0.3$ between the incline and A. The wheels roll without slipping at B.

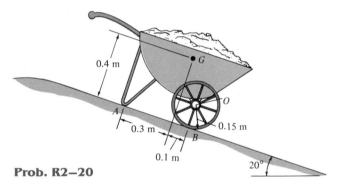

Prob. R2–20

R2–21. The links AB and BC each have a mass of 6 kg. An elastic cable is attached to A and C and has an unstretched length of 0.2 m. If the links are extended out into the horizontal position, then released from rest ($\theta \approx 0°$) so that they move out of equilibrium, compute the velocity of C at the instant $\theta = 45°$. Neglect the mass of the roller.

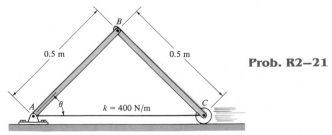

Prob. R2–21

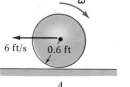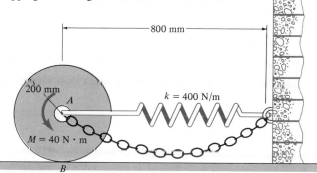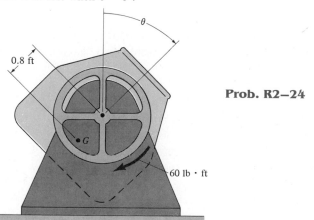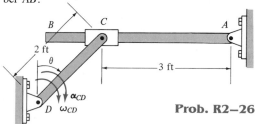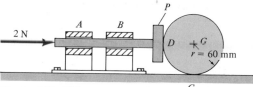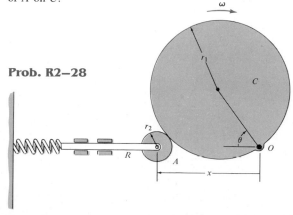

R2–22. If the ball has a weight of 15 lb and is thrown onto a *rough surface* with a velocity of 6 ft/s parallel to the surface, determine the amount of backspin, ω, it must be given so that it stops spinning at the same instant that its forward velocity is zero. It is not necessary to know the coefficient of kinetic friction at A for the calculation.

Prob. R2–22

R2–23. The wheel has a mass of 20 kg and a radius of gyration of $k_A = 0.13$ m. The attached spring is unstretched and is restricted from stretching more than 300 mm due to the chain. If the wheel is subjected to a torque of $M = 40$ N·m, determine its angular velocity just before the chain becomes taut. Assume no slipping at B. Neglect the mass of the chain.

Prob. R2–23

*R2–24. The tub of the mixer has a weight of 70 lb and a radius of gyration about its center of gravity of $k_G = 1.3$ ft. If a constant torque of 60 lb·ft is applied to the dumping wheel, determine the angular velocity of the tub when it has turned 90°. Originally the tub is at rest when $\theta = 0°$.

Prob. R2–24

R2–25. A flywheel has its angular speed increased uniformly from 15 rad/s to 60 rad/s in 120 s. If the diameter of the wheel is 2 ft, determine the normal and tangential components of acceleration of a point on the rim of the wheel when $t = 120$ s, and the total distance the point travels during the time period.

R2–26. At the instant $\theta = 45°$ link CD has an angular velocity of $\omega_{CD} = 4$ rad/s and an angular acceleration of $\alpha_{CD} = 2$ rad/s^2. Determine the angular velocity and angular acceleration of member AB.

Prob. R2–26

R2–27. The 3-kg ball rests on a horizontal surface and is being pushed forward by moving the piston P with a force of 2 N. If the ball rolls without slipping at C and the coefficient of kinetic friction between the piston and the ball at D is $\mu_D = 0.20$, compute the velocity of the ball's center of mass G after the piston moves 50 mm. Neglect the mass of the piston and assume that the guides at A and B are smooth. The ball is originally at rest.

Prob. R2–27

*R2–28. Compute the velocity of rod R for any angle θ of the cam C if the cam rotates with a constant angular velocity ω. The pin connection at O does not cause an interference with the motion of A on C.

Prob. R2–28

405

R2–29. The lawn roller has a mass of 80 kg and a radius of gyration of $k_G = 0.175$ m. If it is pushed forward with a force of 200 N when the handle is at 45°, determine its angular acceleration. The coefficients of static and kinetic friction between the ground and the roller are $\mu_s = 0.12$ and $\mu_k = 0.10$, respectively.

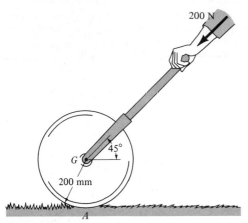

Prob. R2–29

R2–30. A 4-kg disk A is mounted on arm BC, which has a negligible mass. If a constant torque of $M = (5e^{0.5t})$ N · m, where t is in seconds, is applied to the arm C, determine the angular velocity of BC in 2 s starting from rest. Solve the problem assuming that (a) the disk is set in a smooth bearing at B so that it rotates with curvilinear translation and (b) the disk is fixed to the shaft BC.

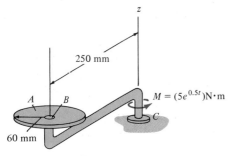

Prob. R2–30

R2–31. The 10-kg rod AB is pin-connected at A and subjected to a couple moment of $M = 15$ N · m. If the rod is released from rest when the spring is unstretched, at $\theta = 30°$, determine the rod's angular velocity at the instant $\theta = 60°$. As the rod rotates, the spring always remains horizontal, because of the roller support at C.

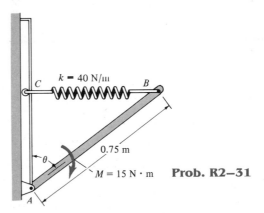

Prob. R2–31

*R2–32. The assembly weighs 10 lb and has a radius of gyration $k_G = 0.6$ ft about its center of mass G. The kinetic energy of the assembly is 31 ft-lb when it is in the position shown. If it is rolling counterclockwise on the surface without slipping, determine its linear momentum at this instant.

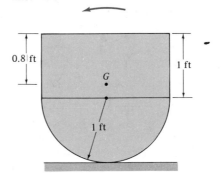

Prob. R2–32

R2–33. The pendulum consists of a 30-lb sphere and a 10-lb slender rod. Compute the reaction at the pin O just after the cord AB is cut.

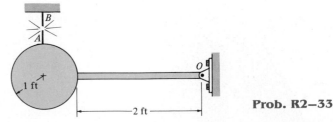

Prob. R2–33

R2–34. The center of the pulley is being lifted vertically with an acceleration of 4 m/s² at the instant it has a velocity of 2 m/s. If the cable does not slip on the pulley's surface, determine the accelerations of the cylinder B and point C on the pulley.

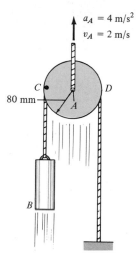

Prob. R2–34

R2–35. The 50-kg cylinder has an angular velocity of 30 rad/s when it is brought into contact with the horizontal surface at C. If the coefficient of kinetic friction is $\mu_k = 0.2$, determine how long it takes for the cylinder to stop spinning. What force is developed at the pin A during this time? The axis of the cylinder is connected to *two* symmetrical links. (Only AB is shown.) For the computation, neglect the weight of the links.

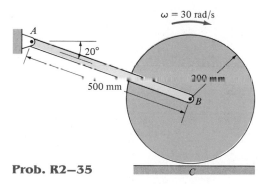

Prob. R2–35

*R2–36. The board rests on the surface of two drums. At the instant shown, it has an acceleration of 0.5 m/s² to the right, while at the same instant points on the outer rim of each drum have an acceleration with a magnitude of 3 m/s². If the board does not slip on the drums, determine its speed due to the motion.

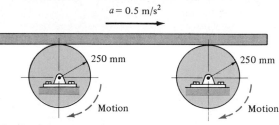

Prob. R2–36

R2–37. The uniform slender rod AB has a mass of 4 kg and a length of 2 m. If it is suspended horizontally by a spring at A and a cord at B, determine the angular acceleration of the rod and the acceleration of the rod's mass center at the instant the cord at B is cut. *Hint:* The stiffness of the spring is not needed for the calculation.

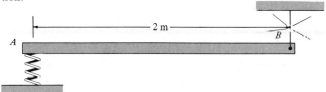

Prob. R2–37

R2–38. The tire has a mass of 9 kg and a radius of gyration $k_O = 225$ mm. If it is released from rest and rolls down the plane without slipping, determine the speed of its center O when $t = 3$ s.

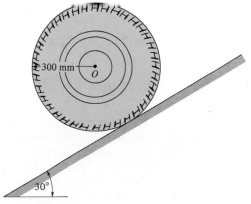

Prob. R2–38

407

R2–39. The double pulley consists of two wheels which are attached to one another and turn at the same rate. The pulley has a mass of 15 kg and a radius of gyration $k_O = 110$ mm. If the block at A has a mass of 40 kg, determine the speed of the block in 3 s after a constant force of 2 kN is applied to the rope wrapped around the inner hub of the pulley. The block is originally at rest. Neglect the mass of the rope.

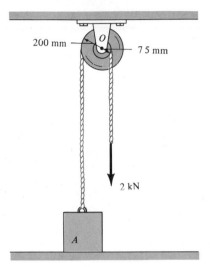

Prob. R2–39

*R2–40. The spool has a weight of 150 lb and a radius of gyration of $k_O = 2.25$ ft. If a cord is wrapped around its inner core and the end is pulled with a horizontal force of $P = 40$ lb, determine the angular velocity of the spool after the center O has moved 10 ft to the right. The spool starts from rest and does not slip as it rolls. Neglect the mass of the cord.

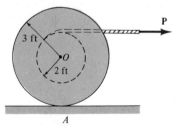

Probs. R2–40/R2–41

R2–41. Determine the magnitude of the horizontal force **P** that must be applied to the cord which is wrapped around the inner core of the spool, so that the spool attains an angular velocity of $\omega = 5$ rad/s when the center O has moved 10 ft to the right. Neglect the mass of the cord. The spool has a weight of 150 lb and a radius of gyration of $k_O = 2.25$ ft. It starts from rest and does not slip as it rolls.

R2–42. At the instant shown, two forces act on the 30-lb slender rod which is pinned at O. Determine the magnitude of force **F** and the initial angular acceleration of the rod so that the horizontal reaction which the *pin exerts on the rod* is 5 lb directed to the right.

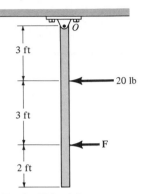

Prob. R2–42

R2–43. The disk A maintains a constant clockwise angular velocity of 30 rad/s. If the 20-kg disk B is initially at rest when it is brought into contact with A, determine the time required for B to attain the same angular velocity as A. The coefficient of kinetic friction between the two disks is $\mu_k = 0.3$. Neglect the mass of bar BC.

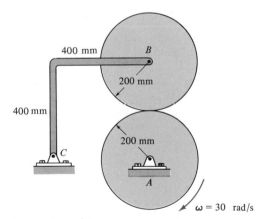

Prob. R2–43

*R2–44. The door has a weight of 200 lb and a center of gravity at G. Determine how far the door moves in 2 s, starting from rest, if a man pushes on it at C with a horizontal force of 30 lb. Also, find the vertical reactions at the rollers A and B.

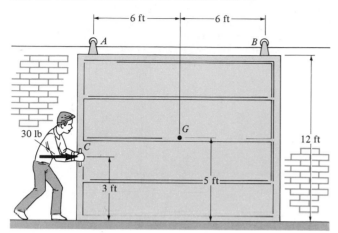

Prob. R2–44

R2–45. The ring has a mass of 10 kg, a center of mass at G, and a radius of gyration $k_G = 135$ mm. If its angular velocity is $\omega_1 = 2$ rad/s when it is in the position shown, compute its angular acceleration at this instant and the normal force of the ring on the ground. Assume that slipping does not occur.

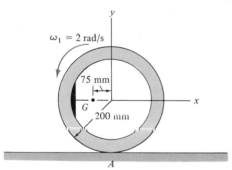

Prob. R2–45

R2–46. The uniform 150-lb stone (rectangular block) is being turned over on its side by pulling the vertical cable *slowly* upward until the stone begins to tip. If it then falls freely ($\mathbf{T} = \mathbf{0}$) from an essentially balanced at rest position, determine the speed at which the corner A strikes the pad at B. The stone does not slip at its corner C as it falls.

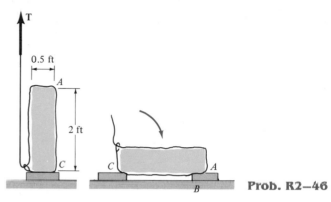

Prob. R2–46

R2–47. The cylinder having a mass of 5 kg is initially at rest when it is placed into contact with the wall B and the rotor at A. If the rotor always maintains a constant clockwise angular velocity $\omega = 6$ rad/s, determine the initial angular acceleration of the cylinder. The coefficient of kinetic friction at the contacting surfaces B and C is $\mu_k = 0.2$.

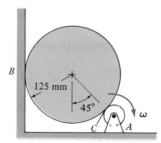

Prob. R2–47

*R2–48. A concrete solid is formed by rotating the shaded area about the y axis. Determine the moment of inertia I_y. The density of material is $\rho = 150$ lb/ft^3.

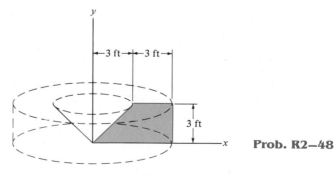

Prob. R2–48

R2–49. A uniform rod having a weight of 10 lb is pin-supported at A from a roller which rides on a horizontal track. If the rod is originally at rest, and a horizontal force of $F = 15$ lb is applied to the roller, determine the acceleration of the roller. Neglect the mass of the roller and its size d in the computations.

R2–50. Solve Prob. R2–49 assuming that the roller at A is replaced by a slider block having a negligible mass. The coefficient of kinetic friction between the block and the track is $\mu_k = 0.2$. Neglect the dimension d and the size of the block in the computations.

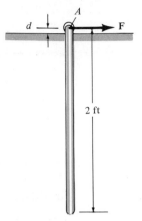

Probs. R2–49/R2–50

20

Three-Dimensional Kinematics
of a Rigid Body

Three types of planar rigid-body motion have been presented in Chapter 16: translation, rotation about a fixed axis, and general plane motion. In this chapter the three-dimensional motion of a rigid body, which consists of rotation about a fixed point and general motion, will be discussed. This is followed by a general study of the motion of particles and rigid bodies using coordinate systems which both translate and rotate. The analysis of these motions is more complex than planar-motion analysis since the angular acceleration of the body will measure a change in both the *magnitude* and *direction* of the body's angular velocity. In order to simplify the motion's three-dimensional aspects, throughout the chapter we will make complete use of vector analysis.*

*20.1 Rotation About a Fixed Point

When a rigid body rotates about a fixed point, the distance *r* from the point to a particle *P* located in the body is the *same* for *any position* of the body. Thus, the path of motion for the particle lies on the *surface of a sphere* having a radius *r* and centered at the fixed point. Since motion along this path occurs only from a series of rotations made during a finite time interval, we will first develop a familiarity with some of the properties of rotational displacements.

*A brief review of vector analysis is given in Appendix C.

Euler's Theorem

This theorem states that two ''component'' rotations about different axes passing through a point are equivalent to a single resultant rotation about an axis passing through the point. If more than two rotations are applied, they can be combined into pairs, and each pair can be further reduced to combine into one rotation.

Finite Rotations

If the component rotations used in Euler's theorem are *finite,* it is important that the *order* in which they are applied be maintained. This is because finite rotations do *not* obey the law of vector addition, and hence they cannot be classified as vector quantities. To show this, consider the two finite rotations $\theta_1 + \theta_2$ applied to the block in Fig. 20–1*a*. Each rotation has a magnitude of 90° and a direction defined by the right-hand rule, as indicated by the arrow. The resultant orientation of the block is shown at the right. When these two rotations are applied in the order $\theta_2 + \theta_1$, as shown in Fig. 20–1*b*, the resultant position of the block is *not* the same as it is in Fig. 20–1*a*. Consequently, *finite rotations* do not obey the commutative law of addition ($\theta_1 + \theta_2 \neq \theta_2 + \theta_1$), and therefore *they cannot be classified as vectors*. If smaller, yet finite, rotations had been used to illustrate this point, e.g., 10° instead of 90°, the *resultant* orientation of the block after each combination of rotations would also be different; however, in this case, only by a small amount.

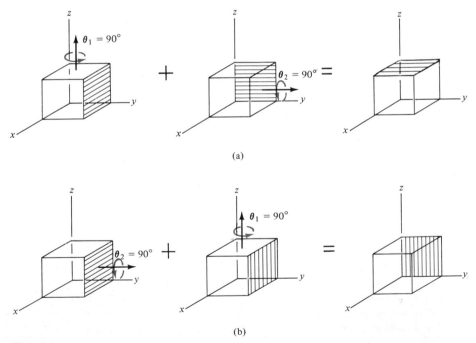

(a)

(b)

Fig. 20–1

Infinitesimal Rotations

When defining the angular motions of a body subjected to three-dimensional motion, only rotations which are *infinitesimally small* will be considered. *Such rotations may be classified as vectors, since they can be added vectorially in any manner.* For example, let us for purposes of simplicity consider the rigid body itself to be a sphere which is allowed to rotate about its central fixed point O, Fig. 20–2a. If we impose two infinitesimal rotations $d\boldsymbol{\theta}_1 + d\boldsymbol{\theta}_2$ on the body, it is seen that point P moves along the path $d\boldsymbol{\theta}_1 \times \mathbf{r} + d\boldsymbol{\theta}_2 \times \mathbf{r}$ and ends up at P'. Had the two successive rotations occurred in the order $d\boldsymbol{\theta}_2 + d\boldsymbol{\theta}_1$, then the resultant displacements of P would have been $d\boldsymbol{\theta}_2 \times \mathbf{r} + d\boldsymbol{\theta}_1 \times \mathbf{r}$. Since the vector cross product obeys the distributive law, by comparison $(d\boldsymbol{\theta}_1 + d\boldsymbol{\theta}_2) \times \mathbf{r} = (d\boldsymbol{\theta}_2 + d\boldsymbol{\theta}_1) \times \mathbf{r}$. Hence infinitesimal rotations $d\boldsymbol{\theta}$ are vectors, since these quantities have both a magnitude and direction for which the order of (vector) addition is not important, i.e., $d\boldsymbol{\theta}_1 + d\boldsymbol{\theta}_2 = d\boldsymbol{\theta}_2 + d\boldsymbol{\theta}_1$. Furthermore, as shown in Fig. 20–2a, the two "component" rotations $d\boldsymbol{\theta}_1$ and $d\boldsymbol{\theta}_2$ are equivalent to a single resultant rotation $d\boldsymbol{\theta} = d\boldsymbol{\theta}_1 + d\boldsymbol{\theta}_2$, a consequence of Euler's theorem.

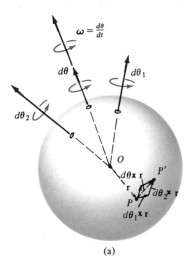

(a)

Angular Velocity

If the body is subjected to an angular rotation $d\boldsymbol{\theta}$ about a fixed point, the angular velocity of the body is defined by the time derivative,

$$\boldsymbol{\omega} = \dot{\boldsymbol{\theta}} \qquad (20\text{–}1)$$

The line specifying the direction of $\boldsymbol{\omega}$, which is collinear with $d\boldsymbol{\theta}$, is referred to as the *instantaneous axis of rotation,* Fig. 20–2b. In general, this axis changes direction during each instant of time. Since $\boldsymbol{\omega}$ is a vector quantity, it follows from vector addition that if the body is subjected to two component angular motions, $\boldsymbol{\omega}_1 = \dot{\boldsymbol{\theta}}_1$ and $\boldsymbol{\omega}_2 = \dot{\boldsymbol{\theta}}_2$, the resultant angular velocity is $\boldsymbol{\omega} = \boldsymbol{\omega}_1 + \boldsymbol{\omega}_2$.

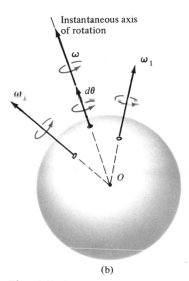

(b)

Fig. 20–2

Angular Acceleration

The body's angular acceleration is determined from the time derivative of the angular velocity, i.e.,

$$\boldsymbol{\alpha} = \dot{\boldsymbol{\omega}} \qquad (20\text{–}2)$$

For motion about a fixed point, $\boldsymbol{\alpha}$ must account for a change in *both* the magnitude and direction of $\boldsymbol{\omega}$, so that, in general, $\boldsymbol{\alpha}$ is not directed along the instantaneous axis of rotation, Fig. 20–3.

As the direction of the instantaneous axis of rotation (or the line of action of $\boldsymbol{\omega}$) changes in space, the locus of points defined by the axis generates a fixed *space cone.* If the change in this axis is viewed with respect to the rotating

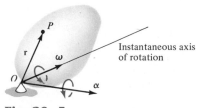

Fig. 20–3

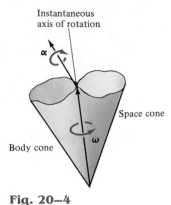

Instantaneous
axis of rotation

Space cone

Body cone

Fig. 20–4

body, the locus of the axis generates a *body cone*, Fig. 20–4. At any given instant, these cones are tangent along the instantaneous axis of rotation, and when the body is in motion, the body cone appears to roll either on the inside or the outside surface of the fixed space cone. Provided the paths defined by the open ends of the cones are described by the head of the $\boldsymbol{\omega}$ vector, $\boldsymbol{\alpha}$ must act tangent to these paths at any given instant, since the time rate of change of $\boldsymbol{\omega}$ is equal to $\boldsymbol{\alpha}$, Fig. 20–4.

Velocity

Once $\boldsymbol{\omega}$ is specified, the velocity of any point P on a body rotating about a fixed point can be determined using the same methods as for a body rotating about a fixed axis (Sec. 16.3). Hence, by the cross product,

$$\boxed{\mathbf{v} = \boldsymbol{\omega} \times \mathbf{r}} \qquad (20\text{–}3)$$

Here $\mathbf{r}$ defines the position of P measured from the fixed point O, Fig. 20–3.

Acceleration

If $\boldsymbol{\omega}$ and $\boldsymbol{\alpha}$ are known at a given instant, the acceleration of any point P on the body can be obtained by time differentiation of Eq. 20–3, which yields

$$\boxed{\mathbf{a} = \boldsymbol{\alpha} \times \mathbf{r} + \boldsymbol{\omega} \times (\boldsymbol{\omega} \times \mathbf{r})} \qquad (20\text{–}4)$$

The form of this equation is the same as that developed in Sec. 16.3, which defines the acceleration of a point located on a body subjected to rotation about a fixed axis.

*20.2 The Time Derivative of a Vector Measured from a Fixed and Translating-Rotating System

In many types of problems involving the motion of a body about a fixed point, the angular velocity $\boldsymbol{\omega}$ is specified in terms of its component angular motions. For example, the disk in Fig. 20–5 spins about the horizontal y axis at $\boldsymbol{\omega}_s$ while it rotates or precesses about the vertical z axis at $\boldsymbol{\omega}_p$. Therefore, its resultant angular velocity is $\boldsymbol{\omega} = \boldsymbol{\omega}_s + \boldsymbol{\omega}_p$. If the angular acceleration $\boldsymbol{\alpha}$ of such a body is to be determined, it is sometimes easier to compute the time derivative of $\boldsymbol{\omega}$, Eq. 20–2, by using a coordinate system which has a *rotation* defined by one or more of the components of $\boldsymbol{\omega}$.* For this reason, and for other uses later, an equation will presently be derived that relates the time

---

*In the case of the spinning disk, Fig. 20–5, the x, y, z axes may be given an angular velocity of $\boldsymbol{\omega}_p$.

derivative of any vector **A** defined from a translating-rotating reference to its derivative defined from a fixed reference.

Consider the x, y, z axes of the moving frame of reference to have an angular velocity $\mathbf{\Omega}$ which is measured from the fixed X, Y, Z axes, Fig. 20–6a. In the following discussion, it will be convenient to express vector **A** in terms of its **i, j, k** components, which define the directions of the moving axes. Hence,

$$\mathbf{A} = A_x\mathbf{i} + A_y\mathbf{j} + A_z\mathbf{k}$$

In general, the time derivative of **A** must account for the change in both the vector's magnitude and direction. However, if this derivative is taken *with respect to the moving frame of reference,* only a change in the magnitudes of the components of **A** must be accounted for, since the directions of the components do not change with respect to the moving reference. Hence,

$$(\dot{\mathbf{A}})_{xyz} = \dot{A}_x\mathbf{i} + \dot{A}_y\mathbf{j} + \dot{A}_z\mathbf{k} \qquad (20\text{–}5)$$

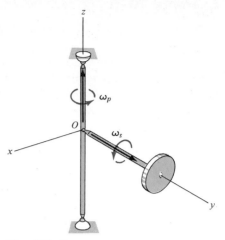

Fig. 20–5

When the time derivative of **A** is taken *with respect to the fixed frame of reference,* the *directions* of **i, j,** and **k** change only on account of the rotation $\mathbf{\Omega}$ of the axes and not their translation. Hence, in general,

$$(\dot{\mathbf{A}})_{XYZ} = \dot{A}_x\mathbf{i} + \dot{A}_y\mathbf{j} + \dot{A}_z\mathbf{k} + A_x\dot{\mathbf{i}} + A_y\dot{\mathbf{j}} + A_z\dot{\mathbf{k}}$$

The time derivatives of the unit vectors will now be considered. For example, $\dot{\mathbf{i}} = d\mathbf{i}/dt$ represents only a change in the direction of **i** with respect to time, since **i** has a fixed magnitude of 1 unit. As shown in Fig. 20–6b, the change, $d\mathbf{i}$, is *tangent to the path* described by the head of **i** as **i** moves due to the rotation $\mathbf{\Omega}$. Accounting for both the magnitude and direction of $d\mathbf{i}$, we can therefore define $\dot{\mathbf{i}}$ using the cross product, $\dot{\mathbf{i}} = \mathbf{\Omega} \times \mathbf{i}$. In general,

$$\dot{\mathbf{i}} = \mathbf{\Omega} \times \mathbf{i} \qquad \dot{\mathbf{j}} = \mathbf{\Omega} \times \mathbf{j} \qquad \dot{\mathbf{k}} = \mathbf{\Omega} \times \mathbf{k}$$

These formulations were also developed in Sec. 16.8, regarding planar motion of the axes. Substituting the results into the above equation and using Eq. 20–5 yields

$$\boxed{(\dot{\mathbf{A}})_{XYZ} = (\dot{\mathbf{A}})_{xyz} + \mathbf{\Omega} \times \mathbf{A}} \qquad (20\text{–}6)$$

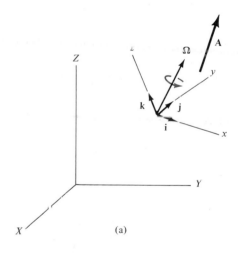

(a)

This result is rather important and it will be used throughout Sec. 20.4 and Chapter 21. In words, it states that *the time derivative of* **A** *as observed from the fixed X, Y, Z frame of reference is equal to the time rate of change of* **A** *as observed from the x, y, z translating-rotating frame of reference, Eq. 20–5, plus* $\mathbf{\Omega} \times \mathbf{A}$, *the change of* **A** *caused by the rotation of the x, y, z frame.*

The following two example problems numerically illustrate the use of this equation for obtaining the angular acceleration of a body rotating about a fixed point.

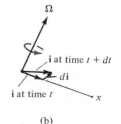

(b)

Fig. 20–6

415

Example 20–1

The disk shown in Fig. 20–7 is spinning about its horizontal axis with a constant angular velocity $\omega_s = 3$ rad/s, while the horizontal platform upon which the disk is mounted is rotating about the vertical axis at a constant rate $\omega_p = 1$ rad/s. Determine the angular acceleration of the disk and the velocity and acceleration of point A on the disk when it is in the position shown.

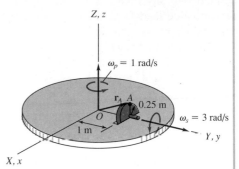

SOLUTION

Point O represents a fixed point of rotation for the disk if one considers a hypothetical extension of the disk to this point. To determine the velocity and acceleration of point A, it is first necessary to determine the resultant angular velocity $\boldsymbol{\omega}$ and angular acceleration $\boldsymbol{\alpha}$ of the disk, since these vectors are used in Eqs. 20–3 and 20–4.

Fig. 20–7

Angular Velocity. The angular velocity is simply the vector addition of the two component motions. Thus,

$$\boldsymbol{\omega} = \boldsymbol{\omega}_s + \boldsymbol{\omega}_p = \{3\mathbf{j} + 1\mathbf{k}\} \text{ rad/s}$$

Angular Acceleration. Since the magnitude of $\boldsymbol{\omega}$ is constant, only a change in its direction, as seen from a fixed reference, creates the angular acceleration $\boldsymbol{\alpha}$ of the disk. One way to obtain $\boldsymbol{\alpha}$ is to compute the time derivative of *each of the two components* of $\boldsymbol{\omega}$ using Eq. 20–6. At the instant shown in Fig. 20–7, imagine the fixed X, Y, Z and a rotating x, y, z frame to be in coincidence. If the rotating x, y, z frame is chosen to have an angular velocity of $\boldsymbol{\Omega} = \boldsymbol{\omega}_p = \{1\mathbf{k}\}$ rad/s, then $\boldsymbol{\omega}_s$ will *always* be directed along the y (not Y) axis, and the time rate of change of $\boldsymbol{\omega}_s$ as seen from x, y, z is *zero;* i.e., $(\dot{\boldsymbol{\omega}}_s)_{xyz} = \mathbf{0}$ (the magnitude of $\boldsymbol{\omega}_s$ is constant). Consequently,

$$(\dot{\boldsymbol{\omega}}_s)_{XYZ} = (\dot{\boldsymbol{\omega}}_s)_{xyz} + \boldsymbol{\omega}_p \times \boldsymbol{\omega}_s = \mathbf{0} + (1\mathbf{k}) \times (3\mathbf{j}) = \{-3\mathbf{i}\} \text{ rad/s}^2$$

By the same choice of axes rotation, $\boldsymbol{\Omega} = \boldsymbol{\omega}_p$, or even with $\boldsymbol{\Omega} = \mathbf{0}$, the time derivative $(\dot{\boldsymbol{\omega}}_p)_{xyz} = \mathbf{0}$, since $\boldsymbol{\omega}_p$ is *always* directed along the z (or Z) axis and has a constant magnitude. Hence,

$$(\dot{\boldsymbol{\omega}}_p)_{XYZ} = (\dot{\boldsymbol{\omega}}_p)_{xyz} + \boldsymbol{\omega}_p \times \boldsymbol{\omega}_p = \mathbf{0} + \mathbf{0} = \mathbf{0}$$

The angular acceleration is therefore

$$\boldsymbol{\alpha} = (\dot{\boldsymbol{\omega}})_{XYZ} = (\dot{\boldsymbol{\omega}}_s)_{XYZ} + (\dot{\boldsymbol{\omega}}_p)_{XYZ} = \{-3\mathbf{i}\} \text{ rad/s}^2 \qquad \textit{Ans.}$$

Velocity and Acceleration. Since $\boldsymbol{\omega}$ and $\boldsymbol{\alpha}$ have been determined, the velocity and acceleration of point A can be computed using Eqs. 20–3 and 20–4. Realizing that $\mathbf{r}_A = \{1\mathbf{j} + 0.25\mathbf{k}\}$ m, Fig. 20–7, we have

$$\mathbf{v}_A = \boldsymbol{\omega} \times \mathbf{r}_A = (3\mathbf{j} + 1\mathbf{k}) \times (1\mathbf{j} + 0.25\mathbf{k}) = \{-0.25\mathbf{i}\} \text{ m/s} \qquad \textit{Ans.}$$

$$\begin{aligned}
\mathbf{a}_A &= \boldsymbol{\alpha} \times \mathbf{r}_A + \boldsymbol{\omega} \times (\boldsymbol{\omega} \times \mathbf{r}_A) \\
&= (-3\mathbf{i}) \times (1\mathbf{j} + 0.25\mathbf{k}) + (3\mathbf{j} + 1\mathbf{k}) \times [(3\mathbf{j} + 1\mathbf{k}) \times (1\mathbf{j} + 0.25\mathbf{k})] \\
&= \{0.5\mathbf{j} - 2.25\mathbf{k}\} \text{ m/s}^2 \qquad \textit{Ans.}
\end{aligned}$$

Example 20–2

At the instant $\theta = 60°$, the gyrotop in Fig. 20–8 has three components of angular motion directed as shown and having magnitudes which are defined as follows:

> *spin:* $\omega_s = 10$ rad/s, increasing at the rate of 6 rad/s^2
> *nutation:* $\omega_n = 3$ rad/s, increasing at the rate of 2 rad/s^2
> *precession:* $\omega_p = 5$ rad/s, increasing at the rate of 4 rad/s^2

Determine the angular velocity and angular acceleration of the top.

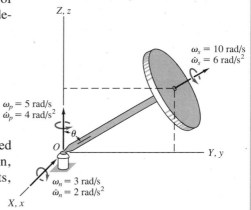

$\omega_s = 10$ rad/s
$\dot{\omega}_s = 6$ rad/s^2

$\omega_p = 5$ rad/s
$\dot{\omega}_p = 4$ rad/s^2

$\omega_n = 3$ rad/s
$\dot{\omega}_n = 2$ rad/s^2

Fig. 20–8

SOLUTION

Angular Velocity. The top is rotating about the fixed point O. If the fixed and rotating frames are considered to be coincident at the instant shown, then the angular velocity can be expressed in terms of $\mathbf{i}$, $\mathbf{j}$, $\mathbf{k}$ components, appropriate to the x, y, z frame, i.e.,

$$\boldsymbol{\omega} = -\omega_n\mathbf{i} + \omega_s \sin \theta\mathbf{j} + (\omega_p + \omega_s \cos \theta)\mathbf{k}$$
$$= -3\mathbf{i} + 10 \sin 60°\mathbf{j} + (5 + 10 \cos 60°)\mathbf{k}$$
$$= \{-3\mathbf{i} + 8.66\mathbf{j} + 10\mathbf{k}\} \text{ rad/s} \qquad\qquad \textit{Ans.}$$

Angular Acceleration. As in the solution of Example 20–1, the angular acceleration $\boldsymbol{\alpha}$ will be determined by investigating separately the time rate of change of *each of the angular velocity components* as observed from the fixed reference X, Y, Z. *This is done by using Eq. 20–6 and choosing an $\boldsymbol{\Omega}$ for the x, y, z reference so that the component of $\boldsymbol{\omega}$ which is being considered is viewed as having a constant direction when observed from x, y, z.*

Careful examination of the motion of the top reveals that $\boldsymbol{\omega}_s$ has a *constant direction* relative to x, y, z if these axes rotate at $\boldsymbol{\Omega} = \boldsymbol{\omega}_n + \boldsymbol{\omega}_p$. Since $\boldsymbol{\omega}_n$ *always* lies in the fixed X-Y plane, this vector has a *constant direction* if the motion is viewed from axes x, y, z having a rotation of $\boldsymbol{\Omega} = \boldsymbol{\omega}_p$ (not $\boldsymbol{\Omega} = \boldsymbol{\omega}_s + \boldsymbol{\omega}_p$). Finally, the component $\boldsymbol{\omega}_p$ is *always directed* along the Z axis so that here it is not necessary to think of x, y, z as rotating, i.e., $\boldsymbol{\Omega} = \mathbf{0}$. Expressing the data in terms of the $\mathbf{i}$, $\mathbf{j}$, $\mathbf{k}$ components, we therefore have

For $\boldsymbol{\omega}_s$, $\boldsymbol{\Omega} = \boldsymbol{\omega}_n + \boldsymbol{\omega}_p$;

$$(\dot{\boldsymbol{\omega}}_s)_{XYZ} = (\dot{\boldsymbol{\omega}}_s)_{xyz} + (\boldsymbol{\omega}_n + \boldsymbol{\omega}_p) \times \boldsymbol{\omega}_s$$
$$= (6 \sin 60°\mathbf{j} + 6 \cos 60°\mathbf{k}) + (-3\mathbf{i} + 5\mathbf{k}) \times (10 \sin 60°\mathbf{j} + 10 \cos 60°\mathbf{k})$$
$$= \{-43.30\mathbf{i} + 20.20\mathbf{j} - 22.98\mathbf{k}\} \text{ rad/s}^2$$

For $\boldsymbol{\omega}_n$, $\boldsymbol{\Omega} = \boldsymbol{\omega}_p$;

$$(\dot{\boldsymbol{\omega}}_n)_{XYZ} = (\dot{\boldsymbol{\omega}}_n)_{xyz} + \boldsymbol{\omega}_p \times \boldsymbol{\omega}_n = -2\mathbf{i} + (5\mathbf{k}) \times (-3\mathbf{i}) \doteq \{-2\mathbf{i} - 15\mathbf{j}\} \text{ rad/s}^2$$

For $\boldsymbol{\omega}_p$, $\boldsymbol{\Omega} = \mathbf{0}$;

$$(\dot{\boldsymbol{\omega}}_p)_{XYZ} = (\dot{\boldsymbol{\omega}}_p)_{xyz} + \mathbf{0} \times \boldsymbol{\omega}_p = \{4\mathbf{k}\} \text{ rad/s}^2$$

Thus, the angular acceleration of the top is

$$\boldsymbol{\alpha} = (\dot{\boldsymbol{\omega}}_s)_{XYZ} + (\dot{\boldsymbol{\omega}}_n)_{XYZ} + (\dot{\boldsymbol{\omega}}_p)_{XYZ}$$
$$= \{-45.3\mathbf{i} + 5.20\mathbf{j} - 19.0\mathbf{k}\} \text{ rad/s}^2 \qquad\qquad \textit{Ans.}$$

*20.3 General Motion

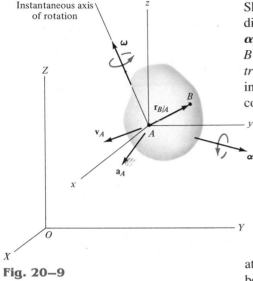

Instantaneous axis of rotation

Fig. 20–9

Shown in Fig. 20–9 is a rigid body subjected to general motion in three dimensions for which the angular velocity is $\boldsymbol{\omega}$ and the angular acceleration is $\boldsymbol{\alpha}$. If point A has a known motion of $\mathbf{v}_A$ and $\mathbf{a}_A$, the motion of any other point B may be determined by using a relative-motion analysis. In this section a *translating coordinate system* will be used to define the relative motion, and in the next section a reference that is both rotating and translating will be considered.

If the origin of the translating coordinate system x, y, z ($\boldsymbol{\Omega} = \mathbf{0}$) is located at the "base point" A, then, at the instant shown, the motion of the body may be regarded as the sum of an instantaneous translation of the body having a motion of $\mathbf{v}_A$ and $\mathbf{a}_A$ and a rotation of the body about an instantaneous axis passing through the base point. Since the body is rigid, the motion of point B measured by an observer located at A is the same as *motion of the body about a fixed point*. This relative motion occurs about the instantaneous axis of rotation and is defined by $\mathbf{v}_{B/A} = \boldsymbol{\omega} \times \mathbf{r}_{B/A}$, Eq. 20–3, and $\mathbf{a}_{B/A} = \boldsymbol{\alpha} \times \mathbf{r}_{B/A} + \boldsymbol{\omega} \times (\boldsymbol{\omega} \times \mathbf{r}_{B/A})$, Eq. 20–4. For translating axes the relative motions are related to absolute motions by $\mathbf{v}_B = \mathbf{v}_A + \mathbf{v}_{B/A}$ and $\mathbf{a}_B = \mathbf{a}_A + \mathbf{a}_{B/A}$, Eqs. 16–14 and 16–16, so that the absolute velocity and acceleration of point B can be determined from the equations

$$\mathbf{v}_B = \mathbf{v}_A + \boldsymbol{\omega} \times \mathbf{r}_{B/A} \qquad (20\text{–}7)$$

and

$$\mathbf{a}_B = \mathbf{a}_A + \boldsymbol{\alpha} \times \mathbf{r}_{B/A} + \boldsymbol{\omega} \times (\boldsymbol{\omega} \times \mathbf{r}_{B/A}) \qquad (20\text{–}8)$$

These two equations are identical to those describing the general plane motion of a rigid body, Eqs. 16–15 and 16–17. However, difficulty in application arises for three-dimensional motion, because $\boldsymbol{\alpha}$ measures the change in *both* the magnitude and direction of $\boldsymbol{\omega}$. (Recall that for general plane motion $\boldsymbol{\alpha}$ and $\boldsymbol{\omega}$ are always parallel or perpendicular to the plane of motion, and therefore $\boldsymbol{\alpha}$ measures only a change in the magnitude of $\boldsymbol{\omega}$.) In some problems the constraints or connections of a body will require that the directions of the angular motions or displacement paths of points on the body be defined. As illustrated in the following example, this information is useful for obtaining some of the terms in the above equations.

Example 20-3

One end of the rigid bar CD shown in Fig. 20–10a slides along the grooved wall slot AB, and the other end slides along the vertical member EF. If the collar at C is moving toward B at a speed of 3 m/s, determine the velocity of the collar at D and the angular velocity of the bar at the instant shown. The bar is connected to the collars at its end points by ball-and-socket joints.

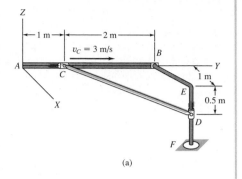

(a)

SOLUTION

Bar CD is subjected to general motion. Why? The velocity of point D on the bar may be related to the velocity of point C by the equation

$$\mathbf{v}_D = \mathbf{v}_C + \boldsymbol{\omega} \times \mathbf{r}_{D/C}$$

The fixed and translating frames of reference are assumed to coincide at the instant considered, Fig. 20–10b. We have

$$\mathbf{v}_D = -v_D\mathbf{k} \qquad \mathbf{v}_C = \{3\mathbf{j}\} \text{ m/s}$$
$$\mathbf{r}_{D/C} = \{1\mathbf{i} + 2\mathbf{j} - 0.5\mathbf{k}\} \text{ m} \qquad \boldsymbol{\omega} = \omega_x\mathbf{i} + \omega_y\mathbf{j} + \omega_z\mathbf{k}$$

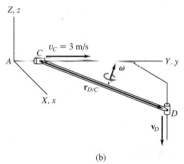

(b)

Fig. 20–10

Substituting these quantities into the above equation gives

$$-v_D\mathbf{k} = 3\mathbf{j} + \begin{vmatrix} \mathbf{i} & \mathbf{j} & \mathbf{k} \\ \omega_x & \omega_y & \omega_z \\ 1 & 2 & -0.5 \end{vmatrix}$$

Expanding and equating the respective $\mathbf{i}, \mathbf{j}, \mathbf{k}$ components yields

$$-0.5\omega_y - 2\omega_z - 0 \qquad (1)$$
$$0.5\omega_x + 1\omega_z + 3 = 0 \qquad (2)$$
$$2\omega_x - 1\omega_y + v_D = 0 \qquad (3)$$

These equations contain four unknowns.* A fourth equation can be written if the direction of $\boldsymbol{\omega}$ is specified. In particular, any component of $\boldsymbol{\omega}$ acting along the bar's axis has no effect on moving the collars. This is because the bar is *free to rotate* about its axis. Therefore, if $\boldsymbol{\omega}$ is specified as acting *perpendicular* to the axis of the bar, then $\boldsymbol{\omega}$ must have a unique magnitude to satisfy the above equations. Perpendicularity is guaranteed provided the dot product of $\boldsymbol{\omega}$ and $\mathbf{r}_{D/C}$ is zero (see Eq. C–14). Hence,

$$\boldsymbol{\omega} \cdot \mathbf{r}_{D/C} = (\omega_x\mathbf{i} + \omega_y\mathbf{j} + \omega_z\mathbf{k}) \cdot (1\mathbf{i} + 2\mathbf{j} - 0.5\mathbf{k}) = 0$$
$$1\omega_x + 2\omega_y - 0.5\omega_z = 0 \qquad (4)$$

Solving Eqs. (1) to (4) simultaneously yields

$$\omega_x = -4.86 \text{ rad/s} \quad \omega_y = 2.29 \text{ rad/s} \quad \omega_z = -0.571 \text{ rad/s} \quad Ans.$$
$$v_D = 12.0 \text{ m/s} \downarrow \qquad Ans.$$

*Although this is the case the magnitude of $\mathbf{v}_D$ can be obtained. For example, solve Eqs. (1) and (2) for ω_y and ω_x in terms of ω_z and substitute into Eq. (3). It will be noted that ω_z will *cancel out*, which will allow a solution for v_D.

PROBLEMS

20–1. The motion of the top is such that at the instant shown it is rotating about the z axis at $\omega_1 = 0.6$ rad/s, while it is spinning at $\omega_2 = 8$ rad/s. Determine the angular velocity and angular acceleration of the top at this instant. Express the result as a Cartesian vector.

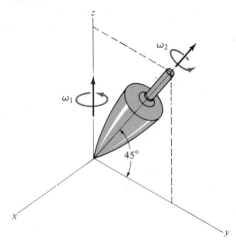

Prob. 20–1

20–2. The ladder of the fire truck rotates about the z axis with an angular velocity $\omega_1 = 0.15$ rad/s, which is increasing at 0.8 rad/s². At the same instant it is rotating upward at a constant rate $\omega_2 = 0.6$ rad/s. Compute the velocity and acceleration of point A located at the top of the ladder at this instant.

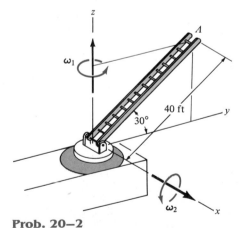

Prob. 20–2

20–3. Gear A is *fixed* while gear B is free to rotate on the shaft S. If the shaft is turning about the z axis at $\omega_z = 5$ rad/s, which is increasing at 2 rad/s², determine the velocity and acceleration of point P at the instant shown. The face of gear B lies in a vertical plane.

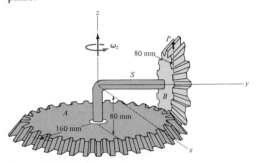

Prob. 20–3

*20–4. At the instant shown, the radar dish is rotating about the z axis at $\omega_1 = 3$ rad/s, which is increasing at 4 rad/s². Also, at this instant the angle of tilt $\phi = 30°$, and $\dot\phi = 2$ rad/s, $\ddot\phi = 6$ rad/s². Determine the angular velocity and angular acceleration of the dish at this instant.

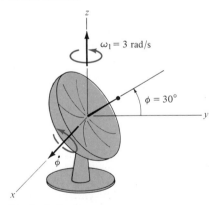

Prob. 20–4

20–5. The drill pipe P turns at a constant angular rate $\omega_P = 4$ rad/s. Determine the angular velocity and angular acceleration of the conical rock bit, which rolls without slipping. Also, what are the velocity and acceleration of point A? Establish x outward, y to the right, and z upwards.

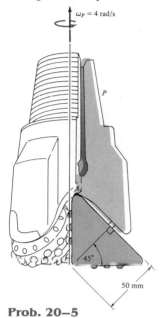

$\omega_P = 4$ rad/s

P

A

$45°$

50 mm

Prob. 20–5

20–6. The tower crane is rotating about the z axis with a constant rate $\omega_1 = 0.25$ rad/s, while the boom OA is rotating downward with a constant rate $\omega_2 = 0.4$ rad/s. Compute the velocity and acceleration of point A located at the top of the boom at the instant shown.

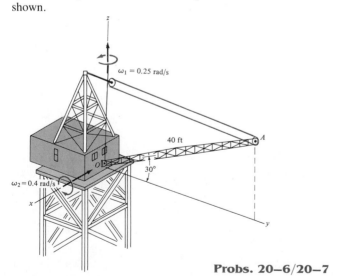

z

$\omega_1 = 0.25$ rad/s

40 ft

A

$30°$

$\omega_2 = 0.4$ rad/s

x

y

Probs. 20–6/20–7

20–7. At the instant shown, the tower crane is rotating about the z axis with an angular velocity $\omega_1 = 0.25$ rad/s, which is increasing at 0.6 rad/s². The boom OA is rotating downward with an angular velocity $\omega_2 = 0.4$ rad/s, which is increasing at 0.8 rad/s². Compute the velocity and acceleration of point A located at the top of the boom at this instant.

*20–8. The truncated cone rotates about the z axis at a constant rate $\omega_z = 0.4$ rad/s without slipping on the horizontal plane. Determine the velocity and acceleration of point A on the cone.

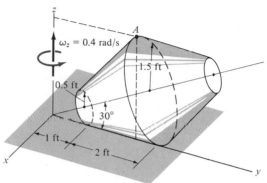

z

A

$\omega_z = 0.4$ rad/s

1.5 ft

0.5 ft

$30°$

1 ft

2 ft

x

y

Probs. 20–8/20–9

20–9. The truncated cone rotates about the z axis at $\omega_z = 0.4$ rad/s without slipping on the horizontal plane. If at this same instant ω_z is increasing at $\dot{\omega}_z = 0.5$ rad/s², determine the velocity and acceleration of point A on the cone.

20–10. The cone A rolls without slipping such that at the instant shown, $\omega = 4$ rad/s and $\alpha = 3$ rad/s². Determine the velocity and acceleration of point A at this instant.

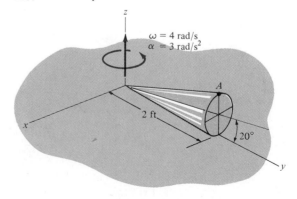

z

$\omega = 4$ rad/s

$\alpha = 3$ rad/s²

A

2 ft

$20°$

x

y

Prob. 20–10

20–11. The differential of an automobile allows the two rear wheels to rotate at different speeds when the automobile travels along a curve. For operation, the rear axles are attached to the wheels at one end and have beveled gears A and B on their other ends. The differential case D is placed over the left axle, but can rotate about C, independent of the axle. The case supports a pinion gear E on a shaft, which meshes with gears A and B. Finally, a ring gear G is *fixed* to the differential case so that the case rotates with the ring gear when the latter is driven by the drive pinion H. This gear, like the differential case, is free to rotate about the left wheel axle. If the drive pinion is turning at $\omega_H = 100$ rad/s and the pinion gear E is spinning about its shaft at $\omega_E = 30$ rad/s, compute the angular velocity, ω_A and ω_B, of each axle.

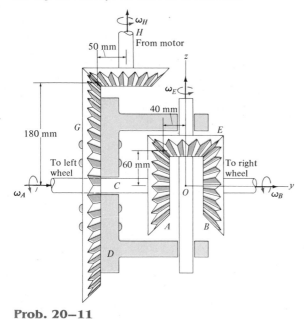

Prob. 20–11

*20–12. Gear B is connected to the rotating shaft, while the plate gear A is fixed. If the shaft is turning at a constant rate of $\omega_z = 10$ rad/s about the z axis, determine the magnitudes of the angular velocity and the angular acceleration of gear B. Also, determine the magnitudes of the velocity and acceleration of point P.

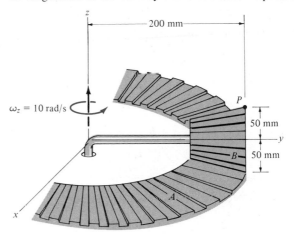

Prob. 20–12

20–13. If the rod is attached to smooth collars A and B at its end points by ball-and-socket joints, determine the speed of B if A is moving downward at a constant speed of 8 ft/s. Also, determine the angular velocity of the rod if it is directed perpendicular to the axis of the rod.

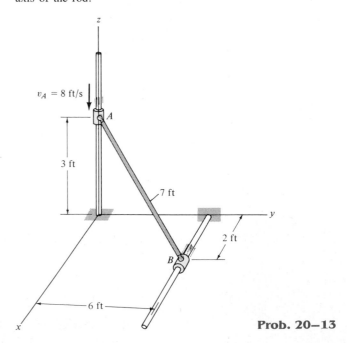

Prob. 20–13

20–14. The rod *AB* is attached to collars at its ends by ball-and-socket joints. If collar *A* has a speed $v_A = 20$ ft/s at the instant shown, determine the speed of collar *B*.

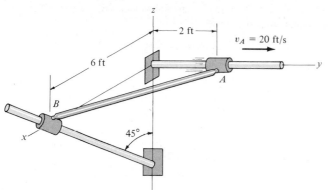

Prob. 20–14

20–15. The rod is attached to smooth collars *A* and *B* at its ends using ball-and-socket joints. Determine the speed of *B* if *A* is moving at $v_A = 6$ ft/s. Also, determine the angular velocity of the rod if it is directed perpendicular to the axis of the rod.

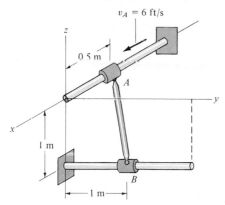

Prob. 20–15

***20–16.** Rod *AB* is attached to collars at its ends by using ball-and-socket joints. If collar *A* moves along the fixed rod with a velocity of $v_A = 8$ ft/s, determine the angular velocity of the rod and the velocity of collar *B* at the instant shown. Assume that the rod's angular velocity is directed perpendicular to the axis of the rod.

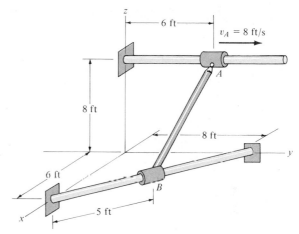

Probs. 20–16/20–17

20–17. Rod *AB* is attached to collars at its ends by using ball-and-socket joints. If collar *A* moves along the fixed rod with a velocity of $v_A = 8$ ft/s and has an acceleration of $a_A = 4$ ft/s² at the instant shown, determine the angular acceleration of the rod and the acceleration of collar *B* at this instant. Assume that the rod's angular velocity and angular acceleration are directed perpendicular to the axis of the rod.

20–18. Rod *AB* is attached to a disk and a collar by ball-and-socket joints. If the disk is rotating at a constant angular velocity of 2 rad/s, determine the velocity and acceleration of the collar at *A* at the instant shown. Assume that the angular velocity of the rod is directed perpendicular to the axis of the rod.

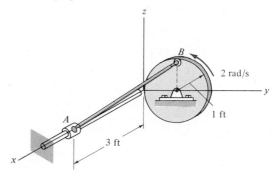

Prob. 20–18

423

20–19. The triangular plate ABC is supported at A by a ball-and-socket and at point C by the x-z plane. The side AB lies in the x-y plane. At the instant $\theta = 60°$, $\dot{\theta} = 2$ rad/s and point C has the coordinates shown. Determine the angular velocity of the plate and the velocity of point C at this instant.

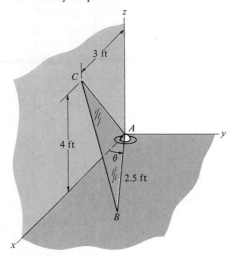

Prob. 20–19

*20–20. The rod ABC is supported at A by a ball-and-socket and at point C by the y-z plane. The segment AB lies in the x-y plane. At the instant $\theta = 60°$, $\dot{\theta} = 2$ rad/s, and point C has the coordinates shown. Determine the angular velocity of the rod and the velocity of point C at this instant.

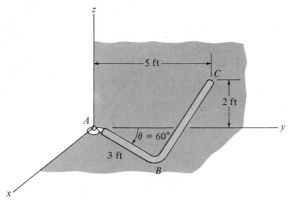

Prob. 20–20

20–21. The rod AB is attached to collars at its ends by ball-and-socket joints. If collar A has a velocity of $v_A = 5$ ft/s, determine the angular velocity of the rod and the velocity of collar B at the instant shown. Assume the angular velocity of the rod is directed perpendicular to the rod.

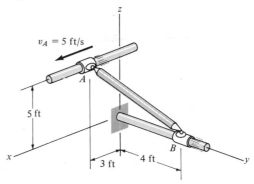

Probs. 20–21/20–22

20–22. The rod AB is attached to the collars at its ends by ball-and-socket joints. If collar A has an acceleration $\mathbf{a}_A = \{6\mathbf{i}\}$ ft/s^2, and a velocity $\mathbf{v}_A = \{5\mathbf{i}\}$ ft/s, determine the angular acceleration of the rod and the acceleration of collar B at the instant shown. Assume the angular velocity and angular acceleration of the rod are both directed perpendicular to the rod.

20–23. Rod AB is attached to the rotating arm using ball-and-socket joints. If AC is rotating with a constant angular velocity of 8 rad/s about the pin at C, determine the angular velocity of link BD at the instant shown.

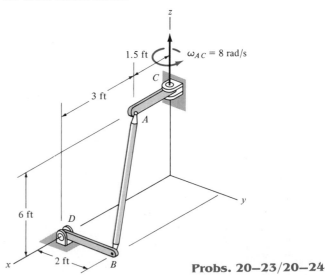

Probs. 20–23/20–24

*20–24. Rod *AB* is attached to the rotating arm using ball-and-socket joints. If *AC* is rotating about point *C* with an angular velocity of 8 rad/s and has an angular acceleration of 6 rad/s² at the instant shown, determine the angular velocity and angular acceleration of link *BD* at this instant.

20–25. Disk *A* is rotating at a constant angular velocity of 10 rad/s. If rod *BC* is joined to the disk and a collar by ball-and-socket joints, determine the velocity of collar *B* at the instant shown, and the rod's angular velocity if ω_{BC} is directed perpendicular to the axis of the rod.

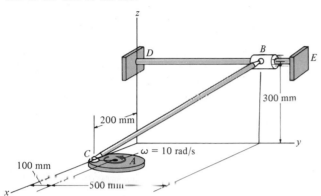

Probs. 20–25/20–26

20–26. If the disk in Prob. 20–25 has an angular acceleration $\alpha = 5$ rad/s², determine the acceleration of the collar *B* at the instant shown, when $\omega = 10$ rad/s. Also, what is the rod's angular acceleration α_{BC} if it is directed perpendicular to the axis of the rod?

20–27. Solve Prob. 20–25 if the connection at *B* consists of a pin as shown in the figure below, rather than a ball-and-socket joint. *Hint:* The constraint allows rotation of the rod both along bar *DE* (**j** direction) and along the axis of the pin (**n** direction). Since there is no rotational component in the **u** direction, i.e., perpendicular to **n** and **j** where $\mathbf{u} = \mathbf{j} \times \mathbf{n}$, an additional equation for solution can be obtained from $\boldsymbol{\omega} \cdot \mathbf{u} = 0$. The vector **n** is in the same direction as $\mathbf{r}_{B/C} \times \mathbf{r}_{D/C}$.

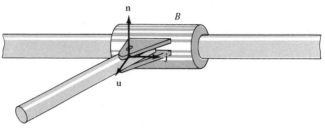

Prob. 20–27

*20–28. The rod assembly is supported at *B* by a ball-and-socket joint and at *A* by a clevis connection. If the collar at *B* moves in the *x-z* plane with a speed $v_B = 5$ ft/s, determine the velocity of points *A* and *C* on the rod assembly. *Hint:* See Prob. 20–27.

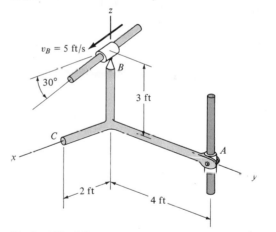

Prob. 20–28

*20.4 Relative-Motion Analysis Using Translating and Rotating Axes

The most general way to analyze the three-dimensional motion of a rigid body requires the use of a system of x, y, z axes which both translate and rotate relative to a second frame X, Y, Z. This analysis also provides a means for determining the motions of two points located on separate members of a mechanism, and for determining the relative motion of one particle with respect to another when one or both particles are moving along *rotating paths*. In this section two equations will be developed which relate the velocities and accelerations of two points A and B, of which one point moves relative to a frame of reference subjected to both translation and rotation. Because of the generality in the derivation, A and B may represent either two particles moving independently of one another or two points located in the same or different rigid bodies.

As shown in Fig. 20–11, the locations of points A and B are specified relative to the X, Y, Z frame of reference by position vectors $\mathbf{r}_A$ and $\mathbf{r}_B$. The base point A represents the origin of the x, y, z coordinate system, which is translating and rotating with respect to X, Y, Z. At the instant considered, the velocity and acceleration of point A are $\mathbf{v}_A$ and $\mathbf{a}_A$, respectively, and the angular velocity and angular acceleration of the x, y, z axes are $\mathbf{\Omega}$ and $\dot{\mathbf{\Omega}} = d\mathbf{\Omega}/dt$, respectively. All these vectors are *measured* with respect to the X, Y, Z frame of reference, although they may be expressed in Cartesian component form along either set of axes.

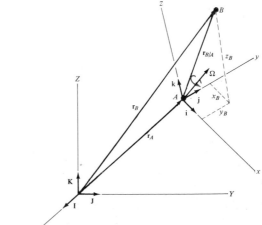

Fig. 20–11

Position

If the position of "B with respect to A" is specified by the *relative-position vector* $\mathbf{r}_{B/A}$, Fig. 20–11, then, by vector addition,

$$\boxed{\mathbf{r}_B = \mathbf{r}_A + \mathbf{r}_{B/A}} \qquad (20\text{–}9)$$

where

$$\mathbf{r}_B = \text{position of } B$$
$$\mathbf{r}_A = \text{position of the origin } A$$
$$\mathbf{r}_{B/A} = \text{relative position of "}B \text{ with respect to } A\text{"}$$

Velocity

The velocity of point B measured from X, Y, Z is determined by taking the time derivative of Eq. 20–9, which yields

$$\boxed{(\dot{\mathbf{r}}_B)_{XYZ} = (\dot{\mathbf{r}}_A)_{XYZ} + (\dot{\mathbf{r}}_{B/A})_{XYZ}}$$

The first two terms represent $\mathbf{v}_B$ and $\mathbf{v}_A$. The last term is evaluated by applying Eq. 20 6, since $\mathbf{r}_{B/A}$ is measured between two points in a rotating reference. Hence,

$$(\dot{\mathbf{r}}_{B/A})_{XYZ} = (\dot{\mathbf{r}}_{B/A})_{xyz} + \mathbf{\Omega} \times \mathbf{r}_{B/A} = (\mathbf{v}_{B/A})_{\text{rel}} + \mathbf{\Omega} \times \mathbf{r}_{B/A} \qquad (20\text{–}10)$$

Here $(\mathbf{v}_{B/A})_{\text{rel}}$ is the relative velocity of B with respect to A measured from x, y, z. Thus,

$$\boxed{\mathbf{v}_B = \mathbf{v}_A + \mathbf{\Omega} \times \mathbf{r}_{B/A} + (\mathbf{v}_{B/A})_{\text{rel}}} \qquad (20 \ 11)$$

where

$$\mathbf{v}_B = \text{velocity of } B$$
$$\mathbf{v}_A = \text{velocity of the origin } A \text{ of the } x, y, z \text{ frame of reference}$$
$$(\mathbf{v}_{B/A})_{\text{rel}} = \text{relative velocity of "}B \text{ with respect to } A\text{" as measured by an observer attached to the rotating } x, y, z \text{ frame of reference}$$
$$\mathbf{\Omega} = \text{angular velocity of the } x, y, z \text{ frame of reference}$$
$$\mathbf{r}_{B/A} = \text{relative position of "}B \text{ with respect to } A\text{"}$$

Acceleration

The acceleration of point B measured from X, Y, Z is determined by taking the time derivative of Eq. 20–11, which yields

$$(\dot{\mathbf{v}}_B)_{XYZ} = (\dot{\mathbf{v}}_A)_{XYZ} + (\dot{\boldsymbol{\Omega}})_{XYZ} \times \mathbf{r}_{B/A} + \boldsymbol{\Omega} \times (\dot{\mathbf{r}}_{B/A})_{XYZ} + (\dot{\mathbf{v}}_{B/A})_{XYZ}$$

The time derivatives defined in the first and second terms represent $\mathbf{a}_B$ and $\mathbf{a}_A$ respectively. The fourth term is evaluated using Eq. 20–10, and the last term is evaluated by applying Eq. 20–6, which yields

$$(\dot{\mathbf{v}}_{B/A})_{XYZ} = (\dot{\mathbf{v}}_{B/A})_{xyz} + \boldsymbol{\Omega} \times (\mathbf{v}_{B/A})_{\text{rel}}$$

$$= (\mathbf{a}_{B/A})_{\text{rel}} + \boldsymbol{\Omega} \times (\mathbf{v}_{B/A})_{\text{rel}}$$

Here $(\mathbf{a}_{B/A})_{\text{rel}}$ is the relative acceleration of B with respect to A measured from x, y, z. Substituting this result and Eq. 20–10 into the above equation and simplifying, we have

$$\boxed{\mathbf{a}_B = \mathbf{a}_A + \dot{\boldsymbol{\Omega}} \times \mathbf{r}_{B/A} + \boldsymbol{\Omega} \times (\boldsymbol{\Omega} \times \mathbf{r}_{B/A}) + 2\boldsymbol{\Omega} \times (\mathbf{v}_{B/A})_{\text{rel}} + (\mathbf{a}_{B/A})_{\text{rel}}}$$

$$(20\text{–}12)$$

where

$\mathbf{a}_B$ = acceleration of B

$\mathbf{a}_A$ = acceleration of the origin A of the x, y, z frame of reference

$(\mathbf{a}_{B/A})_{\text{rel}}$, $(\mathbf{v}_{B/A})_{\text{rel}}$ = relative acceleration and relative velocity of "B with respect to A" as measured by an observer attached to the rotating x, y, z frame of reference

$\dot{\boldsymbol{\Omega}}$, $\boldsymbol{\Omega}$ = angular acceleration and angular velocity of the x, y, z frame of reference

$\mathbf{r}_{B/A}$ = relative position of "B with respect to A"

Equations 20–11 and 20–12 are identical to those used in Sec. 16.8 for analyzing relative plane motion.* In that case, however, application was simplified since $\boldsymbol{\Omega}$ and $\dot{\boldsymbol{\Omega}}$ have a *constant direction* which is always perpendicular to the plane of motion. For three-dimensional motion, $\dot{\boldsymbol{\Omega}}$ must be computed by using Eq. 20–6, since it depends upon the change in both the magnitude and direction of $\boldsymbol{\Omega}$. Furthermore, in some problems, calculation of $\mathbf{v}_A$, $\mathbf{a}_A$ and $(\mathbf{v}_{B/A})_{\text{rel}}$, $(\mathbf{a}_{B/A})_{\text{rel}}$ must also be performed by using Eq. 20–6, since these quantities depend upon the angular rate at which $\mathbf{r}_A$ and $\mathbf{r}_{B/A}$ are "swinging" as measured from their respective frames of reference.

---

*Refer to Sec. 16.8 for an interpretation of the terms.

PROCEDURE FOR ANALYSIS

The following procedure provides a method for applying Eqs. 20–11 and 20–12 to solve problems involving the three-dimensional motion of particles or rigid bodies.

Coordinate Axes. Define the location and orientation of the X, Y, Z and x, y, z coordinate axes. Most often solutions are easily obtained if at the instant considered: (1) the origins are *coincident,* (2) the axes are collinear, and/or (3) the axes are parallel. Since several components of angular velocity may be involved in a problem, the calculations will be reduced if the x, y, z axes are selected such that only one component of angular velocity is observed in this frame (Ω_{rel}) and the frame rotates with Ω defined by the other components of angular velocity.

Kinematic Equations. After the origin of the moving reference, A, is defined and the moving point B is specified, Eqs. 20–11 and 20–12 should be written in symbolic form as

$$\mathbf{v}_B = \mathbf{v}_A + \mathbf{\Omega} \times \mathbf{r}_{B/A} + (\mathbf{v}_{B/A})_{rel}$$

$$\mathbf{a}_B = \mathbf{a}_A + \dot{\mathbf{\Omega}} \times \mathbf{r}_{B/A} + \mathbf{\Omega} \times (\mathbf{\Omega} \times \mathbf{r}_{B/A}) + 2\mathbf{\Omega} \times (\mathbf{v}_{B/A})_{rel} + (\mathbf{a}_{B/A})_{rel}$$

Vectors $\mathbf{\Omega}$, $\mathbf{r}_A$, and $\mathbf{r}_{B/A}$ should be defined from the problem data and represented in Cartesian form. *Motion of the moving reference* ($\mathbf{v}_A$, $\mathbf{a}_A$, and $\dot{\mathbf{\Omega}}$) is determined by applying Eq. 20–6 to compute the time derivatives of $\mathbf{r}_A$ and $\mathbf{\Omega}$. In a similar manner, if $\mathbf{r}_{B/A}$ has an angular motion $\mathbf{\Omega}_{rel}$ when observed from the moving reference, then the *motion of B with respect to the moving reference,* $(\mathbf{v}_{B/A})_{rel}$ and $(\mathbf{a}_{B/A})_{rel}$, must be determined by applying Eq. 20–6 to compute the time derivatives of $\mathbf{r}_{B/A}$. In all cases, the kinematic quantities which are involved in computing the time derivatives of $\mathbf{r}_A$, $\mathbf{r}_{B/A}$, and $\mathbf{\Omega}$ should be treated as *variables.* After the final forms of $\mathbf{v}_A$, $\mathbf{a}_A$, $(\mathbf{v}_{B/A})_{rel}$, $(\mathbf{a}_{B/A})_{rel}$ and $\mathbf{\Omega}$ are obtained, numerical problem data may be substituted and the kinematic terms evaluated. The components of all these vectors may be selected either along the X, Y, Z axes or along x, y, z. The choice is arbitrary, provided a consistent set of unit vectors is used.

Finally, substitute the data into the kinematic equations and perform the vector operations.

The following two examples numerically illustrate this procedure.

Example 20–4

A motor M and attached rod AB have the angular motion shown in Fig. 20–12. A collar C on the rod is located 0.25 m from A, and is moving downward along the rod with a velocity of 3 m/s and an acceleration of 2 m/s². Determine the velocity and acceleration of C at this instant.

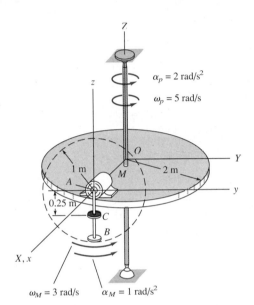

Fig. 20–12

$\alpha_p = 2$ rad/s²

$\omega_p = 5$ rad/s

$\omega_M = 3$ rad/s $\alpha_M = 1$ rad/s²

SOLUTION

Coordinate Axes. The origin of the fixed X, Y, Z reference is chosen at the center of the platform, and the origin of the moving x, y, z frame at point A, Fig. 20–12. Since the collar is subjected to two components of angular motion, $\boldsymbol{\omega}_p$ and $\boldsymbol{\omega}_M$, it is viewed as having an angular velocity of $\boldsymbol{\Omega}_{rel} = \boldsymbol{\omega}_M$ in x, y, z and the x, y, z axes are attached to the platform so that $\boldsymbol{\Omega} = \boldsymbol{\omega}_p$.

Kinematic Equations. Equations 20–11 and 20–12, applied to points C and A, become

$$\mathbf{v}_C = \mathbf{v}_A + \boldsymbol{\Omega} \times \mathbf{r}_{C/A} + (\mathbf{v}_{C/A})_{rel} \qquad (1)$$

$$\mathbf{a}_C = \mathbf{a}_A + \dot{\boldsymbol{\Omega}} \times \mathbf{r}_{C/A} + \boldsymbol{\Omega} \times (\boldsymbol{\Omega} \times \mathbf{r}_{C/A}) + 2\boldsymbol{\Omega} \times (\mathbf{v}_{C/A})_{rel} + (\mathbf{a}_{C/A})_{rel} \qquad (2)$$

Motion of Moving Reference

$$\boldsymbol{\Omega} = \boldsymbol{\omega}_p = 5\mathbf{k}$$

$$\dot{\boldsymbol{\Omega}} = (\dot{\boldsymbol{\omega}}_p)_{xyz} + \boldsymbol{\omega}_p \times \boldsymbol{\omega}_p = 2\mathbf{k} + 5\mathbf{k} \times 5\mathbf{k} = 2\mathbf{k}$$

$$\mathbf{r}_A = 2\mathbf{i}$$

$$\mathbf{v}_A = \dot{\mathbf{r}}_A = (\dot{\mathbf{r}}_A)_{xyz} + \boldsymbol{\omega}_p \times \mathbf{r}_A = 0 + 5\mathbf{k} \times 2\mathbf{i} = 10\mathbf{j}$$

$$\mathbf{a}_A = \ddot{\mathbf{r}}_A = [(\ddot{\mathbf{r}}_A)_{xyz} + \boldsymbol{\omega}_p \times (\dot{\mathbf{r}}_A)_{xyz}] + \dot{\boldsymbol{\omega}}_p \times \mathbf{r}_A + \boldsymbol{\omega}_p \times \dot{\mathbf{r}}_A$$

$$= 0 + 0 + 2\mathbf{k} \times 2\mathbf{i} + 5\mathbf{k} \times 10\mathbf{j} = -50\mathbf{i} + 4\mathbf{j}$$

Since point A moves in a circular path lying in the X-Y plane we can also apply Eqs. 16–15 and 16–17 to obtain these same results.

Motion of C with Respect to Moving Reference

$$\boldsymbol{\Omega}_{rel} = \boldsymbol{\omega}_M = 3\mathbf{i} \qquad \dot{\boldsymbol{\Omega}}_{rel} = \dot{\boldsymbol{\omega}}_M = 1\mathbf{i}\}$$

Note that $\boldsymbol{\omega}_M$ does not change direction relative to x, y, z.

$$\mathbf{r}_{C/A} = -0.25\mathbf{k}$$

$$(\mathbf{v}_{C/A})_{rel} = \dot{\mathbf{r}}_{C/A} = (\dot{\mathbf{r}}_{C/A})_{xyz} + \boldsymbol{\omega}_M \times \mathbf{r}_{C/A}$$

$$= -3\mathbf{k} + [3\mathbf{i} \times (-0.25\mathbf{k})] = 0.75\mathbf{j} - 3\mathbf{k}$$

$$(\mathbf{a}_{C/A})_{rel} = \ddot{\mathbf{r}}_{C/A} = [(\ddot{\mathbf{r}}_{C/A})_{xyz} + \boldsymbol{\omega}_M \times (\dot{\mathbf{r}}_{C/A})_{xyz}] + \dot{\boldsymbol{\omega}}_M \times \mathbf{r}_{C/A} + \boldsymbol{\omega}_M \times \dot{\mathbf{r}}_{C/A}$$

$$= [-2\mathbf{k} + 3\mathbf{i} \times (-3\mathbf{k})] + [(1\mathbf{i}) \times (-0.25\mathbf{k})] + [(3\mathbf{i}) \times (0.75\mathbf{j} - 3\mathbf{k})]$$

$$= 18.25\mathbf{j} + 0.25\mathbf{k}$$

Try to obtain these same results by applying Eqs. 20–11 and 20–12 between the x, y, z axes and coincident (rotating) x', y', z' axes having $\boldsymbol{\Omega} = 3\mathbf{i}$, $\dot{\boldsymbol{\Omega}} = 1\mathbf{i}$.

Substituting the data into Eqs. (1) and (2) yields

$$\mathbf{v}_C = \mathbf{v}_A + \boldsymbol{\Omega} \times \mathbf{r}_{C/A} + (\mathbf{v}_{C/A})_{rel}$$

$$= 10\mathbf{j} + [5\mathbf{k} \times (-0.25\mathbf{k})] + (0.75\mathbf{j} - 3\mathbf{k})$$

$$= \{10.8\mathbf{j} - 3\mathbf{k}\} \text{ m/s} \qquad \textit{Ans.}$$

$$\mathbf{a}_C = \mathbf{a}_A + \dot{\boldsymbol{\Omega}} \times \mathbf{r}_{C/A} + \boldsymbol{\Omega} \times (\boldsymbol{\Omega} \times \mathbf{r}_{C/A}) + 2\boldsymbol{\Omega} \times (\mathbf{v}_{C/A})_{rel} + (\mathbf{a}_{C/A})_{rel}$$

$$= (-50\mathbf{i} + 4\mathbf{j}) + [2\mathbf{k} \times (-0.25\mathbf{k})] + 5\mathbf{k} \times [5\mathbf{k} \times (-0.25\mathbf{k})]$$

$$+ 2[5\mathbf{k} \times (0.75\mathbf{j} - 3\mathbf{k})] + (18.25\mathbf{j} + 0.25\mathbf{k})$$

$$= \{-57.5\mathbf{i} + 22.2\mathbf{j} + 0.25\mathbf{k}\} \text{ m/s}^2 \qquad \textit{Ans.}$$

Example 20–5

The pendulum shown in Fig. 20–13 consists of two rods. AB is pin-supported at A and swings only in the Y-Z plane, whereas a bearing at B allows the attached rod BD to spin about rod AB. At a given instant, the rods have the angular motions shown. If a collar C, located 0.2 m from B, has a velocity of 3 m/s and an acceleration of 2 m/s^2 along the rod, determine the velocity and acceleration of the collar at this instant.

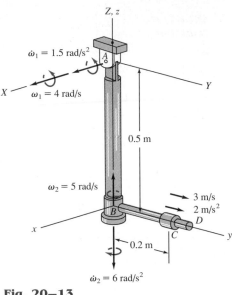

SOLUTION I $(\mathbf{\Omega} = \boldsymbol{\omega}_1, \mathbf{\Omega}_{rel} = \boldsymbol{\omega}_2)$

Coordinate Axes. Since the rod is rotating about the fixed point A, the origin of the X, Y, Z frame will be chosen at A. Motion of the collar is conveniently observed from B, so the origin of the x, y, z frame is located at this point.

Kinematic Equations

$$\mathbf{v}_C = \mathbf{v}_B + \mathbf{\Omega} \times \mathbf{r}_{C/B} + (\mathbf{v}_{C/B})_{rel} \qquad (1)$$

$$\mathbf{a}_C = \mathbf{a}_B + \dot{\mathbf{\Omega}} \times \mathbf{r}_{C/B} + \mathbf{\Omega} \times (\mathbf{\Omega} \times \mathbf{r}_{C/B}) + 2\mathbf{\Omega} \times (\mathbf{v}_{C/B})_{rel} + (\mathbf{a}_{C/B})_{rel} \qquad (2)$$

Fig. 20–13

Motion of Moving Reference

$$\mathbf{\Omega} = \boldsymbol{\omega}_1 = 4\mathbf{i}, \quad \dot{\mathbf{\Omega}} = (\dot{\boldsymbol{\omega}}_1)_{xyz} + \boldsymbol{\omega}_1 \times \boldsymbol{\omega}_1 = 1.5\mathbf{i} + \mathbf{0} = 1.5\mathbf{i}$$

$$\mathbf{r}_B = -0.5\mathbf{k}$$

$$\left.\begin{aligned}
\mathbf{v}_B = \dot{\mathbf{r}}_B &= (\dot{\mathbf{r}}_B)_{xyz} + \boldsymbol{\omega}_1 \times \mathbf{r}_B = \mathbf{0} + 4\mathbf{i} \times (-0.5\mathbf{k}) = 2\mathbf{j} \\
\mathbf{a}_B = \ddot{\mathbf{r}}_B &= [(\ddot{\mathbf{r}}_B)_{xyz} + \boldsymbol{\omega}_1 \times (\dot{\mathbf{r}}_B)_{xyz}] + \dot{\boldsymbol{\omega}}_1 \times \mathbf{r}_B + \boldsymbol{\omega}_1 \times \dot{\mathbf{r}}_B \\
&= [0 + 0] + [1.5\mathbf{i} \times (-0.5\mathbf{k})] + 4\mathbf{i} \times 2\mathbf{j} = 0.75\mathbf{j} + 8\mathbf{k}
\end{aligned}\right\}$$

This is the same as applying Eqs. 16–15 and 16–17 since B has circular motion in the X-Y plane.

Motion of C with Respect to Moving Reference

$$\left.\mathbf{\Omega}_{rel} = \boldsymbol{\omega}_2 = 5\mathbf{k}, \quad \dot{\mathbf{\Omega}}_{rel} = \dot{\boldsymbol{\omega}}_2 = -6\mathbf{k}\right\}$$ Note that $\boldsymbol{\omega}_2$ does not change direction relative to x, y, z.

$$\mathbf{r}_{C/B} = 0.2\mathbf{j}$$

$$\left.\begin{aligned}
(\mathbf{v}_{C/B})_{rel} = \dot{\mathbf{r}}_{C/B} &= (\dot{\mathbf{r}}_{C/B})_{xyz} + \boldsymbol{\omega}_2 \times \mathbf{r}_{C/B} = 3\mathbf{j} + 5\mathbf{k} \times 0.2\mathbf{j} = -1\mathbf{i} + 3\mathbf{j} \\
(\mathbf{a}_{C/B})_{rel} = \ddot{\mathbf{r}}_{C/B} &= [(\ddot{\mathbf{r}}_{C/B})_{xyz} + \boldsymbol{\omega}_2 \times (\dot{\mathbf{r}}_{C/B})_{xyz}] + \dot{\boldsymbol{\omega}}_2 \times \mathbf{r}_{C/B} + \boldsymbol{\omega}_2 \times \dot{\mathbf{r}}_{C/B} \\
&\quad (2\mathbf{j} + 5\mathbf{k} \times 3\mathbf{j}) + (-6\mathbf{k} \times 0.2\mathbf{j}) + [5\mathbf{k} \times (-1\mathbf{i} + 3\mathbf{j})] \\
&= -28.8\mathbf{i} - 3\mathbf{j}
\end{aligned}\right\}$$

The same results can be obtained by applying Eqs. 20–11 and 20–12 between the x, y, z axes and coincident (rotating) x', y', z' axes having $\mathbf{\Omega} = 5\mathbf{k}$, $\dot{\mathbf{\Omega}} = -6\mathbf{k}$.

Substituting the data into Eqs. (1) and (2) yields

$$\mathbf{v}_C = \mathbf{v}_B + \mathbf{\Omega} \times \mathbf{r}_{C/B} + (\mathbf{v}_{C/B})_{rel} = 2\mathbf{j} + 4\mathbf{i} \times 0.2\mathbf{j} + (-1\mathbf{i} + 3\mathbf{j})$$

$$= \{-1\mathbf{i} + 5\mathbf{j} + 0.8\mathbf{k}\} \text{ m/s} \qquad \textit{Ans.}$$

$$\mathbf{a}_C = \mathbf{a}_B + \dot{\mathbf{\Omega}} \times \mathbf{r}_{C/B} + \mathbf{\Omega} \times (\mathbf{\Omega} \times \mathbf{r}_{C/B}) + 2\mathbf{\Omega} \times (\mathbf{v}_{C/B})_{rel} + (\mathbf{a}_{C/B})_{rel}$$

$$= (0.75\mathbf{j} + 8\mathbf{k}) + (1.5\mathbf{i} \times 0.2\mathbf{j}) + [4\mathbf{i} \times (4\mathbf{i} \times 0.2\mathbf{j})]$$

$$+ 2[4\mathbf{i} \times (-1\mathbf{i} + 3\mathbf{j})] + (-28.8\mathbf{i} - 3\mathbf{j})$$

$$= \{-28.8\mathbf{i} - 5.45\mathbf{j} + 32.3\mathbf{k}\} \text{ m/s}^2 \qquad \textit{Ans.}$$

SOLUTION II $(\Omega = \omega_1 + \omega_2,\ \Omega_{\text{rel}} = 0)$

Motion of Moving Reference

$$\Omega = \omega_1 + \omega_2 = 4\mathbf{i} + 5\mathbf{k}$$

From the constraints of the problem ω_1 does not change direction; however, the direction of ω_2 is changed by ω_1. Thus, a simple way of obtaining $\dot{\Omega}$ is to consider x', y', z' axes coincident with the X, Y, Z axes at A, such that the primed axes have an angular velocity ω_1. Then

$$\dot{\Omega} = \dot{\omega}_1 + \dot{\omega}_2 = [(\dot{\omega}_1)_{x'y'z'} + \omega_1 \times \omega_1] + [(\dot{\omega}_2)_{x'y'z'} + \omega_1 \times \omega_2]$$
$$= (1.5\mathbf{i} + 0) + (-6\mathbf{k} + 4\mathbf{i} \times 5\mathbf{k}) = 1.5\mathbf{i} - 20\mathbf{j} - 6\mathbf{k}$$

Also,
$$\mathbf{r}_B = -0.5\mathbf{k}$$

Here again only ω_1 changes the direction of $\mathbf{r}_B$ so that the time derivatives of $\mathbf{r}_B$ can be computed from the primed axes used above, which are rotating at ω_1. Hence,

$$\mathbf{v}_B = \dot{\mathbf{r}}_B = (\dot{\mathbf{r}}_B)_{x'y'z'} + \omega_1 \times \mathbf{r}_B$$
$$= 0 + 4\mathbf{i} \times (-0.5\mathbf{k}) = 2\mathbf{j}$$
$$\mathbf{a}_B = \ddot{\mathbf{r}}_B = [(\ddot{\mathbf{r}}_B)_{x'y'z'} + \omega_1 \times (\dot{\mathbf{r}}_B)_{x'y'z'}] + \dot{\omega}_1 \times \mathbf{r}_B + \omega_1 \times \dot{\mathbf{r}}_B$$
$$= [0 + 0] + [1.5\mathbf{i} \times (-0.5\mathbf{k})] + 4\mathbf{i} \times 2\mathbf{j} = 0.75\mathbf{j} + 8\mathbf{k}$$

Motion of C with Respect to Moving Reference

$$\Omega_{\text{rel}} = 0$$
$$\dot{\Omega}_{\text{rel}} = 0$$
$$\mathbf{r}_{C/B} = 0.2\mathbf{j}$$
$$(\mathbf{v}_{C/B})_{\text{rel}} = \dot{\mathbf{r}}_{C/B} = (\dot{\mathbf{r}}_{C/B})_{xyz} + \Omega_{\text{rel}} \times \mathbf{r}_{C/B}$$
$$= 3\mathbf{j} + 0 = 3\mathbf{j}$$
$$(\mathbf{a}_{C/B})_{\text{rel}} = \ddot{\mathbf{r}}_{C/B} = [(\ddot{\mathbf{r}}_{C/B})_{xyz} + \Omega_{\text{rel}} \times (\dot{\mathbf{r}}_{C/B})_{xyz}] + \dot{\Omega}_{\text{rel}} \times \mathbf{r}_{C/B} + \Omega_{\text{rel}} \times \dot{\mathbf{r}}_{C/B}$$
$$= [2\mathbf{j} + 0] + 0 + 0 = 2\mathbf{j}$$

Substituting the data into Eqs. (1) and (2) yields

$$\mathbf{v}_C = \mathbf{v}_B + \Omega \times \mathbf{r}_{C/B} + (\mathbf{v}_{C/B})_{\text{rel}}$$
$$= 2\mathbf{j} + [(4\mathbf{i} + 5\mathbf{k}) \times (0.2\mathbf{j})] + 3\mathbf{j}$$
$$= \{-1\mathbf{i} + 5\mathbf{j} + 0.8\mathbf{k}\} \text{ m/s} \qquad\qquad Ans.$$

$$\mathbf{a}_C = \mathbf{a}_B + \dot{\Omega} \times \mathbf{r}_{C/B} + \Omega \times (\Omega \times \mathbf{r}_{C/B}) + 2\Omega \times (\mathbf{v}_{C/B})_{\text{rel}} + (\mathbf{a}_{C/B})_{\text{rel}}$$
$$= (0.75\mathbf{j} + 8\mathbf{k}) + [(1.5\mathbf{i} - 20\mathbf{j} - 6\mathbf{k}) \times (0.2\mathbf{j})]$$
$$\qquad\qquad + (4\mathbf{i} + 5\mathbf{k}) \times [(4\mathbf{i} + 5\mathbf{k}) \times 0.2\mathbf{j}] + 2[(4\mathbf{i} + 5\mathbf{k}) \times 3\mathbf{j}] + 2\mathbf{j}$$
$$= \{-28.8\mathbf{i} - 5.45\mathbf{j} + 32.3\mathbf{k}\} \text{ m/s}^2 \qquad\qquad Ans.$$

PROBLEMS

20–29. Solve Example 20–5 by fixing x, y, z axes to rod BD so that $\Omega = \omega_1 + \omega_2$. In this case the collar appears only to move radially outward along BD, hence $\Omega_{C/B} = 0$.

20–30. Solve Example 20–5 such that the x, y, z axes move with curvilinear translation, $\Omega = 0$, in which case the collar appears to have both an angular velocity $\Omega_{C/B} = \omega_1 + \omega_2$ and radial motion.

20–31. At a given instant, the antenna has an angular motion $\omega_1 = 3$ rad/s and $\dot{\omega}_1 = 2$ rad/s^2 about the z axis. At this same instant $\theta = 30°$, the angular motion about the x axis is $\omega_2 = 1.5$ rad/s, and $\dot{\omega}_2 = 4$ rad/s^2. Determine the velocity and acceleration of the signal horn A at this instant. The distance from O to A is $d = 3$ ft.

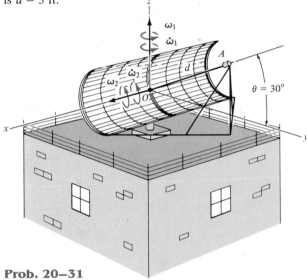

Prob. 20–31

*20–32. At a given instant, rod BD is rotating about the x axis with an angular velocity $\omega_{BD} = 2$ rad/s and an angular acceleration $\alpha_{BD} = 5$ rad/s^2. Also, $\theta = 45°$ and link AC is rotating downward such that $\dot{\theta} = 4$ rad/s and $\ddot{\theta} = 2$ rad/s^2. Determine the velocity and acceleration of point A on the link at this instant.

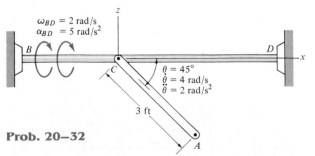

Prob. 20–32

20–33. At a given instant, the crane is moving along the track with a velocity $v_{CD} = 8$ m/s and acceleration of 9 m/s^2. Simultaneously, it has the angular motions shown. If the trolley T is moving outwards along the boom AB with a relative speed of 3 ft/s and relative acceleration of 5 ft/s^2, determine the velocity and acceleration of the trolley.

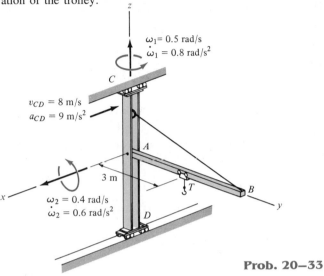

Prob. 20–33

20–34. At a given instant, the rod has the angular motions shown, while the collar C is moving down *relative* to the rod with a velocity of 6 ft/s and an acceleration of 2 ft/s^2. Determine the collar's velocity and acceleration at this instant.

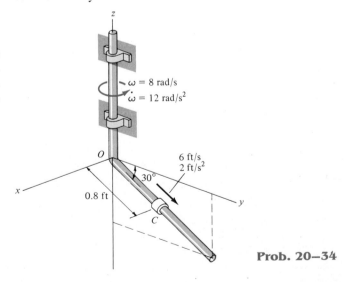

Prob. 20–34

20–35. The boom AB of the crane is rotating about the z axis with an angular velocity $\omega_z = 0.75$ rad/s and an angular acceleration of $\dot{\omega}_z = 2$ rad/s². At the same instant, $\theta = 60°$ and the boom is rotating upward at a constant rate $\dot{\theta} = 0.5$ rad/s. Determine the velocity and acceleration of the tip B of the boom at this instant.

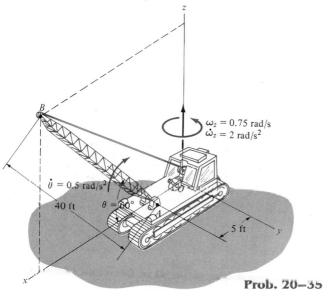

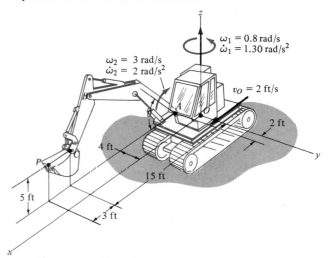

20–37. At the instant shown, the tractor is traveling forward at a constant speed of $v_O = 2$ ft/s, and the boom ABC is rotating about the z axis with an angular velocity of $\omega_1 = 0.8$ rad/s and an angular acceleration $\dot{\omega}_1 = 1.30$ rad/s². At this same instant the boom is rotating with $\omega_2 = 3$ rad/s when $\dot{\omega}_2 = 2$ rad/s², both measured relative to the tractor frame. Determine the velocity and acceleration of point P on the bucket at this instant.

Prob. 20–35

Prob. 20–37

20–36. At the instant shown, the arm AB is rotating about the fixed pin A with an angular velocity $\omega_1 = 4$ rad/s and angular acceleration $\dot{\omega}_1 = 3$ rad/s². At this same instant, rod BD is rotating relative to rod AB with an angular velocity $\omega_2 = 5$ rad/s which is increasing at $\dot{\omega}_2 = 7$ rad/s². Also, the collar C is moving along rod BD with a velocity of 3 m/s and an acceleration of 2 m/s², both measured relative to the rod. Determine the velocity and acceleration of the collar at this instant.

20–38. At the instant shown, the arm AB is rotating about the fixed bearing with an angular velocity $\omega_1 = 2$ rad/s and angular acceleration $\dot{\omega}_1 = 6$ rad/s². At the same instant, rod BD is rotating relative to rod AB at $\omega_2 = 7$ rad/s, which is increasing at $\dot{\omega}_2 = 1$ rad/s². Also, the collar C is moving along rod BD with a velocity $\dot{r} = 2$ ft/s and a *deceleration* $\ddot{r} = -0.5$ ft/s², both measured relative to the rod. Determine the velocity and acceleration of the collar at this instant.

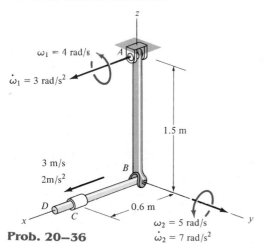

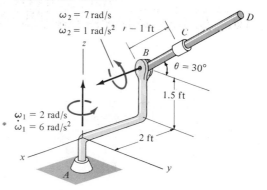

Prob. 20–36

Prob. 20–38

20–39. At the given instant, the rod is spinning about the z axis with an angular velocity $\omega_1 = 3$ rad/s and angular acceleration $\dot{\omega}_1 = 4$ rad/s^2. At this same instant, the disk is spinning with $\omega_2 = 2$ rad/s when $\dot{\omega}_2 = 1$ rad/s^2, both measured *relative* to the rod. Determine the velocity and acceleration of point P on the disk at this instant.

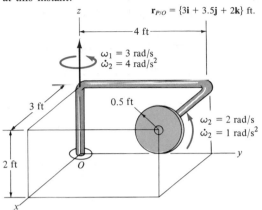

$\mathbf{r}_{P/O} = \{3\mathbf{i} + 3.5\mathbf{j} + 2\mathbf{k}\}$ ft.

Prob. 20–39

*20–40.** At the instant shown the portal crane is moving forward with a velocity $v_C = 6$ ft/s and an acceleration $a_C = 2$ ft/s^2. Simultaneously, it has the relative angular motions shown at $\theta = 60°$. Determine the velocity and acceleration of point A at the top of boom AB.

Prob. 20–40

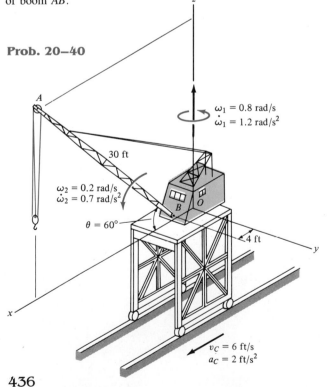

20–41. At the instant shown, the rod AB is rotating about the z axis with an angular velocity $\omega_1 = 4$ rad/s and an angular acceleration $\dot{\omega}_1 = 3$ rad/s^2. At this same instant, the circular rod has an angular motion relative to the rod as shown in the figure. If the collar C is moving down around the circular rod with a speed of 3 in./s, which is increasing at 8 in./s^2, both measured relative to the rod, determine the collar's velocity and acceleration at this instant.

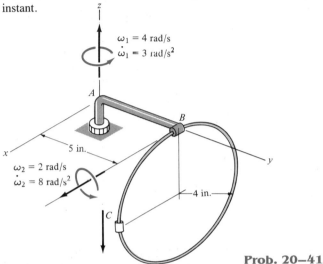

Prob. 20–41

20–42. At the instant shown, the arm OA of the conveyor belt is rotating about the z axis with a constant angular velocity $\omega_1 = 6$ rad/s, while at the same instant the arm is rotating upward at a constant rate $\omega_2 = 4$ rad/s. If the conveyor is running at a constant rate $\dot{r} = 5$ ft/s, determine the velocity and acceleration of the package P at the instant shown. Neglect the size of the package.

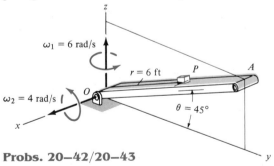

Probs. 20–42/20–43

20–43. At the instant shown, the arm OA of the conveyor belt is rotating about the z axis with a constant angular velocity $\omega_1 = 6$ rad/s, while at the same instant the arm is rotating upward at a constant rate $\omega_2 = 4$ rad/s. If the conveyor is running at a rate $\dot{r} = 5$ ft/s, which is increasing at $\ddot{r} = 8$ ft/s^2, determine the velocity and acceleration of the package P at the instant shown. Neglect the size of the package.

*20–44. At the instant shown, the industrial manipulator is rotating about the z axis at $\omega_1 = 5$ rad/s, and about joint B at $\omega_2 = 2$ rad/s. Determine the velocity and acceleration of the grip A at this instant when $\phi = 30°$, $\theta = 45°$, and $r = 1.6$ m.

20–45. At the instant shown, the industrial manipulator is rotating about the z axis at $\omega_1 = 5$ rad/s, which is increasing at the rate $\dot{\omega}_1 = 7$ rad/s². Simultaneously, the arm BA is rotating about joint B at $\omega_2 = 2$ rad/s, which is increasing at the rate $\dot{\omega}_2 = 6$ rad/s², and the arm BA is telescoping outward (elongating) at $\dot{r} = 0.5$ m/s, which is increasing at $\ddot{r} = 1.3$ m/s². Determine the velocity and acceleration of the grip A at this instant when $\phi = 30°$, $\theta = 45°$, and $r = 1.6$ m.

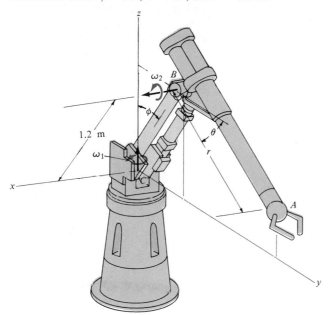

Probs. 20–44/20–45

21

Three-Dimensional Kinetics of a Rigid Body

In general, as a body moves through space it has a simultaneous translation and rotation at a given instant. The kinetic aspects of the *translation* have been discussed in Chapter 17, where it was shown that a system of external forces acting on the body may be related to the acceleration of the body's mass center by the equation $\Sigma \mathbf{F} = m\mathbf{a}_G$. In this chapter emphasis is placed primarily upon the *rotational* aspects of rigid-body motion, since motion of the body's mass center, defined by $\Sigma \mathbf{F} = m\mathbf{a}_G$, is treated in the same manner as particle motion.

The rotational equations of motion relate the body's components of angular motion to the moment components created by the external forces about some point located either on or off the body. To apply these equations, it is first necessary to formulate the moments and products of inertia of the body and to compute the body's angular momentum. Afterward, the principles of impulse and momentum and work and energy will be at our disposal for solving problems in three dimensions. The rotational equations of motion are developed in Sec. 21.4. Of special interest are problems involving the motion of an unsymmetrical body about a fixed axis; motion of a gyroscope, Sec. 21.5; and torque-free motion, Sec. 21.6.

*21.1 Moments and Products of Inertia

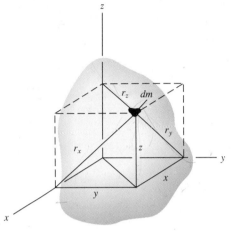

Fig. 21–1

When studying the planar kinetics of a body, it was necessary to introduce the moment of inertia I_G, which was computed about an axis perpendicular to the plane of motion and passing through the mass center G. For the kinetic analysis of three-dimensional motion it will sometimes be necessary to calculate six inertial quantities. These terms, called the moments and products of inertia, describe in a particular way the distribution of mass for a body relative to a given coordinate system that has a specified orientation and point of origin.

Moment of Inertia

Consider the rigid body shown in Fig. 21–1. The *moment of inertia* for a differential element dm of the body about any one of the three coordinate axes is defined as the product of the mass of the element and the square of the shortest distance from the axis to the element. For example, as noted in the figure, $r_x = \sqrt{y^2 + z^2}$, so that the mass moment of inertia of dm about the x axis is

$$dI_{xx} = r_x^2\, dm = (y^2 + z^2)\, dm$$

The moment of inertia I_{xx} for the body is determined by integrating this expression over the entire mass of the body. Hence, for each of the axes, we may write

$$
\begin{aligned}
I_{xx} &= \int_m r_x^2\, dm = \int_m (y^2 + z^2)\, dm \\[2mm]
I_{yy} &= \int_m r_y^2\, dm = \int_m (x^2 + z^2)\, dm \\[2mm]
I_{zz} &= \int_m r_z^2\, dm = \int_m (x^2 + y^2)\, dm
\end{aligned}
\qquad (21\text{–}1)
$$

Here it is seen that the moment of inertia is *always a positive quantity,* since it is the summation of the product of the mass $dm,$ which is always positive, and distances squared.

Product of Inertia

The *product of inertia* for a differential element dm is defined with respect to a set of *two orthogonal planes* as the product of the mass of the element and the perpendicular (or shortest) distances from the planes to the element. For example, with respect to the x-z and y-z planes, the product of inertia dI_{xy} for the element dm shown in Fig. 21–1 is

$$dI_{xy} = xy\, dm$$

Note also that $dI_{yx} = dI_{xy}$. By integrating over the entire mass, the product of inertia of the body for each combination of planes may be expressed as

$$I_{xy} = I_{yx} = \int_m xy \, dm$$

$$I_{yz} = I_{zy} = \int_m yz \, dm \qquad\qquad (21\text{--}2)$$

$$I_{xz} = I_{zx} = \int_m xz \, dm$$

Unlike the moment of inertia, which is always positive, the product of inertia may be positive, negative, or zero. The result depends upon the signs of the two defining coordinates, which vary independently from one another. In particular, if either one or both of the orthogonal planes are *planes of symmetry* for the mass, the *product of inertia* with respect to these planes will be *zero*. In such cases, elements of mass will occur in *pairs,* located on each side of the plane of symmetry. On one side of the plane the product of inertia for the element will be positive, while on the other side the product of inertia for the corresponding element will be negative, the sum therefore yielding zero. Examples of this are shown in Fig. 21–2. In the first case, Fig. 21–2a, the y-z plane is a plane of symmetry, and hence for point O, $I_{xz} = I_{xy} = 0$. Computation for I_{yz} will yield a *positive* result, since all elements of mass are located using only positive y and z coordinates. For the cylinder, with the coordinate axes located as shown in Fig. 21–2b, the x-z and y-z planes are both planes of symmetry. Thus, for point O, $I_{zx} = I_{yz} = I_{xy} = 0$.

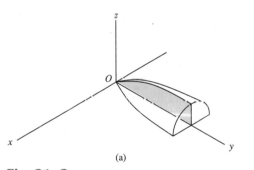

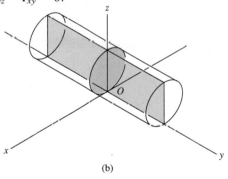

(a)   (b)

Fig. 21–2

Parallel-Axis and Parallel-Plane Theorems

The techniques of integration which are used to determine the moment of inertia of a body were described in Sec. 17.1. Also discussed were methods to determine the moment of inertia of a composite body, i.e., a body that is composed of simpler segments, as tabulated on the inside back cover of the book. In both of these cases the *parallel-axis theorem* is often used for the calculations. This theorem, which was developed in Sec. 17.1, is used to transfer the moment of inertia of a body from an axis passing through its mass center G to a parallel axis passing through some other point. In this regard, if

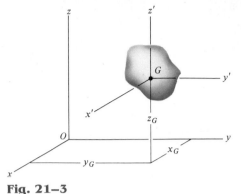

Fig. 21-3

G has coordinates x_G, y_G, z_G defined from the x, y, z axes, Fig. 21–3, then the parallel-axis equations used to calculate the moments of inertia about the x, y, z axes are

$$\begin{aligned} I_{xx} &= (I_{x'x'})_G + m(y_G^2 + z_G^2) \\ I_{yy} &= (I_{y'y'})_G + m(x_G^2 + z_G^2) \\ I_{zz} &= (I_{z'z'})_G + m(x_G^2 + y_G^2) \end{aligned}$$

(21–3)

The products of inertia of a body or a composite are computed in the same manner as the body's moments of inertia. Here, however, the *parallel-plane theorem* is important. This theorem is used to transfer the products of inertia of the body from a set of three orthogonal planes passing through the body's mass center to a corresponding set of three parallel planes passing through some other point O. Defining the perpendicular distances between the planes as x_G, y_G, and z_G, Fig. 21–3, the parallel-plane equations can be written as

$$\begin{aligned} I_{xy} &= (I_{x'y'})_G + m x_G y_G \\ I_{yz} &= (I_{y'z'})_G + m y_G z_G \\ I_{zx} &= (I_{z'x'})_G + m z_G x_G \end{aligned}$$

(21–4)

The derivation of these formulas is similar to that given for the parallel-axis equation, Sec. 17.1.

Inertia Tensor

The inertial properties of a body are completely characterized by nine terms, six of which are independent of one another. This set of terms is defined using Eqs. 21–1 and 21–2 and can be written as

$$\begin{pmatrix} I_{xx} & -I_{xy} & -I_{xz} \\ -I_{yx} & I_{yy} & -I_{yz} \\ -I_{zx} & -I_{zy} & I_{zz} \end{pmatrix}$$

This array is called an *inertia tensor*. It has a unique set of values for a body when it is computed for each location of the origin O and orientation of the coordinate axes.

For point O we can specify a unique axes inclination for which the products of inertia for the body are zero when computed with respect to these axes. When this is done, the inertia tensor is said to be "diagonalized" and may be written in the simplified form

$$\begin{pmatrix} I_x & 0 & 0 \\ 0 & I_y & 0 \\ 0 & 0 & I_z \end{pmatrix}$$

Here $I_x = I_{xx}$, $I_y = I_{yy}$, and $I_z = I_{zz}$ are termed the *principal moments of inertia* for the body, which are computed from the *principal axes of inertia*. Of these three principal moments of inertia, one will be a maximum and another a minimum of the body's moment of inertia.

Mathematical determination of the directions of principal axes of inertia will not be discussed here (see Prob. 21–18). There are many cases, however, in which the principal axes may be determined by inspection. From the previous discussion it was noted that if the coordinate axes are oriented such that *two* of the three orthogonal planes containing the axes are planes of *symmetry* for the body, then all the products of inertia for the body are zero with respect to the coordinate planes, and hence the coordinate axes are principal axes of inertia. For example, the x, y, z axes shown in Fig. 21–2b represent the principal axes of inertia for the cylinder at point O.

Moment of Inertia About an Arbitrary Axis

Consider the body shown in Fig. 21–4, where the nine elements of the inertia tensor have been computed for the x, y, z axes having an origin at O. Here we wish to compute the moment of inertia of the body about the Oa axis, for which the direction is defined by the unit vector $\mathbf{u}_a$. By definition $I_{aa} = \int b^2 \, dm$, where b is the *perpendicular distance* from dm to Oa. If the position of dm is located using $\mathbf{r}$, then $b = r \sin \theta$, which represents the *magnitude* of the cross product $\mathbf{u}_a \times \mathbf{r}$. Hence, the moment of inertia can be expressed as

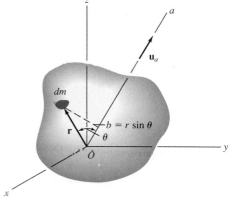

Fig. 21–4

$$I_{aa} = \int_m |(\mathbf{u}_a \times \mathbf{r})|^2 \, dm = \int_m (\mathbf{u}_a \times \mathbf{r}) \cdot (\mathbf{u}_a \times \mathbf{r}) \, dm$$

Provided $\mathbf{u}_a = u_x\mathbf{i} + u_y\mathbf{j} + u_z\mathbf{k}$ and $\mathbf{r} = x\mathbf{i} + y\mathbf{j} + z\mathbf{k}$, so that $\mathbf{u}_a \times \mathbf{r} = (u_y z - u_z y)\mathbf{i} + (u_z x - u_x z)\mathbf{j} + (u_x y - u_y x)\mathbf{k}$, then, after substituting and performing the dot-product operation, we can write the moment of inertia as

$$I_{aa} = \int_m [(u_y z - u_z y)^2 + (u_z x - u_x z)^2 + (u_x y - u_y x)^2] \, dm$$

$$= u_x^2 \int_m (y^2 + z^2) \, dm + u_y^2 \int_m (z^2 + x^2) \, dm + u_z^2 \int_m (x^2 + y^2) \, dm$$

$$- 2u_x u_y \int_m xy \, dm - 2u_y u_z \int_m yz \, dm - 2u_z u_x \int_m zx \, dm$$

Recognizing the integrals to be the moments and products of inertia of the body, Eqs. 21–1 and 21–2, we have

$$\boxed{I_{aa} = I_{xx}u_x^2 + I_{yy}u_y^2 + I_{zz}u_z^2 - 2I_{xy}u_x u_y - 2I_{yz}u_y u_z - 2I_{zx}u_z u_x} \quad (21\text{–}5)$$

Thus, if the inertia tensor is specified for the x, y, z axes, the moment of inertia of the body about the inclined Oa axis can be computed by using Eq. 21–5. For the calculation the direction cosines u_x, u_y, u_z of the axes must be determined. These terms specify the cosines of the coordinate direction angles α, β, γ made between the Oa axis and the x, y, z axes, respectively (see Appendix C).

Example 21–1

Determine the moment of inertia of the bent rod shown in Fig. 21–5a about the *Aa* axis. The mass of each of the three segments is shown in the figure.

SOLUTION

The moment of inertia I_{Aa} can be computed by using Eq. 21–5. It is first necessary, however, to determine the moments and products of inertia of the rod about the *x, y, z* axes. This is done using the formula for the moment of inertia of a slender rod, $I = \frac{1}{12}ml^2$, and the parallel-axis and parallel-plane theorems, Eqs. 21–3 and 21–4. Dividing the rod into three parts and locating the mass center of each segment, Fig. 21–5b, we have

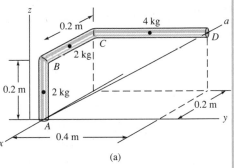

$$I_{xx} = \left[\frac{1}{12}(2)(0.2)^2 + 2(0.1)^2\right] + [0 + 2(0.2)^2]$$

$$+ \left[\frac{1}{12}(4)(0.4)^2 + 4((0.2)^2 + (0.2)^2)\right] = 0.480 \text{ kg} \cdot \text{m}^2$$

$$I_{yy} = \left[\frac{1}{12}(2)(0.2)^2 + 2(0.1)^2\right] + \left[\frac{1}{12}(2)(0.2)^2 + 2((-0.1)^2 + (0.2)^2)\right]$$

$$+ [0 + 4((-0.2)^2 + (0.2)^2)] = 0.453 \text{ kg} \cdot \text{m}^2$$

$$I_{zz} = [0 + 0] + \left[\frac{1}{12}(2)(0.2)^2 + 2(0.1)^2\right]$$

$$+ \left[\frac{1}{12}(4)(0.4)^2 + 4((-0.2)^2 + (0.2)^2)\right] = 0.400 \text{ kg} \cdot \text{m}^2$$

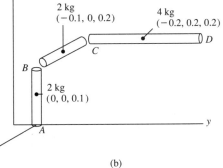

$$I_{xy} = [0 + 0] + [0 + 0] + [0 + 4(-0.2)(0.2)] = -0.160 \text{ kg} \cdot \text{m}^2$$

$$I_{yz} = [0 + 0] + [0 + 0] + [0 + 4(0.2)(0.2)] = 0.160 \text{ kg} \cdot \text{m}^2$$

$$I_{zx} = [0 + 0] + [0 + 2(0.2)(-0.1)] + [0 + 4(0.2)(-0.2)]$$

$$= -0.200 \text{ kg} \cdot \text{m}^2$$

Fig. 21–5

The *Aa* axis is defined by the unit vector

$$\mathbf{u}_{Aa} = \frac{\mathbf{r}_D}{r_D} = \frac{-0.2\mathbf{i} + 0.4\mathbf{j} + 0.2\mathbf{k}}{\sqrt{(-0.2)^2 + (0.4)^2 + (0.2)^2}} = -0.408\mathbf{i} + 0.816\mathbf{j} + 0.408\mathbf{k}$$

Thus,

$$u_x = -0.408, \qquad u_y = 0.816, \qquad u_z = 0.408$$

Substituting the computed data into Eq. 21–5 yields

$$I_{Aa} = I_{xx}u_x^2 + I_{yy}u_y^2 + I_{zz}u_z^2 - 2I_{xy}u_xu_y - 2I_{yz}u_yu_z - 2I_{zx}u_zu_x$$

$$= 0.480(-0.408)^2 + (0.453)(0.816)^2 + 0.400(0.408)^2 - 2(-0.160)(-0.408)(0.816)$$

$$- 2(0.160)(0.816)(0.408) - 2(-0.200)(0.408)(-0.408)$$

$$= 0.168 \text{ kg} \cdot \text{m}^2 \qquad \qquad \qquad \qquad \qquad \qquad \qquad \textit{Ans.}$$

PROBLEMS

21–1. Show that the sum of the moments of inertia of a body, $I_{xx} + I_{yy} + I_{zz}$, is independent of the orientation of the x, y, z axes and thus depends only on the location of the origin.

21–2. Determine the moments of inertia I_x and I_y of the paraboloid of revolution. The mass of the paraboloid is 20 slug.

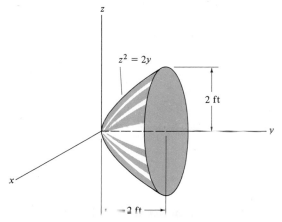

$z^2 = 2y$

2 ft

2 ft

Prob. 21–2

21–3. Determine the moment of inertia I_y and the product of inertia I_{xy} of the body formed by revolving the shaded area about the line $x = 5$ ft. Express the result in terms of the density of the material, ρ.

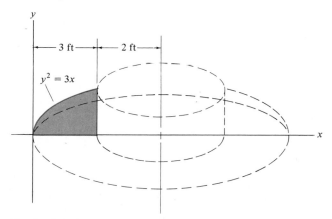

3 ft — 2 ft

$y^2 = 3x$

Prob. 21–3

***21–4.** Determine the radii of gyration k_x and k_y for the solid formed by revolving the shaded area about the y axis. The density of the material is ρ.

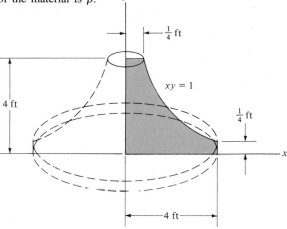

$\frac{1}{4}$ ft

4 ft

$xy = 1$

$\frac{1}{4}$ ft

4 ft

Prob. 21–4

21–5. Determine the product of inertia I_{yz} for the homogeneous prism. The density of the material is ρ. Express the result in terms of the total mass m of the prism.

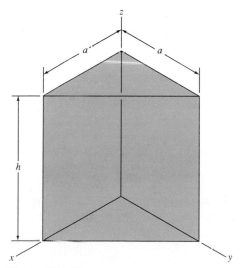

a a

h

Prob. 21–5

21–6. Determine by direct integration the product of inertia I_{yz} for the homogeneous tetrahedron. The mass density of the material is ρ. Express the result in terms of the total mass m of the solid.

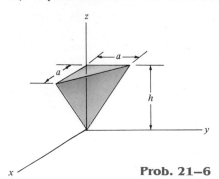

Prob. 21–6

21–7. Determine the moment of inertia of the cone about the z' axis. The weight of the cone is 15 lb, the *height* is $h = 1.5$ ft, and the radius is $r = 0.5$ ft.

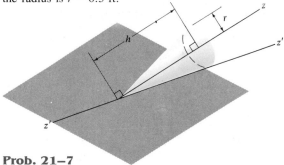

Prob. 21–7

*21–8. Determine the moment of inertia I_{xx} and the product of inertia I_{xy} for the bent rod. The rod has a density of 2 kg/m.

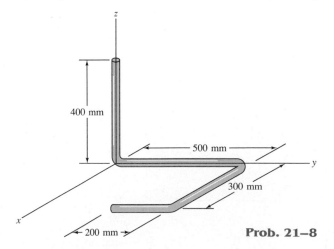

Prob. 21–8

21–9. The bent rod has a weight of 1.5 lb/ft. Locate the center of gravity $G(\bar{x}, \bar{y})$ and determine the principal moments of inertia $I_{x'}$, $I_{y'}$, and $I_{z'}$ of the rod with respect to the x', y', z' axes.

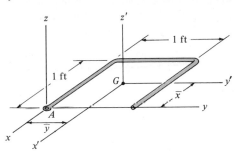

Prob. 21–9

21–10. Rod AB has a weight of 6 lb and each sphere has a weight of 8 lb. Determine the moment of inertia of the assembly about the x axis.

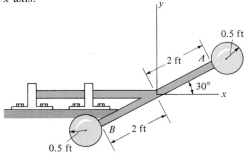

Prob. 21–10

21–11. Determine the moment of inertia of the composite body about the aa axis. The cylinder weighs 20 lb, and each hemisphere weighs 10 lb.

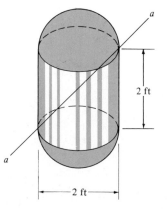

Prob. 21–11

446

*21–12. Compute the moment of inertia of the rod-and-disk assembly about the xx axis. The disks each have a weight of 12 lb. The two rods each have a weight of 4 lb and their ends extend to the rim of the disks.

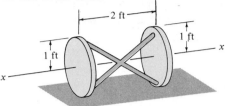

Prob. 21–12

21–13. Determine the moment of inertia of the 4-kg circular plate about the axis of rod OA.

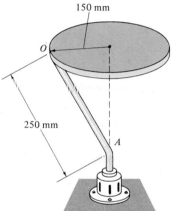

Prob. 21–13

21 14. Compute the moment of inertia of the rod-and-thin-ring assembly about the z axis. The rods and ring have a mass of 2 kg/m.

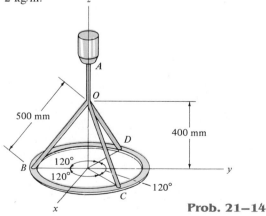

Prob. 21–14

21–15. The assembly consists of two square plates A and B which have a mass of 3 kg each and a rectangular plate C which has a mass of 4.5 kg. Determine the moments of inertia with respect to the principal x, y, z axes.

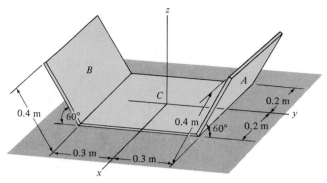

Prob. 21–15

*21–16. Determine the moments of inertia for the homogeneous cylinder of mass m with respect to the x', y', z' coordinate axes.

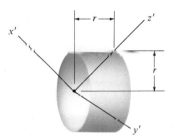

Prob. 21–16

21–17. Determine the elements of the inertia tensor for the cube with respect to the xyz coordinate system. The mass of the cube is m.

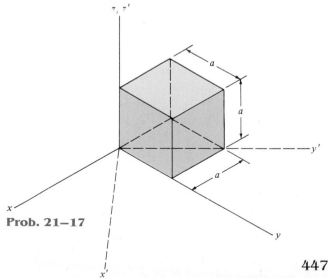

Prob. 21–17

447

*21.2 Angular Momentum

In this section we will develop the necessary equations used to determine the angular momentum of a rigid body about an arbitrary point. This formulation will provide the necessary means for developing both the principle of impulse and momentum and the equations of rotational motion for a rigid body.

Consider the rigid body in Fig. 21–6, which has a total mass m and center of mass located at G. The X, Y, Z coordinate system represents an inertial frame of reference, and hence, its axes are fixed or translate with a constant velocity. The angular momentum as measured from this reference will be computed relative to the arbitrary point A. The position vectors $\mathbf{r}_A$ and $\boldsymbol{\rho}_A$ are drawn from the origin of coordinates to point A and from A to the ith particle of the body. If the particle's mass is m_i, the angular momentum about point A is

$$(\mathbf{H}_A)_i = \boldsymbol{\rho}_A \times m_i \mathbf{v}_i$$

where $\mathbf{v}_i$ represents the particle's velocity as measured from the X, Y, Z coordinate system. If the body has an angular velocity $\boldsymbol{\omega}$ at the instant considered, $\mathbf{v}_i$ may be related to the velocity of A by applying Eq. 20–7, i.e.,

$$\mathbf{v}_i = \mathbf{v}_A + \boldsymbol{\omega} \times \boldsymbol{\rho}_A$$

Thus,

$$(\mathbf{H}_A)_i = \boldsymbol{\rho}_A \times m_i(\mathbf{v}_A + \boldsymbol{\omega} \times \boldsymbol{\rho}_A)$$
$$= (\boldsymbol{\rho}_A \, m_i) \times \mathbf{v}_A + \boldsymbol{\rho}_A \times (\boldsymbol{\omega} \times \boldsymbol{\rho}_A) \, m_i$$

For the entire body, summing all the particles of the body requires an integration, i.e., $m_i \rightarrow dm$ so that

$$\mathbf{H}_A = \left(\int_m \boldsymbol{\rho}_A \, dm \right) \times \mathbf{v}_A + \int_m \boldsymbol{\rho}_A \times (\boldsymbol{\omega} \times \boldsymbol{\rho}_A) \, dm \qquad (21\text{–}6)$$

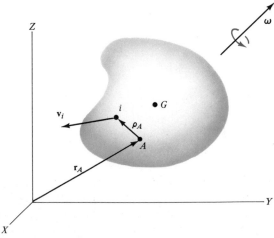

Fig. 21–6

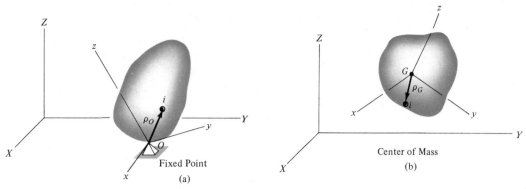

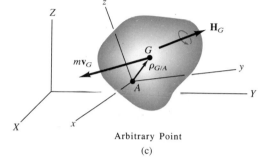

Fig. 21–7

Fixed Point O

If A becomes a *fixed point* O in the body, Fig. 21–7a, then $\mathbf{v}_A = \mathbf{0}$ and Eq. 21–6 reduces to

$$\mathbf{H}_O = \int_m \boldsymbol{\rho}_O \times (\boldsymbol{\omega} \times \boldsymbol{\rho}_O)\, dm \qquad (21\text{–}7)$$

Center of Mass G

If A is located at the *center of mass* G of the body, Fig. 21–7b, then $\displaystyle\int_m \boldsymbol{\rho}_A\, dm = \mathbf{0}$ and

$$\mathbf{H}_G = \int_m \boldsymbol{\rho}_G \times (\boldsymbol{\omega} \times \boldsymbol{\rho}_G)\, dm \qquad (21\text{–}8)$$

Arbitrary Point A

In general, A may be some point other than O or G, Fig. 21–7c, in which case Eq. 21–6 may nevertheless be simplified to the following form (see Prob. 21–19).

$$\mathbf{H}_A = \boldsymbol{\rho}_{G/A} \times m\mathbf{v}_G + \mathbf{H}_G \qquad (21\text{–}9)$$

Here it can be seen that the angular momentum consists of two parts—the moment of the linear momentum $m\mathbf{v}_G$ of the body about point A added (vectorially) to the angular momentum $\mathbf{H}_G$. Equation 21–9 may also be used for computing the angular momentum of the body about a fixed point O; the results, of course, will be the same as those computed using the more convenient Eq. 21–7.

Rectangular Components of **H**

To make practical use of Eqs. 21–7 to 21–9, the angular momentum must be expressed in terms of its scalar components. For this purpose, it is convenient to choose a second set of x, y, z axes having an arbitrary orientation relative to the X, Y, Z axes, Fig. 21–7. For a general explanation, note that Eqs. 21–7 to 21–9 all contain the form

$$\mathbf{H} = \int_m \boldsymbol{\rho} \times (\boldsymbol{\omega} \times \boldsymbol{\rho}) \, dm$$

Expressing **H**, $\boldsymbol{\rho}$, and $\boldsymbol{\omega}$ in terms of x, y, and z components, we have

$$H_x\mathbf{i} + H_y\mathbf{j} + H_z\mathbf{k} = \int_m (x\mathbf{i} + y\mathbf{j} + z\mathbf{k}) \times [(\omega_x\mathbf{i} + \omega_y\mathbf{j} + \omega_z\mathbf{k})$$

$$\times \ (x\mathbf{i} + y\mathbf{j} + z\mathbf{k})] \, dm$$

Expanding the cross products and combining terms yields

$$H_x\mathbf{i} + H_y\mathbf{j} + H_z\mathbf{k} = \left[\omega_x \int_m (y^2 + z^2) \, dm - \omega_y \int_m xy \, dm - \omega_z \int_m xz \, dm \right]\mathbf{i}$$

$$+ \left[-\omega_x \int_m xy \, dm + \omega_y \int_m (x^2 + z^2) \, dm - \omega_z \int_m yz \, dm \right]\mathbf{j}$$

$$+ \left[-\omega_x \int_m zx \, dm - \omega_y \int_m yz \, dm + \omega_z \int_m (x^2 + y^2) \, dm \right]\mathbf{k}$$

Equating the respective **i, j, k** components and recognizing that the integrals represent the moments and products of inertia, we obtain

$$\begin{aligned} H_x &= I_{xx}\omega_x - I_{xy}\omega_y - I_{xz}\omega_z \\ H_y &= -I_{yx}\omega_x + I_{yy}\omega_y - I_{yz}\omega_z \\ H_z &= -I_{zx}\omega_x - I_{zy}\omega_y + I_{zz}\omega_z \end{aligned} \tag{21–10}$$

These three equations represent the scalar form of the **i, j,** and **k** components of $\mathbf{H}_O$ or $\mathbf{H}_G$ (given in vector form by Eqs. 21–7 and 21–8). The angular momentum of the body about the arbitrary point A, other than the fixed point O or the center of mass G, may also be expressed in scalar form. Here it is necessary to use Eq. 21–9 and to represent $\boldsymbol{\rho}_{G/A}$ and $\mathbf{v}_G$ as Cartesian vectors, carry out the cross-product operation, and substitute the components, Eqs. 21–10, for $\mathbf{H}_G$.

Equations 21–10 may be simplified further if the x, y, z coordinate axes are oriented such that they become *principal axes of inertia* for the body at the point. When these axes are used, computation of the products of inertia $I_{xy} = I_{yz} = I_{zx} = 0$, and if the principal moments of inertia about the x, y, z axes are represented as $I_x = I_{xx}$, $I_y = I_{yy}$, and $I_z = I_{zz}$, the three components of angular momentum become

$$H_x = I_x\omega_x \qquad H_y = I_y\omega_y \qquad H_z = I_z\omega_z \tag{21–11}$$

Principle of Impulse and Momentum

Now that the means for computing the angular momentum for a body have been presented, the *principle of impulse and momentum*, as discussed in Sec. 19.2, may be used to solve kinetics problems which involve *force, velocity, and time*. For this case, the following two vector equations are available:

$$m(\mathbf{v}_G)_1 + \Sigma \int_{t_1}^{t_2} \mathbf{F} \, dt = m(\mathbf{v}_G)_2 \qquad (21\text{–}12)$$

$$\mathbf{H}_1 + \Sigma \int_{t_1}^{t_2} \mathbf{M}_O \, dt = \mathbf{H}_2 \qquad (21\text{–}13)$$

In three dimensions each vector term can be represented by three scalar components, and therefore a total of *six scalar equations* can be written. Three equations relate the linear impulse and momentum in the x, y, z directions, and three equations relate the body's angular impulse and momentum about the x, y, z axes. Before applying Eqs. 21–12 and 21–13 to the solution of problems, the material in Secs. 19.2 and 19.3 should be reviewed.

*21.3 Kinetic Energy

In order to apply the principle of work and energy to the solution of problems involving general rigid-body motion, it is first necessary to formulate expressions for the kinetic energy of the body. In this regard, consider the rigid body shown in Fig. 21–8, which has a total mass m and center of mass located at G. The kinetic energy of the ith particle of the body having a mass m_i and velocity $\mathbf{v}_i$, measured relative to the inertial X, Y, Z frame of reference, is

$$T_i = \tfrac{1}{2} m_i \, v_i^2 = \tfrac{1}{2} m_i \, (\mathbf{v}_i \cdot \mathbf{v}_i)$$

Provided the velocity of an arbitrary point A in the body is known, $\mathbf{v}_i$ may be related to $\mathbf{v}_A$ by the equation $\mathbf{v}_i = \mathbf{v}_A + \boldsymbol{\omega} \times \boldsymbol{\rho}_A$, where $\boldsymbol{\omega}$ is the instantaneous angular velocity of the body, measured from the X, Y, Z coordinate system, and $\boldsymbol{\rho}_A$ is a position vector drawn from A to i. Using this expression for $\mathbf{v}_i$, the kinetic energy for the particle may be written as

$$T_i = \tfrac{1}{2} m_i \, (\mathbf{v}_A + \boldsymbol{\omega} \times \boldsymbol{\rho}_A) \cdot (\mathbf{v}_A + \boldsymbol{\omega} \times \boldsymbol{\rho}_A)$$
$$= \tfrac{1}{2}(\mathbf{v}_A \cdot \mathbf{v}_A) \, m_i + \mathbf{v}_A \cdot (\boldsymbol{\omega} \times \boldsymbol{\rho}_A) \, m_i + \tfrac{1}{2}(\boldsymbol{\omega} \times \boldsymbol{\rho}_A) \cdot (\boldsymbol{\omega} \times \boldsymbol{\rho}_A) \, m_i$$

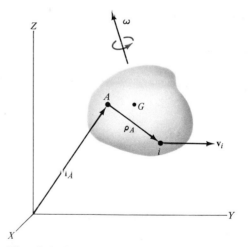

Fig. 21–8

The kinetic energy for the entire body is obtained by summing the kinetic energies of all the particles of the body. This requires an integration, since $m_i \rightarrow dm$, so that

$$T = \tfrac{1}{2} m(\mathbf{v}_A \cdot \mathbf{v}_A) + \mathbf{v}_A \cdot \left(\boldsymbol{\omega} \times \int_m \boldsymbol{\rho}_A \, dm\right) + \tfrac{1}{2} \int_m (\boldsymbol{\omega} \times \boldsymbol{\rho}_A) \cdot (\boldsymbol{\omega} \times \boldsymbol{\rho}_A) \, dm$$

The last term on the right may be rewritten using the vector identity $\mathbf{a} \times \mathbf{b} \cdot \mathbf{c} = \mathbf{a} \cdot \mathbf{b} \times \mathbf{c}$, where $\mathbf{a} = \boldsymbol{\omega}$, $\mathbf{b} = \boldsymbol{\rho}_A$, and $\mathbf{c} = \boldsymbol{\omega} \times \boldsymbol{\rho}_A$. The final result is therefore,

$$T = \tfrac{1}{2}m(\mathbf{v}_A \cdot \mathbf{v}_A) + \mathbf{v}_A \cdot \left(\boldsymbol{\omega} \times \int_m \boldsymbol{\rho}_A \, dm\right) + \tfrac{1}{2}\boldsymbol{\omega} \cdot \int_m \boldsymbol{\rho}_A \times (\boldsymbol{\omega} \times \boldsymbol{\rho}_A) \, dm$$

$$(21-14)$$

This equation is rarely used because of the computations involving the integrals. Simplification occurs, however, if the reference point A is either a fixed point O or the center of mass G.

Fixed Point O

If A is a *fixed point* O in the body, Fig. 21–7a, then $\mathbf{v}_A = \mathbf{0}$, and using Eq. 21–7, we can express Eq. 21–14 as

$$T = \tfrac{1}{2}\boldsymbol{\omega} \cdot \mathbf{H}_O$$

If the x, y, z axes represent the principal axes of inertia for the body, then $\boldsymbol{\omega} = \omega_x\mathbf{i} + \omega_y\mathbf{j} + \omega_z\mathbf{k}$ and $\mathbf{H}_O = I_x\omega_x\mathbf{i} + I_y\omega_y\mathbf{j} + I_z\omega_z\mathbf{k}$. Substituting into the above equation and performing the dot product operations yields

$$\boxed{T = \tfrac{1}{2}I_x\omega_x^2 + \tfrac{1}{2}I_y\omega_y^2 + \tfrac{1}{2}I_z\omega_z^2}$$

$$(21-15)$$

Center of Mass G

If A is located at the *center of mass G* of the body, Fig. 21–7b, then $\int_m \boldsymbol{\rho}_A \, dm = \mathbf{0}$ and, using Eq. 21–8, we can write Eq. 21–14 as

$$T = \tfrac{1}{2}mv_G^2 + \tfrac{1}{2}\boldsymbol{\omega} \cdot \mathbf{H}_G$$

In a manner similar to that for a fixed point, the last term on the right side may be represented in scalar form, in which case

$$\boxed{T = \tfrac{1}{2}mv_G^2 + \tfrac{1}{2}I_x\omega_x^2 + \tfrac{1}{2}I_y\omega_y^2 + \tfrac{1}{2}I_z\omega_z^2}$$

$$(21-16)$$

Here it is seen that the kinetic energy consists of two parts; namely, the translational kinetic energy of the mass center, $\tfrac{1}{2}mv_G^2$, and the body's rotational kinetic energy.

Principle of Work and Energy

Using one of the above expressions for computing the kinetic energy of a body, the *principle of work and energy* may be applied to solve kinetics problems which involve *force, velocity, and displacement*. For this case only one scalar equation can be written for each body, i.e.,

$$\boxed{T_1 + \Sigma U_{1-2} = T_2}$$

$$(21-17)$$

Before applying this equation to the solution of problems the material in Chapter 18 should be reviewed.

Example 21–2

The rod in Fig. 21–9a has a weight of 1.5 lb/ft. Determine its angular velocity just after the end A falls onto the hook at E. The hook provides a permanent connection for the rod due to the spring-lock mechanism S. Just before striking the hook the rod is falling downward with a speed $(v_G)_1 = 10$ ft/s.

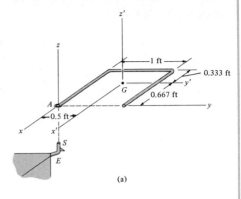

(a)

SOLUTION

The principle of impulse and momentum will be used since impact occurs.

Impulse and Momentum Diagrams. Fig. 21–9b. During the short time Δt, the impulsive force $\mathbf{F}$ acting at A changes the momentum of the rod. (The impulse created by the rod's weight $\mathbf{W}$ during this time is small compared to $\int \mathbf{F}\, dt$, so that it is neglected, i.e., the weight is a nonimpulsive force.) Hence, the angular momentum of the rod is *conserved* about point A since the moment of $\int \mathbf{F}\, dt$ about A is zero.

Conservation of Angular Momentum. Equation 21–9 must be used for computing the angular momentum of the rod, since A does not become a *fixed point* until *after* the impulsive interaction with the hook. Thus, with reference to Fig. 21–9b, $(\mathbf{H}_A)_1 = (\mathbf{H}_A)_2$, or

$$\mathbf{r}_{G/A} \times m(\mathbf{v}_G)_1 = \mathbf{r}_{G/A} \times m(\mathbf{v}_G)_2 + (\mathbf{H}_G)_2 \qquad (1)$$

From Fig. 21–9a, $\mathbf{r}_{G/A} = \{-0.667\mathbf{i} + 0.5\mathbf{j}\}$ ft. Furthermore, the primed axes are principal axes of inertia for the rod because $I_{x'y'} = I_{x'z'} = I_{z'y'} = 0$. Hence, from Eqs. 21–11, $(\mathbf{H}_G)_2 = I_{x'}\omega_x\mathbf{i} + I_{y'}\omega_y\mathbf{j} + I_{z'}\omega_z\mathbf{k}$. The principal moments of inertia are $I_{x'} = I_{y'} = 0.0155$ slug·ft², $I_{z'} = 0.0401$ slug·ft² (see Prob. 21–9). Substituting into Eq. (1), we have

$$(-0.667\mathbf{i} + 0.5\mathbf{j}) \times \left[\left(\frac{4.5}{32.2}\right)(-10\mathbf{k})\right] = (-0.667\mathbf{i} + 0.5\mathbf{j}) \times \left[\left(\frac{4.5}{32.2}\right)(-v_G)_2\mathbf{k}\right]$$
$$+\, 0.0155\omega_x\mathbf{i} + 0.0155\omega_y\mathbf{j} + 0.0401\omega_z\mathbf{k}$$

Expanding and equating the respective $\mathbf{i}$, $\mathbf{j}$, and $\mathbf{k}$ components yields

$$-0.699 = -0.0699v_G + 0.0155\omega_x \qquad (2)$$
$$-0.932 = -0.0932v_G + 0.0155\omega_y \qquad (3)$$
$$0 = 0.0401\omega_z \qquad (4)$$

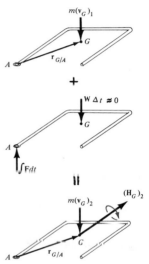

(b)

Fig. 21–9

Kinematics. There are four unknowns in the above equations; however, another equation may be obtained by relating $\boldsymbol{\omega}$ to $(\mathbf{v}_G)_2$ using *kinematics*. Since $\omega_z = 0$ (Eq. 4) and after impact the rod rotates about the fixed point A, Eq. 20–3 may be applied, in which case $\mathbf{v}_G = \boldsymbol{\omega} \times \mathbf{r}_{G/A}$, or

$$-(v_G)_2\mathbf{k} = (\omega_x\mathbf{i} + \omega_y\mathbf{j}) \times (-0.667\mathbf{i} + 0.5\mathbf{j})$$
$$-(v_G)_2 = 0.5\omega_x + 0.667\omega_y \qquad (5)$$

Solving Eqs. (2) to (5) simultaneously yields

$$(\mathbf{v}_G)_2 = \{-8.62\mathbf{k}\} \text{ ft/s} \qquad \boldsymbol{\omega} = \{-6.21\mathbf{i} - 8.29\mathbf{j}\} \text{ rad/s} \qquad \textit{Ans.}$$

Example 21–3

A 5-N · m torque is applied to the vertical shaft CD shown in Fig. 21–10a, which allows the 10-kg gear A to turn freely about CE. Assuming that gear A starts from rest, determine the angular velocity of CD after it has turned two revolutions. Neglect the mass of shaft CD and axle CE and assume that gear A is approximated by a thin disk. Gear B is fixed.

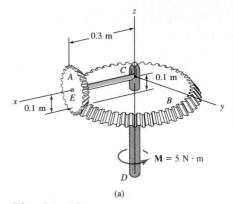

SOLUTION

The principle of work and energy may be used for the solution. Why?

Work. If shaft CD, the axle CE, and gear A are considered as a system of connected bodies, only the applied torque $\mathbf{M}$ does work. For two revolutions of CD, this work is $\Sigma U_{1-2} = (5 \text{ N} \cdot \text{m})(4\pi \text{ rad}) = 62.83 \text{ J}$.

Fig. 21–10a

Kinetic Energy. Since the gear is initially at rest, its initial kinetic energy is zero. A kinematic diagram for the gear is shown in Fig. 21–10b. If the angular velocity of CD is taken as $\boldsymbol{\omega}_{CD}$, then the angular velocity of gear A is $\boldsymbol{\omega}_A = \boldsymbol{\omega}_{CD} + \boldsymbol{\omega}_{CE}$. The gear may be imagined as a portion of a massless extended body which is rotating about the *fixed point C*. The instantaneous axis of rotation for this body is along line CH, because both points C and H on the body (gear) have zero velocity and must therefore lie on this axis. This requires that the components $\boldsymbol{\omega}_{CD}$ and $\boldsymbol{\omega}_{CE}$ be related by the equation $\omega_{CD}/0.1 \text{ m} = \omega_{CE}/0.3 \text{ m}$ or $\omega_{CE} = 3\omega_{CD}$. Thus,

$$\boldsymbol{\omega}_A = -\omega_{CE}\mathbf{i} + \omega_{CD}\mathbf{k} = -3\omega_{CD}\mathbf{i} + \omega_{CD}\mathbf{k} \qquad (1)$$

The x, y, z axes in Fig. 21–10a represent *principal axes of inertia* at C for the gear. Since point C is a fixed point of rotation, Eq. 21–15 may be applied to determine the kinetic energy, i.e.,

$$T = \tfrac{1}{2}I_x\omega_x^2 + \tfrac{1}{2}I_y\omega_y^2 + \tfrac{1}{2}I_z\omega_z^2 \qquad (2)$$

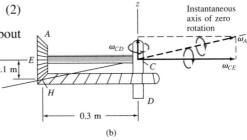

Using the parallel-axis theorem, the moments of inertia of the gear about point C are as follows:

$$I_x = \frac{1}{2}(10)(0.1)^2 = 0.05 \text{ kg} \cdot \text{m}^2$$

$$I_y = I_z = \frac{1}{4}(10)(0.1)^2 + 10(0.3)^2 = 0.925 \text{ kg} \cdot \text{m}^2$$

Fig. 21–10b

Since $\omega_x = -3\omega_{CD}$, $\omega_y = 0$, $\omega_z = \omega_{CD}$, Eq. (2) becomes

$$T_A = \frac{1}{2}(0.05)(-3\omega_{CD})^2 + 0 + \frac{1}{2}(0.925)(\omega_{CD})^2 = 0.6875\omega_{CD}^2$$

Principle of Work and Energy. Applying the principle of work and energy, we obtain

$$T_1 + \Sigma U_{1-2} = T_2 \qquad 0 + 62.83 = 0.6875\omega_{CD}^2$$

$$\omega_{CD} = 9.56 \text{ rad/s} \qquad\qquad Ans.$$

PROBLEMS

21–18. If a body contains *no planes of symmetry*, the principal moments of inertia can be computed mathematically. To show how this is done, consider the rigid body which is spinning with an angular velocity $\boldsymbol{\omega}$ directed along one of its principal axes of inertia. If the principal moment of inertia about this axis is I, the angular momentum can be expressed as $\mathbf{H} = I\boldsymbol{\omega} = I\omega_x\mathbf{i} + I\omega_y\mathbf{j} + I\omega_z\mathbf{k}$. The components of $\mathbf{H}$ may also be expressed by Eqs. 21–10, where the inertia tensor is assumed to be known. Equate the $\mathbf{i}$, $\mathbf{j}$, and $\mathbf{k}$ components of both expressions for $\mathbf{H}$ and consider ω_x, ω_y, and ω_z to be unknown. The solution of these three equations is obtained provided the determinant of the coefficients is zero. Show that this determinant, when expanded, yields the cubic equation

$$I^3 - (I_{xx} + I_{yy} + I_{zz})I^2 + (I_{xx}I_{yy} + I_{yy}I_{zz} + I_{zz}I_{xx}$$
$$- I_{xy}^2 - I_{yz}^2 - I_{zx}^2)I - (I_{xx}I_{yy}I_{zz} - 2I_{xy}I_{yz}I_{zx}$$
$$- I_{xx}I_{yz}^2 - I_{yy}I_{zx}^2 - I_{zz}I_{xy}^2) = 0$$

The three positive roots of I, obtained from the solution of this equation, represent the principal moments of inertia I_x, I_y, and I_z.

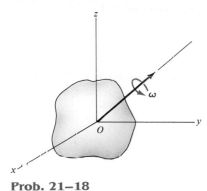

Prob. 21–18

21–19. Show that if the angular momentum of a body is computed with respect to an arbitrary point A, then $\mathbf{H}_A$ can be expressed by Eq. 21–9. This requires substituting $\boldsymbol{\rho}_A = \boldsymbol{\rho}_G + \boldsymbol{\rho}_{G/A}$ into Eq. 21–6 and expanding, noting that $\int \boldsymbol{\rho}_G \, dm = \mathbf{0}$ by definition of the mass center and $\mathbf{v}_G = \mathbf{v}_A + \boldsymbol{\omega} \times \boldsymbol{\rho}_{G/A}$.

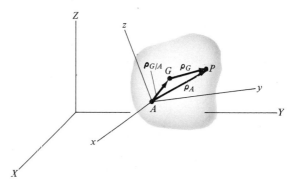

Prob. 21–19

*21–20. Determine the kinetic energy of the 7-kg disk and 1.5-kg rod when the assembly is rotating about the z axis at $\omega = 5$ rad/s.

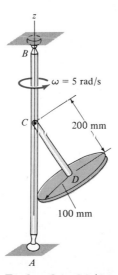

Probs. 21–22/21–23

21–21. Determine the magnitude of the angular momentum of the 7-kg disk and 1.5-kg rod about point C. The shaft BA is rotating at $\omega = 5$ rad/s.

21–22. Rod *AB* has a weight of 6 lb and is attached to two smooth collars at its end points by ball-and-socket joints. If collar *A* is moving downward at a speed of 8 ft/s, determine the kinetic energy of the rod at the instant shown. Assume that at this instant the angular velocity of the rod is directed perpendicular to the rod's axis. *Note:* The motion has been specified in Prob. 20–13.

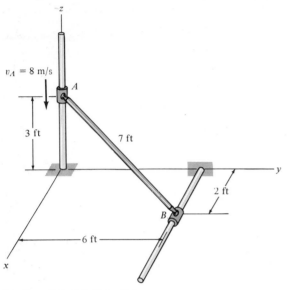

Probs. 21–20/21–21

21–23. Compute the magnitude of the angular momentum of rod *AB* in Prob. 21–22 about the mass center *G* at the instant shown. *Note:* The motion has been specified in Prob. 20–13.

*21–24. The 2-kg gear *A* rolls on the fixed plate gear *B*. Determine the angular velocity of rod *AO* about the *z* axis after it rotates one revolution about the *z* axis starting from rest. The rod is acted upon by the constant moment of $M = 5$ N · m. Neglect the mass of rod *OB*. Assume that gear *A* is a uniform disk having a radius of 100 mm.

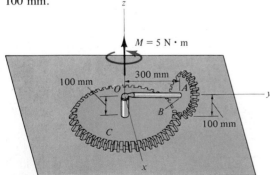

Prob. 21–24

21–25. The 4-lb rod *AB* is attached to the disk and collar using ball-and-socket joints. If the disk has a constant angular velocity of 2 rad/s, compute the kinetic energy of the rod when it is in the position shown. Assume the angular velocity of the rod is directed perpendicular to the axis of the rod. *Hint:* The motion has been specified in Prob. 20–18.

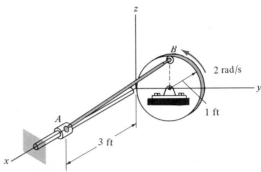

Probs. 21–25/21–26

21–26. Compute the angular momentum of rod *AB* in Prob. 21–25 at the instant shown. *Hint:* The motion has been specified in Prob. 20–18.

21–27. The rod assembly has a mass of 2.5 kg/m and is rotating with a constant angular velocity of $\omega = \{2\mathbf{k}\}$ rad/s when the looped end at *C* encounters a hook at *S*, which provides a permanent connection. Determine the angular velocity of the assembly immediately after impact.

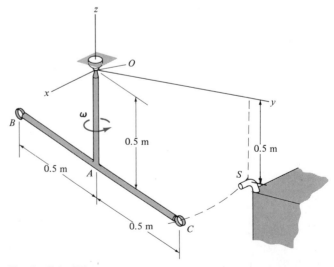

Prob. 21–27

*21–28. The 25-lb thin plate is suspended from a ball-and-socket joint at O. A 0.2-lb bullet is fired with a velocity of $\mathbf{v} = \{-300\mathbf{i} - 250\mathbf{j} + 300\mathbf{k}\}$ ft/s into the plate and becomes embedded in the plate at point A. Determine the angular momentum and the instantaneous axis of rotation of the plate just after impact.

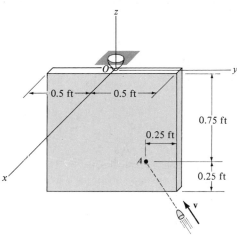

Prob. 21–28

21–29. The 15-lb plate is subjected to a force $F = 8$ lb which is always directed perpendicular to the face of the plate. If the plate is originally at rest, determine its angular velocity after it has rotated one revolution (360°). The plate is supported by ball-and-socket joints at A and B.

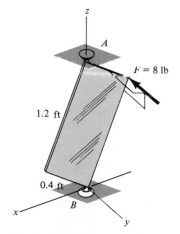

Prob. 21–29

21–30. The 4-lb rod AB is attached to the 1-lb collar at A and a 2-lb link BC using ball-and-socket joints. If the rod is released from rest in the position shown, determine the angular velocity of the link after it has rotated 180°.

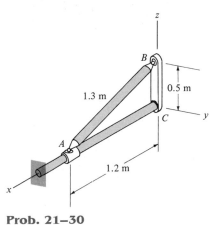

Prob. 21–30

21–31. Rod AB has a weight of 6 lb and is attached to two smooth collars at its ends by ball-and-socket joints. If collar A is moving downward with a speed of 8 ft/s when $z = 3$ ft, determine the speed of collar A at the instant $z = 0$. The spring has an unstretched length of 2 ft. Neglect the mass of the collars. Assume the angular velocity of rod AB is perpendicular to its axis. See Prob. 20–11.

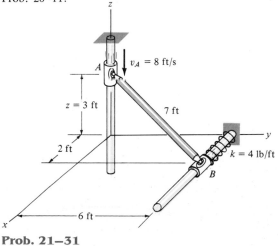

Prob. 21–31

457

*21–32. The 5-kg thin plate is suspended at O using a ball-and-socket joint. It is rotating with a constant angular velocity $\boldsymbol{\omega} = \{2\mathbf{k}\}$ rad/s when the corner A strikes the hook at S, which provides a permanent connection. Determine the angular velocity of the plate immediately after impact.

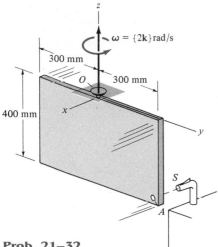

Prob. 21–32

21–33. The 15-kg rectangular plate is free to rotate about the y axis because of the bearing supports at A and B. When the plate is balanced in the vertical plane, a 3-g bullet is fired into it, perpendicular to its surface, with a velocity $\mathbf{v} = \{-2000\mathbf{i}\}$ m/s. Compute the angular velocity of the plate at the instant it has rotated $180°$. If the bullet strikes corner D with the same velocity $\mathbf{v}$, instead of at C, does the angular velocity remain the same? Why or why not?

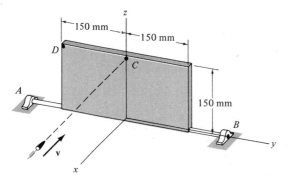

Prob. 21–33

21–34. The circular disk has a weight of 15 lb and is mounted on the shaft AB at an angle of $45°$ with the horizontal. Determine the angular velocity of the shaft when $t = 2$ s if a torque $M = (4e^{0.1t})$ lb · ft, where t is in seconds, is applied to the shaft. The shaft is originally spinning at $\omega_1 = 8$ rad/s when the torque is applied.

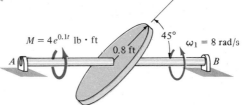

Prob. 21–34

21–35. Each of the two disks has a weight of 10 lb. The axle AB weighs 3 lb. If the assembly is rotating about the z axis at $\omega_z = 6$ rad/s, determine its angular momentum about the z axis and its kinetic energy. The disks roll without slipping.

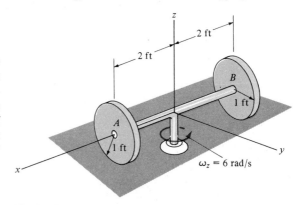

Prob. 21–35

*21–36. The rod assembly is supported at G by a ball-and-socket joint. Each rod has a mass of 0.5 kg/m. If the assembly is originally at rest and an impulse of $\mathbf{I} = \{-8\mathbf{k}\}$ N · s is applied at D, determine the angular velocity of the assembly just after the impact.

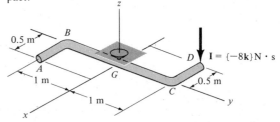

Prob. 21–36

21–37. The assembly consists of a 4-kg rod AB which is connected to link OA and the collar at B by ball-and-socket joints. When $\theta = 0°$, the system is at rest, the spring is unstretched, and a 7-N · m couple moment is applied to the link at O. Determine the angular velocity of the link at the instant $\theta = 90°$. Neglect the mass of the link.

21–38. The space capsule has a mass of 3.5 Mg and the radii of gyration are $k_x = k_z = 0.8$ m and $k_y = 0.5$ m. If it is traveling with a velocity $\mathbf{v}_G = \{600\mathbf{j}\}$ m/s, compute its angular velocity just after it is struck by a meteoroid having a mass of 0.60 kg and a velocity of $\mathbf{v}_m = \{-200\mathbf{i} - 400\mathbf{j} + 200\mathbf{k}\}$ m/s. Assume that the meteoroid embeds itself into the capsule at point A and that the capsule initially has no angular velocity.

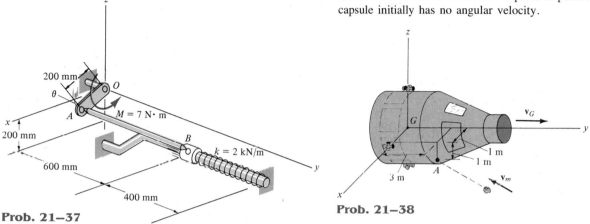

Prob. 21–37

Prob. 21–38

*21.4 Equations of Motion

Having become familiar with the techniques used to describe both the inertial properties and the angular momentum of a body, we can now write the equations which describe the motion of the body in their most useful forms.

Equations of Translational Motion

The *translational motion* of a body is defined in terms of the acceleration of the body's mass center, which is measured from an inertial X, Y, Z reference. The equation of translational motion for the body can be written in vector form as

$$\Sigma \mathbf{F} = m\mathbf{a}_G$$

(21–18)

or by the three scalar equations

$$\begin{aligned} \Sigma F_x &= m(a_G)_x \\ \Sigma F_y &= m(a_G)_y \\ \Sigma F_z &= m(a_G)_z \end{aligned}$$

(21-19)

Here, $\Sigma \mathbf{F} = \Sigma F_x\mathbf{i} + \Sigma F_y\mathbf{j} + \Sigma F_z\mathbf{k}$ represents the sum of all the external forces acting on the body.

Equations of Rotational Motion

In Sec. 15.6, we developed Eq. 15–17, namely,

$$\Sigma \mathbf{M}_O = \dot{\mathbf{H}}_O \qquad (21\text{–}20)$$

which states that the sum of the moments about a fixed point O of all the external forces acting on a system of particles (contained in a rigid body) is equal to the time rate of change of the total angular momentum of the body about point O. When moments of the external forces acting on the particles are summed about the system's *mass center* G, one again obtains the same simple form of Eq. 21–20, relating the moment summation $\Sigma \mathbf{M}_G$ to the angular momentum $\mathbf{H}_G$. To show this, consider the system of particles in Fig. 21–11, where x, y, z represents an inertial frame of reference and the x', y', z' axes, with origin at G, *translate* with respect to this frame. In general, G is *accelerating*, so by definition the translating frame is *not* an inertial reference. The angular momentum of the ith particle with respect to this frame is, however,

$$(\mathbf{H}_i)_G = \mathbf{r}_{i/G} \times m_i \mathbf{v}_{i/G}$$

where $\mathbf{r}_{i/G}$ and $\mathbf{v}_{i/G}$ represent the relative position and relative velocity of the ith particle with respect to G. Taking the time derivative gives

$$(\dot{\mathbf{H}}_i)_G = \dot{\mathbf{r}}_{i/G} \times m_i \mathbf{v}_{i/G} + \mathbf{r}_{i/G} \times m_i \dot{\mathbf{v}}_{i/G}$$

By definition, $\mathbf{v}_{i/G} = \dot{\mathbf{r}}_{i/G}$. Thus, the first term on the right side is zero since the cross product of equal vectors is zero. Also, $\mathbf{a}_{i/G} = \dot{\mathbf{v}}_{i/G}$, so that

$$(\dot{\mathbf{H}}_i)_G = \mathbf{r}_{i/G} \times m_i \mathbf{a}_{i/G}$$

Similar expressions can be written for the other particles of the system. When the results are summed for all the particles, we get

$$\dot{\mathbf{H}}_G = \Sigma(\mathbf{r}_{i/G} \times m_i \mathbf{a}_{i/G})$$

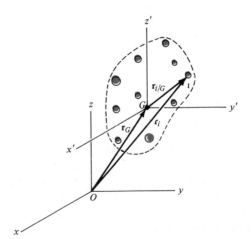

Fig. 21–11

Here $\dot{\mathbf{H}}_G$ is the time rate of change of the total angular momentum of the body computed relative to point G.

The relative acceleration for the ith particle is defined by the equation $\mathbf{a}_{i/G} = \mathbf{a}_i - \mathbf{a}_G$, where $\mathbf{a}_i$ and $\mathbf{a}_G$ represent, respectively, the accelerations of the ith particle and point G measured with respect to the *inertial frame of reference*. Substituting and expanding, using the distributive property of the vector cross product, yields

$$\dot{\mathbf{H}}_G = \Sigma(\mathbf{r}_{i/G} \times m_i\mathbf{a}_i) - (\Sigma m_i\mathbf{r}_{i/G}) \times \mathbf{a}_G$$

By definition of the mass center, the sum $(\Sigma m_i\mathbf{r}_{i/G}) = (\Sigma m_i)\bar{\mathbf{r}}$ is equal to zero, since the position vector $\bar{\mathbf{r}}$ relative to G is zero. Hence, the last term in the above equation is zero. Using the equation of motion, the product $m_i\mathbf{a}_i$ may be replaced by the resultant *external force* $\mathbf{F}_i$ acting on the ith particle. Denoting $\Sigma\mathbf{M}_G = \Sigma\mathbf{r}_{i/G} \times \mathbf{F}_i$, the final result may be written as

$$\Sigma\mathbf{M}_G = \dot{\mathbf{H}}_G \tag{21–21}$$

The rotational equation of motion for the body will now be developed from either Eq. 21–20 or 21–21. In this regard, the scalar components of the angular momentum $\mathbf{H}_O$ or $\mathbf{H}_G$ are defined by Eqs. 21–10 or, if principal axes of inertia are used either at point O or G, by Eqs. 21–11. If these components are computed about x, y, z axes that are *rotating* with an angular velocity $\mathbf{\Omega}$, which may be *different* from the body's angular velocity $\boldsymbol{\omega}$, then the time derivative $\dot{\mathbf{H}} = d\mathbf{H}/dt$, as used in Eqs. 21–20 and 21–21, must account for the rotation of the x, y, z axes as measured from the inertial X, Y, Z axes. Hence, the time derivative of $\mathbf{H}$ must be determined from Eq. 20–6, in which case Eqs. 21–20 and 21–21 become

$$\boxed{\begin{aligned} \Sigma\mathbf{M}_O &= (\dot{\mathbf{H}}_O)_{xyz} + \mathbf{\Omega} \times \mathbf{H}_O \\ \Sigma\mathbf{M}_G &= (\dot{\mathbf{H}}_G)_{xyz} + \mathbf{\Omega} \times \mathbf{H}_G \end{aligned}} \tag{21–22}$$

Here $(\dot{\mathbf{H}})_{xyz}$ is the time rate of change of $\mathbf{H}$ measured from the x, y, z reference.

There are three ways in which one can define the motion of the x, y, z axes. Obviously, motion of this reference should be chosen to yield the simplest set of moment equations for the solution of a particular problem.

x, y, z Axes Having Motion $\mathbf{\Omega} = \mathbf{0}$. If the body has general motion, the x, y, z axes may be chosen with origin at G, such that the axes only *translate* relative to the inertial X, Y, Z frame of reference. Doing this would certainly simplify Eq. 21–22, since $\mathbf{\Omega} = \mathbf{0}$. However, the body may have a rotation $\boldsymbol{\omega}$ about these axes, and therefore the moments and products of inertia of the body would have to be expressed as *functions of time*. In most cases this would be a difficult task, so that such a choice of axes has restricted value.

x, y, z Axes Having Motion $\boldsymbol{\Omega} = \boldsymbol{\omega}$. The x, y, z axes may be chosen such that they are *fixed in and move with the body.* The moments and products of inertia of the body relative to these axes will be *constant* during the motion. Since $\boldsymbol{\Omega} = \boldsymbol{\omega}$, Eqs. 21–22 become

$$\Sigma\mathbf{M}_O = (\dot{\mathbf{H}}_O)_{xyz} + \boldsymbol{\omega} \times \mathbf{H}_O$$
$$\Sigma\mathbf{M}_G = (\dot{\mathbf{H}}_G)_{xyz} + \boldsymbol{\omega} \times \mathbf{H}_G$$

$$(21–23)$$

We may express each of these vector equations as three scalar equations using Eqs. 21–10. Neglecting the subscripts O and G yields

$$\Sigma M_x = I_{xx}\,\dot{\omega}_x - (I_{yy} - I_{zz})\omega_y\omega_z - I_{xy}(\dot{\omega}_y - \omega_z\omega_x) - I_{yz}(\omega_y^2 - \omega_z^2)$$
$$- I_{zx}\,(\dot{\omega}_z + \omega_x\omega_y)$$
$$\Sigma M_y = I_{yy}\,\dot{\omega}_y - (I_{zz} - I_{xx})\omega_z\omega_x - I_{yz}\,(\dot{\omega}_z - \omega_x\omega_y) - I_{zx}(\omega_z^2 - \omega_x^2)$$
$$- I_{xy}\,(\dot{\omega}_x + \omega_y\omega_z) \qquad (21–24)$$
$$\Sigma M_z = I_{zz}\,\dot{\omega}_z - (I_{xx} - I_{yy})\omega_x\omega_y - I_{zx}\,(\dot{\omega}_x - \omega_y\omega_z) - I_{xy}(\omega_x^2 - \omega_y^2)$$
$$- I_{yz}(\dot{\omega}_y + \omega_z\omega_x)$$

Notice that for a rigid body symmetric with respect to the *xy* reference plane, and undergoing general plane motion in this plane, $I_{xz} = I_{yz} = 0$, and $\omega_x = \omega_y = d\omega_x/dt = d\omega_y/dt = 0$. Equations 21–24 reduce to the form $\Sigma M_x = \Sigma M_y = 0$, and $\Sigma M_z = I_{zz}\alpha_z$ (where $\alpha_z = \dot{\omega}_z$), which is essentially the third of Eqs. 17–11 or 17–16 depending upon the choice of point G or O for summing moments.

If the x, y, and z axes are chosen as *principal axes of inertia,* the products of inertia are zero, $I_{xx} = I_x$, etc., and Eqs. 21–24 reduce to the form

$$\Sigma M_x = I_x\dot{\omega}_x - (I_y - I_z)\omega_y\omega_z$$
$$\Sigma M_y = I_y\dot{\omega}_y - (I_z - I_x)\omega_z\omega_x$$
$$\Sigma M_z = I_z\dot{\omega}_z - (I_x - I_y)\omega_x\omega_y$$

$$(21–25)$$

This set of equations is known historically as the *Euler equations of motion,* named after the Swiss mathematician Leonhard Euler (1707–1783). They *only* apply for moments summed either about point O or G.

When applying these equations it should be realized that $\dot{\omega}_x$, $\dot{\omega}_y$, and $\dot{\omega}_z$ represent the time derivatives of the magnitudes of the x, y, z components of $\boldsymbol{\omega}$. Since the x, y, z axes are rotating at $\boldsymbol{\Omega} = \boldsymbol{\omega}$, then, from Eq. 20–6, it may be noted that $(\dot{\boldsymbol{\omega}})_{XYZ} = (\dot{\boldsymbol{\omega}})_{xyz} + \boldsymbol{\omega} \times \boldsymbol{\omega}$. Since $\boldsymbol{\omega} \times \boldsymbol{\omega} = \mathbf{0}$, $(\dot{\boldsymbol{\omega}})_{XYZ} = (\dot{\boldsymbol{\omega}})_{xyz}$. This important result indicates that the required time derivative of $\boldsymbol{\omega}$ can be obtained either by first finding the components of $\boldsymbol{\omega}$ along the x, y, z axes and then taking the time derivative of the magnitudes of these components, i.e., $(\dot{\boldsymbol{\omega}})_{xyz}$, or by first finding the time derivative of $\boldsymbol{\omega}$ with respect to the X, Y, Z axes, i.e., $(\dot{\boldsymbol{\omega}})_{XYZ}$, and then determining the components $\dot{\omega}_x$, $\dot{\omega}_y$, and $\dot{\omega}_z$. In practice, it is generally easier to compute $\dot{\omega}_x$, $\dot{\omega}_y$, and $\dot{\omega}_z$ on the basis of finding $(\dot{\omega})_{XYZ}$.

x, y, z Axes Having Motion $\Omega \neq \omega$. To simplify the calculations for the time derivative of $\boldsymbol{\omega}$, it is often convenient to choose the x, y, z axes having an angular velocity $\boldsymbol{\Omega}$ which is different from the angular velocity $\boldsymbol{\omega}$ of the body. This is particularly suitable for the analysis of spinning tops and gyroscopes which are *symmetrical* about their spinning axis.* When this is the case, the moments and products of inertia remain constant during the motion.

Equations 21–22 are applicable for such a set of chosen axes. Each of these two vector equations may be reduced to a set of three scalar equations which are derived in a manner similar to Eqs. 21–25,† i.e.,

$$\Sigma M_x = I_x\dot{\omega}_x - I_y\Omega_z\omega_y + I_z\Omega_y\omega_z$$
$$\Sigma M_y = I_y\dot{\omega}_y - I_z\Omega_x\omega_z + I_x\Omega_z\omega_x$$
$$\Sigma M_z = I_z\dot{\omega}_z - I_x\Omega_y\omega_x + I_y\Omega_x\omega_y$$

$$(21\text{–}26)$$

Here Ω_x, Ω_y, Ω_z represent the x, y, z components of $\boldsymbol{\Omega}$, measured from the inertial frame of reference.

Any one of these sets of moment equations, Eqs. 21–24, 21–25, or 21–26, represents a series of three first-order nonlinear differential equations. These equations are "coupled," since the angular-velocity components are present in all the terms. Success in determining the solution for a particular problem therefore depends upon what is unknown in these equations. Difficulty certainly arises when one attempts to solve for the unknown components of $\boldsymbol{\omega}$, given the external moments as functions of time. Further complications can arise if the moment equations are coupled to the three scalar equations of translation, Eqs. 21–19. This can happen because of the existence of kinematic constraints which relate the rotation of the body to the translation of its mass center, as in the case of a hoop which rolls without slipping. Problems necessitating the simultaneous solution of differential equations generally require application of numerical methods with the aid of a computer. In many engineering problems, however, one is required to determine the applied moments acting on the body, given information about the motion of the body. Fortunately, many of these types of problems have direct solutions, so that there is no need to resort to computer techniques.

---

* A detailed discussion of such devices is given in Sec. 21.5.

† See Prob. 21–41.

PROCEDURE FOR ANALYSIS

The following procedure provides a method for solving problems involving the three-dimensional motion of a rigid body.

Free-Body Diagram. Draw a *free-body diagram* of the body at the instant considered and specify the x, y, z coordinate system. The origin of this reference must be located either at the body's mass center G, or at point O, considered fixed in an inertial reference frame and located either in the body or on a massless extension of the body. Depending upon the nature of the problem, a decision should be made as to what type of rotational motion $\boldsymbol{\Omega}$ this coordinate system should have, i.e., $\boldsymbol{\Omega} = \mathbf{0}$, $\boldsymbol{\Omega} = \boldsymbol{\omega}$, or $\boldsymbol{\Omega} \neq \boldsymbol{\omega}$. When deciding, one should keep in mind that the moment equations are simplified when the axes move in such a manner that they represent principal axes of inertia for the body at all times.

Kinematics. Compute the necessary moments and products of inertia and the components of the body's angular velocity and angular acceleration. In some cases a *kinematic diagram* might be helpful, since it provides a graphical aid for determining the acceleration of the body's mass center and for computing the angular acceleration. The angular velocity can usually be determined from the constraints of the body or from the given problem data. If this vector is changing direction, Eq. 20–6 $[(\dot{\boldsymbol{\omega}})_{XYZ} = (\dot{\boldsymbol{\omega}})_{xyz} + \boldsymbol{\Omega} \times \boldsymbol{\omega}]$ must be used to determine the components of angular acceleration. In particular, note that if $\boldsymbol{\Omega} = \boldsymbol{\omega}$ then $(\dot{\boldsymbol{\omega}})_{XYZ} = (\dot{\boldsymbol{\omega}})_{xyz}$.

Equations of Motion. Apply either the two vector equations 21–18 and 21–22, or the six scalar component equations appropriate for the x, y, z coordinate axes chosen for the problem.

The following example problems numerically illustrate application of this procedure.

Example 21–4

The gear shown in Fig. 21–12a has a mass of 10 kg and is mounted at an angle of 10° with a rotating shaft having negligible mass. If $I_z = 0.1$ kg $\cdot$ m², $I_x = I_y = 0.05$ kg $\cdot$ m², and the shaft is rotating with a constant angular velocity of $\omega_{AB} = 30$ rad/s, determine the reactions that the bearing supports A and B exert on the shaft at the instant shown.

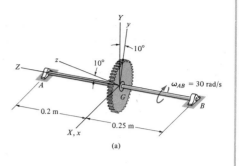

(a)

SOLUTION

Free-Body Diagram Fig. 21–12b. The origin of the x, y, z coordinate system is located at the gear's center of mass G, which is also a fixed point. The axes are fixed in and rotate with the gear; hence $\boldsymbol{\Omega} = \boldsymbol{\omega}$.

Kinematics. As shown in Fig. 21–12c, the angular velocity $\boldsymbol{\omega}$ of the gear is constant in magnitude and is always directed along the axis of the shaft AB. Since this vector is measured from the X, Y, Z inertial frame of reference, for any position of the x, y, z axes $\boldsymbol{\omega}$ has x, y, z components of

$$\omega_x = 0, \qquad \omega_y = -30 \sin 10°, \qquad \omega_z = 30 \cos 10°$$

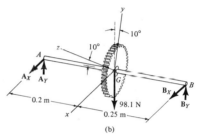

(b)

Since $\boldsymbol{\omega}$ has a constant direction and magnitude when observed from X, Y, Z, $(\dot{\boldsymbol{\omega}})_{XYZ} = (\dot{\boldsymbol{\omega}})_{xyz} = \mathbf{0}$. Hence, $\dot{\omega}_x = \dot{\omega}_y = \dot{\omega}_z = 0$. Also, since G is a fixed point, $(a_G)_x = (a_G)_y = (a_G)_z = 0$.

Equations of Motion. Applying Eqs. 21–25 yields

$$\Sigma M_x = I_x \dot{\omega}_x - (I_y - I_z)\omega_y\omega_z$$
$$- (A_Y)(0.2) + (B_Y)(0.25) = 0 - (0.05 - 0.1)(-30 \sin 10°)(30 \cos 10°)$$
$$- 0.2A_Y + 0.25B_Y = -7.70 \qquad (1)$$

$$\Sigma M_y = I_y \dot{\omega}_y - (I_z - I_x)\omega_z\omega_x$$
$$A_X(0.2) \cos 10° - B_X(0.25) \cos 10° = 0 + 0$$
$$A_X = 1.25B_X \qquad (2)$$

$$\Sigma M_z = I_z \dot{\omega}_z - (I_x - I_y)\omega_x\omega_y$$
$$A_X(0.2) \sin 10° - B_X(0.25) \sin 10° = 0$$
$$A_X = 1.25B_X$$

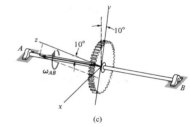

(c)

Fig. 21–12

Applying Eqs. 21–19, we have

$$\Sigma F_X = m(a_G)_X; \qquad A_X + B_X = 0 \qquad (3)$$
$$\Sigma F_Y = m(a_G)_Y; \qquad A_Y + B_Y - 98.1 = 0 \qquad (4)$$
$$\Sigma F_Z = m(a_G)_Z; \qquad 0 = 0$$

Solving Eqs. (1) to (4) simultaneously gives

$$A_X = B_X = 0, \quad A_Y = 71.6 \text{ N}, \quad B_Y = 26.4 \text{ N} \qquad \textit{Ans.}$$

Example 21–5

The airplane shown in Fig. 21–13a is in the process of making a steady *horizontal* turn at the rate of ω_p. During this motion, the airplane's propeller is spinning at the rate of ω_s. If the propeller has two blades, determine the moments which the propeller shaft exerts on the propeller when the blades are in the vertical position. For simplicity, assume the blades to be a uniform slender bar having a moment of inertia I about an axis perpendicular to the blades and passing through their center, and having zero moment of inertia about a longitudinal axis.

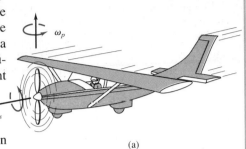

(a)

SOLUTION

Free-Body Diagram Fig. 21–13b. The effect of the connecting shaft on the propeller is indicated by the resultants $\mathbf{F}_R$ and $\mathbf{M}_R$. (The propeller's weight is assumed to be negligible.) The x, y, z axes are taken to be *fixed to the propeller* so that $\boldsymbol{\Omega} = \boldsymbol{\omega} = \boldsymbol{\omega}_s + \boldsymbol{\omega}_p$. Furthermore, the axes so chosen represent the principal axes of inertia for the propeller.

Kinematics. The moments of inertia I_x and I_y are equal ($I_x = I_y = I$) and $I_z = 0$. The angular velocity at the instant shown, as measured from a fixed X, Y, Z frame coincident with x, y, z, Fig. 21–13c, is $\boldsymbol{\omega} = \boldsymbol{\omega}_s + \boldsymbol{\omega}_p = \omega_s\mathbf{i} + \omega_p\mathbf{k}$, so that the components of $\boldsymbol{\omega}$ are

$$\omega_x = \omega_s, \qquad \omega_y = 0, \qquad \omega_z = \omega_p$$

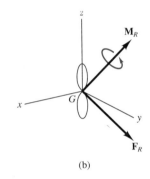

(b)

Since $\boldsymbol{\omega}$ is *changing direction,* the time derivative of $\boldsymbol{\omega}$ will be computed with respect to the fixed X, Y, Z axes and *then* $(\dot{\boldsymbol{\omega}})_{XYZ}$ will be resolved into components along the moving x, y, z axes. To simplify this calculation, a third coordinate system x', y', z' will be chosen which has an angular velocity $\boldsymbol{\Omega}' = \boldsymbol{\omega}_p$ and is coincident with the X, Y, Z axes at the instant shown. Thus

$$(\dot{\boldsymbol{\omega}})_{XYZ} = (\dot{\boldsymbol{\omega}})_{x'y'z'} + \boldsymbol{\Omega}' \times \boldsymbol{\omega} \qquad (1)$$
$$= (\dot{\boldsymbol{\omega}}_s)_{x'y'z'} + (\dot{\boldsymbol{\omega}}_p)_{x'y'z'} + \boldsymbol{\omega}_p \times (\boldsymbol{\omega}_s + \boldsymbol{\omega}_p)$$
$$= \mathbf{0} + \mathbf{0} + \boldsymbol{\omega}_p \times \boldsymbol{\omega}_s + \boldsymbol{\omega}_p \times \boldsymbol{\omega}_p$$
$$= \mathbf{0} + \mathbf{0} + \omega_p\mathbf{k} \times \omega_s\mathbf{i} + \mathbf{0} = \omega_p\omega_s\mathbf{j}$$

Since $\boldsymbol{\Omega} = \boldsymbol{\omega}$, $(\dot{\boldsymbol{\omega}})_{XYZ} = (\dot{\boldsymbol{\omega}})_{xyz}$ and therefore

$$\dot{\omega}_x = 0, \qquad \dot{\omega}_y = \omega_p\omega_s, \qquad \dot{\omega}_z = 0$$

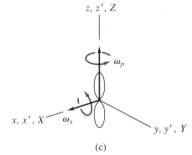

(c)

Fig. 21–13

Equations of Motion. Using Eqs. 21–25, we have

$$\Sigma M_x = I_x\dot{\omega}_x - (I_y - I_z)\omega_y\omega_z = I(0) - (I - 0)(0)\omega_p$$
$$M_x = 0 \qquad\qquad\qquad\qquad\qquad \textit{Ans.}$$
$$\Sigma M_y = I_y\dot{\omega}_y - (I_z - I_x)\omega_z\omega_x = I(\omega_p\omega_s) - (0 - I)\omega_p\omega_s$$
$$M_y = 2I\omega_p\omega_s \qquad\qquad\qquad \textit{Ans.}$$
$$\Sigma M_z = I_z\dot{\omega}_z - (I_x - I_y)\omega_x\omega_y = 0(0) - (I - I)(\omega_s)(0)$$
$$M_z = 0 \qquad\qquad\qquad\qquad\qquad \textit{Ans.}$$

PROBLEMS

21–39. Derive the scalar form of the rotational equation of motion along the *x* axis, when $\Omega \neq \omega$ and the moments and products of inertia of the body are *not constant* with respect to time.

*21–40. Derive the scalar form of the rotational equation of motion along the *x* axis, when $\Omega \neq \omega$ and the moments and the products of inertia of the body are *constant* with respect to time.

21–41. Derive the Euler equations of motion for $\Omega \neq \omega$, i.e., Eqs. 21–26.

21–42. The 40-kg flywheel (disk) is mounted 20 mm off its true center at *G*. If the shaft is rotating at a constant speed $\omega = 8$ rad/s, determine the maximum reactions exerted on the journal bearings at *A* and *B*.

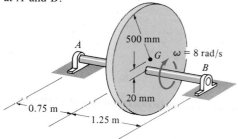

Prob. 21–42

21–43. The 4-lb bar rests along the smooth corners of an open box. At the instant shown the box has a velocity of $\mathbf{v} = \{5\mathbf{k}\}$ ft/s and an acceleration $\mathbf{a} = \{2\mathbf{k}\}$ ft/s². Determine the *x*, *y*, *z* components of force which the corners exert on the bar.

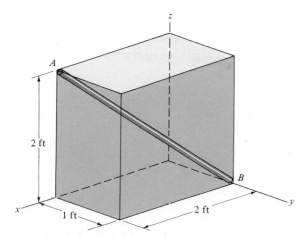

Probs. 21–43/21–44

*21–44. The 4-lb bar rests along the smooth corners of an open box. At the instant shown the box has a velocity of $\mathbf{v} = \{3\mathbf{j}\}$ ft/s and an acceleration of $\mathbf{a} = \{-6\mathbf{j}\}$ ft/s². Determine the *x*, *y*, *z* components of force which the corners exert on the bar.

21–45. The 20-lb plate is mounted on the shaft *AB* so that the plane of the plate makes an angle of $\theta = 30°$ with the vertical. If the shaft is turning with an angular velocity of 25 rad/s, determine the vertical reactions at the bearing supports *A* and *B* when the plate is in the position shown.

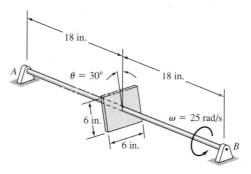

Prob. 21–45

21–46. The 25-lb disk *A* is *fixed* to rod *BCD*, which has negligible mass. Determine the torque **T** which must be applied to the vertical shaft so that the shaft has an angular acceleration of $\alpha = 6$ rad/s².

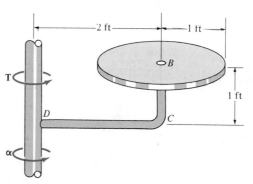

Probs. 21–46/21–47

21–47. Solve Prob. 21–46, assuming rod *BCD* has a weight of 2 lb/ft.

*21–48. The rod assembly has a weight of 5 lb/ft. It is supported at B by a smooth journal bearing, which develops x and y force reactions, and at A by a smooth thrust bearing, which develops x, y, and z force reactions. If a 50-lb · ft torque is applied along rod AB, determine the components of reaction at the bearing at the instant the assembly has an angular velocity $\omega = 10$ rad/s.

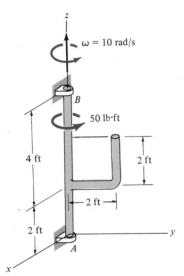

Prob. 21–48

21–49. The 20-lb disk is mounted on the horizontal shaft AB such that its plane forms an angle of 10° with the vertical. If the shaft rotates with an angular velocity of 3 rad/s, determine the vertical reactions developed at the bearings when the disk is in the position shown.

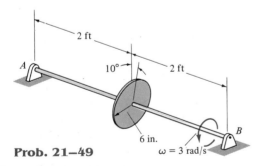

Prob. 21–49

21–50. The crankshaft is constructed from a rod which has a weight of 3 lb/ft. Determine the x and z components of reaction at bearings A and B at the instant the shaft is in the position shown.

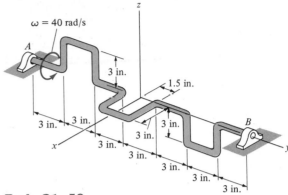

Prob. 21–50

21–51. The rod assembly has a mass of 1.5 kg/m. A resultant frictional moment **M,** developed at the journal bearings A and B, causes the shaft to decelerate at 1 rad/s² when it has an angular velocity $\omega = 6$ rad/s. Determine the reactions at the bearings at the instant shown. What is the frictional moment **M** causing the deceleration? Establish the axes at G with $+y$ upward, and $+z$ directed along the shaft axis towards A.

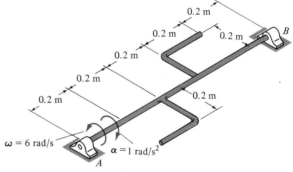

Prob. 21–51

*21–52. The 5-kg rod AB is supported by a rotating arm. The support at A is a journal bearing, which develops reactions normal to the rod. The support at B is a thrust bearing, which develops reactions both normal to the rod and along the axis of the rod. Neglecting friction, determine the x, y, z components of reaction at these supports when the frame rotates with a constant angular velocity of $\omega = 10$ rad/s.

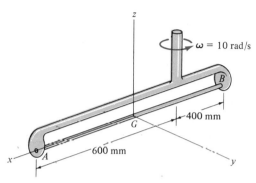

Prob. 21–52

21–53. The rod assembly is supported by bearings at A and B. The mass of each rod is 1.5 kg/m. If the shaft AB is rotating with a speed $\omega = 5$ rad/s, determine the reactions at the journal bearings when the assembly is in the position shown. Also, what is the shaft's angular acceleration?

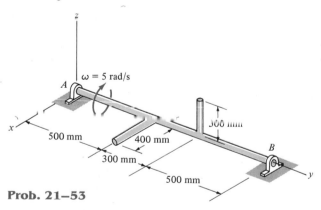

Prob. 21–53

21–54. The rod assembly is supported by a ball-and-socket joint at C and a journal bearing support at D, which develops only x and y force reactions. The rods have a mass of 0.75 kg/m. Determine the angular acceleration of the rods and the components of reaction at the supports at the instant $\omega = 8$ rad/s as shown.

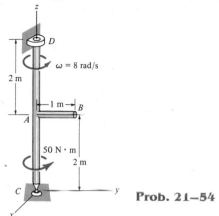

Prob. 21–54

21–55. The cylinder has a mass of 30 kg and is mounted on an axle which is supported by bearings at A and B. If the axle is turning at a constant rate of 40 rad/s, determine the components of force acting at the bearings at the instant shown.

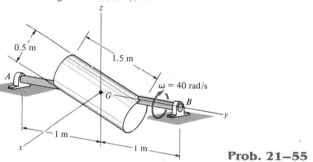

Prob. 21–55

*21–56. The uniform hatch door, having a mass of 15 kg and a mass center at G, is supported in the horizontal plane by bearings at A and B. If a vertical force $F = 300$ N is applied to the door as shown, determine the components of reaction at the bearings and the angular acceleration of the door. The bearing at A will resist a component of force in the y direction, whereas the bearing at B will not. For the calculation, assume the door to be a thin plate and neglect the size of the bearing. The door is originally at rest. Set $a = 100$ mm, $b = 200$ mm, $c = 150$ mm, and $d = 30$ mm.

Prob. 21–56

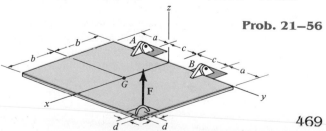

469

21–57. The man sits on a swivel chair which is rotating with a constant angular velocity of 3 rad/s. He holds the uniform 5-lb rod AB horizontal. He suddenly gives it an angular acceleration of 2 rad/s^2, measured relative to him, as shown. Determine the required force and moment components at the grip, A, necessary to do this. Establish axes at G with $+z$ upward, and $+y$ directed along the axis of the rod towards A.

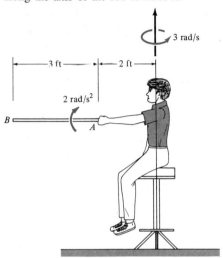

Prob. 21–57

21–58. The 4-kg slender rod AB is pinned at A and held at B by a cord. The axle CD is supported at its ends by ball-and-socket joints and is rotating with a constant angular velocity of 2 rad/s. Determine the tension developed in the cord and the magnitude of force developed at the pin A.

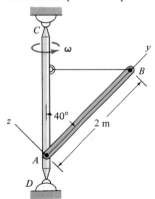

Probs. 21–58/21–59

21–59. Determine the tension in the cord and the magnitude of force at the pin A in Prob. 21–58 if the supporting axle has an angular acceleration $\alpha = 1.5$ rad/s^2 at the instant $\omega = 2$ rad/s. Assume that $\boldsymbol{\alpha}$ acts in the same direction as $\boldsymbol{\omega}$ shown in Prob. 21–58.

*21–60. The bent uniform rod ACD has a weight of 5 lb/ft and is supported at A by a pin and at B by a cord. If the vertical shaft rotates with a constant angular velocity of $\omega = 20$ rad/s, determine the x, y, z components of force and moment developed at A and the tension in the cord.

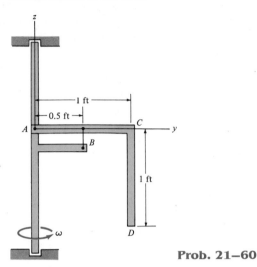

Prob. 21–60

21–61. The rod AB supports the 10-lb sphere. If the rod is pinned at A to the vertical shaft which is rotating at a constant rate of $\omega = \{7\mathbf{k}\}$ rad/s, determine the angle θ of the rod during the motion. Neglect the mass of the rod in the calculation.

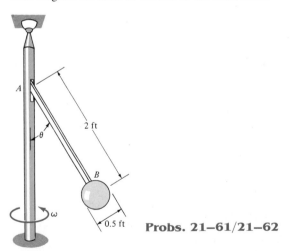

Probs. 21–61/21–62

21–62. The rod AB supports the 10 lb sphere. If the rod is pinned at A to the vertical shaft which is rotating with an angular acceleration of $\alpha = \{2\mathbf{k}\}$ rad/s^2 and at the instant shown the shaft has an angular velocity of $\omega = \{7\mathbf{k}\}$ rad/s, determine the angle θ of the rod at this instant. Neglect the mass of the rod in the calculation.

*21.5 Gyroscopic Motion

In this section the equations used for analyzing the motion of a body (or top) which is symmetrical with respect to an axis and moving about a fixed point lying on the axis will be developed. These equations will then be applied to study the motion of a particularly interesting device, the gyroscope.

The body's motion will be analyzed using *Euler angles* ϕ, θ, ψ (phi, theta, psi). To illustrate how these angles are formed, reference is made to the top shown in Fig. 21–14a. The top is attached to point O and has an orientation relative to the fixed X, Y, Z axes at some instant of time as shown in Fig. 21–14d. To define this final position, a second set of x, y, z axes will be needed. For purposes of discussion, assume that this reference is fixed in the top. Starting with the X, Y, Z and x, y, z axes in coincidence, Fig. 21–14a, the final position of the top is determined using the following three orderly steps:

1. Rotate the top about the Z (or z) axis through an angle ϕ ($0 \leq \phi < 2\pi$), Fig. 21–14b.
2. Rotate the top about the x axis through an angle θ ($0 \leq \theta \leq \pi$), Fig. 21–14c.
3. Rotate the top about the z axis through an angle ψ ($0 \leq \psi < 2\pi$) to obtain the final position, Fig. 20–14d.

When the top is in motion, in general each of the Euler angles is changing with time. Although the three finite rotations defined above are not vectors, the infinitesimal rotations $d\phi$, $d\theta$, and $d\psi$ are vectors, and thus the angular velocity $\boldsymbol{\omega}$ of the top can be expressed in terms of the time derivatives of the Euler angles. The angular-velocity components $\dot{\phi}$, $\dot{\theta}$, and $\dot{\psi}$ are known as the

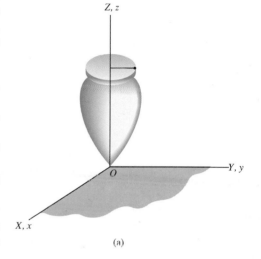

(a)

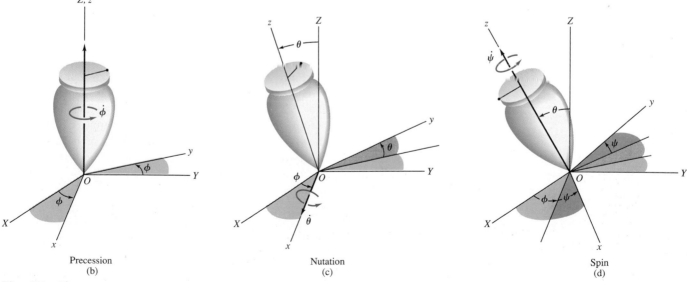

Precession
(b)

Nutation
(c)

Spin
(d)

Fig. 21–14

471

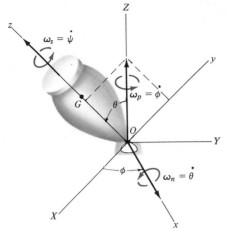

Fig. 21–15

precession, nutation, and *spin,* respectively. Their positive directions are shown in Fig. 21–14. It is seen that these vectors are not all perpendicular to one another; however, $\boldsymbol{\omega}$ of the top can still be expressed in terms of these three components.

In our case the body (top) is symmetric with respect to the z or spin axis. If we consider the top in Fig. 21–15, for which the x axis is oriented such that at the instant considered the spin angle $\psi = 0$ and *the x, y, z axes follow the motion of the body only in nutation and precession,* i.e., $\boldsymbol{\Omega} = \boldsymbol{\omega}_p + \boldsymbol{\omega}_n$, then the nutation and spin are always directed along the x and z axes, respectively. Hence, the angular velocity of the body is specified only in terms of the Euler angle θ, i.e.,

$$\boldsymbol{\omega} = \omega_x \mathbf{i} + \omega_y \mathbf{j} + \omega_z \mathbf{k}$$
$$= \dot{\theta}\mathbf{i} + (\dot{\phi}\sin\theta)\mathbf{j} + (\dot{\phi}\cos\theta + \dot{\psi})\mathbf{k} \qquad (21\text{–}27)$$

Since motion of the axes is not affected by the spin component,

$$\boldsymbol{\Omega} = \Omega_x \mathbf{i} + \Omega_y \mathbf{j} + \Omega_z \mathbf{k}$$
$$= \dot{\theta}\mathbf{i} + (\dot{\phi}\sin\theta)\mathbf{j} + (\dot{\phi}\cos\theta)\mathbf{k} \qquad (21\text{–}28)$$

The x, y, z axes in Fig. 21–15 represent *principal axes of inertia* of the body for *any* spin of the body about these axes. Hence, the moments of inertia are constant and will be represented as $I_{xx} = I_{yy} = I$ and $I_{zz} = I_z$. Since $\boldsymbol{\Omega} \neq \boldsymbol{\omega}$, the Euler Eqs. 21–26 are used to establish the rotational equations of motion. Substituting into these equations the respective angular-velocity components defined by Eqs. 21–27 and 21–28, their corresponding time derivatives, and the moment of inertia components yields

$$\Sigma M_x = I(\ddot{\theta} - \dot{\phi}^2 \sin\theta\cos\theta) + I_z\dot{\phi}\sin\theta(\dot{\phi}\cos\theta + \dot{\psi})$$
$$\Sigma M_y = I(\ddot{\phi}\sin\theta + 2\dot{\phi}\dot{\theta}\cos\theta) - I_z\dot{\theta}(\dot{\phi}\cos\theta + \dot{\psi}) \qquad (21\text{–}29)$$
$$\Sigma M_z = I_z(\ddot{\psi} + \ddot{\phi}\cos\theta - \dot{\phi}\dot{\theta}\sin\theta)$$

Each moment summation applies only at the fixed point O or the center of mass G of the body. Since the equations represent a coupled set of nonlinear second-order differential equations, in general a closed-form solution may not be obtained. Instead, the Euler angles ϕ, θ, and ψ may be obtained graphically as functions of time using numerical analysis and computer techniques.

A special case, however, does exist for which simplification of Eqs. 21–29 is possible. Commonly referred to as *steady precession,* it occurs when the nutation angle θ, precession $\dot{\phi}$, and spin $\dot{\psi}$ all remain *constant.* Equations 21–29 then reduce to the form

$$\boxed{\Sigma M_x = -I\dot{\phi}^2 \sin\theta\cos\theta + I_z\dot{\phi}\sin\theta(\dot{\phi}\cos\theta + \dot{\psi})} \qquad (21\text{–}30)$$

$$\Sigma M_y = 0$$
$$\Sigma M_z = 0$$

Equation 21–30 may be further simplified by noting that, from Eq. 21–27, $\omega_z = \dot{\phi} \cos \theta + \dot{\psi}$, so that

$$\Sigma M_x = -I\dot{\phi}^2 \sin \theta \cos \theta + I_z\dot{\phi}(\sin \theta)\omega_z$$

or

$$\boxed{\Sigma M_x = \dot{\phi} \sin \theta(I_z\omega_z - I\dot{\phi} \cos \theta)} \qquad (21\text{–}31)$$

It is interesting to note what effects the spin $\dot{\psi}$ has on the moment about the x axis. In this regard consider the spinning rotor shown in Fig. 21–16. Here $\theta = 90°$, in which case Eq. 21–30 reduces to the form

$$\Sigma M_x = I_z\dot{\phi}\dot{\psi}$$

or

$$\boxed{\Sigma M_x = I_z\Omega_y\omega_z} \qquad (21\text{–}32)$$

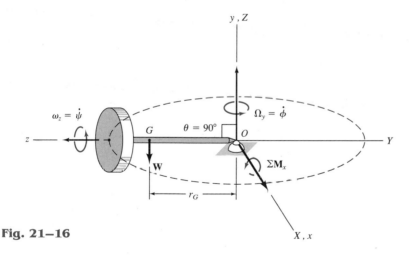

Fig. 21–16

From the figure it is seen that vectors $\Sigma \mathbf{M}_x$, $\mathbf{\Omega}_y$, and $\boldsymbol{\omega}_z$ all act along their respective *positive axes* and therefore they are mutually perpendicular. Instinctively, one would expect the rotor to fall down under the influence of gravity! However, this is not the case at all provided the product $I_z\Omega_y\omega_z$ is correctly chosen to counterbalance the moment $\Sigma M_x = Wr_G$ of the rotor's weight about O. This unusual phenomenon of rigid-body motion is often referred to as the *gyroscopic effect*.

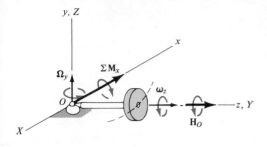

Fig. 21–17

Perhaps a more intriguing demonstration of the gyroscopic effect comes from studying the action of a *gyroscope,* frequently referred to as a *gyro.* A gyro is a rotor which spins at a very high rate about its axis of symmetry. This rate of spin is considerably greater than its precessional rate of rotation about the vertical axis. Hence, for all practical purposes, the angular momentum of the gyro can be assumed directed along its axis of spin. Thus, for the gyro rotor shown in Fig. 21–17, $\omega_z \gg \Omega_y$, and the magnitude of the angular momentum about point O, as computed by Eqs. 21–11, reduces to the form $H_O = I_z\omega_z$. Since both the magnitude and direction of $\mathbf{H}_O$ are constant as observed from x, y, z, direct application of Eq. 21–22 yields

$$\boxed{\Sigma\mathbf{M}_x = \mathbf{\Omega}_y \times \mathbf{H}_O} \qquad (21\text{–}33)$$

Using the right-hand rule applied to the cross product, it is seen that $\mathbf{\Omega}_y$ always swings $\mathbf{H}_O$ (or $\boldsymbol{\omega}_z$) toward the sense of $\Sigma\mathbf{M}_x$. In effect, the *change in direction* of the gyro's angular momentum, $d\mathbf{H}_O$, causes an angular impulse about O, i.e., $d\mathbf{H}_O = \Sigma\mathbf{M}_x\, dt$, Eq. 21–20. Note that since $H_O = I_z\omega_z$ and ΣM_x, Ω_y, and H_O are mutually perpendicular, Eq. 21–33 reduces to Eq. 21–32.

When a gyro is mounted in gimbal rings, Fig. 21–18, it becomes *free* of external moments applied to its base. Thus, in theory, its angular momentum **H** will never precess but, instead, maintain its same fixed orientation along the axis of spin when the base is rotated. This type of gyroscope is called a *free gyro* and is useful as a gyrocompass when the spin axis of the gyro is directed north. In reality, the gimbal mechanism is never completely free of friction, so that such a device is useful only for the local navigation of ships and aircraft. The gyroscopic effect is also useful as a means of stabilizing both the rolling motion of ships at sea and the trajectories of missiles and projectiles. Furthermore, this effect is of significant importance in the design of shafts and bearings for rotors which are subjected to forced precessions.

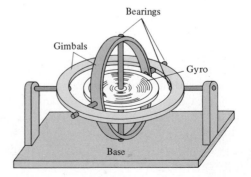

Fig. 21–18

Example 21–6

The top shown in Fig. 21–19a has a mass of 0.5 kg and is precessing about the vertical axis at a constant angle of $\theta = 60°$. If it spins with an angular velocity $\omega_s = 100$ rad/s, determine the precessional velocity ω_p. Assume that the axial and transverse moments of inertia of the top are $4.5(10^{-4})$ kg · m^2 and $12.0(10^{-4})$ kg · m^2, respectively, measured with respect to the fixed point O.

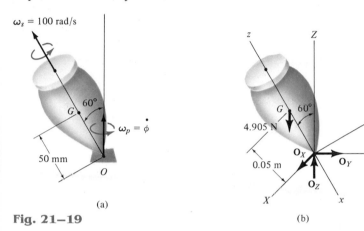

Fig. 21–19

SOLUTION

Equation 21–30 will be used for the solution since the motion is a *steady precession*. As shown on the free-body diagram, Fig. 21–19b, the coordinate axes are established in the usual manner, that is, with the positive z axis in the direction of spin, the positive Z axis in the direction of precession and the positive x axis in the direction of the moment $\Sigma \mathbf{M}_x$ (refer to Fig. 21–15). Thus,

$$\Sigma M_x = -I\dot{\phi}^2 \sin\theta \cos\theta + I_z\dot{\phi} \sin\theta(\dot{\phi}\cos\theta + \dot{\psi})$$

$$4.905(0.05)\sin 60° = -12.0(10^{-4})\dot{\phi}^2 \sin 60° \cos 60°$$
$$+ 4.5(10^{-4})\dot{\phi} \sin 60°(\dot{\phi}\cos 60° + 100)$$

or

$$\dot{\phi}^2 - 120.0\dot{\phi} + 654.0 = 0 \qquad (1)$$

Solving this quadratic equation for the precession gives

$$\dot{\phi} = 114 \text{ rad/s} \qquad \text{(high precession)} \qquad \textit{Ans.}$$

and

$$\dot{\phi} = 5.72 \text{ rad/s} \qquad \text{(low precession)} \qquad \textit{Ans.}$$

In reality low precession of the top would generally be observed since high precession would require a large kinetic energy.

Example 21–7

The 1-kg disk shown in Fig. 21–20a is spinning about its axis with a constant angular velocity $\omega_D = 70$ rad/s. The block at B has a mass of 2 kg, and by adjusting its position s one can change the precession of the disk about its supporting pivot at O. Compute the position s which will enable the disk to have a constant precessional velocity $\omega_p = 0.5$ rad/s about the pivot. Neglect the weight of the shaft.

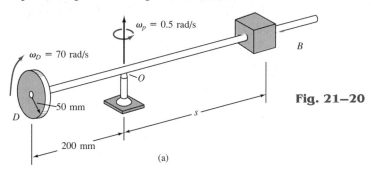

Fig. 21–20

(a)

SOLUTION
The free-body diagram of the disk is shown in Fig. 21–20b, where **F** represents the force reaction of the shaft on the disk. The origin for both the x, y, z and X, Y, Z coordinate systems is located at point O, which represents a *fixed point* for the disk. (Although point O does not lie on the disk, imagine a massless extension of the disk to this point.) In the conventional sense, the Z axis is chosen along the axis of precession, and the z axis is along the axis of spin, so that $\theta = 90°$. Since the precession is *steady,* Eq. 21–31 may be used for the solution. This equation reduces to

$$\Sigma M_x = \dot{\phi} I_z \omega_z$$

which is the same as Eq. 21–32. Substituting the required data gives

$$9.81(0.2) - F(0.2) = 0.5\left[\frac{1}{2}(1)(0.05)^2\right](-70)$$

$$F = 10.0 \text{ N}$$

As shown on the free-body diagram of the shaft and block B, Fig. 21–20c, summing moments about the x axis requires

$$(19.62)s = (10.0)(0.20)$$
$$= 0.102 \text{ m} = 102 \text{ mm} \qquad \textit{Ans.}$$

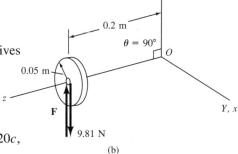

(b)

(c)

*21.6 Torque-Free Motion

When the only external force acting on a body is caused by gravitation, the general motion of the body is referred to as *torque-free motion*. This type of motion is characteristic of planets, artificial satellites, and projectiles—provided the effects of air friction are neglected.

In order to describe the characteristics of this motion, the distribution of the body's mass will be assumed *axisymmetric*. The satellite shown in Fig. 21–21a is an example of such a body, where the z axis represents an axis of symmetry. The origin of the x, y, z coordinates is located at the mass center G, such that $I_{zz} = I_z$ and $I_{xx} = I_{yy} = I$ for the body. If it is assumed that gravitation is the only external force present, the summation of moments about the mass center is zero. From Eq. 21–21, this requires the angular momentum of the body to be constant, i.e.,

$$\mathbf{H}_G = \text{const}$$

At the instant considered, it will be assumed that the inertial frame of reference is oriented such that the positive Z axis is directed along $\mathbf{H}_G$ and the y axis lies in the plane formed by the z and Z axes, Fig. 21–21a. The Euler angle formed between Z and z is θ, and therefore, with this choice of axes the angular momentum may be expressed as

$$\mathbf{H}_G = H_G \sin \theta \mathbf{j} + H_G \cos \theta \mathbf{k}$$

Furthermore, using Eqs. 21–11, we have

$$\mathbf{H}_G = I\omega_x \mathbf{i} + I\omega_y \mathbf{j} + I_z \omega_z \mathbf{k}$$

where ω_x, ω_y, ω_z represent the x, y, z components of the body's angular velocity. Equating the respective $\mathbf{i}$, $\mathbf{j}$, and $\mathbf{k}$ components of the above two equations yields

$$\omega_x = 0, \qquad \omega_y = \frac{H_G \sin \theta}{I}, \qquad \omega_z = \frac{H_G \cos \theta}{I_z} \qquad (21\text{–}34)$$

or

$$\boxed{\boldsymbol{\omega} = \frac{H_G \sin \theta}{I}\mathbf{j} + \frac{H_G \cos \theta}{I_z}\mathbf{k}} \qquad (21\text{–}35)$$

In a similar manner, equating the respective $\mathbf{i}$, $\mathbf{j}$, and $\mathbf{k}$ components of Eq. 21–27 to those of Eq. 21–34, we obtain

$$\dot{\theta} = 0$$

$$\dot{\phi} \sin \theta = \frac{H_G \sin \theta}{I}$$

$$\dot{\phi} \cos \theta + \dot{\psi} = \frac{H_G \cos \theta}{I_z}$$

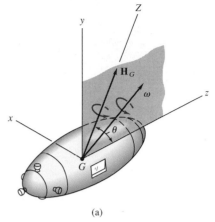

(a)

Fig. 21–21a

477

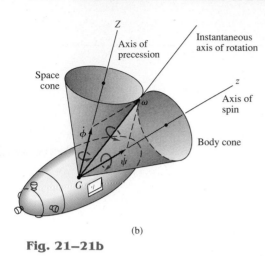

(b)

Fig. 21–21b

Solving, we get

$$
\begin{aligned}
\theta &= \text{const} \\
\dot{\phi} &= \frac{H_G}{I} \\
\dot{\psi} &= \frac{I - I_z}{I I_z} H_G \cos \theta
\end{aligned}
\tag{21–36}
$$

Thus, for torque-free motion of an axisymmetrical body, the angle θ formed between the angular-momentum vector and the spin of the body remains constant. Furthermore, the angular momentum $\mathbf{H}_G$, precession $\dot{\phi}$, and spin $\dot{\psi}$ for the body remain constant at all times during the motion. Eliminating H_G from the second and third of Eqs. 21–36 yields the following relationship between the spin and precession:

$$
\dot{\psi} = \frac{I - I_z}{I_z} (\dot{\phi}) \cos \theta
\tag{21–37}
$$

As shown in Fig. 21–21b, the body precesses about the Z axis, which is fixed in direction, while it spins about the z axis. These two components of angular motion may be analyzed by using a simple cone model, introduced in Sec. 20.1. The *space cone* defining the precession is fixed from rotating, since the precession has a fixed direction, while the *body cone* rotates around the space cone's outer surface without slipping. On this basis, an attempt should be made to imagine the motion. The interior angle of each cone is chosen such that the resultant angular velocity of the body is directed along the line of contact of the two cones. This line of contact represents the instantaneous axis of rotation for the body cone, and hence the angular velocity of both the body cone and the body must be directed along this line. Since the spin is a function of the moments of inertia I and I_z of the body, Eq. 21–36, the cone model in Fig. 21–21b is satisfactory for describing the motion, provided $I > I_z$. Torque-free motion which meets these requirements is called *regular precession*. If $I < I_z$, the spin is negative and the precession positive. This motion is represented by the satellite motion shown in Fig. 21–22 ($I < I_z$). The cone model may again be used to represent the motion; however, to preserve the correct vector addition of spin and precession to obtain the angular velocity $\boldsymbol{\omega}$, the inside surface of the body cone must roll on the outside surface of the (fixed) space cone. This motion is referred to as *retrograde precession*.

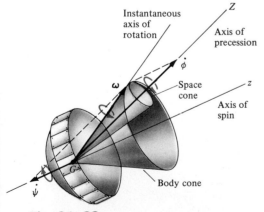

Fig. 21–22

Example 21–8

The motion of a football is observed using a slow-motion projector. From the film, the spin of the football is seen to be directed 30° from the horizontal, as shown in Fig. 21–23a. Also, the football is precessing about the vertical axis at a rate $\dot{\phi} = 3$ rad/s. If the ratio of the axial to transverse moments of inertia of the football is $\frac{1}{3}$, measured with respect to the center of mass, determine the magnitude of the football's spin and its angular velocity. Neglect the effect of air resistance.

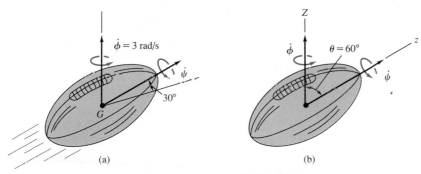

(a)   (b)

Fig. 21–23

SOLUTION

Since the weight of the football is the only force acting, the motion is torque-free. In the conventional sense, if the z axis is established along the axis of spin and the Z axis along the precession axis, as shown in Fig. 21–23b, then the angle $\theta = 60°$. Applying Eq. 21–37 gives

$$\dot{\psi} = \frac{I - I_z}{I_z}(\dot{\phi}) \cos \theta = \frac{I - \frac{1}{3}I}{\frac{1}{3}I}(3) \cos 60°$$

$$= 3 \text{ rad/s} \qquad\qquad\qquad Ans.$$

Using Eqs. 21–34, where $H_G = \dot{\phi}I$ (Eq. 21–36), we have

$$\omega_x = 0$$

$$\omega_y = \frac{H_G \sin \theta}{I} = \frac{3I \sin 60°}{I} = 2.60 \text{ rad/s}$$

$$\omega_z = \frac{H_G \cos \theta}{I_z} = \frac{3I \cos 60°}{\frac{1}{3}I} = 4.50 \text{ rad/s}$$

Thus,

$$\omega = \sqrt{(\omega_x)^2 + (\omega_y)^2 + (\omega_z)^2}$$

$$= \sqrt{(0)^2 + (2.60)^2 + (4.50)^2}$$

$$= 5.20 \text{ rad/s} \qquad\qquad\qquad Ans.$$

PROBLEMS

21–63. Show that the angular velocity of a body, in terms of Euler angles ϕ, θ, and ψ, may be expressed as $\boldsymbol{\omega} = (\dot{\phi} \sin \theta \sin \psi + \dot{\theta} \cos \psi)\mathbf{i} + (\dot{\phi} \sin \theta \cos \psi - \dot{\theta} \sin \psi)\mathbf{j} + (\dot{\phi} \cos \theta + \dot{\psi})\mathbf{k}$, where $\mathbf{i}$, $\mathbf{j}$, and $\mathbf{k}$ are directed along the x, y, z axes shown in Fig. 21–14d.

*21–64. A thin rod is initially coincident with the Z axis when it is given three rotations defined by the Euler angles $\theta = 45°$, $\phi = 30°$, and $\psi = 60°$. If these rotations are given in the order stated, determine the coordinate direction angles of the axis of the rod with respect to the X, Y, and Z axes. Is this direction the same for any order of the rotations? Why?

21–65. The propeller on a single-engine airplane has a mass of 15 kg and a centroidal radius of gyration of 0.3 m computed about the axis of spin. When viewed from the front of the airplane, the propeller is turning clockwise at 350 rad/s about the spin axis. If the airplane enters a vertical curve having a radius $\rho = 80$ m and is traveling at 200 km/h, determine the gyroscopic bending moment which the propeller exerts on the bearings of the engine when the airplane is in its lowest position.

21–66. An airplane descends at a steep angle and then levels off horizontally to land. If the propeller is turning clockwise when observed from the rear of the plane, determine the direction in which the plane tends to turn as caused by the gyroscopic effect.

21–67. The toy gyroscope consists of a rotor R which is attached to the frame of negligible mass. If the rotor is spinning about its axle at an angular speed $\omega_R = 150$ rad/s, determine the constant angular velocity ω_p at which the frame is precessing about the pivot point at O. OA moves in the horizontal plane. The rotor has a mass of 200 g and a radius of gyration $k_{OA} = 20$ mm about OA.

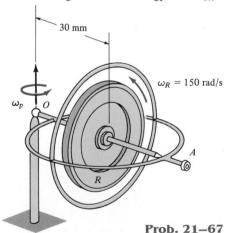

Prob. 21–67

*21–68. The top consists of a thin disk that has a weight of 8 lb and a radius of 0.3 ft. The rod has a negligible mass and a length of 0.5 ft. If the top is spinning with an angular velocity $\omega_s = 300$ rad/s, determine the steady-state precessional angular velocity ω_p of the rod when $\theta = 40°$.

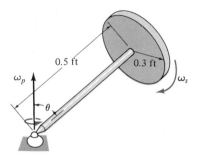

0.5 ft   0.3 ft

ω_p ω_s

θ

Probs. 21–68/21–69

21–69. Solve Prob. 21–68 when $\theta = 90°$.

21–70. The rotor assembly on the engine of a jet airplane consists of the turbine, drive shaft, and compressor. The total mass is 700 kg, the radius of gyration about the shaft axis is $k_{AB} = 0.35$ m, and the mass center is at G. If the rotor has an angular velocity $\omega_{AB} = 1000$ rad/s, and the plane is pulling out of a vertical curve while traveling at 250 m/s, determine the components of reaction at the bearings A and B due to the gyroscopic effect.

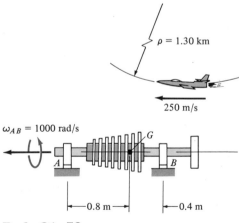

$\rho = 1.30$ km

250 m/s

$\omega_{AB} = 1000$ rad/s G

A B

0.8 m   0.4 m

Prob. 21–70

21–71. The top has a mass of 90 g, a center of mass at G, and a radius of gyration $k = 18$ mm about its axis of symmetry. About any transverse axis acting through point O the radius of gyration is $k_t = 35$ mm. If the top is pinned at O and the precession is $\omega_p = 0.5$ rad/s, determine the spin ω_s.

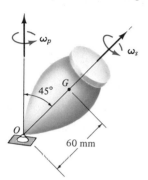

Prob. 21–71

21–73. The 30-lb wheel is rolling without slipping. If it has a radius of gyration $k_{AB} = 1.2$ ft about its axle AB, and the vertical drive shaft is turning at 8 rad/s, determine the normal reaction the wheel exerts on the ground at C.

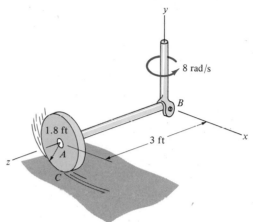

Prob. 21–73

*21–72. The motor weighs 50 lb and has a radius of gyration of 0.2 ft about the z axis. The shaft of the motor is supported by bearings at A and B, and is turning at a constant rate of $\omega_s = \{100\mathbf{k}\}$ rad/s, while the frame has an angular velocity of $\omega_y = \{2\mathbf{j}\}$ rad/s. Determine the moment which the bearing forces at A and B exert on the shaft due to this motion.

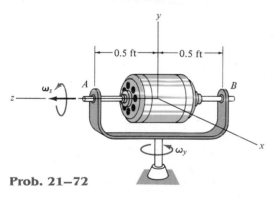

Prob. 21–72

21–74. The car is traveling at $v_C = 100$ km/h around the horizontal curve having a radius of curvature of 80 m. If each wheel has a mass of 16 kg, a radius of gyration $k_G = 300$ mm about its spinning axis, and a diameter of 400 mm, determine the difference between the normal forces of the rear wheels caused by the gyroscopic effect. The distance between the wheels is 1.30 m.

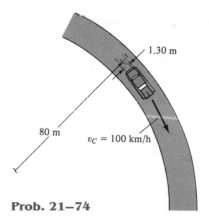

Prob. 21–74

21–75. The 10-lb sphere spins with an angular speed of $\omega_s =$ 1500 rpm about the axis of the horizontal rod. If the counterbalance block weighs 6 lb, determine the precession of rod DC. The cross frame has negligible mass and is supported by ball-and-socket joints at points C and D.

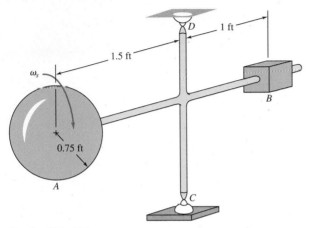

Prob. 21–75

*21–76. The conical top has a mass of 0.8 kg, and the moments of inertia are $I_x = I_y = 3.5(10^{-3})$ kg · m² and $I_z = 0.8(10^{-3})$ kg · m². If it spins freely in the ball-and-socket joint at A with an angular velocity $\omega_s = 750$ rad/s, compute the precession of the top about the axis of the shaft AB.

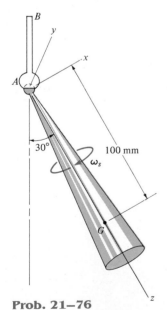

Prob. 21–76

21–77. The projectile shown is subjected to torque-free motion. The transverse and axial moments of inertia are I and I_z, respectively. If θ represents the angle between the precessional axis Z and the axis of symmetry z, and β is the angle between the angular velocity $\boldsymbol{\omega}$ and the z axis, show that β and θ are related by the equation $\tan \theta = (I/I_z) \tan \beta$.

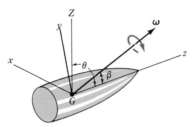

Prob. 21–77

21–78. The radius of gyration about an axis passing through the axis of symmetry of the 2.5-Mg satellite is $k_z = 2.3$ m, and about any transverse axis passing through the center of mass G, $k_t = 3.4$ m. If the satellite has a steady-state precession of two revolutions per hour about the Z axis, determine the rate of spin about the z axis.

Prob. 21–78

482

21–79. While the rocket is in free flight, it has a spin of 3 rad/s and precesses about an axis measured 10° from the axis of spin. If the ratio of the axial to transverse moments of inertia of the rocket is 1/15, computed about axes which pass through the mass center G, determine the angle which the resultant angular velocity makes with the spin axis. Construct the body and space cones used to describe the motion. Is the precession regular or retrograde?

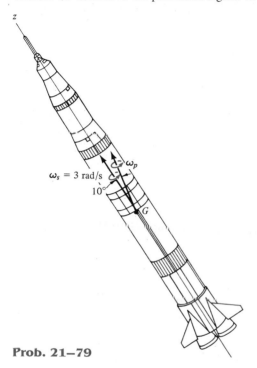

Prob. 21–79

*21–80. The 4-kg disk is thrown in the air with a spin $\omega_z = 6$ rad/s. If the angle θ is measured as 160°, determine the precession about the Z axis.

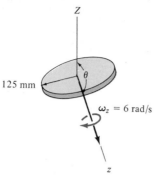

Prob. 21–80

21–81. The satellite has a mass of 1.8 Mg, and about axes passing through the mass center G the axial and transverse radii of gyration are $k_z = 0.8$ m and $k_t = 1.2$ m, respectively. If it is spinning at $\omega_s = 6$ rad/s when it is launched, determine its angular momentum. Precession occurs about the Z axis.

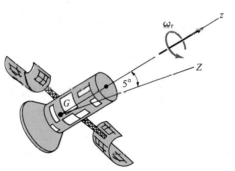

Prob. 21–81

22

Vibrations

A *vibration* is the periodic motion of a body or system of connected bodies displaced from a position of equilibrium. In general, there are two types of vibration, free and forced. *Free vibration* occurs when the motion is maintained by gravitational or elastic restoring forces, such as the swinging motion of a pendulum or the vibration of an elastic rod. *Forced vibration* is caused by an external periodic or intermittent force applied to the system. Both of these types of vibration may be either damped or undamped. *Undamped* vibrations can continue indefinitely because frictional effects are neglected in the analysis. Since in reality both internal and external frictional forces are present, the motion of all vibrating bodies is actually *damped*.

In this chapter we will study the characteristics of the above types of vibrating motion. The analysis will apply to those bodies which are constrained to move only in one direction. These single-degree-of-freedom systems require only one coordinate to specify completely the position of the system at any time. The analysis of multi-degree-of-freedom systems is based on this simplified case and is thoroughly treated in textbooks devoted to vibrational theory.

*22.1 Undamped Free Vibration

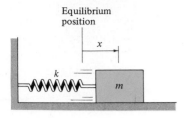

Equilibrium
position

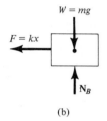

(a)

(b)

Fig. 22–1

The simplest type of vibrating motion is undamped free vibration, represented by the model shown in Fig. 22–1a. The block has a mass m and is attached to a spring having a stiffness k. Vibrating motion occurs when displacing the block a distance x from its equilibrium position and allowing the spring to restore it to its original position. As the spring pulls on the block, the block will attain a velocity such that it will proceed to move out of equilibrium when $x = 0$. Provided the supporting surface is smooth, oscillation will continue indefinitely.

The time-dependent path of motion of the block may be determined by applying the equation of motion to the block when it is in the displaced position x. The free-body diagram is shown in Fig. 22–1b. The elastic restoring force $F = kx$ is always directed toward the equilibrium position, whereas the acceleration **a** is assumed to act in the direction of *positive displacement*. Noting that $a = d^2x/dt^2 = \ddot{x}$, we have

$$\xrightarrow{+}\Sigma F_x = ma_x; \qquad\qquad -kx = m\ddot{x}$$

Here it is seen that the acceleration is proportional to the position. Motion described in this manner is called *simple harmonic motion*. Rearranging the terms into a "standard form" gives

$$\ddot{x} + p^2x = 0 \qquad\qquad (22\text{–}1)$$

The constant p is called the *circular frequency,* expressed in rad/s, and in this case

$$p = \sqrt{\frac{k}{m}} \qquad\qquad (22\text{–}2)$$

Equation 22–1 may also be obtained by considering the block to be suspended, as shown in Fig. 22–2a, and measuring the displacement y from the block's *equilibrium position.* The free-body diagram is shown in Fig. 22–2b. When the block is in equilibrium, the spring exerts an upward force of $F = W = mg$ on the block. Hence, when the block is displaced a distance y downward from this position, the magnitude of the spring force is $F = W + ky$.

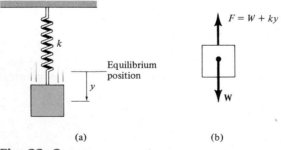

(a)   (b)

Fig. 22–2

Applying the equation of motion gives

$$+\downarrow \Sigma F_y = ma_y; \qquad -W - ky + W = m\ddot{y}$$

or

$$\ddot{y} + p^2 y = 0$$

which is the same form as Eq. 22–1, where p is defined by Eq. 22–2.

Equation 22–1 is a homogeneous, second-order, linear, differential equation with constant coefficients. It can be shown, using the methods of differential equations, that the general solution of this equation is

$$x = A \sin pt + B \cos pt \qquad (22\text{–}3)$$

where A and B represent two constants of integration. The block's velocity and acceleration are determined by taking successive time derivatives, which yields

$$v = \dot{x} = Ap \cos pt - Bp \sin pt \qquad (22\text{–}4)$$

$$a = \ddot{x} = -Ap^2 \sin pt - Bp^2 \cos pt \qquad (22\text{–}5)$$

When Eqs. 22–3 and 22–5 are substituted into Eq. 22–1, the differential equation is indeed satisfied, and therefore Eq. 22–3 represents the true solution to Eq. 22–1.

The integration constants A and B in Eq. 22–3 are generally determined from the initial conditions of the problem. For example, suppose that the block in Fig. 22–1a has been displaced a distance x_1 to the right from its equilibrium position and given an initial (positive) velocity $\mathbf{v}_1$ directed to the right. Substituting $x = x_1$ at $t = 0$ into Eq. 22–3 yields $B = x_1$. Since $v = v_1$ at $t = 0$, using Eq. 22–4 we obtain $A = v_1/p$. If these values are substituted into Eq. 22–3, the equation describing the motion becomes

$$x = \frac{v_1}{p} \sin pt + x_1 \cos pt \qquad (22\text{–}6)$$

Equation 22–3 may also be expressed in terms of simple sinusoidal motion. Let

$$A = C \cos \phi \qquad (22\text{–}7)$$

and

$$B = C \sin \phi \qquad (22\text{–}8)$$

where C and ϕ are new constants to be determined in place of A and B. Substituting into Eq. 22–3 yields

$$x = C \cos \phi \sin pt + C \sin \phi \cos pt$$

Since $\sin(\theta + \phi) = \sin \theta \cos \phi + \cos \theta \sin \phi$, then

$$x = C \sin(pt + \phi) \qquad (22\text{–}9)$$

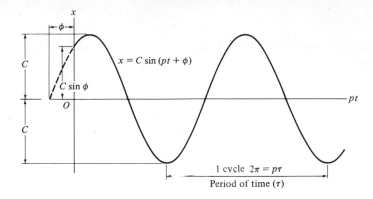

Fig. 22–3

If this equation is plotted on an x versus pt axis, the graph shown in Fig. 22–3 is obtained. The maximum displacement of the block from its equilibrium position is defined as the *amplitude* of vibration. From either the figure or Eq. 22–9 the amplitude is C. The angle ϕ is called the *phase angle* since it represents the amount by which the curve is displaced from the origin when $t = 0$. The constants C and ϕ are related to A and B by Eqs. 22–7 and 22–8. Squaring and adding these two equations, the amplitude becomes

$$C = \sqrt{A^2 + B^2} \qquad (22\text{–}10)$$

If Eq. 22–8 is divided by Eq. 22–7, the phase angle is

$$\phi = \tan^{-1}\frac{B}{A} \qquad (22\text{–}11)$$

Note that the sine curve, Eq. 22–9, completes one *cycle* in time $t = \tau$ (tau) when $p\tau = 2\pi$ or

$$\tau = \frac{2\pi}{p} \qquad (22\text{–}12)$$

This length of time is called a *period*, Fig. 22–3. Using Eq. 22–2, the period may also be represented as

$$\tau = 2\pi\sqrt{\frac{m}{k}} \qquad (22\text{–}13)$$

The *frequency* f is defined as the number of cycles completed per unit of time, which is the reciprocal of the period:

$$f = \frac{1}{\tau} = \frac{p}{2\pi} \qquad (22\text{–}14)$$

or

$$f = \frac{1}{2\pi}\sqrt{\frac{k}{m}} \qquad (22\text{–}15)$$

The frequency is expressed in cycles/s. This ratio of units is called a *hertz* (Hz), where 1 Hz = 1 cycle/s = 2π rad/s.

When a body or system of connected bodies is given an initial displacement from its equilibrium position and released, it will vibrate with a definite frequency known as the *natural frequency*. This type of vibration is called *free vibration,* provided no external forces except gravitational or elastic forces act on the body during the motion. Also, if the *amplitude* of vibration remains *constant,* the motion is said to be *undamped*. The undamped free vibration of a body having a single degree of freedom has the same characteristics as simple harmonic motion of the block and spring discussed above. Consequently, the body's motion is described by a differential equation of the *same form* as Eq. 22–1, i.e.,

$$\ddot{x} + p^2 x = 0 \qquad\qquad (22\text{–}16)$$

Hence, if the circular frequency p of the body is known, the period of vibration τ, natural frequency f, and other vibrating characteristics of the body can be established using the above equations.

PROCEDURE FOR ANALYSIS

As in the case of the block and spring, the circular frequency p of a rigid body or system of connected rigid bodies having a single degree of freedom can be determined using the following procedure:

Free-Body Diagram. Draw the free-body diagram of the body when the body is displaced by a *small amount* from its equilibrium position. Locate the body with respect to its equilibrium position by using an appropriate *inertial coordinate q*. The acceleration of the body's mass center $\mathbf{a}_G$ or the body's angular acceleration $\boldsymbol{\alpha}$ should have a sense which is in the positive direction of the position coordinate. If it is decided that the rotational equation of motion $\Sigma M_P = \Sigma(\mathcal{M}_k)_P$ is to be used, then it may be beneficial to also draw the kinetic diagram since it graphically accounts for the components $m(\mathbf{a}_G)_x$, $m(\mathbf{a}_G)_y$, and $I_G\boldsymbol{\alpha}$, and thereby makes it convenient for visualizing the terms needed in the moment sum $\Sigma(\mathcal{M}_k)_P$.

Equation of Motion. Apply the equation of motion to relate the elastic or gravitational restoring forces and couples acting on the body to the body's accelerated motion.

Kinematics. Using kinematics, express the body's accelerated motion in terms of the second time derivative of the position coordinate, $\ddot{q}$. Substitute this result into the equation of motion and determine p by rearranging the terms so that the resulting equation is of the form $\ddot{q} + p^2 q = 0$.

The following examples illustrate this procedure.

Example 22–1

Determine the period of vibration for the simple pendulum shown in Fig. 22–4a. The bob has a mass m and is attached to a cord of length l. Neglect the size of the bob.

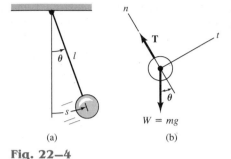

Fig. 22–4

SOLUTION

Free-Body Diagram. Motion of the system will be related to the position coordinate $(q =)\theta$, Fig. 22–4b. When the bob is displaced by an angle θ, the *restoring force* acting on the bob is created by the *weight component* $mg \sin \theta$. Furthermore, $\mathbf{a}_t$ acts in the direction of *increasing s* (or θ).

Equation of Motion. Applying the equation of motion in the *tangential direction,* since it involves the restoring force, yields

$$+ \nearrow \Sigma F_t = ma_t; \qquad -mg \sin \theta = ma_t \qquad (1)$$

Kinematics. $a_t = d^2s/dt^2 = \ddot{s}$. Furthermore, s may be related to θ by the equation $s = l\theta$, so that $a_t = l\ddot{\theta}$. Hence, Eq. (1) reduces to the form

$$\ddot{\theta} + \frac{g}{l} \sin \theta = 0 \qquad (2)$$

The solution of this equation involves the use of an elliptic integral. For *small displacements,* however, $\sin \theta \approx \theta$, in which case

$$\ddot{\theta} + \frac{g}{l} \theta = 0 \qquad (3)$$

Comparing this equation with Eq. 22–16 ($\ddot{x} + p^2x = 0$), which is the "standard form" for simple harmonic motion, it is seen that $p = \sqrt{g/l}$. From Eq. 22–12, the period of time required for the bob to make one complete swing is therefore

$$\tau = \frac{2\pi}{p} = 2\pi \sqrt{\frac{l}{g}} \qquad \qquad Ans.$$

This interesting result, originally discovered by Galileo Galilei through experiment, indicates that the period depends only on the length of the cord and not on the mass of the pendulum bob.

The solution of Eq. (3) is given by Eq. 22–3, where $p = \sqrt{g/l}$ and θ is substituted for x. Like the block and spring, the constants A and B in this problem may be determined if, for example, one knows the displacement and velocity of the bob at a given instant.

Example 22–2

The 10-kg rectangular plate shown in Fig. 22–5a is suspended at its center from a rod having a torsional stiffness $k = 1.5$ N · m/rad. Determine the natural period of vibration of the plate when it is given a small angular displacement θ in the plane of the plate.

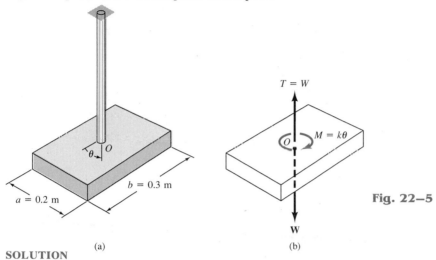

Fig. 22–5

(a)   (b)

SOLUTION

Free-Body Diagram. Fig. 22–5b. Since the plate is displaced in its own plane, the torsional *restoring* moment created by the rod is $M = k\theta$. This moment acts in the direction opposite to the angular displacement θ. The angular acceleration $\ddot{\theta}$ acts in the direction of *positive* θ.

Equation of Motion

$$\Sigma M_O = I_O \alpha; \qquad\qquad -k\theta = I_O \ddot{\theta}$$

or

$$\ddot{\theta} + \frac{k}{I_O}\theta = 0$$

Since this equation is in the "standard form," the circular frequency is $p = \sqrt{k/I_O}$.

The moment of inertia of the plate about an axis coincident with the rod is $I_O = \frac{1}{12}m(a^2 + b^2)$. Hence,

$$I_O = \frac{1}{12}(10)[(0.2)^2 + (0.3)^2] = 0.108 \text{ kg} \cdot \text{m}^2$$

The natural period of vibration is, therefore,

$$\tau = \frac{2\pi}{p} = 2\pi\sqrt{\frac{I_O}{k}} = 2\pi\sqrt{\frac{0.108}{1.5}} = 1.69 \text{ s} \qquad\qquad \textit{Ans.}$$

Example 22–3

The bent rod shown in Fig. 22–6a has a negligible mass and supports a 5-kg collar at its end. Determine the natural period of vibration for the system.

SOLUTION

Free-Body and Kinetic Diagrams. Fig. 22–6b. Here the rod is displaced by a small amount θ from the equilibrium position. Since the spring is subjected to an initial compression of x_{st} for statical equilibrium, then when the displacement $x > x_{st}$ the spring exerts a force of $F_s = kx - kx_{st}$ on the rod. To obtain the "standard form," Eq. 22–16, $5a_y$ acts *upward*, which is in accordance with positive θ displacement.

Equation of Motion. Moments will be summed about point B to eliminate the unknown reaction at this point. Since θ is small,

$$\zeta + \Sigma M_B = \Sigma(\mathcal{M}_k)_B; \quad kx(0.1) - kx_{st}(0.1) + 49.05(0.2) = -5a_y(0.2)$$

The second term on the left side, $-kx_{st}(0.1)$, represents the moment created by the spring force which is necessary to hold the collar in *equilibrium*, i.e., at $x = 0$. Since this moment is equal and opposite to the moment $49.05(0.2)$ created by the weight of the collar, these two terms cancel in the above equation, so that

$$kx(0.1) = -5a_y(0.2) \qquad (1)$$

Kinematics. The positions of the spring and the collar may be related to the angle θ, Fig. 22–6c. Since θ is small, $x = 0.1\theta$ and $y = 0.2\theta$. Therefore, $a_y = \ddot{y} = 0.2\ddot{\theta}$. Substituting into Eq. (1) yields

$$400(0.1\theta)0.1 = -5(0.2\ddot{\theta})0.2$$

Rewriting this equation in standard form gives

$$\ddot{\theta} + 20\theta = 0$$

Compared with $\ddot{x} + p^2x = 0$ (Eq. 22–16), we have

$$p^2 = 20, \qquad p = 4.47 \text{ rad/s}$$

The natural period of vibration is therefore,

$$\tau = \frac{2\pi}{p} = \frac{2\pi}{4.47} = 1.40 \text{ s}$$

Ans.

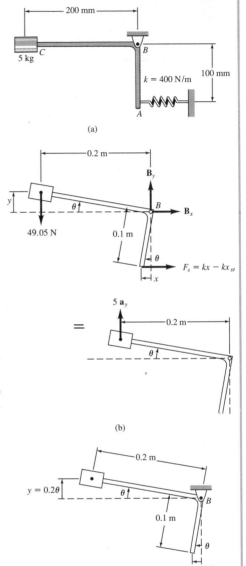

Fig. 22–6

Example 22–4

A 10-lb block is suspended from a cord that passes over a 15-lb disk, as shown in Fig. 22–7a. The spring has a stiffness $k = 200$ lb/ft. Determine the natural period of vibration for the system.

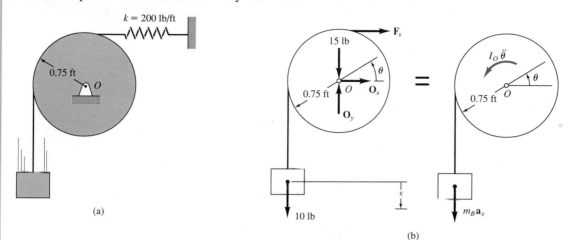

(a)

(b)

Fig. 22–7

SOLUTION

Free-Body and Kinetic Diagrams. Fig. 22–7b. The *system* consists of the disk, which undergoes a rotation defined by the angle θ, and the block, which translates by an amount s. The vector $I_O\ddot{\theta}$ acts in the direction of *positive* θ, and consequently $m_B a_s$ acts downward in the direction of *positive s*.

Equation of Motion. Summing moments about point O to eliminate the reactions $\mathbf{O}_x$ and $\mathbf{O}_y$, realizing that $I_O = \frac{1}{2}mr^2$, yields

$$\zeta + \Sigma M_O = \Sigma(\mathcal{M}_k)_O;$$

$$10(0.75) - F_s(0.75) - \frac{1}{2}\left(\frac{15}{32.2}\right)(0.75)^2\ddot{\theta} + \left(\frac{10}{32.2}\right)a_s(0.75) \quad (1)$$

Kinematics. As shown on the kinematic diagram in Fig. 22–7c, a small positive displacement θ of the disk causes the block to lower by an amount $s = 0.75\theta$; hence, $a_s = \ddot{s} = 0.75\ddot{\theta}$. When $\theta = 0°$, the spring force required for *equilibrium* of the disk is 10 lb, acting to the right. For position θ, the spring force is $F_s = (200\ \text{lb/ft})(0.75\theta\ \text{ft}) + 10$ lb. Substituting these results into Eq. (1) and simplifying yields

$$\ddot{\theta} + 368\theta = 0$$

Hence,

$$p^2 = 368 \qquad p = 19.2\ \text{rad/s}$$

Therefore, the natural period of vibration is

$$\tau = \frac{2\pi}{p} = \frac{2\pi}{19.2} = 0.328\ \text{s} \qquad \textit{Ans.}$$

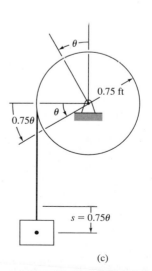

(c)

PROBLEMS

22–1. When a 20-lb weight is suspended from a spring, the spring is stretched a distance of 4 in. Determine the natural frequency and the period of vibration for a 10-lb weight attached to the same spring.

22–2. Determine the frequency of vibration for the block. The springs are originally compressed Δ.

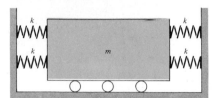

Prob. 22–2

22–3. A spring has a stiffness of 600 N/m. If a 4-kg block is attached to the spring, pushed 50 mm above its equilibrium position, and released from rest, determine the equation which describes the block's motion. Assume that positive displacement is measured downward.

*22–4. When a 3-kg block is suspended from a spring, the spring is stretched a distance of 60 mm. Determine the natural frequency and the period of vibration for a 0.2-kg block attached to the same spring.

22–5. An 8-kg block is suspended from a spring having a stiffness $k = 80$ N/m. If the block is given an upward velocity of 0.4 m/s when it is 90 mm above its equilibrium position, determine the equation which describes the motion and the maximum upward displacement of the block measured from the equilibrium position. Assume that positive displacement is measured downward.

22–6. A 2-lb weight is suspended from a spring having a stiffness of $k = 2$ lb/in. If the weight is pushed 1 in. upward from its equilibrium position and then released from rest, determine the equation which describes the motion. What is the amplitude and the natural frequency of the vibration?

22–7. A 6-lb weight is suspended from a spring having a stiffness of $k = 3$ lb/in. If the weight is given an upward velocity of 20 ft/s when it is 2 in. above its equilibrium position, determine the equation which describes the motion and the maximum upward displacement of the weight, measured from the equilibrium position. Assume positive displacement is downward.

*■22–8. A spring is stretched 175 mm by an 8-kg block. If the block is displaced 100 mm downward from its equilibrium position and given a downward velocity of 1.50 m/s, determine the differential equation which describes the motion. Assume that positive displacement is measured downward. Use the Runge-Kutta method to determine the position of the block, measured from its unstretched position, when $t - 0.22$ s. Use a time increment of $\Delta t = 0.02$ s.

22–9. If the block in Prob. 22–8 is given an upward velocity of 4 m/s when it is displaced downward a distance of 60 mm from its equilibrium position, determine the equation which describes the motion. What is the amplitude of the motion? Assume that positive displacement is measured downward.

22–10. A pendulum has a 0.4-m-long cord and is given a tangential velocity of 0.2 m/s toward the vertical from a position $\theta = 0.3$ rad. Determine the equation which describes the angular motion.

22–11. Determine to the nearest degree the maximum angular displacement of the bob in Prob. 22–10 if it is initially displaced $\theta = 0.2$ rad from the vertical and given a tangential velocity of 0.4 m/s away from the vertical.

*22–12. The square plate has a mass m and is suspended at its corner by the pin O. Determine the period of oscillation if it is displaced a small amount and released.

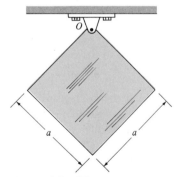

Prob. 22–12

22–13. The body of arbitrary shape has a mass m, mass center at G, and a radius of gyration about G of k_G. If it is displaced by a slight amount θ from its equilibrium position and released, determine the period of vibration.

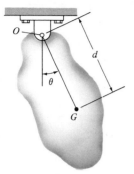

Prob. 22–13

22–14. The uniform beam is supported at its ends by two springs A and B, each having the same stiffness k. When nothing is supported on the beam, it has a period of vertical vibration of 0.83 s. If a 50-kg mass is placed at its center, the period of vertical vibration is 1.52 s. Compute the stiffness of each spring and the mass of the beam.

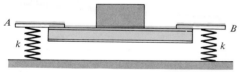

Prob. 22–14

22–15. The thin hoop of mass m is supported by a knife-edge. Determine the period of oscillation for small amplitudes of swing.

Prob. 22–15

*22–16. The semicircular disk weighs 20 lb. Determine the period of oscillation if it is displaced a small amount and released.

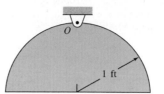

Prob. 22–16

22–17. While standing in an elevator, the man holds a pendulum which consists of an 18-in. cord and a 0.5-lb bob. If the elevator is descending with an acceleration $a = 4 \text{ ft/s}^2$, determine the period of vibration for small amplitudes of swing.

Prob. 22–17

22–18. The disk has a weight of 10 lb and rolls without slipping on the horizontal surface as it oscillates about its equilibrium position. If the disk is displaced, by rolling it counterclockwise 0.4 rad, determine the equation which describes its oscillatory motion when it is released.

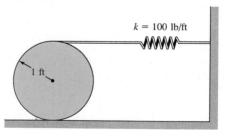

Prob. 22–18

495

22–19. The platform AB when empty has a mass of 400 kg, center of mass at G_1, and period of oscillation $\tau_1 = 2.38$ s. If a car having a mass of 1.2 Mg and center of mass at G_2 is placed on the platform, the period of oscillation becomes $\tau_2 = 3.16$ s. Determine the moment of inertia of the car about an axis passing through G_2.

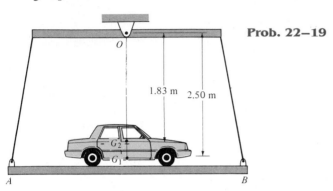

Prob. 22–19

*22–20.** The uniform rod has a mass m and is supported by the pin O. If the rod is given a small displacement and released, determine the period of free vibration. The springs are unstretched when the rod is in the position shown.

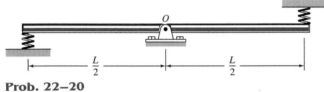

Prob. 22–20

22–21. The pointer on a metronome supports a 0.4-lb slider Λ, which is positioned at a fixed distance from the pivot O of the pointer. When the pointer is displaced, a torsional spring at O exerts a restoring moment on the pointer having a magnitude $M = (1.2\theta)$ lb · ft, where θ represents the angle of displacement from the vertical, measured in radians. Determine the period of vibration when the pointer is displaced a small amount θ and released. Neglect the mass of the pointer.

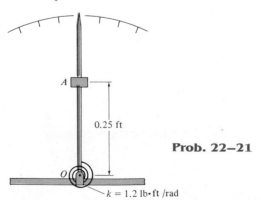

Prob. 22–21

22–22. The 50-lb spool is attached to two springs. If the spool is displaced a small amount and released, determine the period of vibration. The radius of gyration of the spool is $k_G = 1.5$ ft. The spool rolls without slipping.

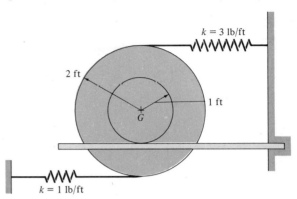

Prob. 22–22

22–23. The 20-lb rectangular plate has a natural period of vibration of $\tau = 0.3$ s, as it oscillates around the axis of rod AB. Determine the torsional stiffness k, measured in lb · ft/rad, of the rod. Neglect the mass of the rod.

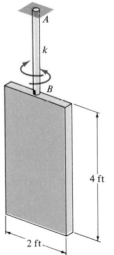

Prob. 22–23

*22–24. The 6-lb weight is attached to the rods of negligible mass. Determine the frequency of vibration of the weight when it is displaced slightly and released.

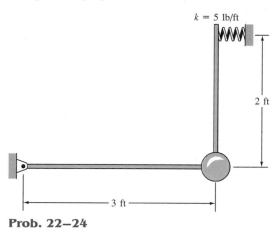

Prob. 22–24

22–26. If the wire *AB* is subjected to a tension force of 20 lb, determine the equation which describes the motion when the 5-lb weight is displaced 2 in. horizontally and released from rest.

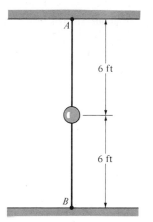

Prob. 22–26

22–25. The 25-lb weight is fixed to the end of the rod assembly. If both springs are unstretched when the assembly is in the position shown, determine the natural period of vibration for the weight when it is displaced slightly and released. Neglect the size of the block and the mass of the rods.

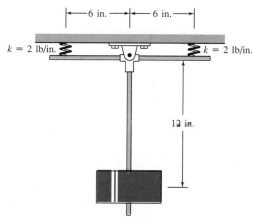

Prob. 22–25

22–27. The plate of mass *m* is supported by three symmetrically placed cords of length *l* as shown. If the plate is given a slight rotation about a vertical axis through its center and released, determine the period of oscillation.

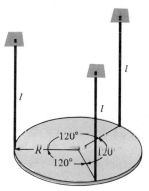

Prob. 22–27

*22–28. The bar has a length l and mass m. It is supported at its ends by rollers of negligible mass. If it is given a small displacement and released, determine the frequency of vibration.

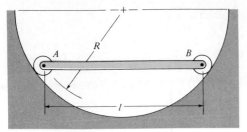

Prob. 22–28

22–29. The cylinder of radius r and mass m is displaced a small amount along the curved surface. If it rolls without slipping, determine the frequency of oscillation when it is released.

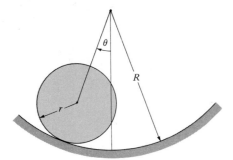

Prob. 22–29

22–30. Determine the frequency for small oscillations of the 10-lb sphere when the rod is displaced a slight distance and released. Neglect the size of the sphere and the mass of the rod. The spring has an unstretched length of 1 ft.

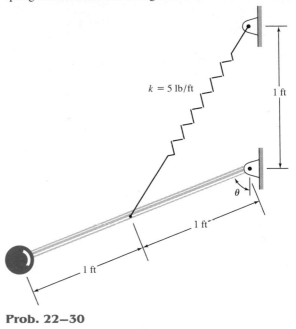

$k = 5$ lb/ft

Prob. 22–30

*22.2 Energy Methods

The simple harmonic motion of a body, discussed in Sec. 22.1, is due only to gravitational and elastic restoring forces acting on the body. Since these types of forces are *conservative,* it is possible to use the conservation of energy equation to obtain the body's natural frequency or period of vibration. To show how to do this, consider the block and spring in Fig. 22–8. When the block is displaced an arbitrary amount x from the equilibrium position, the kinetic energy is $T = \frac{1}{2}mv^2 = \frac{1}{2}m\dot{x}^2$ and the potential energy is $V = \frac{1}{2}kx^2$. By the conservation of energy equation, Eq. 14–21, it is necessary that

$$T + V = \text{const}$$
$$\frac{1}{2}m\dot{x}^2 + \frac{1}{2}kx^2 = \text{const} \qquad (22\text{–}17)$$

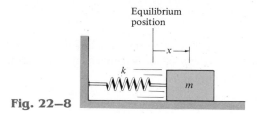

Equilibrium position

Fig. 22–8

The differential equation describing the *accelerated motion* of the block can be obtained by *differentiating* this equation with respect to time, i.e.,

$$m\dot{x}\ddot{x} + kx\dot{x} = 0$$

$$\dot{x}(m\ddot{x} + kx) = 0$$

Since the velocity $\dot{x}$ is not *always* zero in a vibrating system,

$$\ddot{x} + p^2x = 0, \qquad p = \sqrt{k/m}$$

which is the same as Eq. 22–1.

If the energy equation is written for a *system of connected bodies,* the natural frequency or the equation of motion can also be determined by time differentiation. Here it is *not necessary* to dismember the system to account for reactive and connective forces which do no work. Also, by this method, the circular frequency p may be obtained *directly*. For example, consider the total mechanical energy of the block and spring in Fig. 22–8 when the block is at its *maximum displacement*. In this position the block is temporarily at rest, so that the kinetic energy is zero and the potential energy, stored in the spring, is a maximum. Therefore, Eq. 22–17 becomes $\frac{1}{2}kx_{max}^2 = $ const. At the instant the block passes the equilibrium position, the kinetic energy of the block is a maximum and the potential energy of the spring is zero. Hence, Eq. 22–17 becomes $\frac{1}{2}m(\dot{x})_{max}^2 = $ const. Since the vibrating motion of the block is *harmonic*, the *solution* for the displacement and velocity may be written in the form of Eq. 22–9 and its time derivative, i.e.,

$$x = C \sin (pt + \phi), \qquad \dot{x} = Cp \cos (pt + \phi)$$

so that

$$x_{max} = C, \qquad \dot{x}_{max} = Cp$$

Applying the conservation of energy equation ($T + V = $ const) yields

$$V_{max} = T_{max}; \qquad \tfrac{1}{2}kx_{max}^2 = \tfrac{1}{2}m\dot{x}_{max}^2 = \text{const}$$

or

$$kC^2 = mC^2p^2$$

Solving for p yields

$$p = \sqrt{\frac{k}{m}}$$

which is identical to Eq. 22–2.

PROCEDURE FOR ANALYSIS

The following procedure provides a method for determining the circular frequency p of a body or system of connected bodies using the conservation of energy equation.

Energy Equation. Draw the body when it is displaced by a *small amount* from its equilibrium position and define the location of the body from its equilibrium position by an appropriate position coordinate q. Formulate the equation of energy for the body, $T + V = \text{const}$, in terms of the position coordinate.* Recall that, in general, the kinetic energy must account for both the body's translational and rotational motion, $T = \frac{1}{2}mv_G^2 + \frac{1}{2}I_G\omega^2$, Eq. 18–2, and that the potential energy is the sum of the gravitational and elastic potential energies of the body, $V = V_g + V_e$, Eq. 18–15. In particular, V_g should be measured from a datum for which $q = 0$ (equilibrium position).

Time Derivative. Take the time derivative of the energy equation and factor out the common terms. The resultant differential equation represents the equation of motion for the system. The value of p is obtained after rearranging the terms in the standard form $\ddot{q} + p^2q = 0$.

The following examples illustrate this procedure.

*It is suggested that the material in Sec. 18.5 be reviewed.

Example 22–5

The thin hoop shown in Fig. 22–9a is supported by a peg at O. Determine the period of oscillation for small amplitudes of swing. The hoop has a mass m.

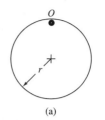

(a)

SOLUTION

Energy Equation. A diagram of the hoop when it is displaced a small amount $(q =)\theta$ from the equilibrium position is shown in Fig. 22–9b. Using the table on the inside back cover of the book and the parallel-axis theorem to determine I_O, we can express the kinetic energy as

$$T = \tfrac{1}{2}I_O\omega^2 = \tfrac{1}{2}[mr^2 + mr^2]\dot{\theta}^2 = mr^2\dot{\theta}^2$$

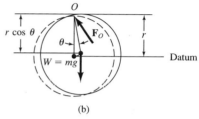

(b)

Fig. 22–9

If a horizontal datum is placed through the center of gravity of the hoop when $\theta = 0$, then the center of gravity moves upward $r(1 - \cos\theta)$ in the displaced position. For *small angles,* $\cos\theta$ may be replaced by the first two terms of its series expansion, $\cos\theta = 1 - \theta^2/2 + \cdots$. Therefore, the potential energy is

$$V = mgr\left[1 - \left(1 - \frac{\theta^2}{2}\right)\right] = mgr\frac{\theta^2}{2}$$

The total energy in the system is

$$T + V = mr^2\dot{\theta}^2 + mgr\frac{\theta^2}{2}$$

Time Derivative

$$mr^2 2\dot{\theta}\ddot{\theta} + mgr\theta\dot{\theta} = 0$$

$$mr\dot{\theta}(2r\ddot{\theta} + g\theta) = 0$$

Since θ is not always equal to zero, from the terms in parentheses

$$\ddot{\theta} + \frac{g}{2r}\theta = 0$$

Hence,

$$p = \sqrt{\frac{g}{2r}}$$

so that

$$\tau = \frac{2\pi}{p} = 2\pi\sqrt{\frac{2r}{g}} \qquad\qquad \textit{Ans.}$$

Example 22–6

A 10-kg block is suspended from a cord wrapped around a 5-kg disk, as shown in Fig. 22–10a. If the spring has a stiffness $k = 200$ N/m, determine the natural period of vibration for the system.

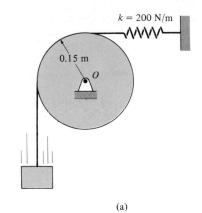

$k = 200$ N/m

0.15 m

O

(a)

SOLUTION

Energy Equation. A diagram of the block and disk when they are displaced by respective amounts s and θ from the equilibrium position is shown in Fig. 22–10b. Since $s = 0.15\theta$, the kinetic energy of the system is

$$T = \tfrac{1}{2}m_b v_b^2 + \tfrac{1}{2}I_O \omega_d^2$$

$$= \frac{1}{2}(10)(0.15\dot{\theta})^2 + \frac{1}{2}\left[\frac{1}{2}(5)(0.15)^2\right](\dot{\theta})^2$$

$$= 0.141(\dot{\theta})^2$$

Establishing the datum at the equilibrium position of the block and realizing that the spring stretches s_{st} for equilibrium, we can write the potential energy as

$$V = \tfrac{1}{2}k(s_{\text{st}} + s)^2 - Ws$$

$$= \frac{1}{2}(200)(s_{\text{st}} + 0.15\theta)^2 - 98.1(0.15\theta)$$

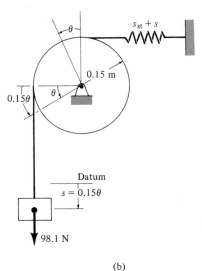

$s_{\text{st}} + s$

0.15 m

θ

0.15θ

Datum

$s = 0.15\theta$

98.1 N

(b)

The total energy for the system is, therefore,

$$T + V = 0.141(\dot{\theta})^2 + 100(s_{\text{st}} + 0.15\theta)^2 - 14.72\theta$$

Time Derivative

$$0.282(\dot{\theta})\ddot{\theta} + 200(s_{\text{st}} + 0.15\theta)0.15\dot{\theta} - 14.72\dot{\theta} = 0$$

Since $s_{\text{st}} = 98.1/200 = 0.4905$ m, the above equation reduces to the standard form

$$\ddot{\theta} + 16\theta = 0$$

so that

$$p = \sqrt{16} = 4 \text{ rad/s}$$

Thus,

$$\tau = \frac{2\pi}{p} = \frac{2\pi}{4} = 1.57 \text{ s} \qquad \textit{Ans.}$$

Fig. 22–10

PROBLEMS

22–31. Solve Prob. 22–12 using energy methods.

*22–32. Solve Prob. 22–13 using energy methods.

22–33. Solve Prob. 22–15 using energy methods.

22–34. Solve Prob. 22–16 using energy methods.

22–35. The machine has a mass m and is uniformly supported by four springs each having a stiffness k. Determine the period of vertical vibration.

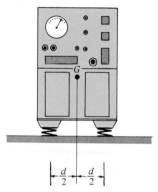

Prob. 22–35

*22–36. Determine the differential equation of motion of the block of mass m when it is displaced slightly and released. Motion occurs in the vertical plane. The springs are originally unstretched and each is attached to the block.

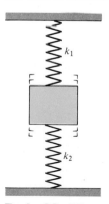

Prob. 22–36

22–37. If the disk has a mass of 8 kg, determine the natural frequency of vibration. The springs are originally unstretched.

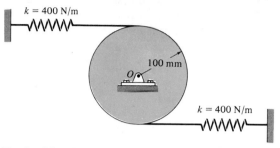

$k = 400$ N/m

100 mm

O

$k = 400$ N/m

Prob. 22–37

22–38. The 7-kg disk is pin-connected at its midpoint. Determine the period of vibration of the disk if the springs have sufficient tension in them to prevent the cord from slipping on the disk as it oscillates. *Hint:* Assume that the initial stretch in each spring is δ_O. This term will cancel out after taking the time derivative of the energy equation.

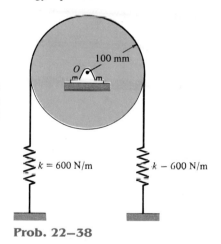

100 mm

O

$k = 600$ N/m $k = 600$ N/m

Prob. 22–38

22–39. Determine the period of vibration of the pendulum. Consider the two rods to be slender, each having a weight $W = 8$ lb/ft.

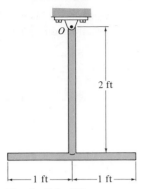

Prob. 22–39

*22–40. Determine the natural frequency of vibration of the 20-lb disk. Assume the disk does not slip on the inclined surface.

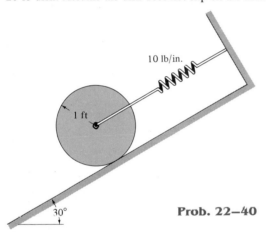

Prob. 22–40

22–41. Determine the period of vibration of the 10-lb semicircular disk.

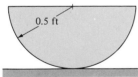

Prob. 22–41

22–42. The 5-lb sphere is attached to a rod of negligible mass. Determine the natural frequency of vibration of the sphere. Neglect the size of the sphere.

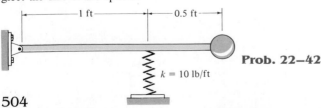

Prob. 22–42

22–43. Determine the period of vibration of the 3-kg sphere. Neglect the mass of the rod and the size of the sphere.

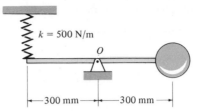

Prob. 22–43

*22–44. The slender rod has a weight $W = 4$ lb/ft. If it is supported in the horizontal plane by a ball-and-socket joint at A and a cable at B, determine the natural frequency of vibration when the end B is given a small horizontal displacement and then released.

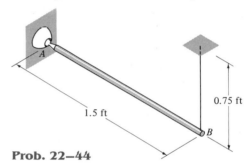

Prob. 22–44

22–45. The bar has a mass of 8 kg and is suspended from two springs such that when it is in equilibrium, the springs make an angle of 45° with the horizontal as shown. Determine the period of vibration if the bar is pulled down a short distance and released. Each spring has a stiffness of $k = 40$ N/m.

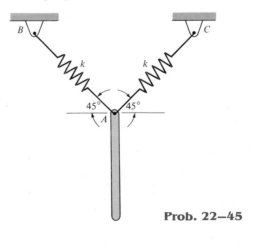

Prob. 22–45

*22.3 Undamped Forced Vibration

Undamped forced vibration is considered to be one of the most important types of vibrating motion in engineering work. The principles which describe the nature of this motion may be applied to the analysis of forces which cause vibration in various types of machines and structures.

Periodic Force

The block and spring shown in Fig. 22–11a provide a convenient "model" which represents the vibrational characteristics of a system subjected to a periodic force $F = F_O \sin \omega t$. This force has a maximum magnitude of F_O and a *forcing frequency* ω. The free-body diagram for the block when it is displaced a distance x is shown in Fig. 22–11b. Applying the equation of motion yields

$$\overset{+}{\rightarrow}\Sigma F_x = ma_x; \qquad F_O \sin \omega t - kx = m\ddot{x}$$

or

$$\ddot{x} + \frac{k}{m}x = \frac{F_O}{m} \sin \omega t \qquad (22\text{–}18)$$

This equation is referred to as a nonhomogeneous second-order differential equation. The general solution consists of a complementary solution, x_c, *plus* a particular solution, x_p.

   The *complementary solution* is determined by setting the term on the right side of Eq. 22–18 equal to zero and solving the resulting homogeneous equation, which is equivalent to Eq. 22–1. The solution is defined by Eq. 22–3, i.e.,

$$x_c = A \sin pt + B \cos pt \qquad (22\text{–}19)$$

where p is the circular frequency, $p = \sqrt{k/m}$, Eq. 22–2.

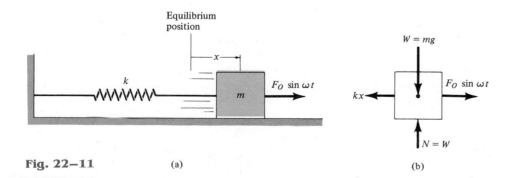

Fig. 22–11   (a)   (b)

Since the motion is periodic, the *particular solution* of Eq. 22–18 may be determined by assuming a solution of the form

$$x_p = C \sin \omega t \qquad (22\text{–}20)$$

where C is a constant. Taking the second time derivative and substituting into Eq. 22–18 yields

$$-C\omega^2 \sin \omega t + \frac{k}{m}(C \sin \omega t) = \frac{F_O}{m} \sin \omega t$$

Factoring out $\sin \omega t$ and solving for C gives

$$C = \frac{F_O/m}{\dfrac{k}{m} - \omega^2} = \frac{F_O/k}{1 - \left(\dfrac{\omega}{p}\right)^2} \qquad (22\text{–}21)$$

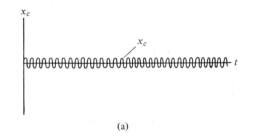

(a)

Substituting into Eq. 22–20, we obtain the particular solution

$$x_p = \frac{F_O/k}{1 - \left(\dfrac{\omega}{p}\right)^2} \sin \omega t \qquad (22\text{–}22)$$

The *general solution* is therefore

$$x = x_c + x_p = A \sin pt + B \cos pt + \frac{F_O/k}{1 - \left(\dfrac{\omega}{p}\right)^2} \sin \omega t \quad (22\text{–}23)$$

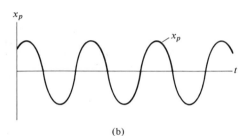

(b)

Here x describes two types of vibrating motion of the block. The *complementary solution* x_c defines the *free vibration,* which depends upon the circular frequency $p = \sqrt{k/m}$ and the constants A and B, Fig. 22–12a. Specific values for A and B are obtained by evaluating Eq. 22–23 at a given instant when the displacement and velocity are known. The *particular solution* x_p describes the *forced vibration* of the block caused by the applied force $F = F_O \sin \omega t$. Fig. 22–12b. The resultant vibration x is shown in Fig. 22–12c. Since all vibrating systems are subject to *friction,* the free vibration, x_c, will in time dampen out. For this reason the free vibration is referred to as *transient,* and the forced vibration is called *steady state,* since it is the only vibration that remains, Fig. 22–12d.

From Eq. 22–21 it is seen that the *amplitude* of forced vibration depends upon the *frequency ratio* ω/p. If the *magnification factor* MF is defined as the ratio of the amplitude of steady-state vibration, $(x_p)_{max}$, to the static deflection F_O/k, which is caused by the amplitude of the periodic force F_O, then, from Eq. 22–22,

$$\text{MF} = \frac{(x_p)_{max}}{F_O/k} = \frac{1}{1 - \left(\dfrac{\omega}{p}\right)^2} \qquad (22\text{–}24)$$

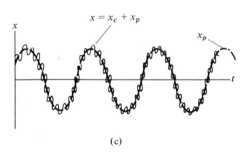

(c)

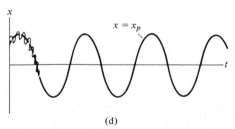

(d)

Fig. 22–12

This equation is graphed in Fig. 22–13, where it is seen that for $\omega \approx 0$, the MF ≈ 1. In this case, because of the very low frequency $\omega \ll p$, the vibration of the block will be in phase with the applied force **F**. If the force or displacement is applied with a frequency close to the natural frequency of the system, i.e., $\omega/p \approx 1$, the amplitude of vibration of the block becomes extremely large. This condition is called *resonance,* and in practice, resonating vibrations can cause tremendous stress and rapid failure of parts. When the cyclic force $F_O \sin \omega t$ is applied at high frequencies ($\omega > p$), the value of the MF becomes negative, indicating that the motion of the block is out of phase with the force. Under these conditions, as the block is displaced to the right, the force acts to the left, and vice versa. For extremely high frequencies ($\omega \gg p$) the block remains almost stationary, and hence the MF is approximately zero.

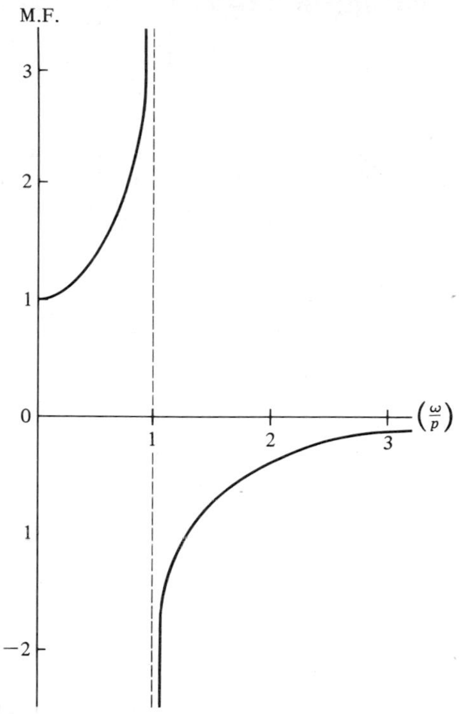

Periodic Support Displacement

Forced vibrations can also arise from the periodic excitation of the support of a system. The model shown in Fig. 22–14a represents the periodic vibration of a block which is caused by harmonic movement $\delta = \delta_O \sin \omega t$ of the support. The free-body diagram for the block in this case is shown in Fig. 22–14b. The coordinate x is measured from the point of zero displacement of the support, i.e., when the radial line OA coincides with OB, Fig. 22–14a. Therefore, general displacement of the spring is $(x - \delta_O \sin \omega t)$. Applying the equation of motion yields

$$\xrightarrow{+}\Sigma F_x = ma_x; \qquad -k(x - \delta_O \sin \omega t) = m\ddot{x}$$

or

$$\ddot{x} + \frac{k}{m}x = \frac{k\delta_O}{m} \sin \omega t \qquad (22\text{–}25)$$

By comparison, this equation is identical to the form of Eq. 22–18, *provided* F_O is *replaced* by $k\delta_O$. If this substitution is made into the solutions defined by Eqs. 22–21 to 22–23, the results are appropriate for describing the motion of the block when subjected to the support displacement $\delta = \delta_O \sin \omega t$.

Fig. 22–13

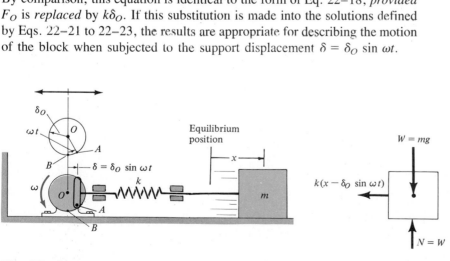

Fig. 22–14   (a)   (b)

507

Example 22–7

The instrument shown in Fig. 22–15 is rigidly attached to a platform P, which in turn is supported by *four* springs, each having a stiffness $k = 800$ N/m. Initially the platform is at rest when the floor is subjected to a displacement $\delta = 10 \sin (8t)$ mm, where t is in seconds. If the instrument is constrained to move vertically, and the total mass of the instrument and platform is 20 kg, determine the vertical displacement y of the platform, measured from the equilibrium position, as a function of time. What floor vibration is required to cause resonance?

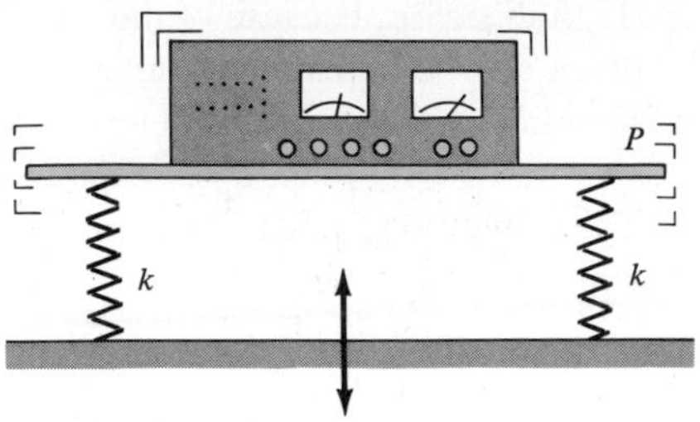

Fig. 22–15

SOLUTION

Since the induced vibration is caused by the displacement of the supports, the motion is described by Eq. 22–23, with F_O replaced by $k\delta_O$, i.e.,

$$y = A \sin pt + B \cos pt + \frac{\delta_O}{1 - \left(\dfrac{\omega}{p}\right)^2} \sin \omega t \qquad (1)$$

Here $\delta = \delta_O \sin \omega t = 10 \sin (8t)$ mm, so that

$$\delta_O = 10 \text{ mm}, \qquad \omega = 8 \text{ rad/s}$$

$$p = \sqrt{\frac{k}{m}} = \sqrt{\frac{4(800)}{20}} = 12.6 \text{ rad/s}$$

From Eq. 22–22, with $k\delta_O$ replacing F_O, the amplitude of vibration caused by the floor displacement is

$$(y_p)_{max} = \frac{\delta_O}{1 - \left(\dfrac{\omega}{p}\right)^2} = \frac{10}{1 - \left(\dfrac{8}{12.6}\right)^2} = 16.7 \text{ mm} \qquad (2)$$

Hence, Eq. (1) and its time derivative become

$$y = A \sin (12.6t) + B \cos (12.6t) + 16.7 \sin (8t)$$
$$\dot{y} = A(12.6) \cos (12.6t) - B(12.6) \sin (12.6t) + 133.3 \cos (8t)$$

The constants A and B are evaluated from these equations. Since $y = \dot{y} = 0$ at $t = 0$, then

$$0 = 0 + B + 0; \qquad\qquad B = 0$$
$$0 = A(12.6) - 0 + 133.3; \qquad A = -10.6$$

The vibrating motion is therefore described by the equation

$$y = -10.6 \sin (12.6t) + 16.7 \sin (8t) \qquad\qquad \textit{Ans.}$$

Resonance will occur when the amplitude of vibration caused by the floor displacement approaches infinity. From Eq. (2), this requires

$$\omega = p = 12.6 \text{ rad/s} \qquad\qquad \textit{Ans.}$$

*22.4 Viscous Damped Free Vibration

The vibration analysis considered thus far has not included the effects of friction or damping in the system, and as a result, the solutions obtained are only in close agreement with the actual motion. Since all vibrations die out in time, the presence of damping forces should be included in the analysis.

In many cases damping is attributed to the resistance created by the substance, such as water, oil, or air, in which the system vibrates. Provided the body moves slowly through this substance, the resistance to motion is directly proportional to the body's speed. The type of force developed under these conditions is called a *viscous damping force*. The magnitude of this force may be expressed by an equation of the form

$$F = c\dot{x} \qquad (22\text{–}26)$$

where the constant c is called the *coefficient of viscous damping* and has units of $N \cdot s/m$ or $lb \cdot s/ft$.

The vibrating motion of a body or system having viscous damping may be characterized by the block and spring shown in Fig. 22–16a. The effect of damping is provided by the *dashpot* connected to the block on the right side. Damping occurs when the piston P moves to the right or left within the enclosed cylinder. The cylinder contains a fluid, and the motion of the piston is retarded since the fluid must flow around or through a small hole in the piston. The dashpot is assumed to have a coefficient of viscous damping c.

If the block is displaced a distance x from its equilibrium position, the resulting free-body diagram is shown in Fig. 22–16b. Both the spring force kx and the damping force $c\dot{x}$ oppose the forward motion of the block, so that applying the equation of motion yields

$$\overset{+}{\rightarrow}\Sigma F_x = ma_x; \qquad\qquad -kx - c\dot{x} = m\ddot{x}$$

or

$$m\ddot{x} + c\dot{x} + kx = 0 \qquad (22\text{–}27)$$

This linear, second-order, homogeneous, differential equation has solutions of the form

$$x = e^{\lambda t}$$

where e is the base of the natural logarithm and λ (lambda) is a constant. The value of λ (lambda) may be obtained by substituting this solution into Eq. 22–27, which yields

$$m\lambda^2 e^{\lambda t} + c\lambda e^{\lambda t} + k e^{\lambda t} = 0$$

or

$$e^{\lambda t}(m\lambda^2 + c\lambda + k) = 0$$

Since $e^{\lambda t}$ is always positive, a solution is possible provided

$$m\lambda^2 + c\lambda + k = 0$$

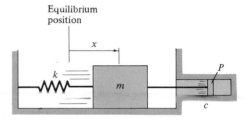

Equilibrium position

(a)

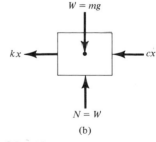

(b)

Fig. 22–16

Hence, by the quadratic formula, the two values of λ are

$$\lambda_1 = -\frac{c}{2m} + \sqrt{\left(\frac{c}{2m}\right)^2 - \frac{k}{m}}$$

$$\lambda_2 = -\frac{c}{2m} - \sqrt{\left(\frac{c}{2m}\right)^2 - \frac{k}{m}}$$

(22–28)

The general solution of Eq. 22–27 is therefore a linear combination of exponentials which involves both of these roots. There are three possible combinations of λ_1 and λ_2 which must be considered for the general solution. Before discussing these combinations, however, it is first necessary to consider the definition of the *critical damping coefficient* c_c as the value of c which makes the radical in Eqs. 22–28 equal to zero; i.e.,

$$\left(\frac{c_c}{2m}\right)^2 - \frac{k}{m} = 0$$

or

$$c_c = 2m\sqrt{\frac{k}{m}} = 2mp$$

(22–29)

Here the value of p is the circular frequency $p = \sqrt{k/m}$, Eq. 22–2.

Overdamped System

When $c > c_c$, the roots λ_1 and λ_2 are both real. The general solution of Eq. 22–27 may then be written as

$$x = Ae^{\lambda_1 t} + Be^{\lambda_2 t}$$

(22–30)

Motion corresponding to this solution is *nonvibrating*. The effect of damping is so strong that when the block is displaced and released, it simply creeps back to its original position without oscillating. The system is said to be *overdamped*.

Critically Damped System

If $c = c_c$, then $\lambda_1 = \lambda_2 = -c_c/2m = -p$. This situation is known as *critical damping,* since it represents a condition where c has the smallest value necessary to cause the system to be nonvibrating. Using the methods of differential equations, it may be shown that the solution to Eq. 22–27 for critical damping is

$$x = (A + Bt)e^{-pt}$$

(22–31)

Underdamped System

Most often $c < c_c$, in which case the system is referred to as *underdamped*. In this case the roots λ_1 and λ_2 are complex numbers and it may be shown that the general solution of Eq. 22–27 can be written as

$$x = D[e^{-(c/2m)t} \sin (p_d t + \phi)] \qquad (22\text{–}32)$$

where D and ϕ are constants generally determined from the initial conditions of the problem. The constant p_d is called the *damped natural frequency* of the system. It has a value of

$$p_d = \sqrt{\frac{k}{m} - \left(\frac{c}{2m}\right)^2} = p\sqrt{1 - \left(\frac{c}{c_c}\right)^2} \qquad (22\text{–}33)$$

where the ratio c/c_c is called the *damping factor*.

The graph of Eq. 22–32 is shown in Fig. 22–17. The initial limit of motion, D, diminishes with each cycle of vibration, since motion is confined within the bounds of the exponential curve. Using the damped natural frequency p_d, the period of damped vibration may be written as

$$\tau_d = \frac{2\pi}{p_d} \qquad (22\text{–}34)$$

Since $p_d < p$, Eq. 22–33, the period of damped vibration, τ_d, will be greater than that of free vibration, $\tau = 2\pi/p$.

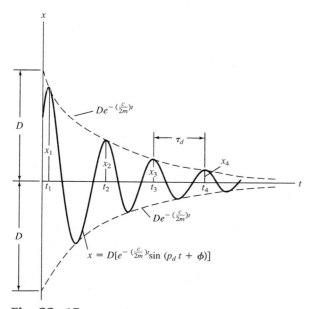

Fig. 22–17

*22.5 Viscous Damped Forced Vibration

The most general case of single-degree-of-freedom vibrating motion occurs when the system includes the effects of forced motion and induced damping. The analysis of this particular type of vibration is of practical value when applied to systems having significant damping characteristics.

If a dashpot is attached to the block and spring shown in Fig. 22–11a, the differential equation which describes the motion becomes

$$m\ddot{x} + c\dot{x} + kx = F_O \sin \omega t \tag{22–35}$$

A similar equation may be written for a block and spring having a periodic support displacement, Fig. 22–14a, which includes the effects of damping. In that case, however, F_O is replaced by $k\delta_O$. Since Eq. 22–35 is nonhomogeneous, the general solution is the sum of a complementary solution, x_c, and a particular solution, x_p. The complementary solution is determined by setting the right side of Eq. 22–35 equal to zero and solving the homogeneous equation, which is equivalent to Eq. 22–27. The solution is therefore given by Eq. 22–30, 22–31, or 22–32, depending upon the values of λ_1 and λ_2. Because all systems contain friction, however, this solution will dampen out with time. Only the particular solution, which describes the *steady-state vibration* of the system, will remain. Since the applied forcing function is harmonic, the steady-state motion will also be harmonic. Consequently, the particular solution will be of the form

$$x_p = A' \sin \omega t + B' \cos \omega t \tag{22–36}$$

The constants A' and B' are determined by taking the necessary time derivatives and substituting them into Eq. 22–35, which after simplification yields

$$[-A'm\omega^2 - cB'\omega + kA'] \sin \omega t + [-B'm\omega^2 + cA'\omega + kB'] \cos \omega t$$
$$= F_O \sin \omega t$$

Since this equation holds for all time, the constant coefficients of $\sin \omega t$ and $\cos \omega t$ may be equated; i.e.,

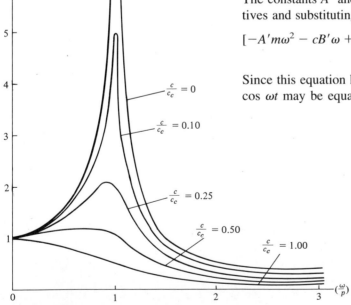

M.F.

$\frac{c}{c_c} = 0$

$\frac{c}{c_c} = 0.10$

$\frac{c}{c_c} = 0.25$

$\frac{c}{c_c} = 0.50$

$\frac{c}{c_c} = 1.00$

$\left(\frac{\omega}{p}\right)$

Fig. 22–18

$$-A'm\omega^2 - cB'\omega + kA' = F_O$$
$$-B'm\omega^2 + cA'\omega + kB' = 0$$

Solving for A' and B', realizing that $p^2 = k/m$, yields

$$A' = \frac{\left(\dfrac{F_O}{m}\right)(p^2 - \omega^2)}{(p^2 - \omega^2)^2 + \left(\dfrac{c\omega}{m}\right)^2}$$

(22–37)

$$B' = \frac{-F_O\dfrac{c\omega}{m^2}}{(p^2 - \omega^2)^2 + \left(\dfrac{c\omega}{m}\right)^2}$$

It is also possible to express Eq. 22–36 in a form similar to Eq. 22–9,

$$x_p = C' \sin(\omega t - \phi')$$

(22 38)

in which case the constants C' and ϕ' are

$$C' = \frac{F_O/k}{\sqrt{\left[1 - \left(\dfrac{\omega}{p}\right)^2\right]^2 + \left(2\dfrac{c}{c_c}\dfrac{\omega}{p}\right)^2}}$$

(22–39)

$$\phi' = \tan^{-1}\left(\frac{c\omega/k}{1 - \left(\dfrac{\omega}{p}\right)^2}\right)$$

The angle ϕ' represents the phase difference between the applied force and the resulting steady-state vibration of the damped system.

The *magnification factor* MF has been defined in Sec. 22.3 as the ratio of the amplitude of deflection caused by the forced vibration to the deflection caused by a static force F_O. From Eq. 22–38, the forced vibration has an amplitude of C'; thus,

$$\text{MF} = \frac{C'}{F_O/k} = \frac{1}{\sqrt{\left[1 - \left(\dfrac{\omega}{p}\right)^2\right]^2 + \left(2\dfrac{c}{c_c}\dfrac{\omega}{p}\right)^2}}$$

(22–40)

The MF is plotted in Fig. 22–18 versus the frequency ratio ω/p for various values of the damping factor c/c_c. It can be seen from this graph that the magnification of the amplitude increases as the damping factor decreases. Resonance obviously occurs only when the damping factor is zero and the frequency ratio equals 1.

Example 22–8

The 30-kg electric motor shown in Fig. 22–19 is supported by *four* springs, each spring having a stiffness of 200 N/m. If the rotor R is unbalanced such that its effect is equivalent to a 4-kg mass located 60 mm from the axis of rotation, determine the amplitude of vibration when the rotor is turning at $\omega = 10$ rad/s. The damping factor is $c/c_c = 0.15$.

SOLUTION

The periodic force which causes the motor to vibrate is the centrifugal force due to the unbalanced rotor. This force has a constant magnitude of

$$F_O = ma_n = mr\omega^2 = 4 \text{ kg}(0.06 \text{ m})(10 \text{ rad/s})^2 = 24 \text{ N}$$

Oscillation in the vertical direction may be expressed in the periodic form $F = F_O \sin \omega t$, where $\omega = 10$ rad/s. Thus,

$$F = 24 \sin 10t$$

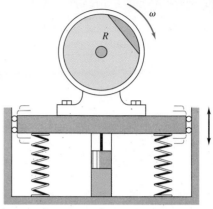

Fig. 22–19

The stiffness of the entire system of four springs is $k = 4(200) = 800$ N/m. Therefore, the circular frequency of vibration is

$$p = \sqrt{\frac{k}{m}} = \sqrt{\frac{800}{30}} = 5.16 \text{ rad/s}$$

Since the damping factor is known, the steady-state amplitude may be determined from the first of Eqs. 22–39, i.e.,

$$C' = \frac{F_O/k}{\sqrt{\left[1 - \left(\dfrac{\omega}{p}\right)^2\right]^2 + \left(2\dfrac{c}{c_c}\dfrac{\omega}{p}\right)^2}}$$

$$= \frac{24/800}{\sqrt{\left[1 - \left(\dfrac{10}{5.16}\right)^2\right]^2 + \left[2(0.15)\dfrac{10}{5.16}\right]^2}}$$

$$= 0.0107 \text{ m} = 10.7 \text{ mm} \qquad\qquad \textit{Ans.}$$

*22.6 Electrical Circuit Analogues

The characteristics of a vibrating mechanical system may be represented by an electric circuit. Consider the circuit shown in Fig. 22–20a, which consists of an inductor L, a resistor R, and a capacitor C. When a voltage $E(t)$ is applied, it causes a current of magnitude i to flow through the circuit. As the current flows past the inductor the voltage drop is $L(di/dt)$, when it flows across the resistor the drop is Ri, and when it arrives at the capacitor the drop is $(1/C) \int i \, dt$. Since current cannot flow past a capacitor, it is only possible to measure the charge q acting on the capacitor. The charge may, however, be related to the current by the equation $i = dq/dt$. Thus, the voltage drops which occur across the inductor, resistor, and capacitor may be written as $L \, d^2q/dt^2$, $R \, dq/dt$, and q/C, respectively. According to Kirchhoff's voltage law, the applied voltage balances the sum of the voltage drops around the circuit. Therefore,

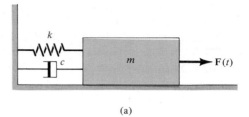

(a)

$$L\frac{d^2q}{dt^2} + R\frac{dq}{dt} + \frac{1}{C}q = E(t) \qquad (22\text{–}41)$$

Consider now the model of a single-degree-of-freedom mechanical system, Fig. 22–20b, which is subjected to both a general forcing function $F(t)$ and damping. The equation of motion for this system was established in the previous section and can be written as

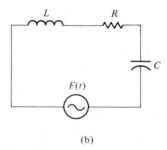

(b)

Fig. 22–20

$$m\frac{d^2x}{dt^2} + c\frac{dx}{dt} + kx = F(t) \qquad (22\text{–}42)$$

By comparison, it is seen that Eqs. 22–41 and 22–42 have the same form, hence mathematically the problem of analyzing an electric circuit is the same as that of analyzing a vibrating mechanical system. The analogues between the two equations are given in Table 22–1.

This analogy has important application to experimental work, for it is much easier to simulate the vibration of a complex mechanical system using an electric circuit, which can be constructed on an analogue computer, than to make an equivalent mechanical spring and dashpot model.

Table 22–1 Electrical-Mechanical Analogues

Electrical		Mechanical	
Electric charge	q	Displacement	x
Electric current	i	Velocity	dx/dt
Voltage	$E(t)$	Applied force	$F(t)$
Inductance	L	Mass	m
Resistance	R	Viscous damping coefficient	c
Reciprocal of capacitance	$1/C$	Spring stiffness	k

515

PROBLEMS

22–46. If the block is subjected to the impressed force $F = F_O \cos \omega t$, show that the differential equation of motion is $\ddot{y} + (k/m)y = (F_O/m) \cos \omega t$, where y is measured from the equilibrium position of the block. What is the general solution of this equation?

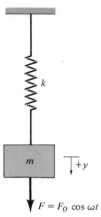

k

m $\quad$ $+y$

$F = F_O \cos \omega t$

Prob. 22–46

22–47. The 20-lb block is attached to a spring having a stiffness of 20 lb/ft. A force of $F = (6 \cos 2t)$ lb, where t is in seconds, is applied to the block. Determine the maximum speed of the block after frictional forces cause the free vibrations to dampen out.

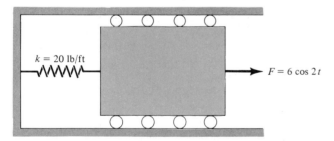

$k = 20$ lb/ft

$F = 6 \cos 2t$

Prob. 22–47

*22–48. A 5-lb weight is suspended from a vertical spring having a stiffness of 50 lb/ft. An impressed force of $F = (0.25 \sin 8t)$ lb, where t is in seconds, is acting on the weight. Determine the equation of motion of the weight when it is pulled down 3 in. from the equilibrium position and released from rest.

22–49. A 5-kg block is suspended from a spring having a stiffness of 300 N/m. If the block is acted upon by a vertical force $F = (7 \sin 8t)$ N, where t is in seconds, determine the equation which describes the motion of the block when it is pulled down 100 mm from the equilibrium position and released from rest at $t = 0$. Assume that positive displacement is measured downward.

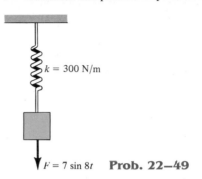

$k = 300$ N/m

$F = 7 \sin 8t$ $\quad$ **Prob. 22–49**

22–50. A 7-kg block is suspended from a spring that has a stiffness of $k = 350$ N/m. The block is drawn downward 70 mm from the equilibrium position and released from rest at $t = 0$. If the support moves with an impressed displacement $\delta = (20 \sin 4t)$ mm, where t is in seconds, determine the equation which describes the vertical motion of the block. Assume that positive displacement is measured downward.

22–51. The electric motor turns an eccentric flywheel which is equivalent to an unbalanced 0.25-lb weight located 10 in. from the axis of rotation. If the static deflection of the beam is 1 in. because of the weight of the motor, determine the angular velocity of the flywheel at which resonance will occur. The motor weighs 150 lb. Neglect the mass of the beam.

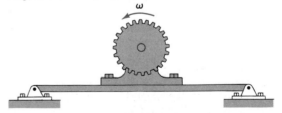

ω

Probs. 22–51/22–52/22–53

*22–52. What will be the amplitude of steady-state vibration of the motor in Prob. 22–51 if the angular velocity of the flywheel is 20 rad/s?

22–53. Determine the angular velocity of the motor in Prob. 22–51 to produce an amplitude of vibration of 0.25 in.

22–54. The fan has a mass of 25 kg and is fixed to the end of a horizontal beam that has a negligible mass. The fan blade is mounted eccentrically on the shaft such that it is equivalent to an unbalanced 3.5-kg mass located 100 mm from the axis of rotation. If the static deflection of the beam is 50 mm as a result of the weight of the fan, determine the angular speed of the fan at which resonance will occur. *Hint:* See the first part of Example 22–8.

Probs. 22–54/22–55/22–56

22–55. What is the amplitude of steady-state vibration of the fan in Prob. 22–54 when the angular speed of the fan is 10 rad/s? *Hint:* See the first part of Example 22–8.

*22–56. What will be the amplitude of steady-state vibration of the fan in Prob. 22–54 if the angular speed of the fan is 18 rad/s? *Hint:* See the first part of Example 22–8.

22–57. The engine is mounted on a foundation block which is spring-supported. Describe the steady state vibration of the system if the block and engine have a total weight $W = 1500$ lb and the engine, when running, creates an impressed force $F = (50 \sin 2t)$ lb, where t is in seconds. Assume that the system vibrates only in the vertical direction, with the positive displacement measured downward, and that the total stiffness of the springs can be represented as $k = 2000$ lb/ft.

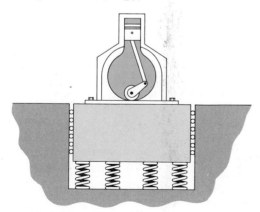

Probs. 22–57/22–58

22–58. Determine the rotational speed ω of the engine in Prob. 22–57 which will cause resonance.

22–59. The 80-lb block is attached to a spring, the end of which is subjected to a periodic support displacement $\delta_A = (0.5 \sin 8t)$ ft, where t is in seconds. Determine the amplitude of the steady-state motion.

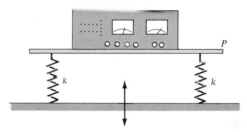

Prob. 22–59

*22–60. The instrument is centered uniformly on a platform P, which in turn is supported by *four* springs, each spring having a stiffness $k = 130$ N/m. If the floor is subjected to a vibration $\omega = 7$ Hz, having a vertical displacement amplitude $\delta_O = 0.17$ ft, determine the vertical displacement amplitude of the platform and instrument. The instrument and the platform have a total weight of 18 lb.

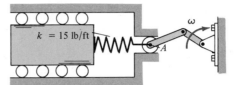

Prob. 22–60

22–61. The electric motor has a mass of 50 kg and is supported by *four springs,* each spring having a stiffness of 100 N/m. If the motor turns a disk D which is mounted eccentrically, 20 mm from the disk's center, determine the angular rotation ω at which resonance occurs. Assume that the motor only vibrates in the vertical direction.

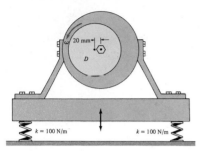

Prob. 22–61

22–62. The small block at A has a mass of 4 kg and is mounted on the bent rod having negligible mass. If the rotor at B causes a harmonic movement $\delta_B = (0.1 \cos 15t)$ m, where t is in seconds, determine the amplitude of vibration of the block.

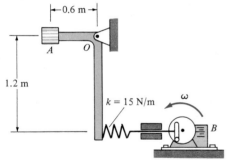

Prob. 22–62

22–63. The 450-kg trailer is pulled with a constant speed over the surface of a bumpy road, which may be approximated by a cosine curve having an amplitude of 50 mm and wave length of 4 m. If the two springs s which support the trailer each have a stiffness of 800 N/m, determine the speed v which will cause the greatest vibration (resonance) of the trailer. Neglect the weight of the wheels.

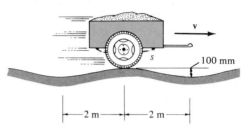

Probs. 22–63/22–64

***22–64.** Determine the amplitude of vibration of the trailer in Prob. 22–63 if the speed $v = 15$ km/h.

22–65. A block having a weight of 7 lb is suspended from a spring having a stiffness $k = 75$ lb/ft. The support to which the spring is attached is given a simple harmonic motion which may be expressed by $\delta = (0.15 \sin 2t)$ ft, where t is in seconds. If the damping factor is $c/c_c = 0.8$, determine the phase angle ϕ of forced vibration.

22–66. Determine the magnification factor of the block, spring, and dashpot combination in Prob. 22–65.

22–67. A block having a mass of 0.5 slug is suspended from a spring having a stiffness of $k = 0.6$ lb/in. If a dashpot provides a damping force of 0.2 lb on the block when the speed of the block is $v = 1$ ft/s, determine the period of free vibration.

***22–68.** The 4-kg circular disk is attached to three springs, each spring having a stiffness of $k = 180$ N/m. If the disk is immersed in a fluid and given a downward velocity of 0.3 m/s at the equilibrium position, determine the equation which describes the motion. Assume that positive displacement is measured downward, and that fluid resistance acting on the disk furnishes a damping force having a magnitude of $F = (60|v|)$ N, where v is in m/s.

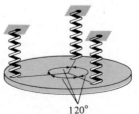

Prob. 22–65

22–69. A 5-lb weight is suspended from a spring having a stiffness of $k = 6$ lb/in. The support to which the spring is attached is given simple harmonic motion which may be expressed by $\delta = (1.5 \sin 2t)$ in. where t is in seconds. If the damping coefficient is $c = 0.08$ lb $\cdot$ s/in., determine the phase angle ϕ of forced vibration.

22–70. Determine the magnification ratio of the spring and dashpot combination in Prob. 22–69.

22–71. The 200-lb electric motor is fastened to the midpoint of the simply supported beam. It is found that the beam deflects 2 in. when the motor is not running. The rotor turns an eccentric flywheel which is equivalent to an unbalanced weight of 1 lb located 5 in. from the axis of rotation. If the rotor is turning at 100 rpm, determine the amplitude of steady-state vibration. The damping factor is $c/c_c = 0.20$. Neglect the mass of the beam.

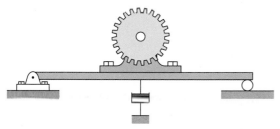

Prob. 22–71

*22–72. A spring has a stiffness of $k = 2$ lb/in. If a 5-lb weight is attached to the spring and is given an upward velocity of 2 ft/s from the equilibrium position, determine the position of the weight as a function of time. Assume that motion takes place in a medium which furnishes a retarding force F (pounds) having a magnitude numerically equal to four times the speed v of the weight, where v is in ft/s.

22–73. A block having a mass of 0.8 kg is suspended from a spring having a stiffness of 120 N/m. If a dashpot provides a damping force of 2.5 N when the speed of the block is 0.2 m/s, determine the period of free vibration.

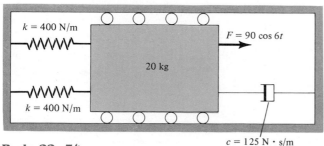

Prob. 22–74

22–74. The 20-kg block is subjected to the action of the harmonic force $F = (90 \cos 6t)$ N, where t is in seconds. Write the equation which describes the steady-state motion.

22–75. The bar has a weight of 6 lb. If the stiffness of the spring is $k = 8$ lb/ft and the dashpot has a damping coefficient $c = 60$ lb $\cdot$ s/ft, determine the differential equation which describes the motion in terms of the angle θ of the bar's rotation.

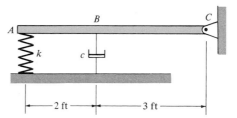

Prob. 22–75

*22–76. The bell-crank mechanism consists of a bent rod, having a negligible mass, and an attached 5-lb weight. Determine the critical damping coefficient c_c and the damping natural frequency for small vibrations about the equilibrium position. Neglect the size of the weight.

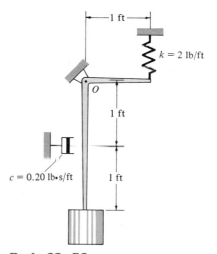

Prob. 22–76

22–77. The damping factor, c/c_c, may be determined experimentally by measuring the successive amplitudes of vibrating motion of a system. If two of these maximum displacements can be approximated by x_1 and x_2, as shown in Fig. 22–17, show that the ratio $\ln x_1/x_2 = 2\pi(c/c_c)/\sqrt{1 - (c/c_c)^2}$. The quantity $\ln x_1/x_2$ is called the *logarithmic decrement*.

22–78. The block shown in Fig. 22–16 has a mass of 20 kg and the spring has a stiffness $k = 600$ N/m. When the block is displaced and released, two successive amplitudes are measured as $x_1 = 150$ mm and $x_2 = 87$ mm. Determine the coefficient of viscous damping, c. *Hint:* See Prob. 22–77.

22–79. The barrel of a cannon has a mass of 700 kg, and after firing it recoils a distance of 0.64 m. If it returns to its original position by means of a single recuperator having a damping coefficient of 2 kN · s/m, determine the required stiffness of each of the two springs fixed to the base and attached to the barrel so that the barrel recuperates without vibration.

*22–80. A block having a mass of 7 kg is suspended from a spring that has a stiffness $k = 600$ N/m. If it is given an upward velocity of 0.6 m/s from its equilibrium position at $t = 0$, determine the position of the block as a function of time. Assume that positive displacement of the block is downward, and that motion takes place in a medium which furnishes a damping force $F = (50|v|)$ N, where v is measured in m/s.

22–81. Determine the differential equation of motion for the damped vibratory system shown. What type of motion occurs?

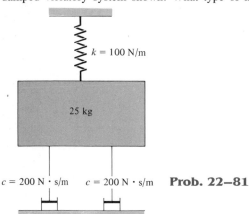

$k = 100$ N/m

25 kg

$c = 200$ N · s/m $c = 200$ N · s/m   **Prob. 22–81**

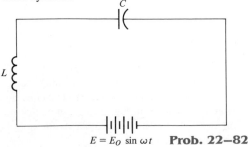

22–82. Determine the mechanical analogue for the electrical circuit. What differential equations describe the mechanical and electrical systems?

C

L

$E = E_O \sin \omega t$ **Prob. 22–82**

22–83. Draw the electrical circuit that is equivalent to the mechanical system shown. What is the differential equation which describes the motion of the current in the circuit?

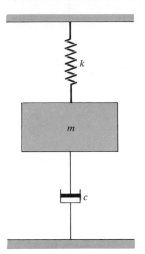

k

m

c

Prob. 22–83

*22–84. Draw the electrical circuit that is equivalent to the mechanical system shown. What is the differential equation which describes the motion of the current in the circuit?

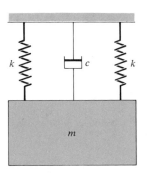

k c k

m

Prob. 22–84

22–85. Draw the electric circuit that is equivalent to the mechanical system shown. Determine the differential equation which describes the motion of the current in the circuit.

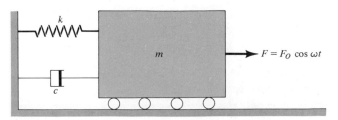

k

m

c

$F = F_O \cos \omega t$

Prob. 22–85

Mathematical Expressions

Quadratic Formula

If $ax^2 + bx + c = 0$, then $x = \dfrac{-b \pm \sqrt{b^2 - 4ac}}{2a}$

Hyperbolic Functions

$\sinh x = \dfrac{e^x - e^{-x}}{2}$, $\cosh x = \dfrac{e^x + e^{-x}}{2}$, $\tanh x = \dfrac{\sinh x}{\cosh x}$

Trigonometric Identities

$\sin^2 \theta + \cos^2 \theta = 1$

$\sin (\theta \pm \phi) = \sin \theta \cos \phi \pm \cos \theta \sin \phi$

$\sin 2\theta = 2 \sin \theta \cos \theta$

$\cos (\theta \pm \phi) = \cos \theta \cos \phi \mp \sin \theta \sin \phi$

$\cos 2\theta = \cos^2 \theta - \sin^2 \theta$

$\cos \theta = \pm\sqrt{\dfrac{1 + \cos 2\theta}{2}}$, $\sin \theta = \pm\sqrt{\dfrac{1 - \cos 2\theta}{2}}$

$\tan \theta = \dfrac{\sin \theta}{\cos \theta}$

$1 + \tan^2 \theta = \sec^2 \theta$

$1 + \operatorname{ctn}^2 \theta = \csc^2 \theta$

Power–Series Expansions

$\sin x = x - \dfrac{x^3}{3!} + \dfrac{x^5}{5!} - \dfrac{x^7}{7!} + \cdots$

$\cos x = 1 - \dfrac{x^2}{2!} + \dfrac{x^4}{4!} - \dfrac{x^6}{6!} + \cdots$

$\sinh x = x + \dfrac{x^3}{3!} + \dfrac{x^5}{5!} + \cdots$

$\cosh x = 1 + \dfrac{x^2}{2!} + \dfrac{x^4}{4!} + \cdots$

Derivatives

$$\frac{d}{dx}(u^n) = nu^{n-1}\frac{du}{dx}$$

$$\frac{d}{dx}(uv) = u\frac{dv}{dx} + v\frac{du}{dx}$$

$$\frac{d}{dx}\left(\frac{u}{v}\right) = \frac{v\dfrac{du}{dx} - u\dfrac{dv}{dx}}{v^2}$$

$$\frac{d}{dx}(\sin u) = \cos u\frac{du}{dx}$$

$$\frac{d}{dx}(\cos u) = -\sin u\frac{du}{dx}$$

$$\frac{d}{dx}(\tan u) = \sec^2 u\frac{du}{dx}$$

$$\frac{d}{dx}(\cot u) = -\csc^2 u\frac{du}{dx}$$

$$\frac{d}{dx}(\sec u) = \tan u \sec u\frac{du}{dx}$$

$$\frac{d}{dx}(\csc u) = -\csc u \cot u\frac{du}{dx}$$

$$\frac{d}{dx}(\sinh u) = \cosh u\frac{du}{dx}$$

$$\frac{d}{dx}(\cosh u) = \sinh u\frac{du}{dx}$$

Integrals

$$\int x^n\, dx = \frac{x^{n+1}}{n+1} + C,\ n \neq -1$$

$$\int \frac{dx}{a+bx} = \frac{1}{b}\ln(a+bx) + C$$

$$\int \frac{dx}{a+bx^2} = \frac{1}{2\sqrt{-ba}}\ln\left[\frac{\sqrt{a}+2\sqrt{-b}}{\sqrt{a}-x\sqrt{-b}}\right] + C,\ a>0,\ b<0$$

$$\int \frac{x\, dx}{a+bx^2} = \frac{1}{2b}\ln(bx^2+a) + C$$

$$\int \frac{x^2\, dx}{a+bx^2} = \frac{x}{b} - \frac{a}{b\sqrt{ab}}\tan^{-1}\frac{x\sqrt{ab}}{a} + C$$

$$\int \frac{dx}{a^2-x^2} = \frac{1}{2a}\ln\left[\frac{a+x}{a-x}\right] + C,\ a^2 > x^2$$

$$\int \sqrt{a+bx}\, dx = \frac{2}{3b}\sqrt{(a+bx)^3} + C$$

$$\int x\sqrt{a+bx}\, dx = \frac{-2(2a-3bx)\sqrt{(a+bx)^3}}{15b^2} + C$$

$$\int x^2\sqrt{a+bx}\, dx = \frac{2(8a^2-12abx+15b^2x^2)\sqrt{(a+bx)^3}}{105b^3} + C$$

$$\int \sqrt{a^2-x^2}\, dx = \frac{1}{2}\left[x\sqrt{a^2-x^2} + a^2\sin^{-1}\frac{x}{a}\right] + C,\ a>0$$

$$\int x\sqrt{a^2-x^2}\, dx = -\frac{1}{3}\sqrt{(a^2-x^2)^3} + C$$

$$\int x^2\sqrt{a^2-x^2}\, dx = -\frac{x}{4}\sqrt{(a^2-x^2)^3} + \frac{a^2}{8}\left(x\sqrt{a^2-x^2} + a^2\sin^{-1}\frac{x}{a}\right) + C,\ a>0$$

$$\int \sqrt{x^2 \pm a^2} \, dx = \frac{1}{2}\left[x\sqrt{x^2 + a^2} \pm a^2 \ln(x + \sqrt{x^2 \pm a^2})\right] + C$$

$$\int x\sqrt{x^2 \pm a^2} \, dx = \frac{1}{3}\sqrt{(x^2 \pm a^2)^3} + C$$

$$\int x^2\sqrt{x^2 \pm a^2} \, dx = \frac{x}{4}\sqrt{(x^2 \pm a^2)^3} \pm \frac{a^2}{8} x\sqrt{x^2 \pm a^2} - \frac{a^4}{8} \ln(x + \sqrt{x^2 \pm a^2}) + C$$

$$\int \frac{dx}{\sqrt{a + bx}} = \frac{2\sqrt{a + bx}}{b} + C$$

$$\int \frac{x \, dx}{\sqrt{x^2 \pm a^2}} = \sqrt{x^2 \pm a^2} + C$$

$$\int \frac{dx}{\sqrt{a + bx + cx^2}} = \frac{1}{\sqrt{c}} \ln\left[\sqrt{a + bx + cx^2} + x\sqrt{c} + \frac{b}{2\sqrt{c}}\right] + C, \, c > 0$$

$$= \frac{1}{\sqrt{-c}} \sin^{-1}\left(\frac{-2cx - b}{\sqrt{b^2 - 4ac}}\right) + C, \, c < 0$$

$$\int \sin x \, dx = -\cos x + C$$

$$\int \cos x \, dx = \sin x + C$$

$$\int x \cos(ax) \, dx = \frac{1}{a^2} \cos(ax) + \frac{x}{a} \sin(ax) + C$$

$$\int x^2 \cos(ax) \, dx = \frac{2x}{a^2} \cos(ax) + \frac{a^2x^2 - 2}{a^3} \sin(ax) + C$$

$$\int e^{ax} \, dx = \frac{1}{a}e^{ax} + C$$

$$\int x e^{ax} \, dx = \frac{e^{ax}}{a^2}(ax - 1) + C$$

$$\int \sinh x \, dx = \cosh x + C$$

$$\int \cosh x \, dx = \sinh x + C$$

B

Numerical and Computer Analysis

Occasionally the application of the laws of mechanics will lead to a system of equations for which a closed-form solution is difficult or impossible to obtain. When confronted with this situation, engineers will often use a numerical method which in most cases can be programmed on a microcomputer or "programmable" pocket calculator. Here we will briefly present a computer program for solving a set of linear algebraic equations and three numerical methods which can be used to solve an algebraic or transcendental equation, evaluate a definite integral, and solve an ordinary differential equation. Application of each method will be explained by example, and an associated computer program written in Microsoft BASIC, which is designed to run on most personal computers, is provided.* A text on numerical analysis should be consulted for further discussion regarding a check of the accuracy of each method and the inherent errors that can develop.

Linear Algebraic Equations

Application of the equations of static equilibrium or the equations of motion sometimes requires solving a set of linear algebraic equations. The computer program listed in Fig. B–1 can be used for this purpose. It is based on the

*Similar types of programs can be written or purchased for programmable pocket calculators.

```
 1 PRINT"Linear system of equations":PRINT
 2 DIM A(10,11)
 3 INPUT"Input number of equations : ",N
 4 PRINT
 5 PRINT"A  coefficients"
 6 FOR I = 1 TO N
 7 FOR J = 1 TO N
 8 PRINT "A(";I;",";J;
 9 INPUT")=",A(I,J)
10 NEXT J
11 NEXT I
12 PRINT
13 PRINT"B  coefficients"
14 FOR I = 1 TO N
15 PRINT "B(";I;
16 INPUT")=",A(I,N+1)
17 NEXT I
18 GOSUB 25
19 PRINT
20 PRINT"Unknowns"
21 FOR I = 1 TO N
22 PRINT "X(";I;")=";A(I,N+1)
23 NEXT I
24 END
25 REM Subroutine Guassian
26 FOR M=1 TO N
27 NP=M
28 BG=ABS(A(M,M))
29 FOR I = M TO N
30 IF ABS(A(I,M))<=BG THEN 33
31 BG=ABS(A(I,M))
32 NP=I
33 NEXT I
34 IF NP=M THEN 40
35 FOR I = M TO N+1
36 TE=A(M,I)
37 A(M,I)=A(NP,I)
38 A(NP,I)=TE
39 NEXT I
40 FOR I = M+1 TO N
41 FC=A(I,M)/A(M,M)
42 FOR J = M+1 TO N+1
43 A(I,J)=A(I,J)-FC*A(M,J)
44 NEXT J
45 NEXT I
46 NEXT M
47 A(N,N+1)=A(N,N+1)/A(N,N)
48 FOR I = N-1 TO 1 STEP -1
49 SM=0
50 FOR J=I+1 TO N
51 SM=SM+A(I,J)*A(J,N+1)
52 NEXT J
53 A(I,N+1)=(A(I,N+1)-SM)/A(I,I)
54 NEXT I
55 RETURN
```

Fig. B—1

method of a Gauss elimination and can solve at most ten equations with ten unknowns. To do so, the equations should first be written in the following general format:

$$A_{11}x_1 + A_{12}x_2 + \cdots + A_{1n}x_n = B_1$$
$$A_{21}x_1 + A_{22}x_2 + \cdots + A_{2n}x_n = B_2$$
$$\cdot$$
$$\cdot$$
$$\cdot$$
$$A_{n1}x_1 + A_{n2}x_2 + \cdots + A_{nn}x_n = B_n$$

The "A" and "B" coefficients are "called" for when running the program. The output presents the unknowns $x_1, \ldots, x_n$.

Example B—1

Solve the two equations
$$3x_1 + x_2 = 4$$
$$2x_1 - x_2 = 10$$

SOLUTION

When the program begins to run, it first calls for the number of equations (2), then the A coefficients in the following sequence: $A_{11} = 3$, $A_{12} = 1$, $A_{21} = 2$, $A_{22} = -1$ and finally the B coefficients $B_1 = 4$, $B_2 = 10$. The output appears as:

Unknowns

$X(1) = 2.8$ *Ans.*

$X(2) = -4.4$ *Ans.*

Simpson's Rule

This numerical method can be used to determine the area under a curve given as a graph or as an explicit function $y = f(x)$. Likewise it can be used to compute the value of a definite integral which involves the function $y = f(x)$. To do so, the area must be subdivided into an *even number* of strips or intervals having a width h. The curve between three consecutive ordinates is approximated by a parabola, and the entire area or definite integral is then determined from the formula

$$\int_{x_0}^{x_n} f(x)\, dx \simeq \frac{h}{3}[y_0 + 4(y_1 + y_3 + \cdots + y_{n-1})$$
$$+ 2(y_2 + y_4 + \cdots + y_{n-2}) + y_n] \quad \text{(B-1)}$$

The computer program for this equation is given in Fig. B–2. For its use, one must first enter the function (line 6), then the upper and lower limits of the integral (lines 7 and 8), and finally the number of intervals (line 9). The value of the integral is then given as the output.

```
1 PRINT"Simpson's rule":PRINT
2 PRINT" To execute this program :":PRINT
3 PRINT"   1- Modify right-hand side of the equation given below,
4 PRINT"      then press RETURN key"
5 PRINT"   2- Type  RUN 6":PRINT:EDIT 6
6 DEF FNF(X)=LOG(X)
7 PRINT:INPUT" Enter Lower Limit = ",A
8 INPUT" Enter Upper Limit = ",B
9 INPUT" Enter Number (even) of Intervals = ",N%
10 H=(B-A)/N%:AR=FNF(A):X=A+H
11 FOR J%=2 TO N%
12 K=2*(2-J%+2*INT(J%/2))
13 AR=AR+K*FNF(X)
14 X=X+H:NEXT J%
15 AR=H*(AR+FNF(B))/3
16 PRINT" Integral = ",AR
17 END
```

Fig. B–2

Example B–2

Evaluate the definite integral $\int_{2}^{5} \ln x \, dx$

SOLUTION

The interval $x_0 = 2$ to $x_6 = 5$ will be divided into six equal parts ($n = 6$), each having a width $h = (5 - 2)/6 = 0.5$. We then compute $y = f(x) = \ln x$ at each point of subdivision.

n	x_n	y_n
0	2	0.693
1	2.5	0.916
2	3	1.099
3	3.5	1.253
4	4	1.386
5	4.5	1.504
6	5	1.609

Thus, Eq. B–1 becomes

$$\int_{2}^{5} \ln x \, dx \simeq \frac{0.5}{3}[0.693 + 4(0.916 + 1.253 + 1.504)$$
$$+ 2(1.099 + 1.386) + 1.609]$$
$$\simeq 3.66 \qquad\qquad Ans.$$

This answer is equivalent to the exact answer to three significant figures. Obviously, accuracy to a larger number of significant figures can be improved by selecting a smaller interval h (or larger n).

Using the computer program, we enter the function $\ln x$, line 6 in Fig. B–2, the upper and lower limits 2 and 5, and the number of intervals $n = 6$. The output appears as

$$\text{Integral} = 3.66082 \qquad\qquad Ans.$$

The Secant Method

This numerical method is used to find the real roots of an algebraic or transcendental equation $f(x) = 0$. The method derives its name from the fact that the formula used is established from the slope of the secant line to the graph $y = f(x)$. This slope is $(f(x_n) - f(x_{n-1}))/(x_n - x_{n-1})$ and the secant formula is

$$x_{n+1} = x_n - f(x_n) \left(\frac{x_n - x_{n-1}}{f(x_n) - f(x_{n-1})} \right) \qquad (B\text{–}2)$$

For application it is necessary to provide two initial guesses, x_0 and x_1, and thereby evaluate x_2 from Eq. B–2 ($n = 1$). One then proceeds to reapply Eq. B–2 with x_1 and the calculated value of x_2 and obtain x_3 ($n = 2$), etc., until the value $x_{n+1} \simeq x_n$. One can see this will occur if x_n is approaching the root of the function $f(x) = 0$ since the correction term on the right of Eq. B–2 will tend toward zero. In particular, the larger the slope the smaller the correction to x_n and the faster the root will be found. Likewise, if the slope is very small in the neighborhood of the root, the method leads to large corrections for x_n and convergence to the root is slow and may even lead to a failure to find it. In such cases other numerical techniques must be used for solution.

A computer program based on Eq. B–2 is listed in Fig. B–3. The user must supply the function (line 7) along with two initial guesses x_0 and x_1 used to approximate the solution. The output specifies the value of the root. If it cannot be determined it is so stated.

```
1 PRINT"Secant method":PRINT
2 PRINT" To execute this program :":PRINT
3 PRINT"    1) Modify right hand side of the equation given below,"
4 PRINT"       then press RETURN key."
5 PRINT"    2) Type  RUN 7"
6 PRINT:EDIT 7
7 DEF FNF(X)=.5*SIN(X)-2*COS(X)+1.3
8 INPUT"Enter point #1 =",X
9 INPUT"Enter point #2 =",X1
10 IF X=X1 THEN 14
11 EP=.00001:TL=2E-20
12 FP=(FNF(X1)-FNF(X))/(X1-X)
13 IF ABS(FP)>TL THEN 15
14 PRINT"Root can not be found.":END
15 DX=FNF(X1)/FP
16 IF ABS(DX)>EP THEN 19
17 PRINT "Root = ";X1;"      Function evaluated at this root = ";FNF(X1)
18 END
19 X=X1:X1=X1-DX
20 GOTO 12
```

Fig. B–3

Example B–3

Determine the root of the equation

$$f(x) = 0.5 \sin x - 2 \cos x + 1.30 = 0$$

SOLUTION

Guesses of the initial roots will be $x_0 = 45°$ and $x_1 = 30°$. Applying Eq. B–2,

$$x_2 = 30° - (-0.1821)\frac{(30° - 45°)}{(-0.1821 - 0.2393)} = 36.48°$$

Using this value in Eq. B–2, along with $x_1 = 30°$, we have

$$x_3 - 36.48° \quad (-0.0108)\frac{36.48° - 30°}{(-0.0108 + 0.1821)} = 36.89°$$

Repeating the process with this value and $x_2 = 36.48°$ yields

$$x_4 = 36.89° - (0.0005)\left(\frac{36.89° - 36.48°}{(0.0005 + 0.0108)}\right) = 36.87°$$

Thus $x = 36.9°$ is appropriate to three significant figures.

If the problem is solved using the computer program, we enter the function, line 7 in Fig. B–3. The first and second guesses must be in radians. Choosing these to be 0.8 rad, and 0.5 rad, the result appears as

Root = 0.6435022.
Function evaluated at this root = 1.66893E–06.

This result converted from radians to degrees is therefore

$$x = 36.9° \qquad\qquad Ans.$$

Runge-Kutta Method

This numerical method is used to solve an ordinary differential equation. It consists of applying a set of formulas which are used to find specific values of y for corresponding incremental values h in x. The formulas given in general form are as follows:

First-Order Equation. To integrate $\dot{x} = f(t, x)$ step-by-step use

$$x_{i+1} = x_i + \frac{1}{6}(k_1 + 2k_2 + 2k_3 + k_4) \qquad \text{(B–3)}$$

where

$$k_1 = hf(t_i, x_i)$$
$$k_2 = hf\left(t_i + \frac{h}{2}, x_i + \frac{k_1}{2}\right)$$
$$k_3 = hf\left(t_i + \frac{h}{2}, x_i + \frac{k_2}{2}\right) \qquad \text{(B–4)}$$
$$k_4 = hf(t_i + h, x_i + k_3)$$

Second-Order Equation. To integrate $\ddot{x} = f(t, x, \dot{x})$ use

$$x_{i+1} = x_i + h\left[\dot{x}_i + \frac{1}{6}(k_1 + k_2 + k_3)\right]$$
$$\dot{x}_{i+1} = \dot{x}_i + \frac{1}{6}(k_1 + 2k_2 + 2k_3 + k_4) \qquad \text{(B–5)}$$

where

$$k_1 = hf(t_i, x_i, \dot{x}_i)$$
$$k_2 = hf\left(t_i + \frac{h}{2}, x_i + \frac{h}{2}\dot{x}_i, \dot{x}_i + \frac{k_1}{2}\right)$$
$$k_3 = hf\left(t_i + \frac{h}{2}, x_i + \frac{h}{2}\dot{x}_i + \frac{h}{4}k_1, \dot{x}_i + \frac{k_2}{2}\right) \qquad \text{(B–6)}$$
$$k_4 = hf\left(t_i + h, x_i + h\dot{x}_i + \frac{h}{2}k_2, \dot{x}_i + k_3\right)$$

To apply these equations, one starts with initial values $t_i = t_0$, $x_i = x_0$ and $\dot{x}_i = \dot{x}_0$ (for the second-order equation). Choosing an increment h for t_0 the four constants k are computed and these results are substituted into Eq. B–3 or B–5 in order to compute $x_{i+1} = x_1$, $\dot{x}_{i+1} = x_1$, corresponding to $t_{i+1} = t_1 = t_0 + h$. Repeating this process using t_1, x_1, $\dot{x}_1$ and h, values for x_2, $\dot{x}_2$ and $t_2 = t_1 + h$ are then computed, etc.

Computer programs which solve first and second-order differential equations by this method are listed in Figs. B–4 and B–5, respectively. In order to use these programs, the operator specifies the function $\dot{x} = f(t, x)$ or $\ddot{x} = f(t, x, \dot{x})$ (line 7), the initial values t_0, x_0, $\dot{x}_0$ (for second-order equation), the final time t_n, and the step size h. The output gives the values of t, x, and $\dot{x}$ for each time increment until t_n is reached.

```
1 PRINT"Runge-Kutta Method for 1-st order Differential Equation":PRINT
2 PRINT" To execute this program :":PRINT
3 PRINT"   1) Modify right hand side of the equation given below,"
4 PRINT"      then Press RETURN key"
5 PRINT"   2) Type  RUN 7"
6 PRINT:EDIT 7
7 DEF FNF(T,X)=5*T+X
8 CLS:PRINT" Initial Conditions":PRINT
9 INPUT"Input  t   = ",T
10 INPUT"       x   = ",X
11 INPUT"Final  t   = ",T1
12 INPUT"step size  = ",H:PRINT
13 PRINT"           t              x"
14 IF T>=T1+H THEN 23
15 PRINT USING"######.#####";T;X
16 K1=H*FNF(T,X)
17 K2=H*FNF(T+.5*H,X+.5*K1)
18 K3=H*FNF(T+.5*H,X+.5*K2)
19 K4=H*FNF(T+H,X+K3)
20 T=T+H
21 X=X+(K1+K2+K2+K3+K3+K4)/6
22 GOTO 14
23 END
```

Fig. B–4

```
1 PRINT"Runge-Kutta Method for 2-nd order Differential Equation":PRINT
2 PRINT" To execute this program :":PRINT
3 PRINT"   1) Modify right hand side of the equation given below,"
4 PRINT"      then Press RETURN key"
5 PRINT"   2) Type  RUN 7"
6 PRINT:EDIT 7
7 DEF FNF(T,X,XD)=
8 INPUT"Input  t   = ",T
9 INPUT"       x   = ",X
10 INPUT"     dx/dt = ",XD
11 INPUT"Final  t   = ",T1
12 INPUT"step size  = ",H:PRINT
13 PRINT"         t          x          dx/dt"
14 IF T>=T1+H THEN 24
15 PRINT USING"######.#####";T;X;XD
16 K1=H*FNF(T,X,XD)
17 K2=H*FNF(T+.5*H,X+.5*H*XD,XD+.5*K1)
18 K3=H*FNF(T+.5*H,X+(.5*H)*(XD+.5*K1),XD+.5*K2)
19 K4=H*FNF(T+H,X+H*XD+.5*H*K2,XD+K3)
20 T=T+H
21 X=X+H*XD+H*(K1+K2+K3)/6
22 XD=XD+(K1+K2+K2+K3+K3+K4)/6
23 GOTO 14
24 END
```

Fig. B–5

Example B–4

Solve the differential equation $\dot{x} = 5t + x$. Obtain the results for two steps using time increments of $h = 0.02$ s. At $t_0 = 0$, $x_0 = 0$.

SOLUTION

This is a first-order equation, so Eqs. B–3 and B–4 apply. Thus, for $t_0 = 0$, $x_0 = 0$, $h = 0.02$, we have

$$k_1 = 0.02[0 + 0] = 0$$

$$k_2 = 0.02[5(0.1) + 0] = 0.001$$

$$k_3 = 0.02[5(0.01) + 0.0005] = 0.00101$$

$$k_4 = 0.02[5(0.02) + 0.00101] = 0.00202$$

$$x_1 = 0 + \frac{1}{6}[0 + 2(0.001) + 2(0.00101) + 0.00202] = 0.00101$$

Using the values $t_1 = 0 + 0.02 = 0.02$ and $x_1 = 0.00101$ with $h = 0.02$, the value for x_2 is now computed from Eqs. B–3 and B–4.

$$k_1 = 0.02[5(0.02) + 0.00101] = 0.00202$$

$$k_2 = 0.02[5(0.03) + 0.00202] = 0.00304$$

$$k_3 = 0.02[5(0.03) + 0.00253] = 0.00305$$

$$k_4 = 0.02[5(0.04) + 0.00406] = 0.00408$$

$$x_2 = 0.001 + \frac{1}{6}[0.00202 + 2(0.00304) + 2(0.00305) + 0.00408]$$

$$x_2 = 0.00405 \qquad \qquad \textit{Ans.}$$

To solve this problem using the computer program in Fig. B–4, the function is first entered on line 7, then the data $t_0 = 0$, $x_0 = 0$, $t_n = 0.04$, and $h = 0.02$ is specified. The results appear as

t	x
0.00000	0.00000
0.02000	0.00101
0.04000	0.00405

Ans.

C

Vector Analysis

The following discussion provides a brief review of the vector analysis used in statics. A more detailed treatment of these topics is given in *Engineering Mechanics: Statics*.

Vector

A vector, **A**, is a quantity which has a magnitude and direction, and adds according to the parallelogram law. As shown in Fig. C–1, **A** = **B** + **C**, where **A** is the *resultant vector* and **B** and **C** are *component vectors*.

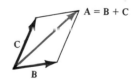

Fig. C–1

Unit Vector

A unit vector, $\mathbf{u}_A$, has a magnitude of one "dimensionless" unit and acts in the same direction as **A**. It is determined by dividing **A** by its magnitude A, i.e.,

$$\mathbf{u}_A = \frac{\mathbf{A}}{A}$$

(C–1)

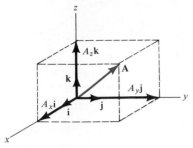

Fig. C–2

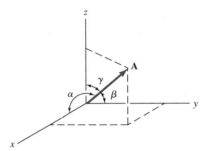

Fig. C–3

Cartesian Vector Notation

The directions of the positive x, y, z axes are defined by the Cartesian unit vectors, **i, j, k,** respectively.

As shown in Fig. C–2, vector **A** is formulated by the addition of its x, y, z components as

$$\mathbf{A} = A_x\mathbf{i} + A_y\mathbf{j} + A_z\mathbf{k} \tag{C–2}$$

The *magnitude* of **A** is determined from

$$A = \sqrt{A_x^2 + A_y^2 + A_z^2} \tag{C–3}$$

The *direction* of **A** is defined in terms of its *coordinate direction angles*, α, β, γ, measured from the *tail* of **A** to the *positive x, y, z axes*, Fig. C–3. These angles are determined from the *direction cosines* which represent the **i, j, k** components of the unit vector $\mathbf{u}_A$; i.e., from Eqs. C–1 and C–2,

$$\mathbf{u}_A = \frac{A_x}{A}\mathbf{i} + \frac{A_y}{A}\mathbf{j} + \frac{A_z}{A}\mathbf{k} \tag{C–4}$$

so that the direction cosines are

$$\cos\alpha = \frac{A_x}{A}, \quad \cos\beta = \frac{A_y}{A}, \quad \cos\gamma = \frac{A_z}{A} \tag{C–5}$$

Hence $\mathbf{u}_A = \cos\alpha\,\mathbf{i} + \cos\beta\,\mathbf{j} + \cos\gamma\,\mathbf{k}$, and using Eq. C–3, it is seen that

$$\cos^2\alpha + \cos^2\beta + \cos^2\gamma = 1 \tag{C–6}$$

The Cross Product

The cross product of two vectors **A** and **B,** which yields the resultant vector **C,** is written as

$$\mathbf{C} = \mathbf{A} \times \mathbf{B} \tag{C–7}$$

and reads **C** equals **A** "cross" **B.** The *magnitude* of **C** is

$$C = AB\sin\theta \tag{C–8}$$

where θ is the angle made between the *tails* of **A** and **B** ($0° \le \theta \le 180°$). The *direction* of **C** is determined by the right-hand rule, whereby the fingers of the right hand are curled *from* **A** *to* **B** and the thumb points in the direction of **C,** Fig. C–4. This vector is perpendicular to the plane containing vectors **A** and **B.**

The vector cross product is *not* commutative, i.e., $\mathbf{A} \times \mathbf{B} \ne \mathbf{B} \times \mathbf{A}$. Rather,

$$\mathbf{A} \times \mathbf{B} = -\mathbf{B} \times \mathbf{A}$$

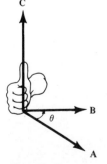

Fig. C–4

The distributive law is valid, i.e.,

$$\mathbf{A} \times (\mathbf{B} + \mathbf{D}) = \mathbf{A} \times \mathbf{B} + \mathbf{A} \times \mathbf{D} \qquad (C\text{-}10)$$

And, the cross product may be multiplied by a scalar m in any manner, i.e.,

$$m(\mathbf{A} \times \mathbf{B}) = (m\mathbf{A}) \times \mathbf{B} = \mathbf{A} \times (m\mathbf{B}) = (\mathbf{A} \times \mathbf{B})m \qquad (C\text{-}11)$$

If $\mathbf{A}$ and $\mathbf{B}$ are expressed in Cartesian component form, then the cross product, Eq. C–7, may be evaluated by expanding the determinant

$$\mathbf{C} = \mathbf{A} \times \mathbf{B} = \begin{vmatrix} \mathbf{i} & \mathbf{j} & \mathbf{k} \\ A_x & A_y & A_z \\ B_x & B_y & B_z \end{vmatrix} \qquad (C\text{-}12)$$

which yields

$$\mathbf{C} = (A_y B_z - A_z B_y)\mathbf{i} - (A_x B_z - A_z B_x)\mathbf{j} + (A_x B_y - A_y B_x)\mathbf{k}$$

Recall that the cross product is used in statics to define the moment of a force $\mathbf{F}$ about point O, in which case

$$\mathbf{M}_O = \mathbf{r} \times \mathbf{F} \qquad (C\text{-}13)$$

where $\mathbf{r}$ is a position vector directed from point O to *any point* on the line of action of $\mathbf{F}$.

The Dot Product

The dot product of two vectors $\mathbf{A}$ and $\mathbf{B}$, which yields a scalar, is defined as

$$\mathbf{A} \cdot \mathbf{B} = AB \cos \theta \qquad (C\text{-}14)$$

and reads $\mathbf{A}$ "dot" $\mathbf{B}$. The angle θ is formed between the *tails* of $\mathbf{A}$ and $\mathbf{B}$ ($0° \le \theta \le 180°$).

The dot product is commutative, i.e.,

$$\mathbf{A} \cdot \mathbf{B} = \mathbf{B} \cdot \mathbf{A} \qquad (C\text{-}15)$$

The distributive law is valid, i.e.,

$$\mathbf{A} \cdot (\mathbf{B} + \mathbf{D}) = \mathbf{A} \cdot \mathbf{B} + \mathbf{A} \cdot \mathbf{D} \qquad (C\text{-}16)$$

And, scalar multiplication can be performed in any manner, i.e.,

$$m(\mathbf{A} \cdot \mathbf{B}) = (m\mathbf{A}) \cdot \mathbf{B} = \mathbf{A} \cdot (m\mathbf{B}) = (\mathbf{A} \cdot \mathbf{B})m \qquad (C\text{-}17)$$

If $\mathbf{A}$ and $\mathbf{B}$ are expressed in Cartesian component form, then the dot product, Eq. C–14, can be determined from

$$\mathbf{A} \cdot \mathbf{B} = A_x B_x + A_y B_y + A_z B_z \qquad (C\text{-}18)$$

The dot product may be used to determine the *angle θ formed between two vectors*. From Eq. C–14,

$$\theta = \cos^{-1}\left(\frac{\mathbf{A} \cdot \mathbf{B}}{AB}\right) \qquad \text{(C–19)}$$

It is also possible to find the *component of a vector in a given direction* using the dot product. For example, the magnitude of the component (or projection) of vector **A** in the direction of **B**, Fig. C–5, is defined by $A \cos \theta$. From Eq. C–14, this magnitude is

$$A \cos \theta = \mathbf{A} \cdot \frac{\mathbf{B}}{B} = \mathbf{A} \cdot \mathbf{u}_B \qquad \text{(C–20)}$$

where $\mathbf{u}_B$ represents a unit vector acting in the direction of **B**, Fig. C–5.

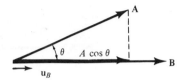

Fig. C–5

Differentiation and Integration of Vector Functions

The rules for differentiation and integration of the sums and products of scalar functions apply as well to vector functions. Consider, for example, the two vector functions $\mathbf{A}(s)$ and $\mathbf{B}(s)$. Provided these functions are smooth and continuous for all s, then

$$\frac{d}{ds}(\mathbf{A} + \mathbf{B}) = \frac{d\mathbf{A}}{ds} + \frac{d\mathbf{B}}{ds} \qquad \text{(C–21)}$$

$$\int (\mathbf{A} + \mathbf{B}) \, ds = \int \mathbf{A} \, ds + \int \mathbf{B} \, ds \qquad \text{(C–22)}$$

For the cross product,

$$\frac{d}{ds}(\mathbf{A} \times \mathbf{B}) = \left(\frac{d\mathbf{A}}{ds} \times \mathbf{B}\right) + \left(\mathbf{A} \times \frac{d\mathbf{B}}{ds}\right) \qquad \text{(C–23)}$$

Similarly, for the dot product,

$$\frac{d}{ds}(\mathbf{A} \cdot \mathbf{B}) = \frac{d\mathbf{A}}{ds} \cdot \mathbf{B} + \mathbf{A} \cdot \frac{d\mathbf{B}}{ds} \qquad \text{(C–24)}$$

Answers

12–1. 12.5 m/s, 2.55 s.

12–2. -1.96 ft/s^2, 28.6 s.

12–3. $v_{avg} = 0.222$ m/s, $v_{sp} = 2.22$ m/s.

12–5. $h_B = 26.1$ m, $h_{max} = 28.0$ m.

12–6. $d = 13.0$ ft, $v_{avg} = -1.00$ ft/s, $v_{sp} = 2.60$ ft/s.

12–7. 54.0 m.

12–9. 5.12 m.

12–10. 0.371 ft/s^2.

12–11. 517 ft, 616 ft.

12–13. $s = 12.5$ m at $t = 2$ s.

12–14. $v = 322$ m/s, $t = 19.3$ s.

12–15. $s = -18.0$ ft, $s_{Tot} = 46.0$ ft.

12–17. $s = -27.0$ ft, $s_{Tot} = 69.0$ ft.

12–18. 11.9 m, 0.250 m/s.

12–19. 16.9 ft.

12–21. 10.3 s, 4.11 km.

12–22. (a) 6.25 m/s. (b) 7.23 m/s. (c) 1.42 m.

12–23. 6.53 m, 3.27 s.

12–25. 31.6 m/s, 16.5 m.

12–26. $\Delta s = 152$ ft, $s_{A\ Tot} = 41.0$ ft, $s_{B\ Tot} = 200$ ft.

12–27. 11.2 km/s.

12–29. $0 \le t < 30$, $a = 0.4$, $s = 0.2t^2$,
$30 < t \le 50$, $a = 0$, $s = 12t - 180$.

12–30. $0 \le t \le 5$, $v = 3t^2$, $5 \le t \le 10$,
$v = t^2 + 20t - 50$, $s = 917$ m.

12–31. $0 \le t < 4$, $a = 1.25$, $s = 0.625t^2$,
$4 < t < 10$, $a = 0$, $s = 5t - 10$,
$10 < t \le 15$, $a = -1$, $s = 15t - 0.5t^2 - 60$, $s = 52.5$ m.

12–33. $0 \le t < 30$, $v = 0.8t$, $a = 0.8$, $30 < t \le 40$, $v = 24$, $a = 0$.

12–34. $v_{max} = 11.25$ ft/s, $0 \le t \le 15$, $s = 0.375t^2$,
$15 \le t \le 51.9$, $s = 11.25t - 84.375$.

12–35. $0 \le t \le 30$, $v = 0.00333t^3$, $s = 0.000833t^4$, $30 \le t \le 60$,
$v = 15t - 360$, $s = 7.5t^2 - 360t + 4725$.

12–37. $0 \le t \le 10$, $s = 0.133t^3$, $10 \le t \le 30$, $s = 0.5t^2 + 30t - 217$,
$v_{avg} = 37.8$ ft/s, $s_{Tot} = 1133$ ft.

12–38. $0 \le t < 4$, $a = 3.5$, $4 < t < 5$, $a = 0$, $5 < t \le 8$, $a = 4$,
$a_{max} = 4$ m/s^2.

12–39. $0 \le t \le 4$, $s = 1.75t^2$, $4 \le t \le 5$, $s = 14t - 28$, $5 \le t \le 8$,
$s = 2t^2 - 6t + 22$, $s_{Tot} = 102$ m.

12–41. $v_{40} = 12.7$ ft/s, $v_{90} = 22.8$ ft/s, $v_{200} = 36.1$ ft/s.

12–42. $v_{max} = 55.0$ ft/s, $t' = 14.6$ s.

12–43. $0 \le s \le 100$, $a = 0.16\ s$, $100 < s \le 200$, $a = 0.01s + 3$,
$a_{100} = 16.0$ m/s^2, $a_{150} = 4.50$ m/s^2.

12–45. $0 \le s \le 300$, $v = 4.90\ s^{1/2}$, $300 \le s \le 450$,
$v = (-0.04s^2 + 48s - 3600)^{1/2}$, $t = 5.77$ s.

12–46. $0 \le t < 13$, $v = 0.75\ t$, $13 \le t \le 160$, $v = 9.77$, $t' = 13.0$ s.

12–47. $s_{Tot} = 9$ km, $\Delta s = 3.61$ km, $\measuredangle\ 56.3°$, $v_{avg} = 2.61$ m/s,
$v_{sp} = 6.52$ m/s.

12–49. $a = 13.5$ m/s^2, $\alpha = 63.7°$, $\beta = 82.4°$, $\gamma = 27.6°$.

12–50. (18.0 ft, 6.20 ft, 4.00 ft).

12–51. 3.05 m/s, 7.48 m.

12–53. 9.68 m/s, 16.8 m/s^2.

12–54. (a) $s_A = 1.40$ m, $s_B = 3$ m, (b) $\mathbf{r}_A = \{1.38\mathbf{i} + 0.195\mathbf{j}\}$ m,
$\mathbf{r}_B = \{-2.82\mathbf{i} + 0.873\mathbf{j}\}$ m, (c) $\Delta r = 4.26$ m.

12–55. $v = 168$ mm/s, $\alpha = 91.0°$, $\beta = 87.7°$, $\gamma = 2.49°$,
$a = 198$ mm/s^2, $\alpha = 114°$, $\beta = 110°$, $\gamma = 32.0°$.

12–57. (9.0 m, 0), $v_A = 27.4$ m/s, $v_B = 10.1$ m/s, $t = 2.0$ s.

12–58. (4 ft, 3 ft), $v_A = 4.47$ ft/s, $v_B = 8.06$ ft/s.

12–59. $y = \left(\dfrac{x}{4}\right)^{1/4}$ $[1 + 8x^{1/2}]$, $v = 162$ in./s, $a = 215$ in./s.

12–61. 2.69 ft/s, 0.02 ft/s².

12–62. 4.22 m/s, 0.130 m/s².

12–63. $v_x = 0.398$ ft/s, $v_y = 3.98$ ft/s.

12–65. 8.34 m, 0.541 m/s².

12–66. 19.8 m/s, ∡ 36.2°.

12–67. 19.4 m/s, 4.54 s.

12–69. 18.2 m/s, 12.7 m.

12–70. 0.400 s, 2.42 ft.

12–71. 166 ft.

12–73. 36.7 ft/s, 11.5 ft.

12–74. 0.242 s, 0.714 m.

12–75. $x = 13.3$ ft, $y = -7.10$ ft.

12–77. $\theta_C = 75.3°$, $\theta_D = 14.7°$, $\Delta t = 1.45$ s.

12–78. 45°, 10.2 m.

12–79. 8.83 m.

12–81. 20.8 ft/s.

12–82. $v_B = 21.6$ ft/s < 23 ft/s, (he can catch the ball).

12–83. $\Delta t = \dfrac{2v_0}{g}\left[\dfrac{\sin(\theta_1 - \theta_2)}{\cos\theta_2 + \cos\theta_1}\right]$.

12–85. 35.0 ft/s², 2.64 (10)³ ft.

12–86. $v_x = \dfrac{v_0}{\sqrt{1 + b^2}}$, $v_y = \dfrac{bv_0}{\sqrt{1 + b^2}}$, $a_n = \dfrac{2cv_0^2}{[1 + b^2]^{3/2}}$.

12–87. (0, −9 ft), 450 ft/s².

12–89. 256 ft/s, 5.25 (10³) ft.

12–90. 12.6 ft/s².

12–91. 21.6 ft/s², 40.5 ft.

12–93. 0.921 m/s².

12–94. 97.2 ft/s, 42.6 ft/s².

12–95. 115 ft/s, 58.4 ft/s².

12–97. 1.96 m/s, 0.930 m/s².

12–98. $a_n = 8.00$ ft/s², $a_t = 24$ ft/s².

12–99. 1.68 s.

12–101. 2.32 m/s².

12–102. $v_r = 0$, $v_\theta = 100$ ft/s, $a_r = -25.0$ ft/s², $a_\theta = -3.2$ ft/s².

12–103. $\dot{\theta} = 0.075$ rad/s, $a = 2.25$ ft/s².

12–105. 24.1 ft/s, 8.17 ft/s².

12–106. 3.03 m/s, 12.6 m/s².

12–107. $v_r = \dfrac{-20}{\sqrt{1 + \theta^2}}$, $v_\theta = \dfrac{20\,\theta}{\sqrt{1 + \theta^2}}$, $v_r = -14.1$ ft/s, $v_\theta = 14.1$ ft/s.

12–109. $v_r = -0.230$ ft/s, $v_\theta = 1.35$ ft/s, $a_r = -1.47$ ft/s², $a_\theta = -0.706$ ft/s².

12–110. $v_r = 0.425$ ft/s, $v_\theta = 0.425$ ft/s, $a_r = 0.425$ ft/s², $a_\theta = 0.595$ ft/s².

12–111. $v_r = 0$, $v_\theta = 4.8$ ft/s, $v_z = -0.664$ ft/s, $a_r = -2.88$ ft/s², $a_\theta = 0$, $a_z = -0.365$ ft/s².

12–113. $v_r = 32$ ft/s, $v_\theta = 50.3$ ft/s, $a_r = -161$ ft/s², $a_\theta = 319$ ft/s².

12–114. $v_r = -93.6$ ft/s, $v_\theta = 27.0$ ft/s, $a_r = -68.7$ ft/s², $a_\theta = -41.3$ ft/s².

12–115. $v_r = -14.1$ m/s, $v_\theta = 14.1$ m/s, $a_r = -1.41$ m/s², $a_\theta = -1.41$ m/s².

12–117. $v_r = a\dot{\theta}$, $v_\theta = a\theta\dot{\theta}$, $a_r = -a\theta\dot{\theta}^2$, $a_\theta = 2a\dot{\theta}^2$.

12–118. $v_r = ake^{k\theta}\dot{\theta}$, $v_\theta = ae^{k\theta}\dot{\theta}$, $a_r = ae^{k\theta}(\dot{\theta})^2[k^2 - 1]$, $a_\theta = 2ake^{k\theta}\dot{\theta}^2$.

12–119. $v_r = 6$ ft/s, $v_\theta = 18.3$ ft/s, $a_r = -67.1$ ft/s², $a_\theta = 66.3$ ft/s².

12–121. 640 m/s, 657 m/s².

12–122. 155 m/s, 67.1 m/s².

12–123. $v_r = 0.9$ ft/s, $v_\theta = 2.46$ ft/s, $a_r = -12.1$ ft/s², $a_\theta = 5.40$ ft/s².

12–125. $v_r = -16.9$ m/s, $v_\theta = 4.87$ m/s, $a_r = -89.4$ m/s², $a_\theta = -53.7$ m/s².

12–126. $\dot{\mathbf{a}} = (\dddot{r} - 3\dot{r}\dot{\theta}^2 - 3r\dot{\theta}\ddot{\theta})\mathbf{u}_r + (3\dot{r}\ddot{\theta} + r\dddot{\theta} + 3\ddot{r}\dot{\theta} - r\dot{\theta}^3)\mathbf{u}_\theta + (\dddot{z})\mathbf{u}_z$.

12–127. $v_r = 0.242$ m/s, $v_\theta = 0.943$ m/s, $a_r = -2.33$ m/s², $a_\theta = 1.74$ m/s².

12–129. $\dot{\mathbf{u}}_b = -0.349\mathbf{i} - 0.0127\mathbf{j} - 0.937\mathbf{k}$

12–130. 1.33 ft ↑ .

12–131. 12.0 ft/s → .

12–133. 8.0 ft/s ↑ .

12–134. 58 ft/s ↑ .

12–135. 0.667 m/s ↑ .

12–137. 1.82 s, 5.29 m/s ↑ .

12–138. 1.5 ft/s ↑ .

12–139. 2 ft/s ↑ .

12–141. $v_B = (-3(y^2 + 16)^{1/2}/y)$ m/s

12–142. 5 m/s, 7 m/s².

12–143. 2.4 ft/s →, 3.85 ft/s² → .

12–145. 0.513 ft/s².

12–146. 525 mi/hr ↘ 19.1°.

12–147. 23.4 mi.

12–149. 21.7 ft/s 18.0° ↘, 36.9 s.

12–150. 70.9 ft/s 7.11° ↗, 22.6 s.

12–151. $v_{A/B} = 6.69$ m/s ↗ 53.3°, $a_{A/B} = 1.52$ m/s² 41.9° ↗ .

12–153. 15.1°.

12–154. 98.4 ft/s 67.6° ↗, 19.8 ft/s² ∡ 57.4°.

12–155. 5.75 m/s, 17.8 m/s, ↘ 76.2°, 9.81 m/s² ↓ .

CHAPTER 13

13–1. 296 N, 8.47 m/s².

13–2. 199 (10)¹⁸ N.

13–3. 35.3 (10²¹) N.

13–5. 0.343 m/s², 0.436 m/s².

13–6. $T_B = T_A = 30$ lb, $a_A = 6.44$ ft/s², $a_B = 4.83$ ft/s².

13–7. 7.23 m/s.

13–9. $T = \dfrac{mg}{2}\sin 2\theta$.

13–10. $T = mg\cos\theta\,[\sin\theta - \mu\cos\theta]$.

13–11. 64.9 ft/s.

13–13. 286 lb.

13–14. 206 lb.

13–15. $0 \leq t < 10$, $F = 652$ lb, $10 < t \leq 30$, $F = 109$ lb.

13–17. 64.4 ft.

13–18. 13.1 lb.

13–19. 3.62 m/s.

13–21. 0.297 m/s².

13–22. 0.0595 m/s².

13–23. 5.30 ft., 1.82 s.

13–25. $a_A = 9.66$ ft/s^2, $a_B = 15.0$ ft/s^2.

13–26. 8.91 m/s.

13–27. 247 mm.

13–29. 1.75 m/s^2.

13–30. $v = \left[\dfrac{mg}{k} - \left(\dfrac{mg}{k} - v_0^2\right)e^{-\frac{2kh}{m}}\right]^{1/2}$, $v = \sqrt{\dfrac{mg}{k}}$.

13–31. $v = \sqrt{\dfrac{mg}{k}}\left(\dfrac{e^{2t\sqrt{\frac{kg}{m}}} - 1}{e^{2t\sqrt{\frac{kg}{m}}} + 1}\right)$, $v_t = \sqrt{\dfrac{mg}{k}}$.

13–33. 39.5 mm.

13–34. (a) $a = \dfrac{P}{m} - 3\mu g$, (b) $a = \dfrac{P}{2m} - 2\mu g$.

13–35. $a_B = 0$, $a_C = 4.11$ m/s^2, $a_D = 0.162$ m/s^2.

13–37. $P = 2mg\left(\dfrac{\mu \cos \theta + \sin \theta}{\cos \theta - \mu \sin \theta}\right)$.

13–38. 78.7°.

13–39. 1.06 s.

13–41. $F = m(a_B + g)\sqrt{4y^2 + d^2}/(4y)$.

13–42. 188 mm.

13–43. $s = \dfrac{evLl}{v_0^2\, wm}$.

13–45. $v_{B/AC} = a_0 (\sin \theta)t$, $s = \frac{1}{2}a_0 (\sin \theta)\, t^2$.

13–46. 1.99 (10^{30}) kg.

13–47. 80.2 ft/s.

13–49. $N = 277$ lb, $F = 13.4$ lb.

13–50. 15.6 ft/s.

13–53. 962 N.

13–54. $T = 3mg \sin \theta$.

13–55. $a_t = 16.1$ ft^2/s, $T = 46.9$ lb.

13–57. 2.10 m/s.

13–58. 2.09 m/s.

13–59. 4.90 lb.

13–61. $T = 361$ N, $a_t = 3.36$ m/s^2.

13–62. $F = 1.63$ kN, $N_s = 720$ N.

13–63. $F = 1.23$ kN, $N_s = 921$ N.

13–65. 7.31 lb, 6.00 ft/s^2.

13–66. 1.95 m/s^2, 647 N.

13–67. 0.969 m/s.

13–69. 29.4 ft/s^2, 69.0 lb.

13–70. $v = \sqrt{9.81 \cos \theta - 5.81}$, $N = 58.9 \cos \theta - 23.2$, $\theta = 0°$.

13–71. $F_r = -92.5$ N, $F_\theta = 317$ N, $F_z = 35.6$ N.

13–73. $F_r = -1.54$ kN, $F_\theta = 0$, $F_z = 2.45$ kN.

13–74. $F_r = -100$ N, $F_\theta = 0$, $F_z = 2.45$ kN.

13–75. $p = 12.6$ lb, $N = 4.22$ lb.

13–77. 0.300 lb.

13–78. $\theta = \tan^{-1}\left(\dfrac{4r_c}{g}\dot{\theta}_0^2\right)$.

13–79. 12.0 m/s^2 5.71° ⟍.

13–81. $N_C = 3.03$ N, $F = 1.81$ N.

13–82. $N_C = 25.6$ N, $F_{OA} = 0$.

13–83. $N_C = 20.7$ N, $F_{OA} = 0$.

13–85. -0.0108 lb.

13–86. 0.163 lb.

13–87. $F_r = -1.35$ kN, $F_\theta = -0.300$ kN, $F_z = 2.94$ kN.

13–89. $F = 5.07$ kN, $N = 2.74$ kN.

13–90. 113 lb.

13–91. 26.7 lb.

13–93. $r = 42.2$ Mm, $v = 3.07$ km/s.

13–94. (a) $r = 194$ (10^3) mi, (b) 392 (10^3) mi,
(c) 194 (10^3) mi $< r < 392$ (10^3) mi, (d) $r > 392$ (10^3) mi.

13–95. $\Delta v = \sqrt{\dfrac{GM_e}{r_0}}(\sqrt{2} - \sqrt{e + 1})$.

13–97. $\Delta v = 2.57$ km/s.

13–98. 9.60 km/s, 6.50 Mm.

13–99. $v_p = 7.82$ km/s, $T = 1.97$ h.

13–101. $v_B = 1.20$ km/s, $\Delta v = 2.17$ km/s.

13–102. $v_p = 55.3$ (10^3) ft/s, $r = 484$ (10^6) mi.

13–105. 11.5 Mm/h, 27.3 Mm.

13–106. $\Delta v_a = 466$ m/s (decrease), $\Delta v_p = 2.27$ km/s (decrease).

CHAPTER 14

14–1. $U_w = 4.12$ kJ, $U_{NF} = -2.44$ kJ.

14–2. 3.80 kN.

14–3. 2.72 in.

14–5. 16.0 ft/s, 7.98 ft.

14–6. 1.05 ft.

14–7. 4.52 m/s.

14–9. 0.365 ft/s.

14–10. 8.79 ft/s.

14–11. 1.13 kN.

14–13. 642 lb/ft, 18.0 ft/s.

14–14. $v_B = 7.22$ ft/s, $N_B = 27.1$ lb, $v_C = 17.0$ ft/s, $N_C = 133$ lb, $v_D = 18.2$ ft/s

14–15. 179 mm.

14–17. $U = [\frac{1}{2}k(y - d) + mg](y - d)$.

14–18. 0.313 ft.

14–19. 30.5 m/s, 1.74 kN.

14–21. 6.89 m/s.

14–22. 3.41 m.

14–23. For A, 6.08 m/s, for B, 4.08 m/s.

14–25. $s_A = 0.255$ m, $s_B = 0.205$ m.

14–26. 2.36 m/s.

14–27. 2.34 m/s.

14–29. 0.0735 ft.

14–30. 0.591 hp.

14–31. 5.20 days.

14–33. 47.2°.

14–34. 95.4 hp.

14–35. 4.12 MW, 10.3 m/s.

14–37. 35.4 kW.

14–38. 0.229 hp.

14–39. $s = \dfrac{mv^3}{3P}$.

14–41. 1.13 kW.

14–42. $P = (6.03t)$ MW.

14–43. $P = \left(\dfrac{6.72(e^{1.79t} - 1)}{(e^{1.79t} + 1)}\right)$ MW.

14–45. 0.211 hp.

14–46. 3.82 m/s.

14–47. 8.79 ft/s.

14–49. $v_B = 7.22$ ft/s, $N_B = 27.1$ lb, $v_C = 17.0$ ft/s, $N_C = 133$ lb, $v_D = 18.2$ ft/s.

14–50. 179 mm.

14–51. $s_A = 0.255$ m, $s_B = 0.205$ m.

14–53. 13.9 ft/s.

14–54. $v_B = 12.7$ ft/s, $v_C = 15.0$ ft/s, $T = 13.5$ lb.

14–55. 2.86 m/s.

14–57. 39.7°.

14–58. 6.97 m/s.

14–61. $v_A = 1.54$ m/s ↑, $v_B = 4.62$ m/s ↓.

14–62. 0.195 ft.

14–63. 20.4 ft/s.

14–65. 27.2 ft/s.

14–66. 1.29 ft.

CHAPTER 15

15–1. 1.53 s, 11.5 m.

15–2. 0.280 s, 15.6 ft/s.

15–3. 467 N, 1.25 kg · m/s ∡ 14.2°.

15–5. (a) 15 kN · s, (b) 15 kN · s.

15–6. 2.61 N · s ∡ 30°.

15–7. $T = 14.9$ kN, $F = 24.8$ kN.

15–9. For A, 7.40 m/s, for B, 5.40 m/s.

15–10. 20.4 m/s.

15–11. 3.64 s.

15–13. 10.5 ft/s →.

15–14. 6.00 ft/s →.

15–15. 21.8 m/s.

15–17. 40.2 ft/s.

15–18. 10.1 ft/s.

15–19. 6.90 m/s.

15–21. 0.5 m/s →, 16.9 kJ.

15–22. 3.29 ft/s.

15–23. 120 m/s.

15–25. $v_A = 3.29$ m/s →, $v_B = 2.19$ m/s ←.

15–26. $v_B = 0.379$ m/s →.

15–27. $v_{B/C} = 0.632$ m/s →.

15–29. $v_A = 1.58$ ft/s →, $v_B = 0.904$ ft/s ←.

15–30. (a) 2.26 ft/s, (b) 1.85 ft/s.

15–31. 0.364 ft ←.

15–33. 3.18 m.

15–34. 1.65 m.

15–35. 145 ft.

15–37. $t = v_0/\mu g\left(1 + \dfrac{m}{M}\right)$, $v = \dfrac{Mv_0}{m + M}$.

15–38. $x_{C/b} = \left| v_0^2/2\mu g\left(1 + \dfrac{m}{M}\right)\right|$.

15–39. 71.4 mm →.

15–41. $(v_B)_2 = 1.73$ m/s ←, $(v_A)_2 = 4.53$ m/s ←.

15–42. 3.24 m/s, 43.9°.

15–45. $(v_A)_2 = 0.720$ m/s, $(v_B)_2 = 1.92$ m/s →, $\Delta d = 0.538$ m.

15–46. 8.42 ft/s.

15–47. 1.31 ft.

15–49. $(v_A)_1 = 19.7$ ft/s, $(v_A)_2 = 11.1$ ft/s, $(v_B)_2 = 17.0$ ft/s, $d = 11.3$ ft.

15–50. $(v_B)_2 = \frac{1}{3}(\sqrt{2gh})(1 + e)$.

15–51. 6.59 ft/s.

15–53. 1.53 m.

15–54. 1.68 kN.

15–55. $t_{\text{Tot}} = \sqrt{\dfrac{2h}{g}}\left(\dfrac{1 + e}{1 - e}\right)$.

15–57. $v_A = 1.35$ m/s →, $v_B = 5.89$ m/s 32.9° ↘.

15–58. 90°.

15–59. $v_A = 2.46$ ft/s, $v_B = 3.09$ ft/s.

15–61. $(v_B)_2 = \sqrt{3}v(1 + e)/4$ ↗ 30, $\Delta T = \frac{3}{16}mv^2(e^2 - 1)$.

15–62. $v_B = \frac{1}{4}(e + 1)v$.

15–63. $(H_A)_O = 72$ kg · m²/s ↓, $(H_B)_O = 59.5$ kg · m²/s ↓.

15–65. $\mathbf{H}_A = \{-1.83\mathbf{i} - 0.913\mathbf{k}\}$ slug · ft²/s.

15–66. $\mathbf{H}_B = \{-5.17\mathbf{i} + 0.304\mathbf{j} - 2.74\mathbf{k}\}$ slug · ft²/s.

15–67. $\mathbf{H}_B = \{108\mathbf{i} + 96\mathbf{j} + 40\mathbf{k}\}$ kg · m²/s.

15–69. 3.47 m/s.

15–70. $v_2 = 2.53$ m/s, $\theta = 27.0°$.

15–71. $v_2 = 9.22$ ft/s 1, $\Sigma U_{1-2} = 3.04$ ft · lb.

15–73. $v_B = 10.2$ km/s, $r_B = 13.8$ Mm.

15–74. $v = 20.2$ ft/s, $h = 6.36$ ft.

15–75. 402 N.

15–77. 242 lb.

15–78. 303 lb.

15–79. $F_x = 19.5$ lb, $F_y = 1.96$ lb.

15–81. 17.5 hp.

15–82. 844 N.

15–83. 3.36(10³) lb.

15–85. 1.42 lb.

15–86. $a_{\text{init}} = 133$ ft/s², $a_{\text{finial}} = 200$ ft/s².

15–87. 2.07 (10³) ft/s.

15–89. 11.5 kN.

15–90. 37.5 ft/s².

15–91. 12.2 m/s.

15–93. $R = (20t + 2.48)$ lb.

15–94. $F = m'v^2$.

15–95. $F = (6t + 0.0559)$ lb.

REVIEW PROBLEMS 1

R1–1. 1.84 m.

R1–2. 946 kW.

R1–3. 23.8 s.

R1–5. 980 m.

R1–6. 744 lb, 14.4 ft · kip.

R1–7. 21.5 ft/s.

R1–9. $a_A = 0.755$ m/s², $a_B = 1.51$ m/s², $T_A = 90.6$ N, $T_B = 45.3$ N.

R1–10. 77.5 ft/s.

R1–11. 0.575 ft/s 71.6° ↘, 0.192 ft · lb.

R1–13. 24.8 m/s.

R1–14. $(v_A)_2 = 1.30$ m/s ∡ 8.47°, $(v_B)_2 = 1.34$ m/s ←.

R1–15. 0.933 m.
R1–17. $F_s = 68$ N, $N_s = 153$ N.
R1–18. 27.3 s.
R1–19. 38.5 N.
R1–21. 20 ft.
R1–22. 51.0 lb.
R1–23. 1.80 hp.
R1–25. $v_{B/A} = 675$ mph, 40.1° ⊾.
R1–26. 38.5 ft/s.
R1–27. $F_r = -39.2$ N, $F_\theta = 0$, $F_z = 392$ N.
R1–29. 0.221 m.
R1–30. 13.1 ft/s.
R1–31. $v_r = 6$ ft/s, $v_\theta = 18.3$ ft/s, $a_r = -67.1$ ft/s², $a_\theta = 66.3$ ft/s².
R1–33. 14.6 m/s.
R1–34. 0.9 m/s, 22.7 kJ.
R1–35. 2.09 s, 50.9 ft/s.
R1–37. 1.90 ft, 16.9 ft/s.
R1–38. 3 11 s.
R1–39. 1.02 s.
R1–41. 24.5 m, $N_B = 0$, $N_C = 9.81$ kN.
R1–42. 1.81 ft/s².
R1–43. $v_B = 55.2$ ft/s, $v_A = 27.6$ ft/s.
R1–45. 47.8 ft/s.
R1–46. 0.5 m/s ↑.
R1–47. 29.0 m/s².
R1–49. 2.22 ft.
R1–50. 13.5 m/s, 147 kN.

CHAPTER 16

16–1. 2 ft/s, 8.02 ft/s².
16–2. $a_t = 0.286$ ft/s², $a_n = 331$ ft/s², $s = 525$ ft.
16–3. 0.838 m/s, 0.890 m/s².
16–5. $v = 52$ ft/s, $a_t = 12$ ft/s², $a_n = 1352$ ft/s².
16–6. $v_B = 22$ ft/s, $a_t = 9$ ft/s², $a_n = 322$ ft/s².
16–7. 211 rad/s.
16–9. $\omega_C = 4.50$ rad/s, $\omega_D = 3.38$ rad/s.
16–10. 10.6 rad/s.
16–11. 1.80 m/s ↓, 0.450 m/s² ↓.
16–13. 10.8 m/s ↑, 16.2 m.
16–14. 465 rad/s.
16–15. 2.89 m.
16–17. 3.52 ft/s, 12.8 ft/s², 1.15 s.
16–18. 48.7 ft/s, 8.54 rev.
16–19. $a = \left(\dfrac{\omega^2}{2\pi}\right)\left(\dfrac{r_2 - r_1}{L}\right)d.$
16–21. 650 mm/s, -1615 mm/s².
16–22. -19.2 rad/s, -183 rad/s².
16–23. 4.17 rad/s.
16–25. $v = 0.480 \cos \theta$ m/s, $a = -1.92 \sin \theta$ m/s².
16–26. 0.416 m/s, -0.752 m/s².
16–27. $\dot{\theta} = -v_B h/(h^2 + x^2)$, $\alpha = 2v_B^2 hx/(h^2 + x^2)^2$.
16–29. $\omega = -rv_A/y\sqrt{y^2 - r^2}$, $\alpha = rv_A^2(2y^2 - r^2)/y^2(y^2 - r^2)^{3/2}$.
16–30. $\dot{\theta} = \dfrac{v_0}{a} \sin^2 \theta$, $\ddot{\theta} = \left(\dfrac{v_0}{a}\right)^2 \sin 2\theta \sin^2 \theta$.

16–31. 1.08 rad/s, 4.39 ft/s.
16–33. $\omega = -\dfrac{v_A \tan \theta}{x}$, $\alpha = \dfrac{v_A^2 \tan \theta}{x^2}(2 + \tan^2 \theta)$.
16–34. $\omega = -\dfrac{v_A \tan \theta}{x}$, $\alpha = \dfrac{\tan \theta}{x^2}[v_A^2(2 + \tan^2 \theta) - a_A x]$.
16–35. 33.3 rad/s, 1 rev.
16–37. $v = lr\omega[r + d \cos \theta/(d + r \cos \theta)^2]$,
$a = lr\omega^2 \sin \theta \, [2r^2 - d^2 + dr \cos \theta/(d + r \cos \theta)^3]$.
16–38. 50 m/s ↓.
16–39. 9.20 m/s →.
16–41. 9.33 in/s ⬉ 16.5°.
16–42. 2.40 m/s →.
16–43. 2.45 rad/s ↰, 2.20 ft/s ←.
16–45. 4 rad/s ↰, 100 mm/s ↓.
16–46. $\omega_A = 4$ rad/s ↓, $\omega_B = 1$ rad/s ↰, $v_D = 60$ mm/s ↑.
16–47. $\omega_{CD} = 10.0$ rad/s ↰, $\omega_{BC} = 1.33$ rad/s ↓.
16–49. 4.00 m/s ⬉ 52.7°.
16–50. 10.3 rad/s ↓.
16–51. $v_C = 2.89$ ft/s ←, $v_E = 9.33$ ft/s ↓.
16–53. 50 m/s ↓.
16–54. 2.40 m/s →.
16–55. 2.45 rad/s ↰, 2.20 ft/s ←.
16–57. $\omega_{CD} = 10.0$ rad/s ↰, $\omega_{BC} = 1.33$ rad/s ↓.
16–59. $v_C = 6$ ft/s →, $v_B = 4$ ft/s →.
16–61. $v_A = 60$ ft/s →, $v_C = 220$ ft/s ←, $v_B = 161$ ft/s ⬊ 60.3°.
16–62. 77.3 in./s ↑
16–63. 1.5 v/r.
16–65. $v_C = 2.50$ ft/s ←, $v_D = 9.43$ ft/s ⬂ 55.8°.
16–66. 6 rad/s ↰.
16–67. 8.50 rad/s ↰.
16–69. $\omega_s = 15$ rad/s ↰, $\omega_R = 3$ rad/s ↰.
16–70. 1.04 m/s →.
16–71. 4.42 in./s 30° ⊾.
16–73. 0.332 rad/s² ↓, 7.88 ft/s² ←.
16–74. 2.25 m/s², ⬉ 32.6°.
16–75. 10.0 m/s², 2.02° ⬈.
16–77. 5.94 m/s², 53.9° ⬈.
16–78. 6.21 m/s², 45.6° ⬊.
16–79. $\alpha_{AB} = 4.81$ rad/s² ↓, $\alpha_{BC} = 5.21$ rad/s² ↰.
16–81. 2.41 m/s² ↓.
16–82. $\omega_{BC} = 5.58$ rad/s ↓, $\omega_{CD} = 5.44$ rad/s ↰, $\alpha_{BC} = 54.4$ rad/s², $\alpha_{CD} = 103$ rad/s².
16–83. 10.7 rad/s² ↰, 14.1 in./s² ⬋ 18.4°.
16–85. 4.73 rad/s ↰, 131 rad/s² ↓.
16–86. $\mathbf{v}_B = \{5\mathbf{i} + 6\mathbf{j}\}$ ft/s, $\mathbf{a}_B = \{-15\mathbf{i} + 40\mathbf{j}\}$ ft/s².
16–87. $\mathbf{v}_A = \{-2.5\mathbf{i} + 2\mathbf{j}\}$ ft/s, $\mathbf{a}_A = \{-3\mathbf{i} + 1.75\mathbf{j}\}$ ft/s².
16–89. $\mathbf{v}_B = \{-12\mathbf{i}\}$ ft/s, $\mathbf{a}_B = \{-20\mathbf{i} - 70\mathbf{j}\}$ ft/s².
16–90. $\mathbf{v}_B = \{-12\mathbf{i} + 5.77\mathbf{j}\}$ ft/s, $\mathbf{a}_B = \{-28.9\mathbf{i} - 70\mathbf{j}\}$ ft/s².
16–91. 2.40 m/s.
16–93. $\mathbf{v}_C = \{-7.00\mathbf{i} + 17.3\mathbf{j}\}$ ft/s, $\mathbf{a}_C = \{-34.6\mathbf{i} - 15.5\mathbf{j}\}$ ft/s².
16–94. $\mathbf{v}_C = \{-7.00\mathbf{i} + 17.3\mathbf{j}\}$ ft/s, $\mathbf{a}_C = \{-38.8\mathbf{i} - 6.84\mathbf{j}\}$ ft/s².
16–95. $\mathbf{v}_C = \{-0.944\mathbf{i} + 2.02\mathbf{j}\}$ m/s, $\mathbf{a}_C = \{-11.2\mathbf{i} - 4.15\mathbf{j}\}$ m/s².
16–97. 10.0 rad/s ↓, 24.0 rad/s² ↓.
16–98. 9.00 rad/s ↓, 249 rad/s² ↓.
16–99. $\omega_{AC} = 0$, $\alpha_{AC} = 14.4$ rad/s² ↓.

Answers

CHAPTER 17

17–1. $I_y = \frac{1}{3}ml^2$.

17–2. $I_z = mR^2$.

17–3. $I_x = \frac{3}{10}mr^2$.

17–5. $I_x = \frac{2}{5}mb^2$.

17–6. $I_x = \frac{2}{5}mr^2$.

17–7. 1.20 in.

17–9. 2.25 slug · ft².

17–10. 2.17 slug · ft².

17–11. 7.67 kg · m².

17–13. 1.53 kg · m².

17–14. 118 slug · ft².

17–15. 293 slug · ft².

17–17. 0.450 (10^{-3}) kg · m², $\bar{y} = 57.7$ mm, 0.150 (10^{-3}) kg · m².

17–18. 0.888 m, 5.61 kg · m².

17–19. $a_G = 13.3$ ft/s², $N_A = 44.0$ lb, $N_B = 61.0$ lb.

17–21. 2.00 kN.

17–22. $a_{max} = 77.3$ ft/s² →, $B_x = 432$ lb, $B_y = 180$ lb.

17–23. $N_A = 120$ lb, $B_x = 73.9$ lb, $B_y = 69.7$ lb.

17–25. $P = 24$ lb, $a_G = 3.22$ ft/s², $N_A = 65.3$ lb, $N_B = 14.8$ lb.

17–26. 53.3 lb.

17–27. 16.1 ft/s².

17–29. 11.2 s.

17–30. 18.6°.

17–31. (a) $a_B = 2.26$ m/s², $N_B = 226$ N, $N_A = 559$ N,
(b) $N_B = 454$ N.

17–33. (a) 82.8 ft, (b) 82.8 ft.

17–34. 5.59 ft/s².

17–35. 20.0 rad/s².

17–37. $a_G|_{h=0.2 \text{ m}} = 1.41$ m/s², $a_G|_{h=0} = 1.38$ m/s².

17–38. impossible, $a_G = 5.89$ m/s² < 39.2 m/s².

17–39. $F_B = 4.50$ kN, $N_A = 1.78$ kN, $N_B = 5.58$ kN.

17–41. 9.67 rad/s².

17–42. $N_B = 4.70$ kN, $N_C = 2.31$ kN, $N_D = 3.78$ kN.

17–43. $A_x = 14.9$ lb, $A_y = 2.50$ lb, $\alpha = 8.05$ rad/s² ↓.

17–45. 20.8 rad/s.

17–46. 2.48 rad/s ↓.

17–47. $A_x = 0$, $A_y = 262$ N.

17–49. $a = (m_B - m_A)g/(\frac{1}{2}M + m_B + m_A)$.

17–50. $N_A = N_B = 346$ N, $\alpha = 2.52$ rad/s².

17–51. 1.06 s.

17–53. 7.28 rad/s².

17–54. $O_x = 54.8$ N, $O_y = 41.2$ N.

17–57. 7.69 ft/s.

17–58. 10.9 rad/s.

17–59. 0.548 m/s.

17–61. $A_x = 0$, $A_y = 289$ N, $\alpha = 23.1$ rad/s².

17–62. $T_A = 13.3$ lb, $T_B = 16.7$ lb.

17–63. $T_A = 18.3$ lb, $T_B = 20.7$ lb.

17–65. $\omega = \sqrt{(3g/L)}\,[1 - \cos\theta]$, $O_x = mg\sin\theta\,[2.25\cos\theta - 1.5]$,
$O_y = mg(1 - 1.5\cos\theta + 1.5\cos^2\theta - 0.75\sin^2\theta)$.

17–66. $F = 2.5\,mg\sin\theta$, $N = 0.25\,mg\cos\theta$, $\theta = \tan^{-1}(0.1\,\mu)$.

17–67. 29.8°

17–69. 16.1 ft/s², 5.80 rad/s².

17–70. 0.860 rad/s², 5.92 ft/s².

17–71. 4.10 rad/s².

17–73. 1.30 rad/s².

17–74. 0.5 rad/s².

17–75. 31.2 rad/s².

17–77. 5.62 rad/s², 196 N.

17–78. $A_x = 0.466$ lb, $A_y = 15$ lb, $a_G = 1.00$ ft/s².

17–79. 13.4 rad/s².

17–81. $s = \dfrac{\omega_1 r}{\mu g}(v_1 - \frac{1}{2}\omega_1 r)$.

17–82. $A_x = 42.2$ lb, $A_y = 22.0$ lb.

17–83. $T_A = \frac{4}{7}W$.

17–85. $\alpha = \dfrac{6P}{mL}$↓, $a_B = \dfrac{2P}{m}$→.

17–86. $A_x = 139$ N, $A_y = 252$ N, $B_y = 180$ N.

17–87. $A_x = 34.3$ N, $A_y = 200$ N, $B_y = 143$ N.

17–89. $B_x = 16.7$ lb, $B_y = 18.8$ lb, $N_C = 31.2$ lb.

17–91. $\alpha = \dfrac{2g}{3r}\sin\theta$, $N = m\left[g\cos\theta + \dfrac{r^2\omega^2}{R - r}\right]$.

17–93. Does not slip, $F = 20.1$ N < μN.

17–94. $N_B = 297$ N ↖ 30°, $A_x = 1.63$ N, $A_y = 344$ N.

17–95. 33.5°.

17–97. $\alpha = \dfrac{3g}{2l}$, $N = \frac{1}{4}mg$.

17–98. $\alpha = \dfrac{1.5g}{l}\cos\theta$.

CHAPTER 18

18–1. $v = 23.8$ m/s, $a_n = 1.88\,(10^3)$ m/s², $a_t = 37.5$ m/s², $A_x = 0$, $A_y = 98.1$ N.

18–2. 7.69 ft/s.

18–3. 10.9 rad/s.

18–5. 14.1 rad/s.

18–6. 8.25 rad/s.

18–7. 55.2 ft/s.

18–9. 2.45 rad/s.

18–10. 5.74 rad/s.

18–11. 11.0 rad/s.

18–13. 2.58 m/s, 141 N.

18–14. 9.75 m.

18–15. 1.32 rad/s.

18–17. 0.661 m.

18–18. 0.859 m.

18–19. 2.58 m/s.

18–21. 10.4 ft/s.

18–22. 8.37 ft/s.

18–23. (a) $\omega = \sqrt{\dfrac{3\pi w_0}{2m}}$, (b) $\omega = \sqrt{\dfrac{3\pi w_0}{2m} + \dfrac{3g}{L}}$.

18–25. 0.891 rev.

18–26. $\omega_{BC} = \omega_{AB} = \sqrt{\dfrac{3g}{L}\sin\theta}$.

18–27. $v_G = 3\sqrt{\dfrac{3gR}{7}}$.

18–29. 0.661 m.

18–30. $\omega_{BC} = \omega_{AB} = \sqrt{\dfrac{3g}{L} \sin \theta}.$

18–31. $v_G = 3\sqrt{\dfrac{3gR}{7}}.$

18–33. 2.58 m/s.

18–34. 19.8 rad/s.

18–35. 7.02 ft.

18–37. 1.80 rad/s.

18–38. 39.3°.

18–39. 2.57 rad/s.

18–41. $\omega_{AB} = \omega_{BC} = 5.28$ rad/s.

18–42. $\omega_{AB} = \omega_{BC} = 2.21$ rad/s.

18–43. 10.5 ft/s.

18–45. 85.1 rad/s.

18–46. 11.6 rad/s.

18–47. 5.48 rad/s.

18–49. 5.04 rad/s.

18–50. 2.49 rad/s.

18–51. 4.00 m/s.

CHAPTER 19

19–1. 20.8 rad/s.

19–2. 2.48 rad/s.

19–3. 1.06 s.

19–9. 18.7 rad/s.

19–10. 119 rad/s, 64.5 ft/s $\measuredangle$ 22.5°.

19–11. $v_A = 8.83$ m/s $\downarrow$, $v_B = 5.30$ m/s $\uparrow$.

19–13. 12.7 rad/s.

19–14. 14.4 rad/s.

19–15. 4.26 ft/s.

19–17. $\omega_A = 1.70$ rad/s, $\omega_B = 5.10$ rad/s.

19–18. 23.4 rad/s.

19–19. $\omega_A = 47.4$ rad/s, $\omega_B = 75.6$ rad/s.

19–21. $\theta = \tan^{-1}(3.5\,\mu).$

19–22. $I = m\sqrt{\dfrac{gL}{6}}.$

19–23. $y = \dfrac{\sqrt{2}}{3}a.$

19–25. $d = \frac{2}{3}L.$

19–26. $\omega = 0.$

19–27. 51.5 rad/s.

19–29. $\omega_2 = \frac{1}{3}\omega_1.$

19–30. 0.0906 rad/s.

19–31. 0.0181 rad/s.

19–33. $\omega' = \frac{11}{3}\omega_1.$

19–34. (a) $\omega = 0.025$ rad/s, (b) $\omega = 0.$

19–35. $\omega_1 < 1.02\sqrt{\dfrac{g}{r}}.$

19–37. $\omega_2 = \left(\frac{2}{7} + \frac{5}{7}\cos\theta\right)\omega_1.$

19–38. 0.500 ft.

19–39. $\theta = \tan^{-1}\left(\sqrt{\dfrac{7e}{5}}\right).$

19–41. 39.8°.

19–42. 39.8°.

19–43. 3.23 rad/s, 32.8°.

REVIEW PROBLEMS 2

R2–1. 20.4 ft/s.

R2–2. $v_C = 25.3$ in./s, $\measuredangle$ 63.4°, $a_C = 73.8$ in./s², 32.5° $\searrow$.

R2–3. 41.8 rad/s.

R2–5. 6.04 ft/s².

R2–6. 5.75 ft.

R2–7. 100 rad/s.

R2–9. 52.0°.

R2–10. 24 ft/s.

R2–11. 24 ft/s.

R2–13. 8.89 rad/s².

R2–14. 65.6 rad/s.

R2–15. 27.8 rad/s.

R2–17. 0.393 m, 0.0885 kg · m².

R2–18. $\omega_{AB} = 6$ rad/s $\downdownarrows$.

R2–19. 216 mm, 0.0264 kg · m².

R2–21. 4.72 m/s.

R2–22. 25 rad/s.

R2–23. 8.59 rad/s.

R2–25. $a_t = 0.375$ ft/s², $a_n = 3600$ ft/s², $s = 4500$ ft.

R2–26. 1.89 rad/s, 15.6 rad/s².

R2–27. 195 mm/s.

R2–29. 5.0 rad/s².

R2–30. (a) 68.7 rad/s, (b) 68.8 rad/s.

R2–31. 4.60 rad/s.

R2–33. 6.14 lb.

R2–34. 8 m/s², $\uparrow$, 50.6 m/s², $\measuredangle$ 9.09°.

R2–35. 1.64 s, 48.7 N.

R2–37. 4.90 m/s², 14.7 rad/s².

R2–38. 9.42 m/s.

R2–39. 24.1 m/s.

R2–41. 49.1 lb.

R2–42. 12.1 rad/s², 30.0 lb.

R2–43. 0.714 s.

R2–45. 12.5 rad/s², 88.8 N.

R2–46. 11.9 ft/s.

R2–47. 14.2 rad/s².

R2–49. 193 ft/s².

R2–50. 167 ft/s².

CHAPTER 20

20–1. $\boldsymbol{\omega} = \{5.66\mathbf{j} + 6.26\mathbf{k}\}$ rad/s, $\boldsymbol{\alpha} = \{-3.39\mathbf{i}\}$ rad/s².

20–2. $\mathbf{v}_A = \{-5.20\mathbf{i} - 12.0\mathbf{j} + 20.8\mathbf{k}\}$ ft/s,
$\mathbf{a}_A = \{-24.1\mathbf{i} - 13.3\mathbf{j} - 7.20\mathbf{k}\}$ ft/s².

20–3. $\mathbf{v}_P = \{-1.6\mathbf{i}\}$ m/s, $\mathbf{a}_P = \{-0.32\mathbf{i} - 12.0\mathbf{j} - 8.00\mathbf{k}\}$ m/s².

20–5. $\boldsymbol{\omega} = \{-4\mathbf{j}\}$ rad/s, $\boldsymbol{\alpha} = \{16\mathbf{i}\}$ rad/s², $\mathbf{v}_A = \{-0.283\mathbf{i}\}$ m/s,
$\mathbf{a}_A = \{-1.13\mathbf{j} - 1.13\mathbf{k}\}$ m/s².

20–6. $\mathbf{v}_A = \{-8.66\mathbf{i} + 8.00\mathbf{j} - 13.9\mathbf{k}\}$ ft/s,
$\mathbf{a}_A = \{-4.00\mathbf{i} - 7.71\mathbf{j} - 3.20\mathbf{k}\}$ ft/s².

20–7. $\mathbf{v}_A = \{-8.66\mathbf{i} + 8.00\mathbf{j} - 13.9\mathbf{k}\}$ ft/s,
$\mathbf{a}_A = \{-24.8\mathbf{i} + 8.29\mathbf{j} - 30.9\mathbf{k}\}$ ft/s².

20–9. $\mathbf{v}_A = \{1.80\mathbf{i}\}$ ft/s, $\mathbf{a}_A = \{-0.750\mathbf{i} - 0.720\mathbf{j} - 0.831\mathbf{k}\}$ ft/s².

20–10. $\mathbf{v}_A = \{-14.1\mathbf{i}\}$ ft/s, $\mathbf{a}_A = \{-4.60\mathbf{i} - 56.5\mathbf{j} - 88.0\mathbf{k}\}$ ft/s².

20–11. $\omega_A = 47.8$ rad/s, $\omega_B = 7.78$ rad/s.

20–13. $v_B = 12.0$ ft/s, $\boldsymbol{\omega} = \{0.980\mathbf{i} - 1.06\mathbf{j} - 1.47\mathbf{k}\}$ rad/s.

20–14. 9.43 ft/s.

20–15. $v_B = 3.00$ ft/s, $\boldsymbol{\omega} = \{1.33\mathbf{i} + 2.67\mathbf{j} + 3.33\mathbf{k}\}$ rad/s.

20–17. $\boldsymbol{\alpha} = \{0.413\mathbf{i} + 0.622\mathbf{j} - 0.412(10^{-3})\mathbf{k}\}$ rad/s^2,
$\mathbf{a}_B = \{-5.98\mathbf{i} + 7.98\mathbf{j}\}$ ft/s^2.

20–18. $\mathbf{v}_A = \{0.667\mathbf{i}\}$ ft/s, $\mathbf{a}_A = \{-0.148\mathbf{i}\}$ ft/s^2.

20–19. $\boldsymbol{\omega} = \{1.50\mathbf{i} + 2.60\mathbf{j} + 2.00\mathbf{k}\}$ rad/s, $\mathbf{v}_C = \{10.4\mathbf{i} - 7.79\mathbf{k}\}$ ft/s.

20–21. $\boldsymbol{\omega} = \{-0.375\mathbf{i} + 0.5\mathbf{j} + 0.625\mathbf{k}\}$ rad/s, $\mathbf{v}_B = \{-3.75\mathbf{j}\}$ ft/s.

20–22. $\boldsymbol{\alpha} = \{-1.43\mathbf{i} + 0.600\mathbf{j} + 1.34\mathbf{k}\}$ rad/s^2, $\mathbf{a}_B = \{-14.3\mathbf{j}\}$ ft/s^2.

20–23. $\boldsymbol{\omega}_{BD} = \{-2\mathbf{i}\}$ rad/s.

20–25. $\mathbf{v}_B = \{-0.333\mathbf{j}\}$, $\boldsymbol{\omega} = \{0.204\mathbf{i} - 0.612\mathbf{j} + 1.36\mathbf{k}\}$ rad/s.

20–26. $\boldsymbol{\alpha}_{BC} = \{1.24\mathbf{i} - 0.306\mathbf{j} + 1.44\mathbf{k}\}$ rad/s^2, $\mathbf{a}_B = \{7.98\mathbf{j}\}$ m/s^2.

20–27. $\boldsymbol{\omega} = \{0.769\mathbf{i} - 2.31\mathbf{j} + 0.513\mathbf{k}\}$ rad/s, $\mathbf{v}_B = \{-0.333\mathbf{j}\}$ m/s.

20–30. $\mathbf{v}_C = \{3\mathbf{i} + 6\mathbf{j} - 3\mathbf{k}\}$ m/s,
$\mathbf{a}_C = \{-13\mathbf{i} + 28.5\mathbf{j} - 10.2\mathbf{k}\}$ m/s^2.

20–31. $\mathbf{v}_A = \{-7.79\mathbf{i} - 2.25\mathbf{j} + 3.90\mathbf{k}\}$ ft/s,
$\mathbf{a}_A = \{8.30\mathbf{i} - 35.2\mathbf{j} + 7.02\mathbf{k}\}$ ft/s^2.

20–33. $\mathbf{v}_T = \{-9.50\mathbf{i} + 3\mathbf{j} + 1.20\mathbf{k}\}$ m/s,
$\mathbf{a}_T = \{-14.4\mathbf{i} - 4.23\mathbf{j} + 4.20\mathbf{k}\}$ m/s^2.

20–34. $\mathbf{v}_C = \{-5.54\mathbf{i} + 5.20\mathbf{j} - 3.00\mathbf{k}\}$ ft/s,
$\mathbf{a}_C = \{-91.5\mathbf{i} - 42.6\mathbf{j} - 1.00\mathbf{k}\}$ ft/s^2.

20–35. $\mathbf{v}_B = \{-17.3\mathbf{i} + 18.8\mathbf{j} + 10\mathbf{k}\}$ ft/s,
$\mathbf{a}_B = \{-19.1\mathbf{i} + 24.0\mathbf{j} - 8.66\mathbf{k}\}$ ft/s^2.

20–37. $\mathbf{v}_P = \{-9.80\mathbf{i} + 14.4\mathbf{j} + 48.0\mathbf{k}\}$ ft/s,
$\mathbf{a}_P = \{-160\mathbf{i} + 5.16\mathbf{j} - 13.0\mathbf{k}\}$ ft/s^2.

20–38. $\mathbf{v}_C = \{-1.73\mathbf{i} - 5.77\mathbf{j} + 7.06\mathbf{k}\}$ ft/s,
$\mathbf{a}_C = \{9.88\mathbf{i} - 72.8\mathbf{j} + 0.365\mathbf{k}\}$ ft/s^2.

20–39. $\mathbf{v}_P = \{-10.5\mathbf{i} + 9.00\mathbf{j} - 1.00\mathbf{k}\}$ ft/s,
$\mathbf{a}_P = \{-41.0\mathbf{i} - 17.5\mathbf{j} - 0.500\mathbf{k}\}$ ft/s^2.

20–41. $\mathbf{v}_C = \{-20.0\mathbf{i} + 24.0\mathbf{j} - 3.00\mathbf{k}\}$ in./s,
$\mathbf{a}_C = \{-145\mathbf{i} - 24.0\mathbf{j} + 8.00\mathbf{k}\}$ in./s^2.

20–42. $\mathbf{v}_P = \{-25.5\mathbf{i} - 13.4\mathbf{j} + 20.5\mathbf{k}\}$ ft/s,
$\mathbf{a}_P = \{161\mathbf{i} - 249\mathbf{j} - 39.6\mathbf{k}\}$ ft/s^2.

20–43. $\mathbf{v}_P = \{-25.5\mathbf{i} - 13.4\mathbf{j} + 20.5\mathbf{k}\}$ ft/s,
$\mathbf{a}_P = \{161\mathbf{i} - 243\mathbf{j} - 33.9\mathbf{k}\}$ ft/s^2.

20–45. $\mathbf{v}_A = \{-8.66\mathbf{i} + 2.62\mathbf{j} + 1.91\mathbf{k}\}$ m/s,
$\mathbf{a}_A = \{-38.3\mathbf{i} - 38.7\mathbf{j} + 11.8\mathbf{k}\}$ m/s^2.

CHAPTER 21

21–2. $I_y = 26.7$ slug-ft^2, $I_x = 53.3$ slug-ft^2.

21–3. $I_y = 4.48 (10^3) \rho$, $I_{xy} = 636 \rho$.

21–5. $I_{yz} = \dfrac{mah}{6}$.

21–6. $I_{yz} = \dfrac{mah}{5}$.

21–7. 0.0961 slug-ft^2.

21–9. $\bar{x} = -0.667$ ft, $\bar{y} = 0.500$ ft, $I_x' = 0.0272$ slug-ft^2,
$I_y' = 0.0155$ slug-ft^2, $I_z' = 0.0427$ slug-ft^2.

21–10. 0.888 slug-ft^2.

21–11. 1.13 slug-ft^2.

21–13. 0.0945 kg $\cdot$ m^2.

21–14. 0.429 kg $\cdot$ m^2.

21–15. $I_x = 1.36$ kg $\cdot$ m^2, $I_y = 0.380$ kg $\cdot$ m^2, $I_z = 1.26$ kg $\cdot$ m^2.

21–17. $I_x = I_y = I_z = \dfrac{2ma^2}{3}$, $I_{xy} = \dfrac{-a^2m}{4}$, $I_{yz} = \dfrac{a^2m}{4}$, $I_{xz} = -\dfrac{a^2m}{4}$.

21–21. 0.727 kg $\cdot$ m^2/s.

21–22. 6.46 ft $\cdot$ lb.

21–23. 1.57 slug $\cdot$ ft^2/s.

21–25. 0.0920 ft $\cdot$ lb.

21–26. $\mathbf{H} = \{0.0207\mathbf{i} - 0.00690\mathbf{j} + 0.0690\mathbf{k}\}$ ft $\cdot$ lb $\cdot$ s.

21–27. $\boldsymbol{\omega} = \{-0.444\mathbf{j} + 0.444\mathbf{k}\}$ rad/s.

21–29. 58.4 rad/s.

21–30. 19.7 rad/s.

21–31. 18.2 ft/s.

21–33. 21.4 rad/s.

21–34. 87.2 rad/s.

21–35. $\mathbf{H} = \{16.6\mathbf{k}\}$ slug-ft^2/s, $T = 72.1$ ft $\cdot$ lb.

21–37. 20.2 rad/s.

21–38. $\boldsymbol{\omega} = \{0.0536\mathbf{i} + 0.0536\mathbf{k}\}$ rad/s.

21–39. $\Sigma M_x = \dfrac{d}{dt}(I_x\omega_x - I_{xy}\omega_y - I_{xz}\omega_z) - \Omega_z(I_y\omega_y - I_{yz}\omega_z - I_{yx}\omega_x) + \Omega_y(I_z\omega_z - I_{zx}\omega_x - I_{zy}\omega_y).$

21–41. $\Sigma M_x = I_x\dot{\omega}_x - I_y\Omega_z\omega_y + I_z\Omega_y\omega_z.$

21–42. $F_A = 277$ N, $F_B = 166$ N.

21–43. $A_x = -2.12$ lb, $A_y = 1.06$ lb, $B_x = 2.12$ lb, $B_y = -1.06$ lb,
$B_z = 4.25$ lb.

21–45. $F_A = 8.83$ lb, $F_B = 11.2$ lb.

21–46. 21.0 lb $\cdot$ ft.

21–47. 23.4 lb $\cdot$ ft.

21–49. $F_B = 5.01$ lb, $F_A = 4.99$ lb.

21–50. $A_x = -8.97$ lb, $B_x = -9.66$ lb, $A_z = -6.07$ lb,
$B_z = 15.2$ lb.

21–51. $M = 0.0320$ N $\cdot$ m, $A_x = -1.08$ N, $A_y = 13.2$ N,
$B_x = 1.08$ N, $B_y = 13.3$ N.

21–53. $\dot{\omega}_y = 25.9$ rad/s^2, $B_x = 9.59$ N, $B_z = 12.3$ N, $A_x = -10.8$ N,
$A_z = 12.3$ N.

21–54. $\dot{\omega}_z = 200$ rad/s^2, $D_x = -37.5$ N, $D_y = -12.9$ N,
$C_x = -37.5$ N, $C_y = -11.1$ N, $C_z = 36.8$ N.

21–55. $A_x = B_x = 0$, $A_y = -1.09$ kN, $B_y = 1.38$ kN.

21–57. $A_x = 0$, $A_y = 4.89$ lb, $A_z = 5.47$ lb, $M_x = -8.43$ lb $\cdot$ ft,
$M_y = M_z = 0$.

21–58. $T = 23.3$ N, $F_A = 51.7$ N.

21–59. $T = 23.3$ N, $F_A = 51.8$ N.

21–61. $70.8°$.

21–62. $70.8°$.

21–65. 328 N $\cdot$ m.

21–66. To the right when viewed from above.

21–67. 4.91 rad/s.

21–69. 1.19 rad/s.

21–70. $A_y = -B_y = 13.7$ kN.

21–71. $3.63 (10^3)$ rad/s.

21–73. 77.7 lb.

21–74. 107 N.

21–75. 0.820 rad/s.

21–78. 2.33 rev/h.

21–79. $0.673°$ regular precession.

21–81. 12.5 Mg $\cdot$ m^2/s.

CHAPTER 22

22–1. $\tau = 0.452$ s, $f = 2.21$ Hz.

22–2. $f = \dfrac{1}{\pi}\sqrt{\dfrac{k}{m}}$.

22–3. $x = -0.05 \cos(12.2t)$ m.

22–5. $x = (-0.126 \sin(3.16t) - 0.09 \cos(3.16t))$ m, $C = 0.155$ m.

22–6. $y = (0.0833 \cos 19.7t)$ ft, $C = 1$ in., $f = 3.13$ Hz.

22–7. $y = (-1.44 \sin 13.9t - 0.167 \cos 13.9t)$ ft, $C = 1.45$ ft.

22–9. $x = (-0.534 \sin(7.49\,t) + 0.06 \cos(7.49t))$ m,
$C = 0.538$ m.

22–10. $\theta = (-0.101 \sin(4.95\,t) + 0.3 \cos(4.95t))$ rad.

22–11. $16.3°$.

22–13. $\tau = 2\pi\sqrt{\dfrac{k_G{}^2 + d^2}{gd}}$.

22–14. 21.2 kg, 609 N/m.

22–15. $\tau = 2\pi\sqrt{\dfrac{2r}{g}}$.

22–17. 1.45 s.

22–18. $\theta = 0.4 \cos 20.7t$.

22–19. 2.50 Mg $\cdot$ m^2.

22–21. 0.167 s.

22–22. 2.67 s.

22–23. 90.8 lb $\cdot$ ft/rad.

22–25. 0.910 s.

22–26. $x = 0.167 \cos 6.55t$.

22–27. $\tau = 2\pi\sqrt{\dfrac{l}{2g}}$.

22–29. $f = \dfrac{1}{2\pi}\sqrt{\dfrac{2}{3}\left(\dfrac{g}{R-r}\right)}$.

22–30. $f = 0.597$ Hz.

22–31. $\tau = 6.10\sqrt{\dfrac{a}{g}}$.

22–33. $\tau = 2\pi\sqrt{\dfrac{2r}{g}}$.

22–34. 1.18 s.

22–35. $\tau = \pi\sqrt{\dfrac{m}{k}}$.

22–37. 2.25 Hz.

22–38. 0.339 s.

22–39. 1.52 s.

22–41. 0.970 s.

22–42. 0.851 Hz.

22–43. 0.487 s.

22–45. 2.81 s.

22–46. $y = A \sin pt + B \cos pt + \dfrac{F_0/m}{(k/m - \omega^2)} \cos \omega t$.

22–47. 0.685 ft/s.

22–49. $y = (361 \sin 7.75t + 100 \cos 7.75t - 350 \sin 8t)$ mm.

22–50. $y = (-16.6 \sin 7.07t + 70 \cos 7.07t + 29.4 \sin 4t)$ mm.

22–51. 19.7 rad/s.

22–53. 19.0 rad/s.

22–57. $x_P = (0.0276 \sin 2t)$ ft.

22–58. 6.55 rad/s.

22–59. 0.625 in.

22–61. 2.83 rad/s.

22–62. 5.95 mm.

22–63. 1.20 m/s.

22–65. $9.89°$.

22–66. 0.997.

22–67. 1.66 s.

22–69. $1.54°$.

22–70. 1.01.

22–71. 0.0269 in.

22–73. 0.666 s.

22–74. $x = 0.119 \cos(6t - 83.9°)$ m.

22–75. $1.55\ddot{\theta} + 340\dot{\theta} + 200\theta = 0$.

22–78. 18.9 N $\cdot$ s/m.

22–79. 714 N/m.

22–81. Overdamped, $\ddot{y} + 16\dot{y} + 4y = 0$.

22–82. $m\ddot{x} + kx = F_0 \sin \omega t$, $L\ddot{q} + \dfrac{1}{c}q = E_0 \sin \omega t$.

22–83. $L\ddot{q} + R\dot{q} + \dfrac{1}{c}q = 0$.

22–85. $L\ddot{q} + R\dot{q} + \dfrac{1}{c}q = E_0 \cos \omega t$.

Index

Geometric Properties of Line and Area Elements

Centroid Location	Centroid Location	Area Moment of Inertia
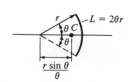 Circular arc segment	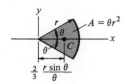 Circular sector area	$I_x = \frac{1}{4}r^4(\theta - \frac{1}{2}\sin 2\theta)$ $I_y = \frac{1}{4}r^4(\theta + \frac{1}{2}\sin 2\theta)$ 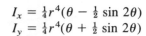
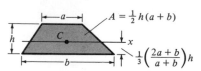 Quarter and semicircular arcs	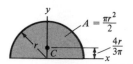 Quarter circular area	$I_x = \frac{1}{16}\pi r^4$ $I_y = \frac{1}{16}\pi r^4$
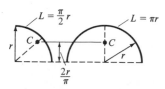 Trapezoidal area	Semicircular area	$I_x = \frac{1}{8}\pi r^4$ $I_y = \frac{1}{8}\pi r^4$
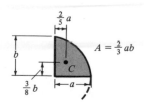 Semiparabolic area	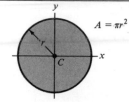 Circular area	$I_x = \frac{1}{4}\pi r^4$ $I_y = \frac{1}{4}\pi r^4$
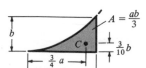 Exparabolic area	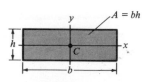 Rectangular area	$I_x = \frac{1}{12}bh^3$ $I_y = \frac{1}{12}hb^3$
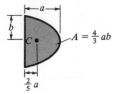 Parabolic area	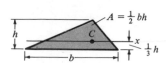 Triangular area	$I_x = \frac{1}{36}bh^3$

Center of Gravity and Mass Moment of Inertia of Homogeneous Solids

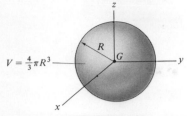

Sphere

$$I_{xx} = I_{yy} = I_{zz} = \tfrac{2}{5}mR^2$$

$$V = \tfrac{4}{3}\pi R^3$$

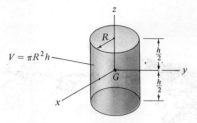

Cylinder

$$I_{xx} = I_{yy} = \tfrac{1}{12}m(3R^2 + h^2) \qquad I_{zz} = \tfrac{1}{2}mR^2$$

$$V = \pi R^2 h$$

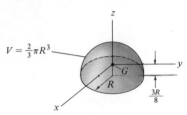

Hemisphere

$$I_{xx} = I_{yy} = 0.259mR^2 \qquad I_{zz} = \tfrac{2}{5}mR^2$$

$$V = \tfrac{2}{3}\pi R^3$$

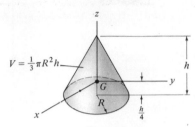

Cone

$$I_{xx} = I_{yy} = \tfrac{3}{80}m(4R^2 + h^2) \qquad I_{zz} = \tfrac{3}{10}mR^2$$

$$V = \tfrac{1}{3}\pi R^2 h$$

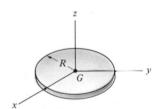

Thin circular disk

$$I_{xx} = I_{yy} = \tfrac{1}{4}mR^2 \qquad I_{zz} = \tfrac{1}{2}mR^2$$

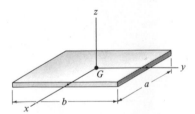

Thin plate

$$I_{xx} = \tfrac{1}{12}mb^2 \qquad I_{yy} = \tfrac{1}{12}ma^2 \qquad I_{zz} = \tfrac{1}{12}m(a^2 + b^2)$$

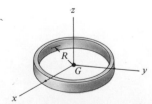

Thin ring

$$I_{xx} = I_{yy} = \tfrac{1}{2}mR^2 \qquad I_{zz} = mR^2$$

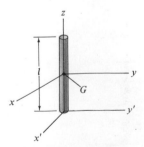

Slender rod

$$I_{xx} = I_{yy} = \tfrac{1}{12}ml^2 \qquad I_{x'x'} = I_{y'y'} = \tfrac{1}{3}ml^2 \qquad I_{zz} = 0$$